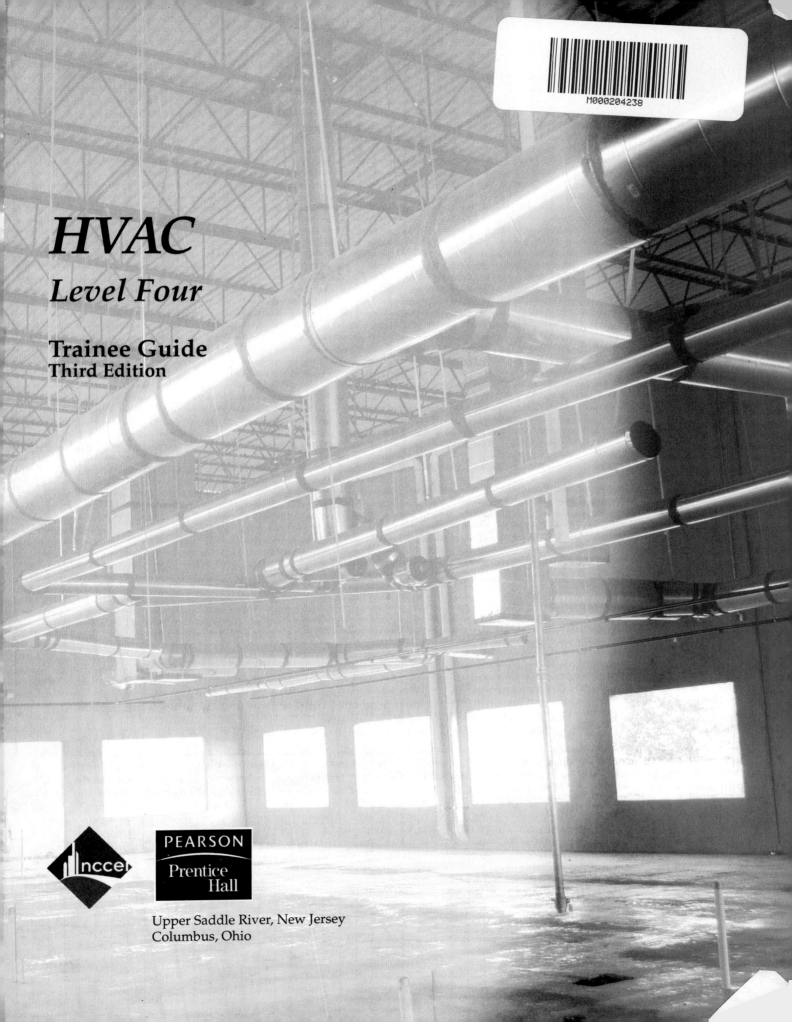

HVAC

Level Four

Trainee Guide
Third Edition

nccer

PEARSON

Prentice Hall

Upper Saddle River, New Jersey
Columbus, Ohio

National Center for Construction Education and Research

President: Don Whyte
Director of Curriculum Revision and Development: Daniele Stacey
HVAC Project Manager: Carla Sly/Tania Domenech
Production Manager: Tim Davis
Quality Assurance Coordinator: Debie Ness
Editors: Rob Richardson, Matt Tischler, and Brendan Coote
Desktop Publishing Coordinator: James McKay
Production Assistant: Brittany Ferguson

NCCER would like to acknowledge the contract service provider for this curriculum:
Topaz Publications, Liverpool, New York.

This information is general in nature and intended for training purposes only. Actual performance of activities described in this manual requires compliance with all applicable operating, service, maintenance, and safety procedures under the direction of qualified personnel. References in this manual to patented or proprietary devices do not constitute a recommendation of their use.

10 9 8 7 6 5 4 3

0-13-604494-8
978-0-13-604494-9

PREFACE

TO THE TRAINEE

Heating and air-conditioning systems (HVAC) regulate the temperature, humidity, and the total air quality in residential, commercial, industrial, and other buildings. This also extends to refrigeration systems used to transport food, medicine, and other perishable items. Other systems may include hydronics (water-based heating systems), solar panels, or commercial refrigeration. HVAC technicians and installers set up, maintain, and repair such systems. As a technician, you must be able to maintain, diagnose, and correct problems throughout the entire system. Diversity of skills and tasks is also significant to this field. You must know how to follow blueprints or other specifications to install any system. You may also need working knowledge of sheet metal practices for the installation of ducts, welding, basic pipefitting, and electrical practices.

Think about it! Nearly all buildings and homes use forms of heating, cooling and/or ventilation. The increasing development of HVAC technology causes employers to recognize the importance of continuous education and keeping up to speed with the latest equipment and skills. Hence, technical school training or apprenticeship programs often provide an advantage and a higher qualification for employment. NCCER's program has been designed by highly-qualified subject matter experts with this in mind. Our four levels present an apprentice approach to the HVAC field, including theoretical and practical skills essential to your success as an HVAC installer or technician.

As the number of buildings increases, so will the demand for HVAC technicians. According to the U.S. Bureau of Labor Statistics, employment of HVAC technicians and installers is projected to increase 18 to 26 percent by 2014. We wish you the best as you begin an exciting and promising career.

WHAT'S NEW IN *HVAC LEVEL FOUR*?

In recognition of the trend toward energy conservation, an *Alternative Heating and Cooling Systems* module was added. Technologies such as in-floor radiant heating, solar heating, and geothermal heat pumps are covered, as well as chilled-beam cooling, ductless split systems, valance cooling systems, and evaporative coolers.

Another change to this level was the expansion of the *Commercial and Industrial Refrigeration Systems* module. Other modules in Level Four were revised to reflect newer technology. The *Building Management Systems* module, for example, was heavily revised to address the current state of the art in BMS software.

The first two levels of HVAC provide you with a broad knowledge of installation and service requirements for residential and commercial systems, while Levels Three and Four provide advanced maintenance, troubleshooting, design, and supervisory skills. Through this course, you will enter the workforce with the knowledge and skills needed to perform productively in either the residential or commercial market.

CONTREN® LEARNING SERIES

The National Center for Construction Education and Research (NCCER) is a not-for-profit 501(c)(3) education foundation established in 1995 by the world's largest and most progressive construction companies and national construction associations. It was founded to address the severe workforce shortage facing the industry and to develop a standardized training process and curricula. Today, NCCER is supported by hundreds of leading construction and maintenance companies, manufacturers, and national associations. The Contren® Learning Series was developed by NCCER in partnership with Pearson Education, Inc., the world's largest educational publisher.

Some features of NCCER's Contren® Learning Series are as follows:

- An industry-proven record of success
- Curricula developed by the industry for the industry
- National standardization providing portability of learned job skills and educational credits
- Compliance with the Office of Apprenticeship requirements for related classroom training (CFR 29:29)
- Well-illustrated, up-to-date, and practical information

NCCER also maintains a National Registry that provides transcripts, certificates, and wallet cards to individuals who have successfully completed modules of NCCER's Contren® Learning Series. *Training programs must be delivered by an NCCER Accredited Training Sponsor in order to receive these credentials.*

Contents

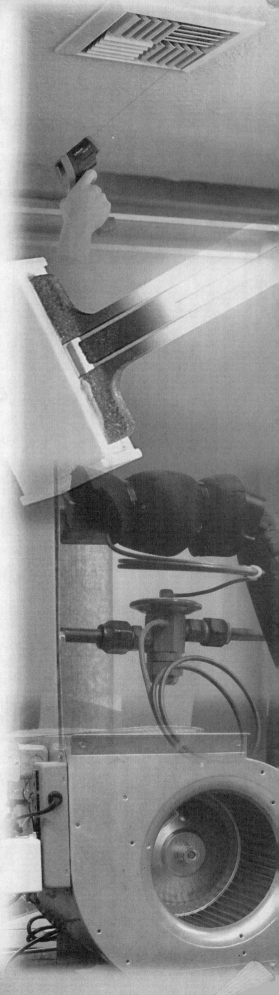

Contren® Curricula

NCCER's training programs comprise over 50 construction, maintenance, and pipeline areas and include skills assessments, safety training, and management education.

Boilermaking
Cabinetmaking
Carpentry
Concrete Finishing
Construction Craft Laborer
Construction Technology
Core Curriculum:
  Introductory Craft Skills
Drywall
Electrical
Electronic Systems Technician
Heating, Ventilating, and
  Air Conditioning
Heavy Equipment Operations
Highway/Heavy Construction
Hydroblasting
Industrial Maintenance
  Electrical and Instrumentation
  Technician
Industrial Maintenance Mechanic
Instrumentation
Insulating
Ironworking
Masonry
Millwright
Mobile Crane Operations
Painting
Painting, Industrial
Pipefitting
Pipelayer
Plumbing
Reinforcing Ironwork
Rigging
Scaffolding
Sheet Metal
Site Layout
Sprinkler Fitting
Welding

Pipeline
Control Center Operations,
  Liquid
Corrosion Control
Electrical and Instrumentation
Field Operations, Liquid
Field Operations, Gas
Maintenance
Mechanical

Safety
Field Safety
Safety Orientation
Safety Technology

Management
Introductory Skills for the
  Crew Leader
Project Management
Project Supervision

Spanish Translations
Andamios
Currículo Básico
  Habilidades Introductorias
  del Oficio
Instalación de Rociadores
  Nivel Uno
Introducción a la Carpintería
  Nivel Uno
Orientación de Seguridad
Principios Básicos de Maniobras
Seguridad de Campo

Supplemental Titles
Applied Construction Math
Careers in Construction

Acknowledgments

This curriculum was revised as a result of the
farsightedness and leadership of the following sponsors:

ABC of Wisconsin
Lincoln Technical Institute
W. B. Guimarin & Co., Inc.
Hunton Trane
Apex Technical School
Entek

This curriculum would not exist were it not for the dedication
and unselfish energy of those volunteers who served on the Authoring Team.
A sincere thanks is extended to the following:

Barry Burkan
Frank Kendall
Daniel Kerkman
Joe Moravek
Troy Staton
Mattew Todd

NCCER PARTNERING ASSOCIATIONS

American Fire Sprinkler Association
Associated Builders and Contractors, Inc.
Associated General Contractors of America
Association for Career and Technical Education
Association for Skilled and Technical Sciences
Carolinas AGC, Inc.
Carolinas Electrical Contractors Association
Center for the Improvement of Construction
    Management and Processes
Construction Industry Institute
Construction Users Roundtable
Design Build Institute of America
Green Advantage
Merit Contractors Association of Canada
Metal Building Manufacturers Association
NACE International
National Association of Minority Contractors
National Association of Women in Construction
National Insulation Association
National Ready Mixed Concrete Association
National Systems Contractors Association
National Technical Honor Society
National Utility Contractors Association
NAWIC Education Foundation
North American Crane Bureau
North American Technician Excellence
Painting & Decorating Contractors of America
Portland Cement Association
SkillsUSA
Steel Erectors Association of America
Texas Gulf Coast Chapter, ABC
U.S. Army Corps of Engineers
University of Florida
Women Construction Owners & Executives, USA

HVAC Level Four

03401-09

Construction Drawings and Specifications

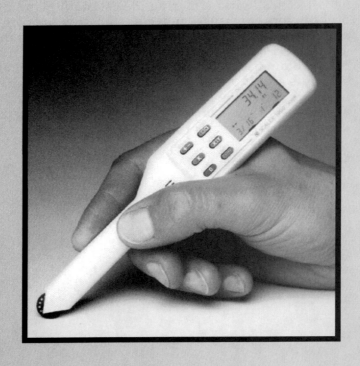

03401-09
Construction Drawings and Specifications

Topics to be presented in this module include:

Overview

Anyone involved in the installation of heating and cooling equipment for new construction must be able to interpret the project drawings and specifications. The drawings show the locations of equipment, duct runs, piping runs, and electrical wiring. During the estimating and planning processes that occur before construction begins, the drawings are used to determine the amount and types of equipment, accessories, and materials needed for the job. Correct interpretation of the drawings is essential in these activities in order to determine the correct price for the job, and to have the correct amounts and types of equipment and materials available to support the installers. Given the importance of the drawings and specifications, a technician or installer who does not learn to interpret them properly is unlikely to advance very far in his or her career.

Objectives

When you have completed this module, you will be able to do the following:

1. Read HVAC drawings and architect's plans and explain their relationships.
2. Compare mechanical plans with the actual installation of duct and pipe runs, fittings, and sections.
3. Interpret specification documents and apply them to the plans.
4. Interpret shop drawings and apply them to the plans and specifications.
5. Describe a submittal, its derivation, routing, and makeup.
6. Develop a field set of as-built drawings.
7. Identify the steps required for transferring design information to component production.
8. Identify, develop, and complete takeoff sheets.
9. List and classify materials most commonly used in HVAC systems.
10. Complete takeoff procedures for HVAC systems.

Trade Terms

Coordination drawings
Cut list
Detail drawing
Elevation view
Floor plan
Longitudinal section
Plan view

Riser diagram
Schedules
Section drawing
Shop drawing
Site plan
Takeoff
Transverse section

Required Trainee Materials

1. Pencil and paper
2. Appropriate personal protective equipment

Prerequisites

Before you begin this module, it is recommended that you successfully complete *Core Curriculum*; *HVAC Level One*; *HVAC Level Two*; and *HVAC Level Three*.

This course map shows all of the modules in the fourth level of the *HVAC* curriculum. The suggested training order begins at the bottom and proceeds up. Skill levels increase as you advance on the course map. The local Training Program Sponsor may adjust the training order.

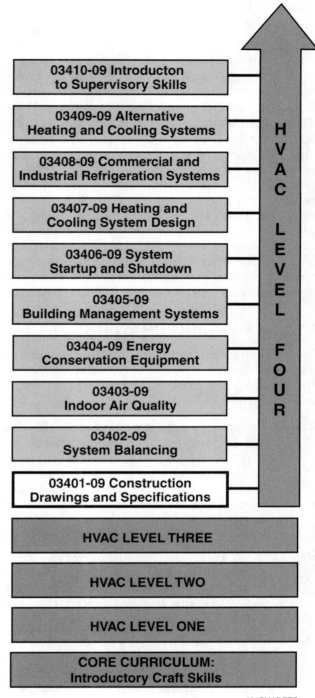

401CMAP.EPS

1.0.0 ◆ INTRODUCTION

This module reviews and builds on the information previously studied in the *Core Curriculum* module *Introduction to Construction Drawings*. It focuses on techniques for reading various types of HVAC-related construction drawings and project specifications. Construction drawings tell the HVAC technician and installer, as well as other skilled tradespeople, how to build a specific building or structure. A specification is a related contractual document used along with the construction drawings. It contains detailed written instructions that supplement the information shown in the set of drawings. As an HVAC technician, you must be able to interpret drawings and specifications correctly. Failure to do so may result in costly rework and unhappy customers. Depending on the severity of a mistake, it can also expose you and your employer to legal liability.

2.0.0 ◆ READING DRAWINGS

The following general procedure is suggested as a method of reading a set of drawings. Use this procedure to familiarize yourself with an available set of drawings:

Step 1 Locate and read the title block. The title block tells you what the drawing is about. It contains critical information about the drawing such as the scale, date of last revision, drawing number, and architect or engineer. If you have to remove a sheet from a set of drawings, be sure to fold the sheet with the title block facing up.

Step 2 Find the north arrow. Always orient yourself to the structure. Knowing where north is enables you to more accurately describe the locations of walls and other parts of the building.

Step 3 Always be aware that drawings work together. The reason the architect or engineer draws plans, elevations, and sections is that it requires more than one type of view to communicate the whole project. Learn how to use more than one drawing when necessary to find the information you need.

Step 4 Check the list of drawings in the set. Note the sequence of the various types of plans. Some drawings have an index on the front cover. Notice that the prints are broken into several categories, as shown in *Figure 1*. However, drawing sets do not have to contain all the categories of drawings shown.

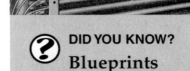

Step 5 Study the **site plan** (plot plan) to observe the location of the building. Also notice that the geographic location of the building may be indicated on the site plan.

Step 6 Check the **floor plan** for the orientation of the building. Observe the location and other details of entries, corridors, offsets, and any special features.

Step 7 Study the features that extend for more than one floor, such as plumbing, vents, stairways, elevator shafts, heating and cooling ductwork, and piping. Determine the location of all main electrical runs and fire sprinkler system lines while looking for possible installation conflicts.

Step 8 Check the floor and wall construction and other details relating to exterior and interior walls.

Step 9 Check the foundation plan for size and types of footings, reinforcing steel, and loadbearing substructures.

Step 10 Study the mechanical plans for heating, cooling, and plumbing details.

Step 11 Observe the electrical entrance and distribution panels, as well as the installation of the lighting and power supplies for special equipment.

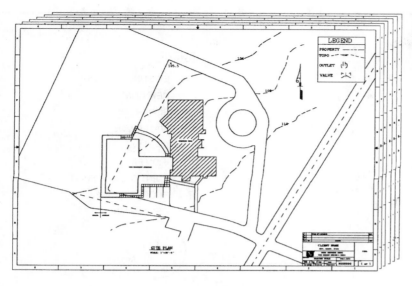

TITLE SHEET(S)
ARCHITECTURAL DRAWINGS
- SITE (PLOT) PLAN
- FOUNDATION PLAN
- FLOOR PLANS
- INTERIOR/EXTERIOR ELEVATIONS
- SECTIONS
- DETAILS
- SCHEDULES

STRUCTURAL DRAWINGS
PLUMBING PLANS
MECHANICAL PLANS
ELECTRICAL PLANS

401F01.EPS

Figure 1 ◆ Typical categories of drawings in a set of construction drawings.

Step 12 Check the notes on the various pages and compare the specifications against the construction details. Look for any variations.

Step 13 Thumb through the sheets of drawings until you are familiar with all the plans and structural details.

Step 14 Recognize applicable symbols and their relative locations in the plans (see *Appendix A*). Note any special construction details or variations that will affect your trade.

2.1.0 Site Plan

The site plan (*Figure 2*) indicates the location of the building on the land site. It may include topographic features such as contour lines, trees, and shrubs. It may also include some construction features such as walks, driveways, curbs, and gutters.

Often the roof plan, if there is one, is also shown on the site plan. General notes pertaining to grading and shrubbery may also be included on the site plan.

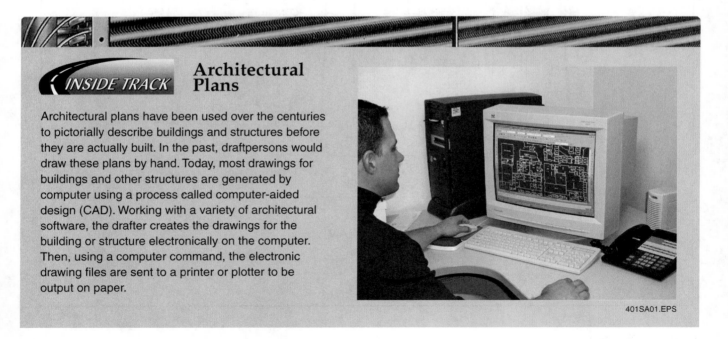

INSIDE TRACK **Architectural Plans**

Architectural plans have been used over the centuries to pictorially describe buildings and structures before they are actually built. In the past, draftpersons would draw these plans by hand. Today, most drawings for buildings and other structures are generated by computer using a process called computer-aided design (CAD). Working with a variety of architectural software, the drafter creates the drawings for the building or structure electronically on the computer. Then, using a computer command, the electronic drawing files are sent to a printer or plotter to be output on paper.

401SA01.EPS

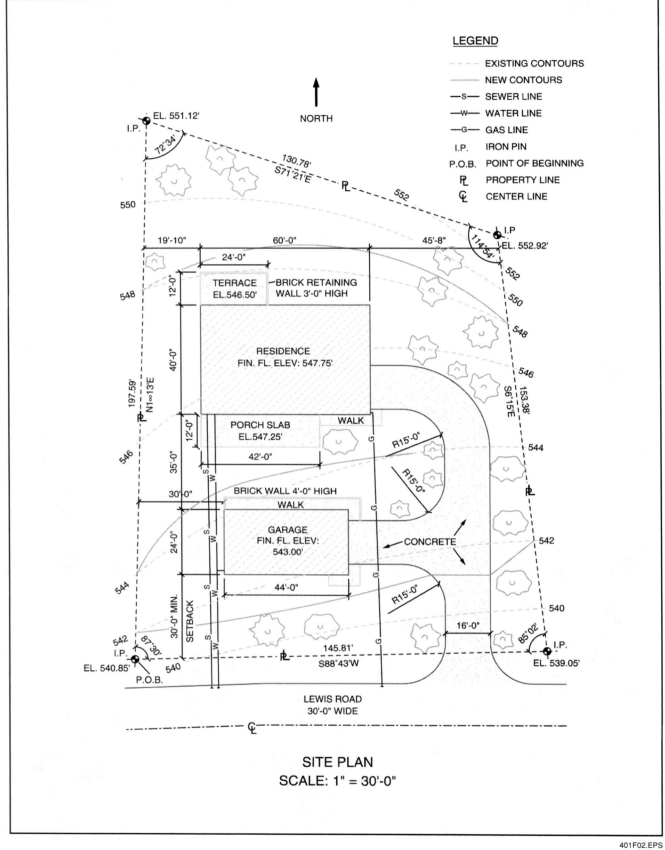

SITE PLAN
SCALE: 1" = 30'-0"

401F02.EPS

Figure 2 ◆ Site plan.

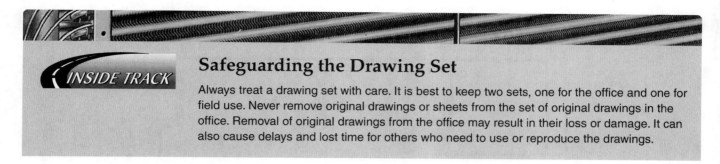
On large commercial jobs, a utility site plan may also be included in the drawing set. It shows the locations for underground facilities such as gas and water pipelines, sanitary sewers, electric power or communication system cables, and other facilities.

2.2.0 Plan Views (Floor, Roof, and Ceiling Plans)

The floor plan (*Figure 3*) is one of the most important working drawings. It is the first drawing done by the designer. It shows the length and width of the building and the location of the rooms and other spaces that the building contains. Each floor of the structure has a different floor plan.

A floor plan is a **plan view** of a horizontal section taken at some distance above the floor, usually midway between the floor and ceiling. However, the cutting plane of the sectional drawing may be offset to a higher or lower level so that it cuts through the desired features such as windows and doors. The cut view crosses all openings for that floor or story. This view gives the dimensions of the window and door openings and indicates which way the doors are to swing.

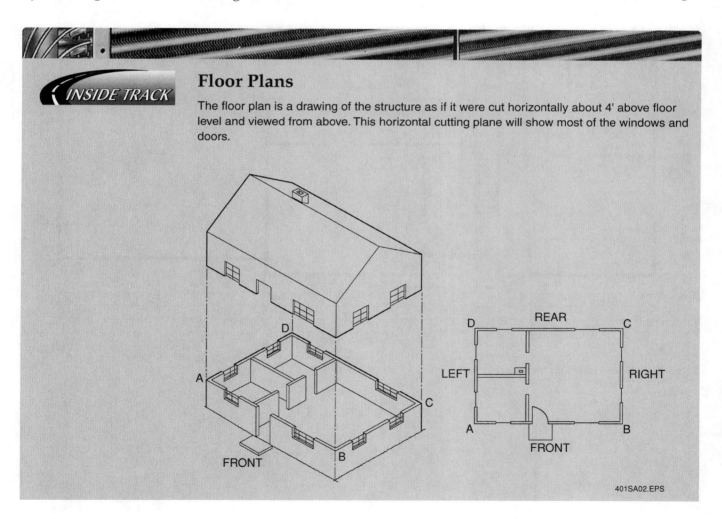

INSIDE TRACK

Floor Plans

The floor plan is a drawing of the structure as if it were cut horizontally about 4' above floor level and viewed from above. This horizontal cutting plane will show most of the windows and doors.

401SA02.EPS

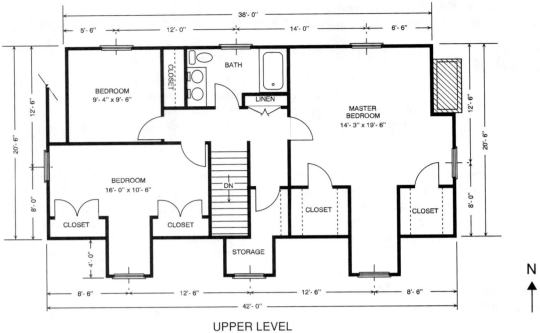

UPPER LEVEL

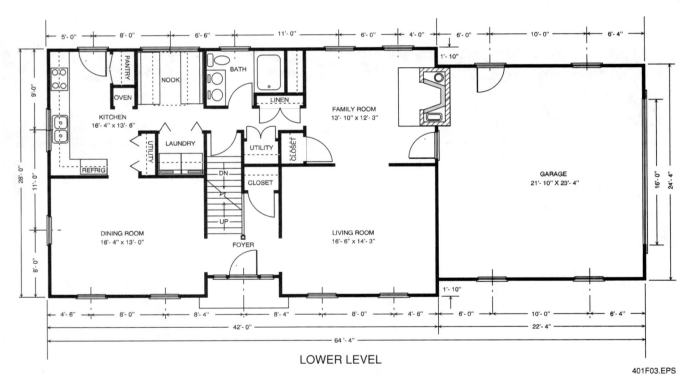

LOWER LEVEL

401F03.EPS

Figure 3 ◆ Floor plans for a building.

When supplied, the contractor consults the roof plan for information pertaining to roof slope, roof drain placement, and other pertinent information regarding ornamental sheet metal work and gutter and downspout technical information. *Figure 4* shows an example of a typical roof plan. Where applicable, the roof plan may also contain information on the location of air conditioning units, exhaust fans, and other ventilation equipment.

Some drawing sets include a reflected ceiling plan (*Figure 5*). This view shows the ceiling as if it were reflected down into a mirror. A reflected ceiling plan is of particular value to the HVAC contractor, providing information that identifies the location of supply diffusers, exhaust grilles, access panels, and other structural components.

2.3.0 Elevation Drawings

The **elevation view** of a structure (*Figure 6*) shows the exterior features of that structure. Unless one or more views are identical, four views are generally used to show each exposure.

With very complex buildings, more than four views may be required. Elevation drawings show the exterior style of the building as well as the placement of doors, windows, chimneys, and decorative trim.

The various views are usually labeled in one of two ways. They may be broken down as front view, right side view, left side view, and rear view, or they may be designated by compass direction. If the front of the building faces east, then this becomes the east elevation. The other elevations are then labeled accordingly: west, south, and north.

Materials used for the exterior finish of a building are also indicated on elevation drawings and described in detail in the specifications, either as part of the drawings or as a separate set of written specifications. Parts of a building hidden from view, foundation walls, and footings are shown by broken lines. Elevation drawings often show elements such as the heights of windows, doors, and porches and the pitches of roofs because all of these measurements cannot be shown conveniently on floor plans.

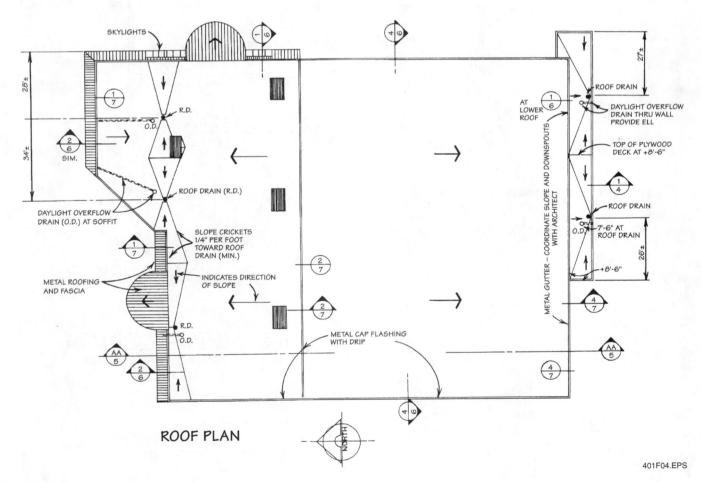

ROOF PLAN

401F04.EPS

Figure 4 ◆ Roof plan.

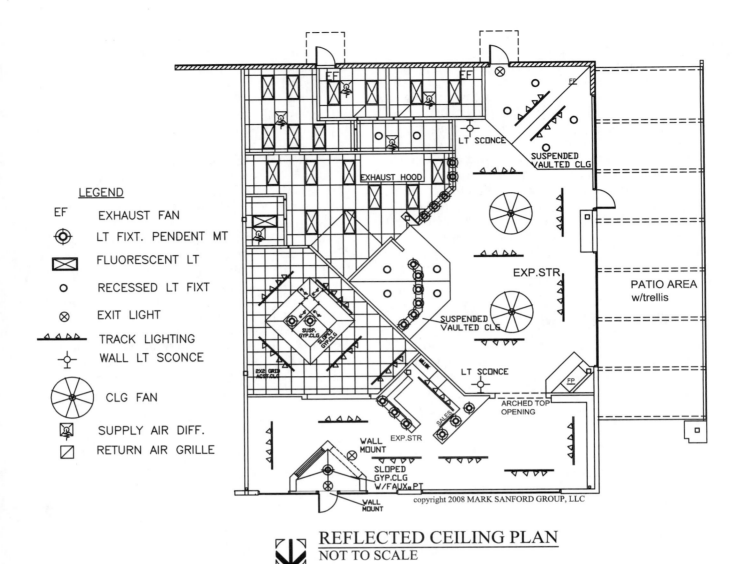

REFLECTED CEILING PLAN
NOT TO SCALE

CEILING PLAN NOTES:

1. LOCATIONS OF DIFFUSSERS ARE APPROXIMATE AND WILL BE DETERMINED BY MENCHANICAL ENGINEER.
2. NEW SUSPENDED CEILING GRID AND PADS TO BE PAINTED BLACK.
3. ALL CCEILINGS ARE TO BE 10'-0" A.F.F.(U.N.O.)

MARK SANFORD GROUP, LLC
1306 N. 162nd ST.
OMAHA, NE 68118
PHONE: 402.598.0442
FAX: 402.493.3567

mark ☐ SANFORD ☐
group

401F05.EPS

Figure 5 ◆ Reflected ceiling plan.

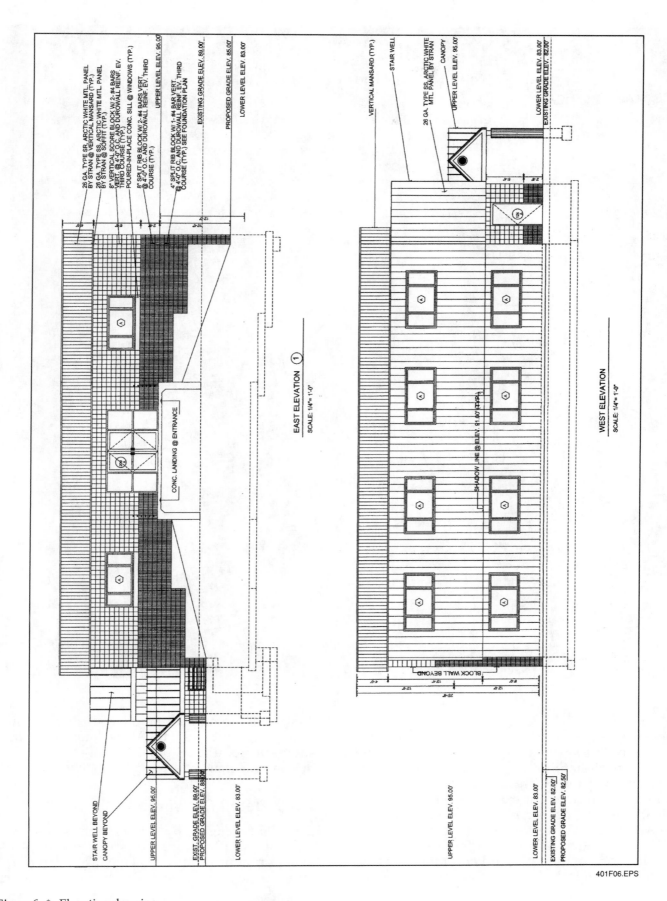

Figure 6 ◆ Elevation drawing.

401F06.EPS

Protection of Underground Facilities

In order to avoid damage or interruptions to underground utilities caused by digging, it is mandatory to contact the various companies for a utility stakeout prior to any digging. Most states have a One Call Notification System center that makes digging notification easy by calling a single phone number or by contacting the center on the Internet. Typically, the call must be made at least two or three working days before the digging is to begin.

A Dig Safely card and other materials are readily available from state Dig Safely notification centers to remind contractors and excavators of this requirement. Shown here is a Dig Safely card available from the Dig Safely New York notification center. In addition to giving the procedure for contacting the center, it shows the American Public Works Association (APWA) universal color codes used for the temporary marking of underground facilities. All other states have cards similar to the one shown for New York.

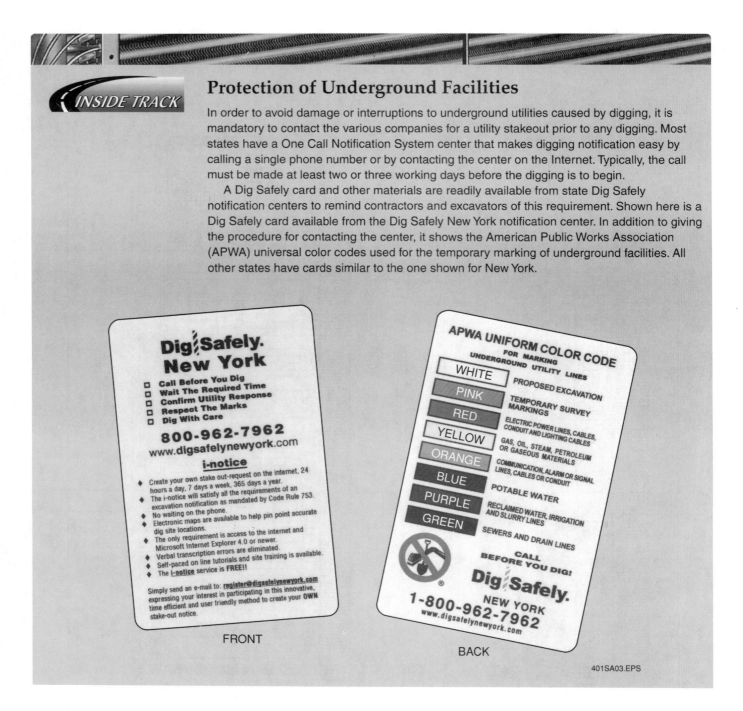

FRONT

BACK

401SA03.EPS

2.4.0 Schedules

Schedules are not drawings. They are tables shown on the various drawings throughout the drawing set that identify the types and sizes of items used by the different trades in the construction of a building. For example, schedules that identify the windows and doors in a building are shown on the floor plans. Of importance to the HVAC technician are the schedules for the mechanical components and equipment shown on the mechanical plans. Schedules shown on the related plumbing plans and electrical plans may also be relevant to the HVAC technician. *Figure 7* shows schedules typical of those found on mechanical plans.

2.5.0 Detail Drawings

Detail drawings show enlargements of special features of a building construction, fixtures, or equipment (*Figure 8*). They are drawn to a larger scale in order to make the details clearer.

CABINET UNIT HEATER SCHEDULE

UNIT HEATER NO.	LOCATION	C.F.M.	FAN MOTOR				MBH	GPM	EWT	EAT	MAX. WATER P.D.	REMARKS
			H.P.	VOLTS	PHASE	Hz						
CUH-1	124	400	1/12	115	1	60	23	2.3	180°F	60°F	2.7	McQUAY #CHF004 SEMI-RECESSED, R.H COIL
CUH-2	137	400	1/12	115	1	60	23	2.3	180°F	60°F	2.7	McQUAY #CHF004 SEMI-RECESSED, L.H COIL
CUH-3	143	400	1/12	115	1	60	23	2.3	180°F	60°F	2.7	McQUAY #CHF004 SEMI-RECESSED, R.H COIL

NOTES:
1. 3 Speed Control
2. Front Discharge
3. With Return Air Filters

PUMP SCHEDULE

UNIT NO.	LOCATION	SERVICE.	GPM	MBH	MOTOR					TYPE	REMARKS
					RPM	H.P.	VOLTS	PHASE	Hz		
P-1	MECH. ROOM	NAVE	45	41'	1750	1 1/2	208/230	3	60	IN-LINE	B/G #60-20T SERVICE 40% GLYCOL SOLUTION
P-2	MECH. ROOM	CHW TO AHU 1-4	65	37'	1750	1 1/2	208/230	3	60	IN-LINE	B/G #60-20T SERVICE 40% GLYCOL SOLUTION
P-3	MECH. ROOM	RECIR. TO TANK	40	17'	1750	1/2	208/230	3	60	IN-LINE	B/G #60-13T SERVICE 40% GLYCOL SOLUTION
P-4	EXIST. MECH. ROOM	HW	73	31'	1750	1 1/2	208/230	3	60	IN-LINE	B/G #60-20T HOT WATER

NOTES:
1. Starters And Disconnects By E.C.

GRILLE, REGISTER AND DIFFUSER SCHEDULE

ITEM	MANUFACTURER	MODEL NO.	QTY.	LOCATION	CFM EACH	AIR PATTERN	SIZE		FINISHES	REMARKS
							FRAME	NECK		
A	BARBER COLMAN	SFSV	8	126,127,128 144	245	4-WAY	12"× 12"	8"Ø	#7 OFF-WHITE	
B	BARBER COLMAN	SFSV	2	142	275	4-WAY	18"× 18"	10"Ø	#7 OFF-WHITE	
C	BARBER COLMAN	SFSV	4	140, 141	240	4-WAY	12"× 12"	8"Ø	#7 OFF-WHITE	
D	BARBER COLMAN	SFSV	2	139	270	4-WAY	18"× 18"	10"Ø	#7 OFF-WHITE	
E	BARBER COLMAN	SFSV	2	138	280	4-WAY	18"× 18"	10"Ø	#7 OFF-WHITE	
F	BARBER COLMAN	SFSV	2	136	250	4-WAY	12"× 12"	6"Ø	#7 OFF-WHITE	
G	BARBER COLMAN	SFSV	2	135	235	4-WAY	12"× 12"	6"Ø	#7 OFF-WHITE	
H	BARBER COLMAN	SFSV	3	134	100	4-WAY	12"× 12"	6"Ø	#7 OFF-WHITE	FIRE DAMPER SEE DETAIL A
I	BARBER COLMAN	SFSV	1	134	190	4-WAY	12"× 12"	8"Ø	#7 OFF-WHITE	FIRE DAMPER SEE DETAIL A
	COLMAN	SFSV	1	134		WAY	12"×			FIRE DAMPER

401F07.EPS

Figure 7 ◆ Mechanical equipment schedules.

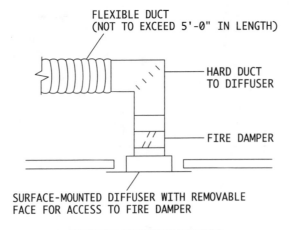

FLEXIBLE DUCT
(NOT TO EXCEED 5'-0" IN LENGTH)

HARD DUCT
TO DIFFUSER

FIRE DAMPER

SURFACE-MOUNTED DIFFUSER WITH REMOVABLE
FACE FOR ACCESS TO FIRE DAMPER

DIFFUSER WITH FIRE DAMPER

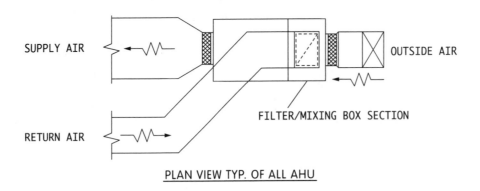

SUPPLY AIR

RETURN AIR

OUTSIDE AIR

FILTER/MIXING BOX SECTION

PLAN VIEW TYP. OF ALL AHU

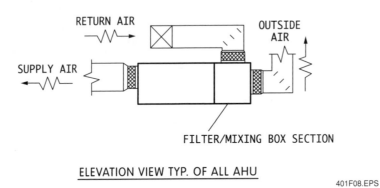

RETURN AIR

OUTSIDE AIR

SUPPLY AIR

FILTER/MIXING BOX SECTION

ELEVATION VIEW TYP. OF ALL AHU

401F08.EPS

Figure 8 ◆ Equipment-related detail drawings.

Note that the scale can vary from one detail view to another on the same sheet. For residential and some commercial projects, the detail drawings are often placed on the same sheet where the feature appears in the plan. However, for large or more complex commercial projects, detailed drawings may be drawn on a different sheet than where the feature appears. When this occurs, the detail drawings are referenced to and from the sheets where they apply.

2.6.0 Section Drawings

Section drawings (*Figure 9*) are cutaway views that allow the viewer to see the inside of a structure or how something is put together internally.

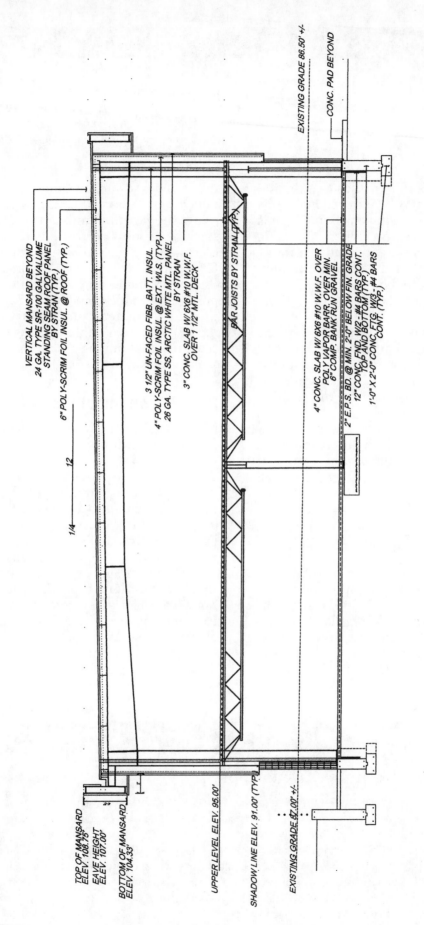

BUILDING SECTION
SCALE: 1/4" = 1'-0"

VERTICAL MANSARD BEYOND
24 GA. TYPE SR-100 GALVALUME STANDING SEAM ROOF PANEL BY STRAN (TYP.)

6" POLY-SCRIM FOIL INSUL. @ ROOF (TYP.)

3 1/2" UN-FACED FIBB. BATT. INSUL.
4" POLY-SCRIM FOIL INSUL. @ EXT. WLS. (TYP.)
26 GA. TYPE SS, ARCTIC WHITE MTL. PANEL BY STRAN
3" CONC. SLAB W/ 6X6 #10 W.W.F. OVER 1 1/2" MTL. DECK

BAR JOISTS BY STRAN (TYP.)

4" CONC. SLAB W/ 6X6 #10 W.W.F. OVER POLY VAPOR BARR. OVER MIN. 6" COMP. BANK RUN GRAVEL

2" E.P.S. BD. @ MIN. 2'-0" BELOW FIN. GRADE
12" CONC. FND. W/2 - #4 BARS CONT. TOP AND BOTTOM (TYP.)
1'-0" X 2'-0" CONC. FTG. W/3 - #4 BARS CONT. (TYP.)

EXISTING GRADE 86.50' +/-

CONC. PAD BEYOND

TOP OF MANSARD ELEV. 108.75'
EAVE HEIGHT ELEV. 107.00'
BOTTOM OF MANSARD ELEV. 104.33'

UPPER LEVEL ELEV. 95.00'

SHADOW LINE ELEV. 91.00' (TYP.)

EXISTING GRADE 87.00' +/-

12
1/4

Figure 9 ◆ Section drawing showing building construction.

401F09.EPS

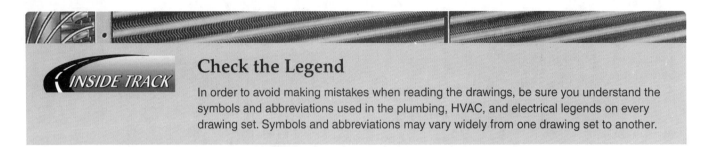

Check the Legend

In order to avoid making mistakes when reading the drawings, be sure you understand the symbols and abbreviations used in the plumbing, HVAC, and electrical legends on every drawing set. Symbols and abbreviations may vary widely from one drawing set to another.

A feature is drawn as if a cut has been made through the middle of it. When a sectional cut is made along the long dimension of a building, it is called a **longitudinal section**. When it is made through the short dimension, it is called a **transverse section**. The point on the plan or elevation drawing showing where the imaginary cut has been made is indicated by the section line, which is usually a dashed line.

Like detail drawings, section drawings are drawn to a larger scale than that used in plan views. Section views are commonly given for the construction of walls, stairs, cabinets, and other building features that require more information than is given on the plan views.

Section views on mechanical plans are typically used to show more information about the installation features of a particular fixture or piece of equipment within a building. The detail views in *Figure 10* show locations for the installation of an air handler unit.

2.7.0 Plumbing Plans

Plumbing plans show the layout of fixtures, water supply lines, natural gas piping, and lines to sewage disposal systems. The plans may be included in the floor plan of a regular construction job or on a separate plan for a large commercial structure. When drawn as a separate plan, the plumbing plan details are usually overlaid on tracings of the various building floor plans from which unnecessary details have been omitted to allow the location and layout of the plumbing systems to show clearly. *Figure 11* is a plumbing plan showing the sanitary plumbing in both plan form and as an isometric drawing.

A plumbing legend shows the various symbols pertaining to the plan (*Figure 12*). Some legends provide tabulated plumbing fixture and equipment schedules. Plumbing plans may also show a schedule of plumbing systems and plumbing system specifications.

2.8.0 Mechanical Plans

Mechanical plans show the heating, ventilation, and air conditioning systems, as well as other mechanical systems for a building. For some residential jobs, the mechanical plan may be combined with the plumbing plan and show very little detailed information other than the locations of the main HVAC system components. This is because the installation location of duct or piping runs is allowed to be determined on the job by the HVAC contractor. For large commercial jobs, the mechanical plans typically show information about the HVAC system installation and the installation of other equipment. Information about the installation of the plumbing and electrical systems is usually shown on separate stand-alone plumbing and electrical plan drawings.

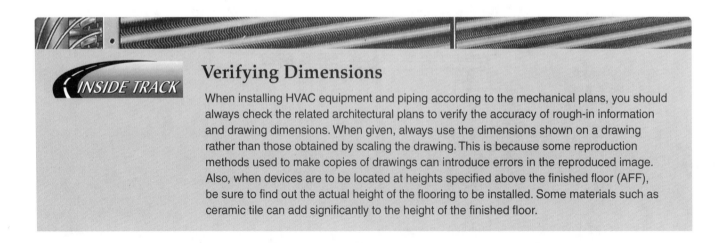

Verifying Dimensions

When installing HVAC equipment and piping according to the mechanical plans, you should always check the related architectural plans to verify the accuracy of rough-in information and drawing dimensions. When given, always use the dimensions shown on a drawing rather than those obtained by scaling the drawing. This is because some reproduction methods used to make copies of drawings can introduce errors in the reproduced image. Also, when devices are to be located at heights specified above the finished floor (AFF), be sure to find out the actual height of the flooring to be installed. Some materials such as ceramic tile can add significantly to the height of the finished floor.

DETAIL SHEET M4-A

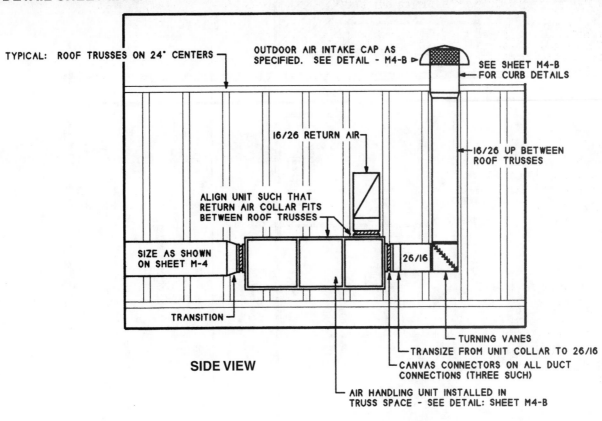

SIDE VIEW

DETAIL SHEET M4-B

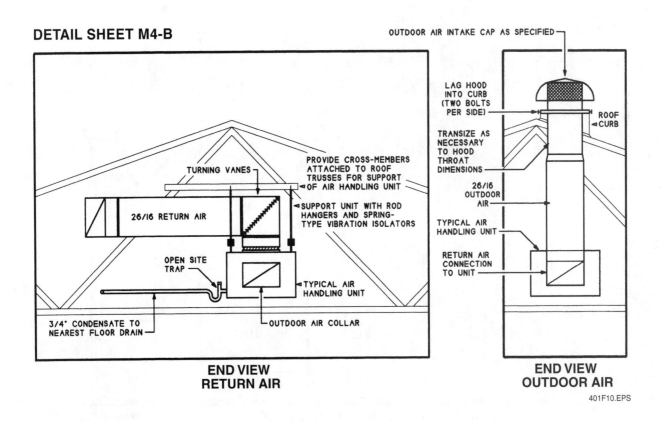

**END VIEW
RETURN AIR**

**END VIEW
OUTDOOR AIR**

401F10.EPS

Figure 10 ◆ Section drawings showing air handling unit installation.

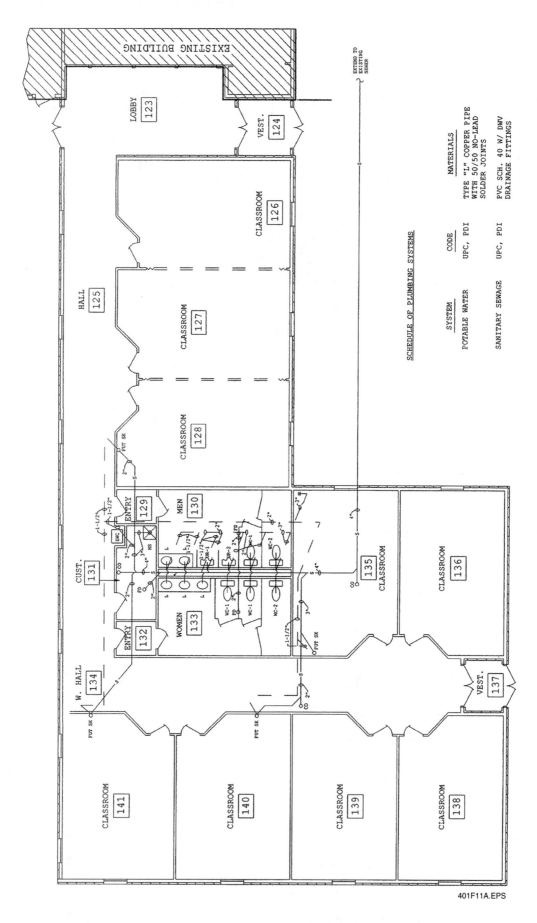

401F11A.EPS

Figure 11 ◆ Sanitary plumbing plan (sheet 1 of 2).

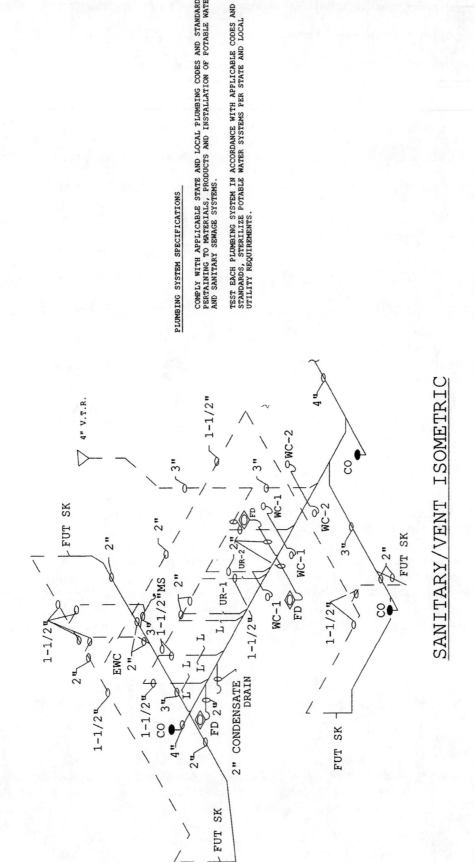

Figure 11 ◆ Sanitary plumbing plan (sheet 2 of 2).

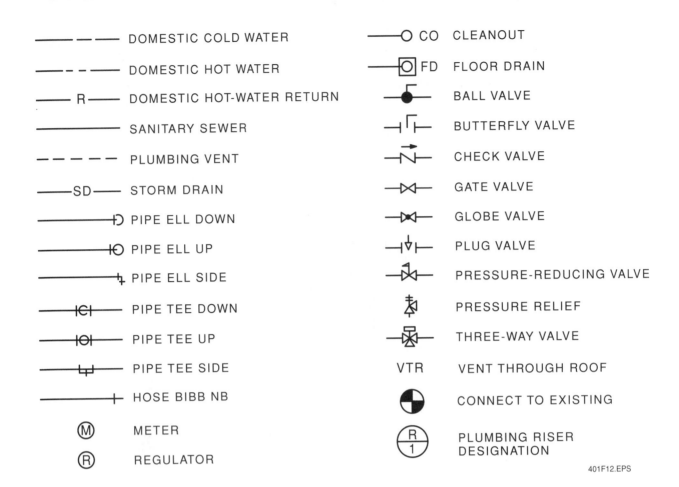

――― – – ―――	DOMESTIC COLD WATER		――○ CO	CLEANOUT
――― - - - ―――	DOMESTIC HOT WATER		――◙ FD	FLOOR DRAIN
――― R ―――	DOMESTIC HOT-WATER RETURN		――●―	BALL VALVE
―――――――	SANITARY SEWER		―⌐ Γ―	BUTTERFLY VALVE
– – – – –	PLUMBING VENT		―◸―	CHECK VALVE
―― SD ――	STORM DRAIN		―◁▷―	GATE VALVE
――――⊃	PIPE ELL DOWN		―◀▶―	GLOBE VALVE
――――○	PIPE ELL UP		―◡―	PLUG VALVE
――――↳	PIPE ELL SIDE		―◁◁―	PRESSURE-REDUCING VALVE
――C―――	PIPE TEE DOWN		◿	PRESSURE RELIEF
――○―――	PIPE TEE UP		―◿―	THREE-WAY VALVE
――⊔―――	PIPE TEE SIDE		VTR	VENT THROUGH ROOF
―――――	HOSE BIBB NB		⊕	CONNECT TO EXISTING
Ⓜ	METER		⊘ R 1	PLUMBING RISER DESIGNATION
Ⓡ	REGULATOR			

401F12.EPS

Figure 12 ◆ Plumbing legend.

As with separate plumbing plans, details of the mechanical plan are usually overlaid on tracings of the various building floor plans from which unnecessary details have been omitted to allow the location and layout of the HVAC system equipment to show clearly. *Figure 13* shows an example of a typical HVAC mechanical plan.

Mechanical plans typically contain tabulated schedules that identify the different items and types of HVAC equipment. As appropriate, detailed views describing the installation of the HVAC equipment are shown. Depending on the nature of the project, these views can include refrigeration piping schematics (*Figure 14*), chilled-water coil and/or hot-water coil piping schematics, and views detailing piping runs and pipe sizes for major items of HVAC equipment.

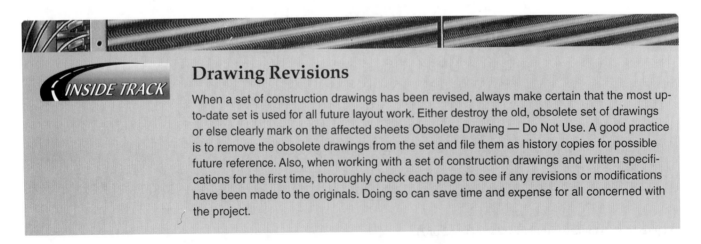

Drawing Revisions

When a set of construction drawings has been revised, always make certain that the most up-to-date set is used for all future layout work. Either destroy the old, obsolete set of drawings or else clearly mark on the affected sheets Obsolete Drawing — Do Not Use. A good practice is to remove the obsolete drawings from the set and file them as history copies for possible future reference. Also, when working with a set of construction drawings and written specifications for the first time, thoroughly check each page to see if any revisions or modifications have been made to the originals. Doing so can save time and expense for all concerned with the project.

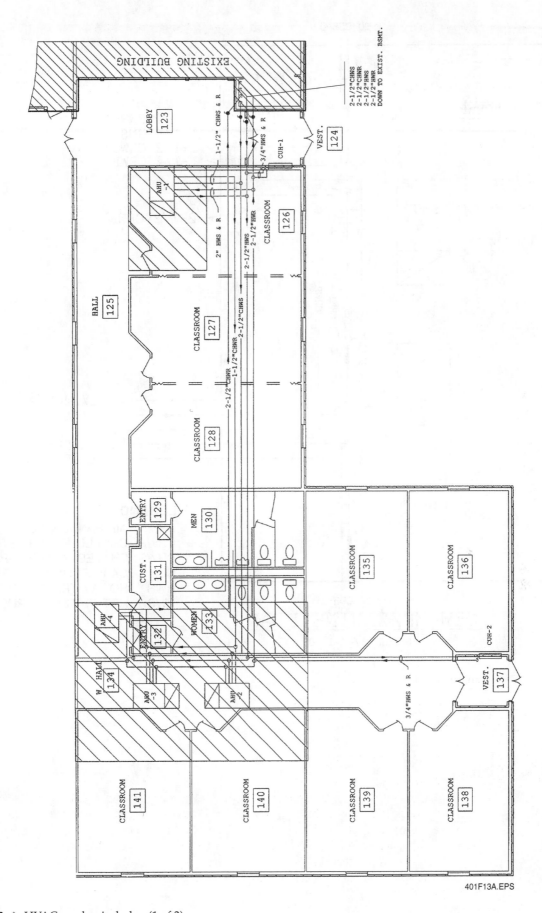

Figure 13 ◆ HVAC mechanical plan (1 of 3).

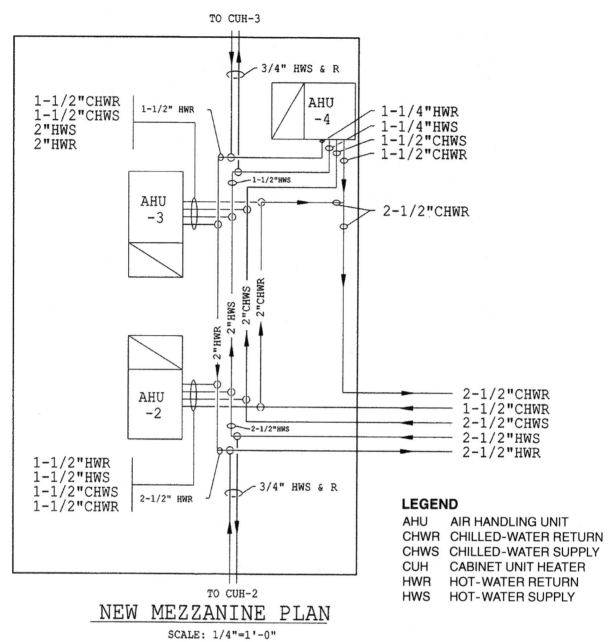

NEW MEZZANINE PLAN

SCALE: 1/4"=1'-0"

LEGEND

AHU	AIR HANDLING UNIT
CHWR	CHILLED-WATER RETURN
CHWS	CHILLED-WATER SUPPLY
CUH	CABINET UNIT HEATER
HWR	HOT-WATER RETURN
HWS	HOT-WATER SUPPLY

401F13B.EPS

Figure 13 ◆ HVAC mechanical plan (2 of 3).

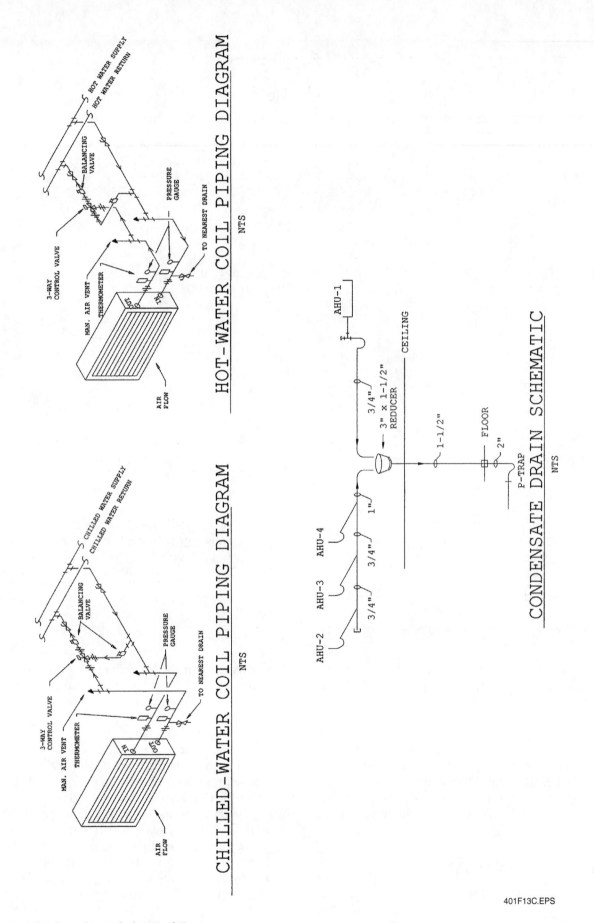

Figure 13 ◆ HVAC mechanical plan (3 of 3).

401F13C.EPS

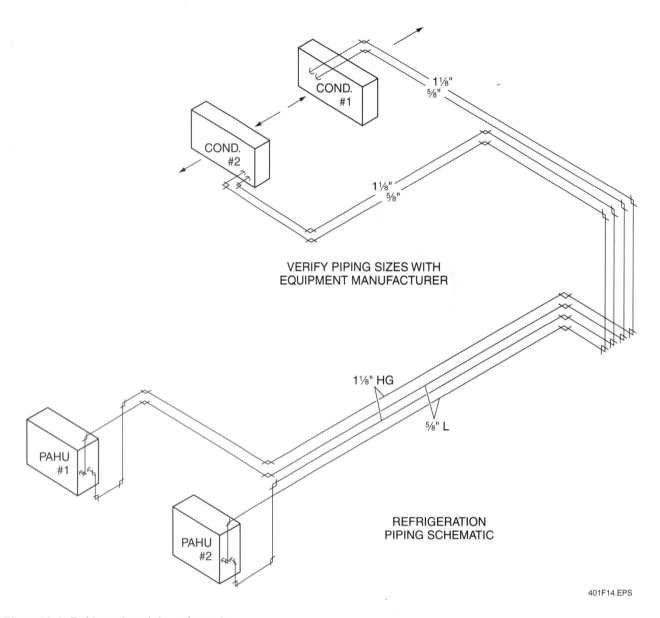

VERIFY PIPING SIZES WITH
EQUIPMENT MANUFACTURER

REFRIGERATION
PIPING SCHEMATIC

401F14.EPS

Figure 14 ◆ Refrigeration piping schematic.

Mechanical plans also normally include an HVAC legend listing the various symbols pertaining to the plan. *Figure 15* shows a partial HVAC legend.

Some mechanical plans also contain a schedule of HVAC systems (*Figure 16*) and information about relevant HVAC system specifications (*Figure 17*). It is important to point out that specifications are job-specific. Always be sure to read the specifications for the particular job you are working on.

2.9.0 Electrical Plans

For smaller construction jobs, the electrical plans are usually shown on the architectural floor plans. For large commercial jobs, the electrical plans are typically stand-alone drawings that show only information about the electrical system installation. Like the separate plumbing and mechanical plans, electrical plans typically overlay tracings of the various building floor plans from which unnecessary details have been omitted to allow the location and layout of the electrical system equipment to show clearly. *Figure 18* shows an example of a typical electrical plan.

Electrical plans show the locations of the meter, distribution panels, light fixtures, switches, and other electrical equipment. Also shown are equipment and fixtures schedules, an electrical legend listing the various symbols pertaining to the plan, specifications for load capacities, and wire sizes. Electrical plans usually have a power **riser diagram** that shows all the major pieces of electrical

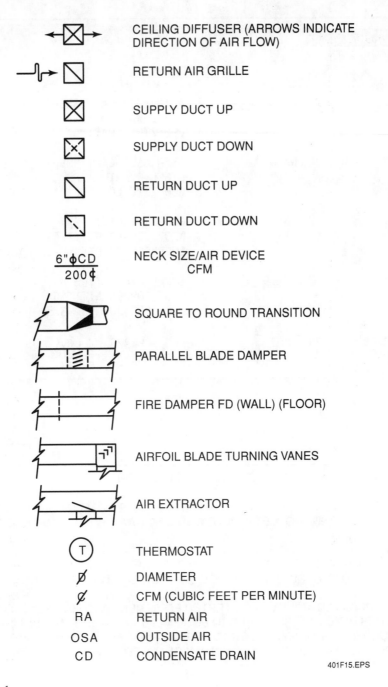

⟷⊠⟷	CEILING DIFFUSER (ARROWS INDICATE DIRECTION OF AIR FLOW)
⌐→ ◻	RETURN AIR GRILLE
⊠	SUPPLY DUCT UP
⊠	SUPPLY DUCT DOWN
◻	RETURN DUCT UP
◻	RETURN DUCT DOWN
6"⌀CD / 200¢	NECK SIZE/AIR DEVICE CFM
	SQUARE TO ROUND TRANSITION
	PARALLEL BLADE DAMPER
	FIRE DAMPER FD (WALL) (FLOOR)
	AIRFOIL BLADE TURNING VANES
	AIR EXTRACTOR
Ⓣ	THERMOSTAT
⌀	DIAMETER
⌀	CFM (CUBIC FEET PER MINUTE)
RA	RETURN AIR
OSA	OUTSIDE AIR
CD	CONDENSATE DRAIN

401F15.EPS

Figure 15 ◆ HVAC legend.

equipment, including HVAC equipment, as well as the connecting lines used to indicate service-entrance conductors and feeders. Electrical plans may also contain information about the electrical specifications.

3.0.0 ◆ REQUEST FOR INFORMATION

The request for information (RFI) is used for clarification. If a discrepancy, conflict, or incomplete information is noted on the plans, an RFI may be issued to the architect or engineer. There is a hierarchy that is usually followed. For example,

should you notice a discrepancy on the plans, you should notify the foreman. The foreman generates the RFI, being as specific as possible and making sure to put the time, date, and company RFI number on it. The foreman gives the RFI to the superintendent or project manager, who passes it on to the general contractor. The general contractor then relays the RFI to the architect or engineer. A sample RFI form is shown in *Figure 19*. The person who finds a discrepancy on the plans should clearly describe the conflicting information and, if possible, suggest how to correct the problem.

SCHEDULE OF HVAC SYSTEMS

SYSTEM	CODE	MATERIALS	INSULATION
METAL DUCTWORK	SMACNA NFPA 90 A & B	ASTM A527 GAL. SHEET STEEL W/ASTM A525 G90 ZINC COATING	1" DUCTLINER TIMA AHC-101
ROUND DUCTWORK	SMACNA NFPA 90 A & B	ASTM A527 GAL. SHEET STEEL W/ASTM A525 G90 ZINC COATING	1½" DUCTWRAP W/ VAPOR JACKET
HOT WATER	ASME 31.9	COPPER TUBE TYPE L WROT COPPER FITTINGS SOLDERED JOINTS 95/5	1" FIBERGLASS
CHILLED WATER	ASME 31.9	COPPER TUBE TYPE L WROT COPPER FITTINGS SOLDERED JOINTS 95/5	1" FIBERGLASS
CONDENSATE	UPC, PDI	PVC SCH 40, W/ DWV DRAINAGE FITTINGS	½" FLEXIBLE UNICELL

401F16.EPS

Figure 16 ◆ Schedule of HVAC systems.

HEATING, VENTILATING & AIR CONDITIONING SYSTEM SPECIFICATIONS

COMPLY WITH APPLICABLE MECHANICAL CODES AND STANDARDS PERTAINING TO MATERIALS, PRODUCTS, AND INSTALLATION OF AIR HANDLING, METAL DUCTWORK, HOT-WATER SYSTEMS, AND CHILLED-WATER SYSTEMS.

SUBMIT MANUFACTURER'S TECHNICAL PRODUCT DATA TAILORED TO THE PROJECT, ASSEMBLY-TYPE SHOP DRAWINGS, WIRING DRAWINGS, AND MAINTENANCE DATA FOR EACH COMPONENT OF EACH HVAC SYSTEM.

PROVIDE FACTORY-FABRICATED AND FACTORY-TESTED EQUIPMENT AND MATERIALS OF SIZES, RATINGS, AND CHARACTERISTICS INDICATED. REFERENCED EQUIPMENT AND MATERIALS INDICATE STYLE AND QUALITY DESIRED. CONTACT ENGINEER PRIOR TO SUBMITTAL OF ANY OTHER MANUFACTURER FOR PRELIMINARY APPROVAL. PROVIDE PROPER QUANTITY OF MATERIAL AND EQUIPMENT AS REQUIRED FOR COMPLETE INSTALLATION OF EACH HVAC SYSTEM.

IDENTIFY EACH HVAC SYSTEM'S COMPONENTS WITH MATERIALS AND DESIGNATIONS AS DIRECTED.

INSTALL EACH HVAC SYSTEM IN ACCORDANCE WITH APPLICABLE MECHANICAL CODES AND STANDARDS, RECOGNIZED INDUSTRY PRACTICES, AND MANUFACTURER'S RECOMMENDATIONS.

TEST AND BALANCE EACH HVAC SYSTEM IN ACCORDANCE WITH APPLICABLE MECHANICAL CODES AND STANDARDS. BALANCE AIR CONDITIONING SYSTEM TO CFM'S SHOWN ON DRAWINGS. REPORT FINDINGS TO ENGINEER USING APPROVED FORMS.

401F17.EPS

Figure 17 ◆ Example of HVAC system specifications information.

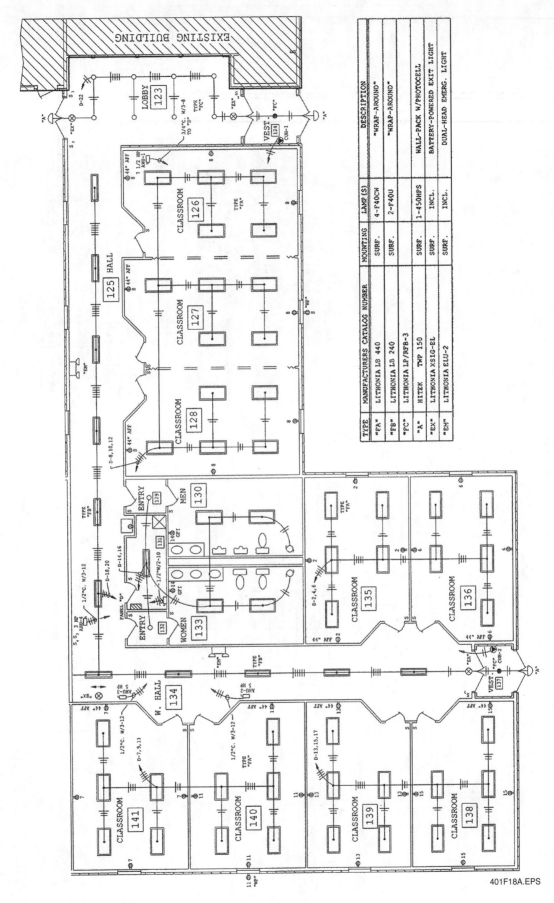

Figure 18 ◆ Electrical plan (1 of 2).

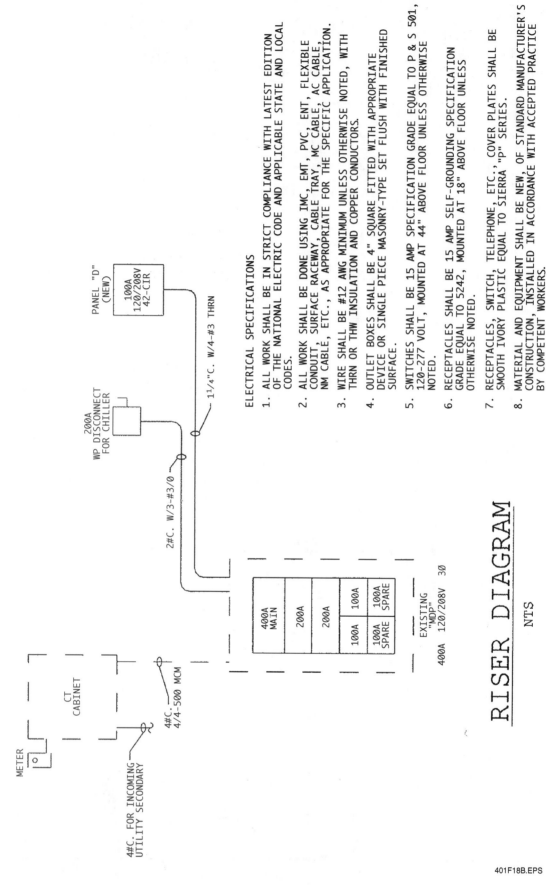

ELECTRICAL SPECIFICATIONS

1. ALL WORK SHALL BE IN STRICT COMPLIANCE WITH LATEST EDITION OF THE NATIONAL ELECTRIC CODE AND APPLICABLE STATE AND LOCAL CODES.

2. ALL WORK SHALL BE DONE USING IMC, EMT, PVC, ENT, FLEXIBLE CONDUIT, SURFACE RACEWAY, CABLE TRAY, MC CABLE, AC CABLE, NM CABLE, ETC., AS APPROPRIATE FOR THE SPECIFIC APPLICATION.

3. WIRE SHALL BE #12 AWG MINIMUM UNLESS OTHERWISE NOTED, WITH THRN OR THW INSULATION AND COPPER CONDUCTORS.

4. OUTLET BOXES SHALL BE 4" SQUARE FITTED WITH APPROPRIATE DEVICE OR SINGLE PIECE MASONRY-TYPE SET FLUSH WITH FINISHED SURFACE.

5. SWITCHES SHALL BE 15 AMP SPECIFICATION GRADE EQUAL TO P & S 501, 120-277 VOLT, MOUNTED AT 44" ABOVE FLOOR UNLESS OTHERWISE NOTED.

6. RECEPTACLES SHALL BE 15 AMP SELF-GROUNDING SPECIFICATION GRADE EQUAL TO 5242, MOUNTED AT 18" ABOVE FLOOR UNLESS OTHERWISE NOTED.

7. RECEPTACLES, SWITCH, TELEPHONE, ETC., COVER PLATES SHALL BE SMOOTH IVORY PLASTIC EQUAL TO SIERRA "P" SERIES.

8. MATERIAL AND EQUIPMENT SHALL BE NEW, OF STANDARD MANUFACTURER'S CONSTRUCTION, INSTALLED IN ACCORDANCE WITH ACCEPTED PRACTICE BY COMPETENT WORKERS.

RISER DIAGRAM

NTS

PANEL "D" (NEW)

100A
120/208V
42-CIR

1¹/₄"C. W/4-#3 THRN

200A
WP DISCONNECT FOR CHILLER

2#C. W/3-#3/0

4#C.
4/4-500 MCM

METER

CT CABINET

4#C. FOR INCOMING UTILITY SECONDARY

400A
MAIN

200A

200A

100A   100A

100A   100A
SPARE   SPARE

EXISTING
"MDP"

400A   120/208V   3Ø

401F18B.EPS

Figure 18 ◆ Electrical plan (2 of 2).

XYZ, Inc
General Contractors
123 Main Street
Bigtown, USA 10001
(111) 444-5555

R.F.I.

Request for Information

XYZ Project #_____
Date: _____
R.F.I. # _____

PROJECT:

TO:

RE:

Specification
Reference:_____

Drawing
Reference:_____

SUBJECT:

REQUIRED:

Date Information
is Required:_____

XYZ, Inc
By: _____

REPLY:

Distribution: Superintendent
Field File

By: _____

Date: _____

401F19.EPS

Figure 19 ◆ Request for information form.

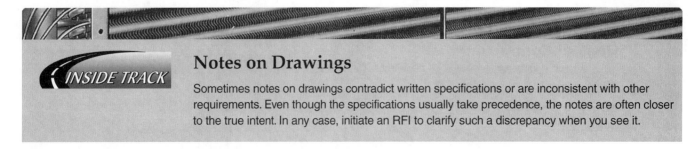

4.0.0 ◆ SPECIFICATIONS

Specifications are written statements provided by the architectural and engineering firm to the general contractor and, consequently, to the subcontractors. Specifications define the quality of work to be done and describe the materials to be used. They are very important to the architect and owner because they guarantee compliance by the contractors to the standards set for the project. Specifications consist of various elements that may differ somewhat for particular construction projects. An example of the abbreviated specifications for a hypothetical construction project is provided in *Appendix B*. This example follows the old CSI format.

4.1.0 Purpose

Specifications have several important purposes:

- They clarify information that cannot be shown on the drawings.
- They identify the work standards, types of materials to be used, and responsibilities of various parties to the contract.
- They provide information on details of construction.
- They serve as a guide for contractors bidding on the construction job.
- They serve as a standard of quality for materials and workmanship.
- They serve as a guide for compliance with building codes and zoning ordinances.

- They serve as the basis of agreement between the owner, architect, and contractors in settling any disputes.

There are two types of information contained in a set of specifications: non-technical (special and general conditions) and technical aspects of construction.

4.2.0 Special and General Conditions

Special and general conditions cover the non-technical aspects of the contractual arrangements. Special conditions cover topics such as safety, temporary construction, **shop drawing(s)** required, and so on. General conditions cover the following points of information:

- Contract terms
- Responsibilities for examining the construction site
- Types and limits of insurance
- Permits and payment of fees
- Use and installation of utilities
- Supervision of construction
- Other pertinent items

The general conditions section is the area of the construction contract where misunderstandings often occur. Therefore, these conditions are usually much more explicit on large, complicated construction projects.

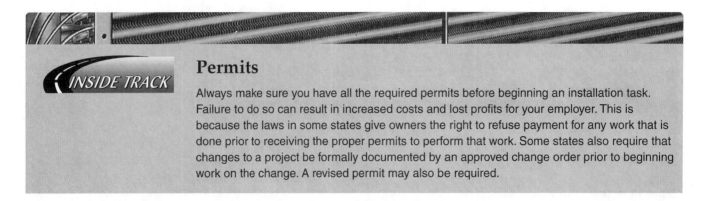

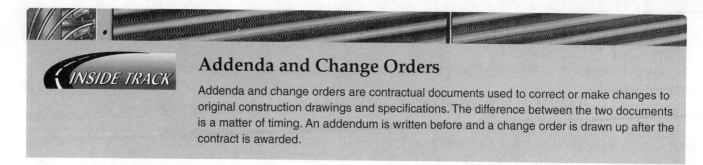

4.3.0 Technical Aspects

The technical aspects section of the specifications covers the work to be done by the major divisions and identifies the standards for each part. The divisions are usually in the order in which the work will be performed; for example, site work is listed before carpentry work.

The technical aspects section includes information on materials that are specified by standard numbers and by standard national testing organizations, such as the American Society of Testing Materials (ASTM). The specifications are usually written around some standard format published by the American Institute of Architects (AIA).

4.4.0 Format

For convenience in writing, speed in estimating work, and ease in reference, the most suitable organization of the specifications is a series of sections dealing successively with the different trades. All the work of each trade should be incorporated into the section devoted to that trade. Those people who use the specifications must be able to find all needed information quickly.

4.4.1 CSI Format

The most commonly used specification-writing format used in North America is the *Master-Format*™. This standard was developed jointly by the Construction Specifications Institute (CSI) and Construction Specifications Canada (CSC). For many years prior to 2004, the organization of construction specifications and suppliers catalogs was based on a standard with 16 sections, otherwise known as divisions, where the divisions and their subsections were individually identified by a five-digit numbering system. The first two digits represented the division number and the next three individual numbers represented succes-

sively lower levels of breakdown. For example, the number 13213 represents division 13, subsection 2, sub-subsection 1 and sub-sub-subsection 3. In this older version of the standard, electrical systems, including any electronic or special electrical systems, were lumped together under Division 16 – *Electrical*. Today, specifications conforming to the 16 division format may still be in use. The older version of the standard contains the following divisions:

- Division 1 – *General Requirements*
- Division 2 – *Site Work*
- Division 3 – *Concrete*
- Division 4 – *Masonry*
- Division 5 – *Metals*
- Division 6 – *Wood and Plastics*
- Division 7 – *Thermal and Moisture Protection*
- Division 8 – *Doors and Windows*
- Division 9 – *Finishes*
- Division 10 – *Specialties*
- Division 11 – *Equipment*
- Division 12 – *Furnishings*
- Division 13 – *Special Construction*
- Division 14 – *Conveying Systems*
- Division 15 – *Mechanical*
- Division 16 – *Electrical*

In 2004, the *MasterFormat*™ standard underwent a major change. The 16 original divisions were expanded to four major groupings and 49 divisions with some divisions reserved for future expansion (*Figure 20*). The first 14 divisions are essentially the same as the old format. Subjects under the old Division 15 – *Mechanical* have been relocated to new divisions 21, 22, and 23. The basic subjects under old Division 16 – *Electrical* have been relocated to new divisions 26 and 27. In addition, the numbering system was changed to 6 digits to allow for more subsections in each division for finer definition.

Division Numbers and Titles

PROCUREMENT AND CONTRACTING REQUIREMENTS GROUP

Division 00   Procurement and Contracting Requirements

SPECIFICATIONS GROUP

GENERAL REQUIREMENTS SUBGROUP

Division 01   General Requirements

FACILITY CONSTRUCTION SUBGROUP

Division 02   Existing Conditions
Division 03   Concrete
Division 04   Masonry
Division 05   Metals
Division 06   Wood, Plastics, and Composites
Division 07   Thermal and Moisture Protection
Division 08   Openings
Division 09   Finishes
Division 10   Specialties
Division 11   Equipment
Division 12   Furnishings
Division 13   Special Construction
Division 14   Conveying Equipment
Division 15   *Reserved*
Division 16   *Reserved*
Division 17   *Reserved*
Division 18   *Reserved*
Division 19   *Reserved*

FACILITY CONSTRUCTION SUBGROUP

Division 20   *Reserved*
Division 21   Fire Suppression
Division 22   Plumbing
Division 23   Heating, Ventilating, and Air Conditioning
Division 24   *Reserved*
Division 25   Integrated Automation
Division 26   Electrical
Division 27   Communications
Division 28   Electronic Safety and Security
Division 29   *Reserved*

SITE AND INFRASTRUCTURE SUBGROUP

Division 30   *Reserved*
Division 31   Earthwork
Division 32   Exterior Improvements
Division 33   Utilities
Division 34   Transportation
Division 35   Waterway and Marine Construction
Division 36   *Reserved*
Division 37   *Reserved*
Division 38   *Reserved*
Division 39   *Reserved*

PROCESS EQUIPMENT SUBGROUP

Division 40   Process Integration
Division 41   Material Processing and Handling Equipment
Division 42   Process Heating, Cooling, and Drying Equipment
Division 43   Process Gas and Liquid Handling, Purification, and Storage Equipment
Division 44   Pollution Control Equipment
Division 45   Industry-Specific Manufacturing Equipment
Division 46   *Reserved*
Division 47   *Reserved*
Division 48   Electrical Power Generation
Division 49   *Reserved*

Div Numbers - 1

401F20.EPS

Figure 20 ◆ 2004 *MasterFormat*™.

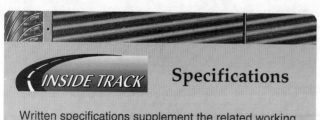

5.0.0 ◆ SHOP DRAWINGS

There are three types of shop drawings. One type is a detail drawing that a drafter creates after the engineer designs the structure. It illustrates the connections used, shows the location of all holes and openings, and provides notes specifying how each part is to be made. Assembly instructions are also included and are used principally for structural steel members.

A second type of shop drawing (or submittal) pertains to the purchase of special items of equipment for installation in a building. This kind of shop drawing is usually prepared by the equipment manufacturers. This drawing shows overall sizes, details of construction, methods for securing the equipment to the structure, and all other pertinent data that the architect and contractor need to know for the final placement and installation of the equipment. The contractor must check the dimensions of a submitted item of equipment to make sure it will fit into the space provided. If the space is not adequate, an RFI form is filled out so that the architect or engineer responsible can correct the problem.

A third type of shop drawing very similar in development to the structural steel shop drawing is the one used by sheet metal fabricators and installers. *Figure 21* shows an example of a shop drawing. This shop drawing involves sheet metal drafting techniques and is developed from the design drawings. This shop drawing is usually drawn to a scale that is several times larger than the design drawing. A sheet metal shop drawing shows the exact layout of the ductwork, the sizes and types of fittings, the types of connectors and hangers required for installation, and notes that will assist in fabrication or installation.

The first and second types of shop drawings usually come from the contractor or fabricator and are submitted to the owner or architect for approval and/or revisions or corrections. The design drawing is often put on the same sheet as the shop drawing. Shop drawings are drawn to a large enough scale so that they are clear, but they must not be crowded. They are usually dimensioned to the nearest $\frac{1}{16}$ of an inch.

One contractor uses the following approach for the development of shop drawings; when the contract is signed by the president of the company, a full set of drawings is given to the drafting department. Upon receipt of the drawings, the drafting department, depending on the workload, immediately begins developing the shop drawings. As the shop drawings are taken from the mechanical prints, the drafting department also researches the plumbing, electrical, HVAC piping, architectural, and structural prints to see if there will be any conflicts. These drawings are commonly called **coordination drawings**.

Coordination drawings are drawings produced for a project by the individual contractors for each trade to prevent a conflict between the trades in the installation of their materials and equipment. They are produced prior to finalizing shop drawings, cut lists, and other drawings and before the installation process begins. Development of these drawings evolves through a series of review and coordination meetings held by the various contractors.

Some contracts mandate that coordination drawings be drawn, while others only recommend it. In the case where one contractor elects to make coordination drawings and another does not, the contractor who made the drawings may be given the installation right-of-way by the presiding authority. As a result, the other contractor may have to bear the expense of removing and reinstalling equipment if the equipment was installed in a space designated for use by the contractor who produced coordination drawings.

5.1.0 Cut Lists

Another function of the shop drawing is to assist the subcontractor in identifying the number and sizes of the fittings and duct run sections that must be fabricated and subsequently installed on the job.

After the shop drawings are complete, or as they are drawn (depending on the workload), the drafter makes a cut sheet of each individual fitting and assigns a fitting number to it that matches the numbers on the shop drawing.

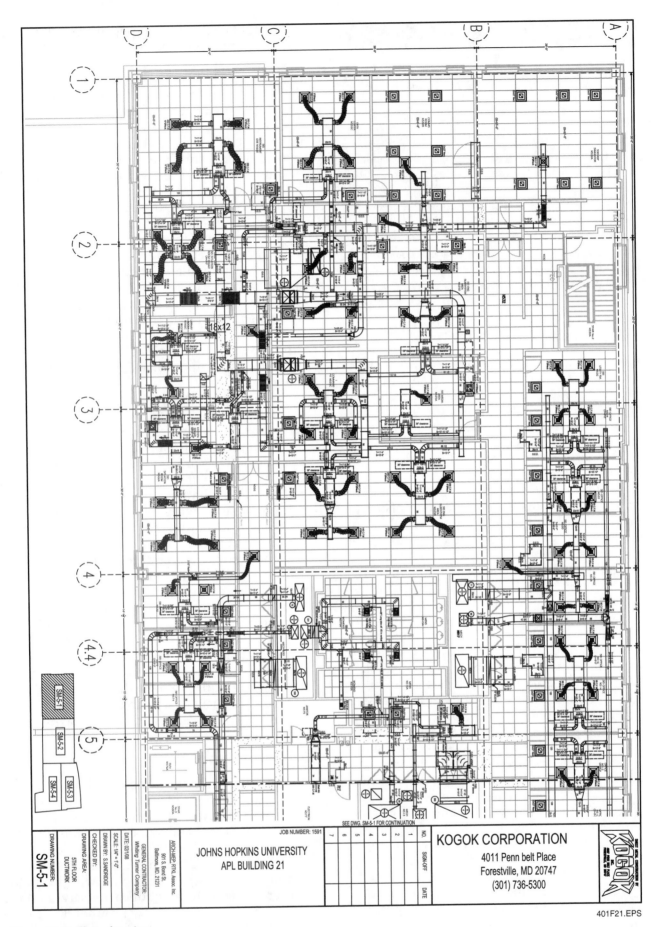

Figure 21 ◆ Shop drawing.

The straight duct sections are given another number that stays the same until the duct size changes. The cut sheets are then given to the fabrication shop. These cut sheets contain the job number, the quantity of ducts required, and the gauge of metal from which they are to be fabricated. Some contractors have their own methods for tracking fabricated duct pertaining to the duct line, delivery date, and job number.

Cut lists are a function of the production phase in the fabrication shop and may either be generated by the drafter (*Figure 22*) or generated by a computer (*Figure 23*). The cut lists identify the fittings and sections by number, type of fitting, amount required, width, depth, length, type of flange or connection to be used, number of parts required for the fitting, cut size, and type and gauge of metal.

5.2.0 General Procedure

In large shops, the sheet metal drafters are usually sheet metal mechanics who have been trained in the use of drawing instruments and layout procedures. In smaller shops, the owner or journeyperson may be required to develop the shop drawings. Freehand sketches from field measurements are often passed on to the drafter to develop a shop drawing. Sometimes, written notes and descriptions are provided, and the drafter must translate that information into shop drawings. A general procedure for producing shop drawings is as follows:

Step 1 Select a scale two to four times larger than the scale used for the design drawing. The scale selected should suit the intended purpose of the drawing. If the drawing is to be used as a coordination drawing, then each contractor should use the same scale in order to make comparisons among the drawings easier. Sometimes the engineers or architects will provide a basic template for the drawing that helps achieve accuracy in the layout.

Step 2 Arrange the layout to be evenly spaced on the sheet; it may be desirable to center the layout.

Step 3 Use the standard symbols on the drawing.

Step 4 Add notes when and where necessary.

Step 5 Draw in partitions, exterior walls, beams, columns, hanging ceilings, and any other obstructions that appear on the architectural plan.

Step 6 Use the design drawing to calculate all measurements needed to properly locate the ductwork.

Step 7 When dimensioning, indicate:
- The measurements from the finished floor to the bottom of the duct
- The duct height
- The clearance from the top of the duct to the bottom of the slab if applicable
- The measurement from the ceiling to the bottom of the slab if a suspended ceiling is part of the construction

Step 8 Properly locate all access doors, volume dampers, fire and smoke dampers, boots, registers, duct linings, thermostats, and other accessories on the drawing.

Step 9 Allow sufficient clearance around all ducts so they can pass through walls easily. Check the specifications for sealing around ducts passing through walls and around ducted fire and smoke dampers.

Step 10 To properly locate ducts, dimension from the centerline of the nearest column.

Step 11 Refer periodically to the design drawing to check for interference with other trades.

Step 12 Number all pieces of ductwork on the shop drawing according to the practices in your shop.

Step 13 Make up a tally sheet or cut sheet that indicates each piece's identification number, the size, description, quantity, type, and gauge of material, and any other pertinent information that will help the fabricator and/or installer.

In addition, consider the following factors when you prepare shop drawings:

- Carefully check the electrical and plumbing mechanical drawings when preparing drawings for ducts.
- Note that the types of connections used for conventional low-pressure or high-pressure duct or for heavy-gauge duct affect the length of the joints and fittings.
- Where necessary, if provided, note on the shop drawing the thickness and type of acoustical lining and external insulation.
- Be aware that duct sizes are usually increased to allow for the thickness of acoustical lining, but plans and specifications should be checked for verification of the designer's intent.

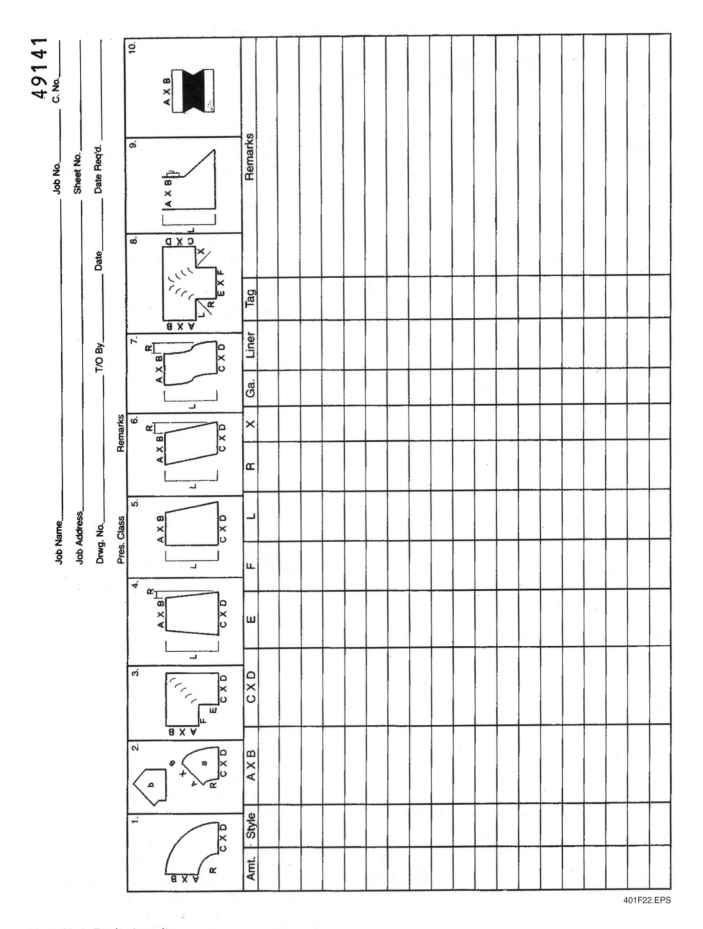

Figure 22 ◆ Drafter's cut list.

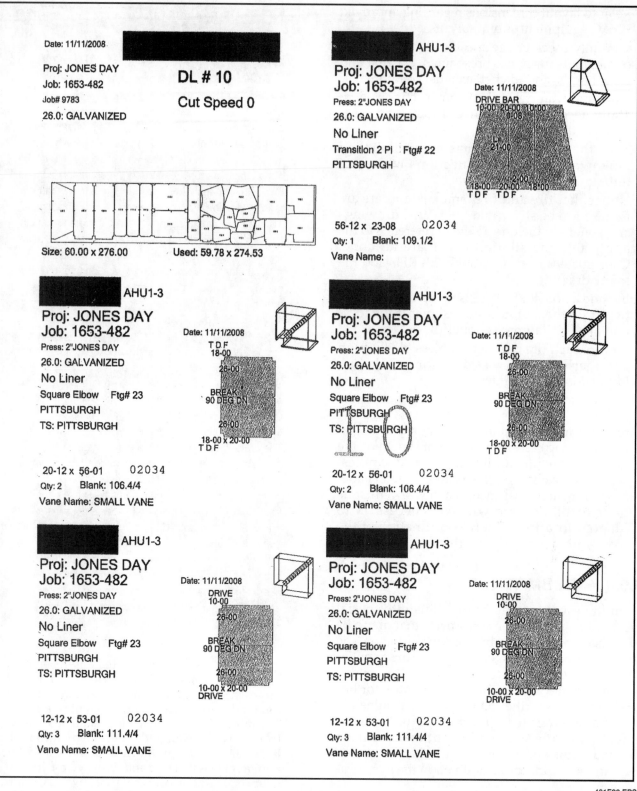

Date: 11/11/2008

Proj: JONES DAY
Job: 1653-482
Job# 9783
26.0: GALVANIZED

DL # 10

Cut Speed 0

Size: 60.00 x 276.00   Used: 59.78 x 274.53

AHU1-3

Proj: JONES DAY
Job: 1653-482
Press: 2"JONES DAY
26.0: GALVANIZED
No Liner
Transition 2 Pi   Ftg# 22
PITTSBURGH

Date: 11/11/2008
DRIVE BAR
10-00 20-00 10-00
0-08
L
21-00
2-00
18-00 20-00 18-00
T D F   T D F

56-12 x 23-08   02034
Qty: 1   Blank: 109.1/2
Vane Name:

AHU1-3

Proj: JONES DAY
Job: 1653-482
Press: 2"JONES DAY
26.0: GALVANIZED
No Liner
Square Elbow   Ftg# 23
PITTSBURGH
TS: PITTSBURGH

Date: 11/11/2008
T D F
18-00
26-00
BREAK
90 DEG DN
26-00
18-00 x 20-00
T D F

20-12 x 56-01   02034
Qty: 2   Blank: 106.4/4
Vane Name: SMALL VANE

AHU1-3

Proj: JONES DAY
Job: 1653-482
Press: 2"JONES DAY
26.0: GALVANIZED
No Liner
Square Elbow   Ftg# 23
PITTSBURGH
TS: PITTSBURGH

Date: 11/11/2008
T D F
18-00
26-00
BREAK
90 DEG DN
26-00
18-00 x 20-00
T D F

20-12 x 56-01   02034
Qty: 2   Blank: 106.4/4
Vane Name: SMALL VANE

AHU1-3

Proj: JONES DAY
Job: 1653-482
Press: 2"JONES DAY
26.0: GALVANIZED
No Liner
Square Elbow   Ftg# 23
PITTSBURGH
TS: PITTSBURGH

Date: 11/11/2008
DRIVE
10-00
26-00
BREAK
90 DEG DN
26-00
10-00 x 20-00
DRIVE

12-12 x 53-01   02034
Qty: 3   Blank: 111.4/4
Vane Name: SMALL VANE

AHU1-3

Proj: JONES DAY
Job: 1653-482
Press: 2"JONES DAY
26.0: GALVANIZED
No Liner
Square Elbow   Ftg# 23
PITTSBURGH
TS: PITTSBURGH

Date: 11/11/2008
DRIVE
10-00
26-00
BREAK
90 DEG DN
26-00
10-00 x 20-00
DRIVE

12-12 x 53-01   02034
Qty: 3   Blank: 111.4/4
Vane Name: SMALL VANE

401F23.EPS

Figure 23 ◆ Computer-generated cut list.

- Refer to layout dimensions regarding approved HVAC equipment submittal cuts.
- Carefully check gauge specifications and types of materials for boiler breechings, exhaust and fume hoods, and kitchen exhaust components.
- Note that watertight duct construction is generally necessary for shower rooms and dishwasher equipment.
- Note that horizontal ducts may be pitched downward to drain connections when run through moist environments.
- Notice that fire and fire/smoke dampers are required to be shown on the HVAC drawings and generally include a note in reference to the applicable installation and material code. Coordinate this information with RFIs and life safety drawings.
- Include a note that requests necessary information from the architect or designer if the location of a fire damper is doubtful.
- Be sure each fire damper is accessible through a properly sized access door for fusible link inspections and maintenance.
- Include additional notes to the attention of the architect or designer if information is necessary to confirm a dimension on the shop drawing in the following cases: when it is not shown on the design or architectural plan; when it is necessary to indicate a specific location that is inadequate for duct clearance; when work must be done by others; and when it is necessary to identify locations where coordination of the work with other trades is critical.

6.0.0 ◆ SUBMITTALS

Submittals are documents that illustrate special pieces of equipment or accessories that are to be furnished and installed by the subcontractor. The submittal document is received from the supplier and submitted by the subcontractor to the general contractor after the bid by the subcontractor has been accepted and the contract has been signed.

The submittal sheet (*Figure 24*) illustrates the accessory or piece of equipment that has been defined in the specifications and that must conform to the standards as outlined in the specifications manual.

For example, the specifications for an in-line centrifugal duct fan may have been stated as follows:

Centrifugal in-line duct fans shall be ACME Company, Model VIDB direct drive or Model VIBA belt drive, as shown on the plans and schedules. Fans shall be constructed of heavy-

Fire and Smoke Dampers

It is critical when reading drawings to know the difference between fire dampers and fire/smoke dampers. Fire dampers are simple spring-return devices. When the lead fusible link melts at the set temperature (usually about 165°F to 185°F), the damper spring is released and slams closed. These dampers must be manually reset once they are tripped, and a new fusible link installed to hold the spring. These dampers are much lower in cost and do not require any other coordination for installation outside of the sheet metal trade.

Combination fire/smoke dampers include the same components as a fire damper, but add the complexity of motorized control. This damper is also a spring damper, but includes a motor, and must be powered open by a signal from the fire alarm system. If that signal is lost, the spring automatically closes the smoke damper whether or not the fusible link in the fire damper is affected. Smoke dampers can be installed separately from fire dampers, but they cannot be separated by any more distance than allowed by code, which is usually about 2 feet, or two to four duct diameters. Check local codes. The added cost of the smoke damper to a system is significant, as several trades are needed for a completed installation (electrical, controls, sheet metal, and sometimes others). The cost and size of the actual fire/smoke damper impacts all aspects of its installation. The cost of fire/smoke dampers may be four to six times that of fire dampers.

Access to fire/smoke dampers is critical. They are supposed to be tested with visual verification of their operation on an annual or semi-annual basis, depending on local fire marshal requirements and recordkeeping.

gauge steel and electro-coated acrylic enamel finish over iron phosphate primer. Wheels 12 inches in diameter and larger shall have median foil blades to assure quiet, efficient operation. The motor drive compartment shall be isolated from the airstream and externally ventilated. Bearings shall be prelubricated and sealed for minimum maintenance and designed for 200,000 hours of operation. Internal parts (wheels, shaft, bearings, motor, and drive) shall be accessible for inspection and repair or replacement without disturbing inlet or outlet ductwork. Fans shall be furnished with a mounted safety disconnect. Single-phase motors shall have integral overload protection. V-belt drives shall be adjustable. Horsepower and noise levels shall not exceed the values shown, and oversized motors will not

SUBMITTAL SHEET   HV 4-8
IN LINE CENTRIFUGAL DUCT FANS

MODEL VIDB – DIRECT DRIVE

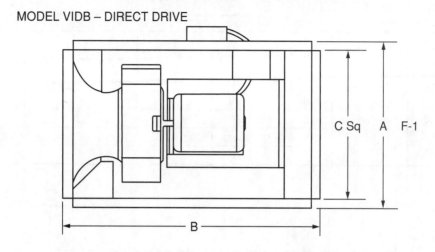

UNIT SIZE	A	B	C	INLET OR OUTLET AREA	WHEEL DIAMETER	WT (lbs)
06	13⅞"	20"	12"	0.979 sq ft	10¾"	30
08	13⅞"	20"	12"	0.979 sq ft	10¾"	40
10	13⅞"	20"	12"	0.979 sq ft	10¾"	40
12	17⅞"	27⅜"	16"	1.750 sq ft	11¹³⁄₁₆"	75
15	*21⅞"	31"	20"	2.740 sq ft	14⅞"	90
18	*21⅞"	33⅜"	26"	4.650 sq ft	17¹³⁄₁₆"	140

* A-1 Larger on access door sides

JOB NAME AND LOCATION	SUBMITTED BY
SECTION 15000 2.4.1	

401F24A.EPS

Figure 24 ◆ Submittal sheet (1 of 2).

be acceptable. Performance ratings shall be certified for air and sound.

The submittal is commonly used to describe the unit specified by the architect or engineer. If the specification allows for substitutions, however, the subcontractor may choose an alternate piece of equipment. In that case, a submittal for the new equipment is acquired by the subcontractor, who submits copies of it to the general contractor, who then submits it to the owner, architect, and any code enforcement authorities. These people either accept or reject the submittal. If the submittal is accepted by the general con-

tractor, owner, and architect, the item may then be installed, as agreed upon, by the subcontractor.

The submittal sheet, as shown, includes the pertinent information that meets the specifications for the construction project. It includes such information as the size of the unit and rough opening, the specifications relating to the size of the inlet or outlet, the cubic feet per minute (cfm), and the sound ratings. The type of mounting may also be stated along with any accessories that would be applicable, such as electronic speed control, spark-resistant wheels, and explosion-proof motors.

SUBMITTAL DATA   TYPE VIDB IN LINE DUCT FANS

MODEL VIDB 10 – DIRECT DRIVE UNITS

SIZE 10	MOTOR HP	RPM	TIP SPEED	STATIC PRESSURE, INCHES W.G.						
				1/8 CFM	1/4 CFM	3/8 CFM	1/2 CFM	5/8 CFM	3/4 CFM	1 CFM
10-D3S	35 MHP	700	1970	200						
with	Standard	800	2250	295						
ES60	Motor	900	2535	365	160					
Speed		1000	2815	430	290					
Control	100W Max.									

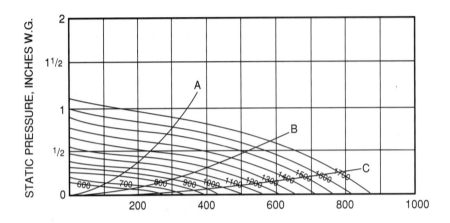

SOUND LEVEL DATA   VIDB SIZE 10

SIZE RPM	*	SOUND POWER LEVEL BY OCTAVE BANDS								LWA	SONE
		1	2	3	4	5	6	7	8		
10-10D 1105	A	62.5	72.5	57.5	54.5	51.0	49.5	48.0	47.0	60.5	5.6
	B	62.5	69.0	58.0	56.5	53.5	50.0	48.0	47.0	60.0	5.2
	C	62.5	68.0	59.0	58.5	57.0	51.5	48.5	47.0	62.0	5.5
10-10H 1710	A	72.0	82.0	67.0	64.0	60.5	59.0	57.5	56.5	69.5	10.7
	B	72.0	78.5	67.5	66.0	63.0	59.5	57.0	56.5	69.5	9.8
	C	72.0	77.5	68.5	68.0	67.0	61.0	58.0	56.5	71.0	10.0

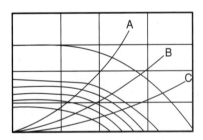

PROJECT:	AGENT:	
	SUBMITTAL DATE:	JOB NO:
The CARNES Company Manufactured Products	DRAWING NO:	
	Date:	Rev:

Figure 24 ◆ Submittal sheet (2 of 2).

When agreed upon by all, the submittal contains the genuine specifications for the unit or accessory with no deviations without approval by a change order from all parties.

Submittal information and shop drawings are usually available from equipment and accessory manufacturers. With this information, the subcontractor has the submittal sheet made up by the drafting department for processing.

7.0.0 ◆ AS-BUILT DRAWINGS

As-built drawings must be made on alteration or addition jobs, on jobs where modifications must be made to make way for other mechanical trades, or to alter the location of a component. In some cases, particularly on additions or alterations, these drawings may be available from the building or plant engineer. These drawings indicate actual installations by the various mechanical trades and must be used for reference by the drafter when called upon to provide a shop drawing for the modified system or components. As-built drawings usually use dashed lines to indicate ducts, piping, and equipment at close proximity to the work. Separate symbols or notes must be used to distinguish ducts that are to be removed and discarded from those that will be relocated.

The as-built drawings are then placed on the architect's plan (sometimes in another color, such

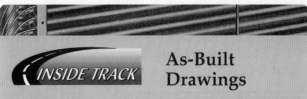
as red), but are more often stamped AS-BUILT. The as-built drawings become a permanent part of the building's drawings. In addition, notes should be made as to the types of connections and existing duct locations that will be reconnected. Duct openings through existing concrete or masonry walls should also be located, checked, and indicated.

8.0.0 ◆ TAKEOFFS

The **takeoff** procedure involves surveying, measuring, and counting all materials and equipment indicated on a set of drawings.

In an HVAC takeoff, materials and equipment should be taken off in the following order:

1. Equipment such as boilers, chillers, pumps, air-handing units, fans, and air cleaners
2. Air devices and air terminals
3. Radiant heating system components
4. Piping and accessory systems
5. Ductwork, dampers, and louvers
6. Insulation for piping, ductwork, and equipment
7. Gauges and thermometers
8. Motor starters, motor control centers, and electrical work to be furnished by the HVAC contractor
9. Temperature control systems
10. Other special systems to be furnished and installed by the HVAC contractor

8.1.0 Takeoff Tools and Materials

Standard takeoff sheets are available for the HVAC technician to use for accurate takeoffs. The sheets are useful because they provide standardization, continuity, and a permanent record. In addition, they reduce the workload as well as the potential for error.

There are several types of takeoff sheets available for use by the HVAC contractor. These sheets include the following:

- *Air devices and equipment takeoff sheets* – These are usually 8½" × 11" standard sheets. They are ruled in four columns, with room to indicate the item number, the classification of the equipment, the quantity, and any necessary remarks.
- *Piping and accessories takeoff sheets* – These often include individual sheets for pipe (*Figure 25*), fittings (*Figure 26*), valves (*Figure 27*), hangers (*Figure 28*), and strainers, traps, and joints.
- *Ductwork takeoff sheets* – The information required includes duct size and length, area per foot of run, total area of duct, maximum duct size, and suggested gauge (*Figure 29*).

PROJECT NAME________					JOB NO. ________		
PIPING SYSTEM ________					MATERIAL ________		
PIPE SIZES (MEASURED IN INCHES)							
FEET	NUMBER OF JOINTS	FEET	NUMBER OF JOINTS	FEET	NUMBER OF JOINTS	FEET	NUMBER OF JOINTS
TOTAL							

401F25.EPS

Figure 25 ◆ Piping takeoff sheet.

PROJECT NAME________					JOB NO. ________	
MATERIAL________					CLASS ________	
FITTING TYPE	FITTING SIZES (MEASURED IN INCHES)					
NO. OF JOINTS						

401F26.EPS

Figure 26 ◆ Piping fitting takeoff sheet.

PROJECT NAME________					JOB NO. ________	
MATERIAL________					CLASS ________	
TYPE OF VALVE DEVICE	VALVE AND DEVICE SIZES (MEASURED IN INCHES)					

401F27.EPS

Figure 27 ◆ Valve takeoff sheet.

PROJECT NAME _____ **JOB NO.** _____

HANGER TYPE	PIPE SIZES (MEASURED IN INCHES)							

401F28.EPS

Figure 28 ◆ Hanger takeoff sheet.

PROJECT NAME _____ **JOB NO.** _____

DUCTWORK SYSTEM _____ **MATERIAL** _____

DUCT SIZE (INCHES)	DUCT SIZE (FEET)	SQ. FT. PER FT. OF RUN	TOTAL AREA OF DUCT (SQUARE FEET)			
			MAXIMUM DUCT SIZE AND SUGGESTED GAUGE			

401F29.EPS

Figure 29 ◆ Ductwork takeoff sheet.

The following materials will make the measuring, counting, and calculating tasks easier:

- Colored pencils for checking off items on drawings as they are taken off. The same color can be used for all the material contained within a system to simplify following up on that system.
- An automatic mechanical counter for counting similar diffusers, grilles, and registers
- An electronic wheel scaler or similar device for measuring duct and piping runs (should have scales of ⅛", ¼", and ½")
- Metallic tape with ⅛" scale on one side and a ¼" scale on the other for measuring duct and piping runs
- Two drafting scales (one architect's and one engineer's)
- An adding machine or calculator
- A magnifying glass for examining details on the drawings
- A large collection of manufacturers' catalogs, technical books and manuals, and previous job files for reference

After the takeoff is complete, you should write up all material and equipment classifications on the job estimating sheets and copy the total quantity of each item as it appears on the takeoff sheets to the job estimating sheets.

8.2.0 Selecting Equipment and Materials

The drawings and specifications are used to identify the types and quantities of equipment and materials required for the project. The selection of heating, ventilating, and air conditioning (HVAC) equipment is based on an analysis of heating and cooling loads performed by the architect or engineer. The purpose of the analysis is to select the equipment that is physically suited to the structure and is sized to meet its heating and cooling loads and air flow requirements. Load estimating and equipment selection are explained in a later module.

The selection of materials for HVAC systems is based on a study of the conditions of operation. The factors to consider include the following:

- Code requirements
- The working fluid in the pipe
- The pressure and temperature of the fluid
- The external environment of the pipe
- The cost of the installed system

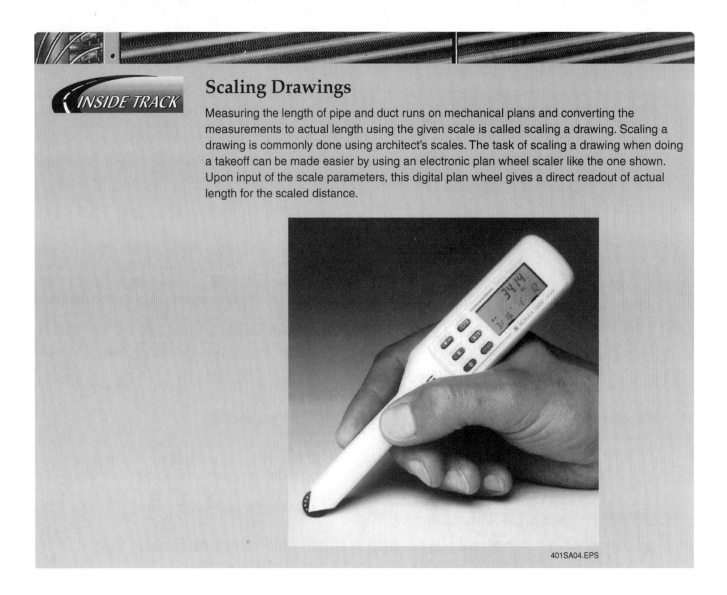
Piping, ductwork, pipe/duct hangers and supports, and valves are also selected from the drawings and specifications. An overview of the criteria that go into the selection process for these items is provided in the sections that follow.

8.2.1 Ductwork

The various elements requiring sheet metal work are the duct, plenum, apparatus casing, and acoustic liner. The function of ductwork is to supply the conditioned air to the space served and to remove the exhaust and return air from that conditioned space. Ducts must be constructed so that they are sufficiently airtight, will not vibrate or breathe when the airstream varies in pressure, and guarantee an even flow of air without undue pressure loss.

The materials commonly used in duct run system construction are the following:

- Galvanized steel sheets
- Hot- or cold-rolled black steel sheets
- Copper
- Stainless steel
- Aluminum sheets
- Transite
- PVC
- Fibrous glass
- Flexible duct

Duct construction is classified in terms of operating pressure and air velocity. The common classifications are shown in *Table 1*.

Table 1 Pressure and Velocity

System	Velocity (fpm)	Static Pressure (in. w.c.)
Low pressure, low velocity	up to 2,000	up to 2
Medium pressure, high velocity	over 2,000	2 to 6
High pressure, high velocity	over 2,000	6 to 10

Ducts can be made in round, rectangular, or flat oval shapes. Round ducts are less expensive to fabricate and to install than the other two. However, when there is a requirement for sound attenuation, there is additional cost if using round duct. This should be considered for new designs or when thinking about converting from rectangular duct to round duct for a retrofit job. However, flat oval ducts combine the advantages of round and rectangular ducts; they have considerably less flat surface that is susceptible to vibration and, therefore, require less reinforcement than corresponding sizes of rectangular duct. Round and flat oval ducts are recommended for medium- and high-velocity duct run systems.

The recommended constructions, gauges, connections, and methods of reinforcement for various types of ducts can be found in applicable Sheet Metal and Air Conditioning Contractors' National Association (SMACNA) duct construction standards manuals.

8.2.2 Duct Hangers

The duct-hanging components consist of three elements: the upper attachment to the overstructure, the hanger itself, and the lower attachment to the duct. Ductwork must not be hung from overhead piping or ceiling hanger irons.

Duct hangers are made of strap iron, rods, angle irons, or a combination of these (trapeze hangers). Rectangular ducts are usually supported by two metal strap hangers that are screwed into the sides and/or bottom of the duct. Round ducts are usually supported by one strap hanger or by two strap or rod hangers that are bolted to the duct bands. Trapeze-type hangers are recommended for large ducts.

All types of hangers are fastened to the supporting structural member by clamps, anchor bolts, metal screws, regular bolts, or nails. The maximum recommended spacing between duct hangers is 10' for round ducts with diameters up to 84 inches and for rectangular ducts with the largest dimension up to 60 inches. For rectangular ducts with the largest dimension over 60 inches, 8-foot spacing is recommended. The size of the hanger depends on the weight and size of the duct. Duct sizes are determined according to airflow rates (cfm), velocity (fpm), and duct friction losses.

8.3.0 Takeoff Procedures

The takeoff procedure for valves, fittings, piping, hangers, and accessories is as follows:

Step 1   Prepare the required takeoff sheets.

Step 2   Read the specifications carefully; then list the materials and joints specified for each system.

Step 3   Take off each system as indicated on the floor plans, section details, and flow diagrams. The takeoff must include piping, fittings, valves, joints, hangers, and all accessories.

Step 4   List all quantities of the same size in one column of the takeoff sheet.

Step 5   Measure pipe lengths in feet; count the fittings, valves, hangers, and so on.

Step 6   After completing the takeoff, add the quantities of each size, and transfer the sum to a master sheet.

The takeoff procedure for ductwork follows:

Step 1   Use a standard takeoff sheet or prepare one.

Step 2   Read the specifications; then list the types of materials, recommended construction, and specified gauges for each system.

Step 3   Include all the takeoff quantities for all the duct runs that have the same material, construction, acoustic lining, and insulation on one takeoff sheet.

Step 4   Take off each system as indicated on the drawings.

Step 5   Place all the findings on a summary sheet.

9.0.0 ◆ BUILDING CODES

Building codes that are national in scope provide minimum standards to guard the life and safety of the public by regulating and controlling the design, construction, and quality of materials used in modern construction. They have also come to govern use and occupancy, location of a type of building, and the on-going maintenance of all buildings and facilities. Once adopted by a local jurisdiction, these model building codes then become law. It is common for localities to change or add new requirements to any model code requirements adopted in order to meet more stringent requirements and/or local needs. The provisions of the model building codes apply to the construction, alteration, movement, demolition, repair, structural maintenance, and use of any building or structure within the local jurisdiction.

The model building codes are the legal instruments that enforce public safety in construction of human habitation and assembly structures. They are used not only in the construction industry but also by the insurance industry for compensation appraisals and claims adjustments, and by the legal industry for court litigation.

Up until 2000, there were three model building codes. The three code writing groups, Building Officials and Code Administrators (BOCA), International Conference of Building Officials (ICBO), and Southern Building Code Congress International (SBCCI), combined into one organization called the International Code Council (ICC) with the purpose of writing one nationally accepted family of building and fire codes. The first edition of the *International Building Code*® was published in 2000, a second edition was published in 2003, and a third in 2006. It is intended to continue on a three-year cycle.

In 2002, the National Fire Protection Association (NFPA) published its own building code, *NFPA 5000*®. There are now two nationally recognized codes competing for adoption by the 50 states.

The format and chapter organization of the two codes differ, but the content and subjects covered are generally the same. Both codes cover all types of occupancies from single-family residences to high-rise office buildings, as well as industrial facilities. They also cover structures, building materials, and building systems, including life safety systems.

When states, counties, and cities adopt a model code as the basis for their own code, they often change it to meet local conditions. They might add further restrictions, or they might only adopt part of the model code. An important general rule to remember about codes is that in almost every case the most stringent local code will apply.

The HVAC technician should be aware of the laws, local building codes, and restrictions that affect the specific job being constructed. This should also include a basic understanding of other codes, such as the NFPA gas and electrical codes.

1. Mechanical plans are architectural drawings.
   a. True
   b. False

2. Site plans are sometimes called _____.
   a. elevations
   b. sections
   c. plot plans
   d. foundation plans

3. The elevation of the finished floor for the garage shown on the site plan in *Figure 2* is _____.
   a. 543.00'
   b. 946.50'
   c. 947.25'
   d. 947.75'

4. The distance between the centers of the north window and door in the garage shown on the floor plan in *Figure 3* is _____.
   a. 6'-0"
   b. 6'-4"
   c. 8'-0"
   d. 10'-0"

5. How many roof drains are called for in the roof plan shown in *Figure 4*?
   a. three
   b. four
   c. five
   d. six

6. The vertical mansard (type of roof) shown in the elevation drawing on *Figure 6* is made of _____.
   a. 26 gauge, Type SR arctic white metal
   b. 26 gauge, Type SS arctic white metal
   c. 8" vertical score block
   d. 8" split rib block

7. Item F listed in the grille, register, and diffuser schedule shown on *Figure 7* identifies a _____ diffuser that delivers _____.
   a. 12" × 12"; 240 cfm
   b. 12" × 12"; 250 cfm
   c. 18" × 18"; 235 cfm
   d. 18" × 18"; 280 cfm

8. Details about air diffusers, such as size and model number, would generally be found on a(n) _____.
   a. elevation drawing
   b. detail drawing
   c. section drawing
   d. schedule

9. The difference in thickness between the concrete slabs for the lower and upper floors of the building shown in *Figure 9* is _____.
   a. 0"
   b. ½"
   c. ¾"
   d. 1"

10. In accordance with the plumbing plan shown in *Figure 11*, _____ urinal(s) must be installed.
    a. one
    b. two
    c. four
    d. six

11. In accordance with the plumbing plan shown in *Figure 11*, _____ sanitary system pipe cleanouts must be installed.
    a. three
    b. four
    c. five
    d. six

12. Piping for natural gas is typically shown on the _____.
    a. floor plan
    b. elevation drawings
    c. HVAC drawings
    d. plumbing drawings

13. What size chilled-water supply piping should be connected to air handling unit–1 shown in the mechanical plan on *Figure 13*?
    a. 1¼"
    b. 1½"
    c. 2"
    d. 2½"

14. The electrical plan in *Figure 18* shows the new power distribution panel, Panel D, to be installed in _____.

    a. hallway 125
    b. lobby 123
    c. room 126
    d. room 131

15. The final recipient of a request for information (RFI) form is the _____.

    a. foreman
    b. architect or engineer
    c. superintendent
    d. general contractor

16. In new CSI-formatted specifications, you would expect to find information on HVAC systems in _____.

    a. Division 7
    b. Division 10
    c. Division 15
    d. Division 23

17. A shop drawing prepared by an equipment manufacturer typically describes _____.

    a. structural steel members
    b. special equipment
    c. sheet metal components
    d. ductwork layout

18. Cut lists are drawn from _____.

    a. takeoffs
    b. as-builts
    c. shop drawings
    d. an RFI

19. In order to secure approval to furnish and install special equipment, a subcontractor must provide the contractor with a(n) _____.

    a. submittal
    b. cut list
    c. as-built
    d. takeoff

20. A drawing that shows alterations or additions made to an original plan is called a(n) _____.

    a. submittal
    b. cut list
    c. as-built
    d. takeoff

21. Types of fittings are usually included in a piping takeoff sheet.

    a. True
    b. False

22. The purpose of an electronic wheel scaler is to _____.

    a. examine details shown on the drawings
    b. measure the diameter of pipe and other circular objects
    c. count the number of grilles and diffusers shown on drawings
    d. measure duct and piping runs shown on drawings

23. Galvanized steel sheets are commonly used in ductwork systems.

    a. True
    b. False

24. Duct sizes are determined according to the airflow rates (cfm), velocity (fpm), and _____.

    a. type of metal
    b. duct friction losses
    c. cost
    d. climate

25. Which of the following statements is true with regard to ductwork?

    a. Ductwork must be hung from overhead piping or ceiling hanger irons.
    b. Rectangular ducts are usually supported by one strap hanger bolted to duct bands.
    c. Trapeze-type hangers are recommended for smaller ducts.
    d. Duct hangers are made of strap iron, rods, angle irons, or a combination.

Summary

Construction drawings and specifications for the HVAC trade contain the information necessary for the layout, fabrication, and installation of duct runs, HVAC piping, and fittings. Specifications are written descriptions of technical design and performance information that must be used when selecting and installing equipment, systems, and construction components. When there is a conflict between the design specifications and the architectural plans, the specifications that are most stringent usually apply. Often the contract or specification will define which document prevails in case of a conflict. Sometimes this is determined by all parties at the contract signing.

Notes

Trade Terms Introduced in This Module

Coordination drawings: Elevation, location, and other drawings produced for a project by the individual contractors for each trade to prevent a conflict between the trades regarding the installation of their materials and equipment. Development of these drawings evolves through a series of review and coordination meetings held by the various contractors.

Cut list: An information sheet that is derived from shop drawings. It is the shop guide for fabricating duct runs and fittings.

Detail drawing: A drawing of a feature that provides more elaborate information than is available on a plan.

Elevation view: A view that depicts a vertical side of a building, usually designated by the direction that side is facing; for example, right, left, east, or west elevation.

Floor plan: A building drawing indicating a plan view of a horizontal section at some distance above the floor, usually midway between the ceiling and the floor.

Longitudinal section: A section drawing where the "cut" is made along the long dimension of the building.

Plan view: The overhead view of an object or structure.

Riser diagram: A one-line schematic depicting the layout, components, and connections of a piping system or electrical system.

Schedules: Tables that describe and specify the types and sizes of items required for the construction of a building.

Section drawing: A drawing that depicts a feature of a building as if there were a cut made through the middle of it.

Shop drawing: A drawing that indicates how to fabricate and install individual components of a construction project. A shop drawing may be drafted from the construction drawings of a project or provided by the manufacturer.

Site plan: A construction drawing that indicates the location of a building on a land site.

Takeoff: The process of surveying, measuring, itemizing, and counting all materials and equipment needed for a construction project, as indicated by the drawings.

Transverse section: A section drawing where the "cut" is made along the short dimension of the building.

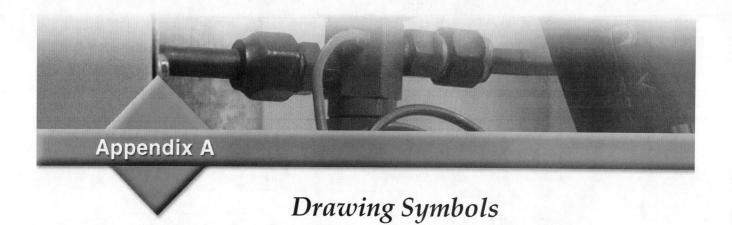

Appendix A

Drawing Symbols

PROPERTY LINE

BOUNDARY LINE (MATCH LINE)

MAIN OBJECT LINE

HIDDEN LINE

CENTER LINE (Used as finished floor line)

DIMENSION AND EXTENSION LINES

2$\frac{1}{8}$"

LONG BREAK LINE

SHORT BREAK LINE

LEADER LINE

SECTION LINE TYP.

A   A'

REF. LINE FOR VARIOUS SECTION TYPES

1
B

1
A

401A01.EPS

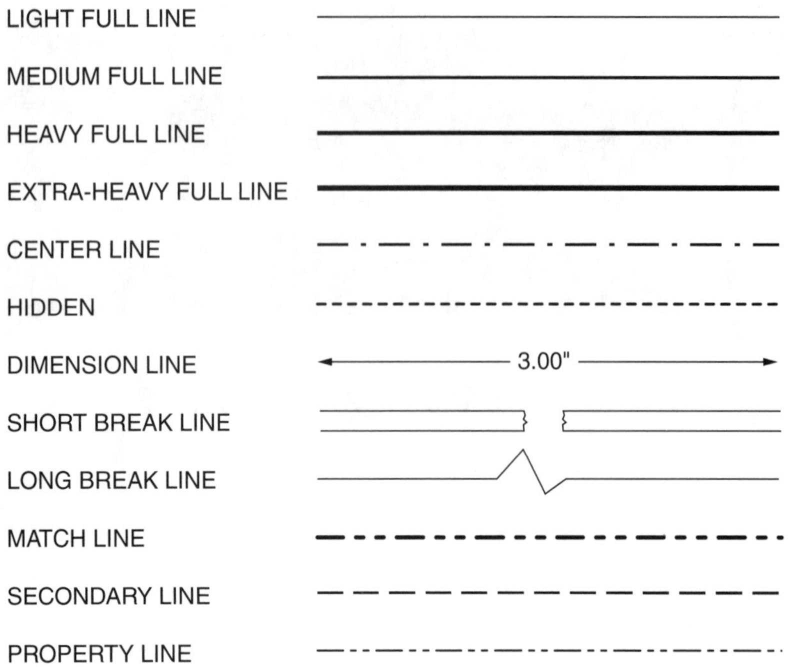

LIGHT FULL LINE	————————————————
MEDIUM FULL LINE	————————————————
HEAVY FULL LINE	————————————————
EXTRA-HEAVY FULL LINE	————————————————
CENTER LINE	— · — · — · — · — · —
HIDDEN	– – – – – – – – – – –
DIMENSION LINE	◄——————— 3.00" ———————►
SHORT BREAK LINE	⸦————⸧ ⸦————⸧
LONG BREAK LINE	————⋀⋁————
MATCH LINE	▬ ▬ ▬ ▬ ▬ ▬ ▬ ▬ ▬
SECONDARY LINE	— — — — — — — —
PROPERTY LINE	— · — · — · — · — ·

401A02.EPS

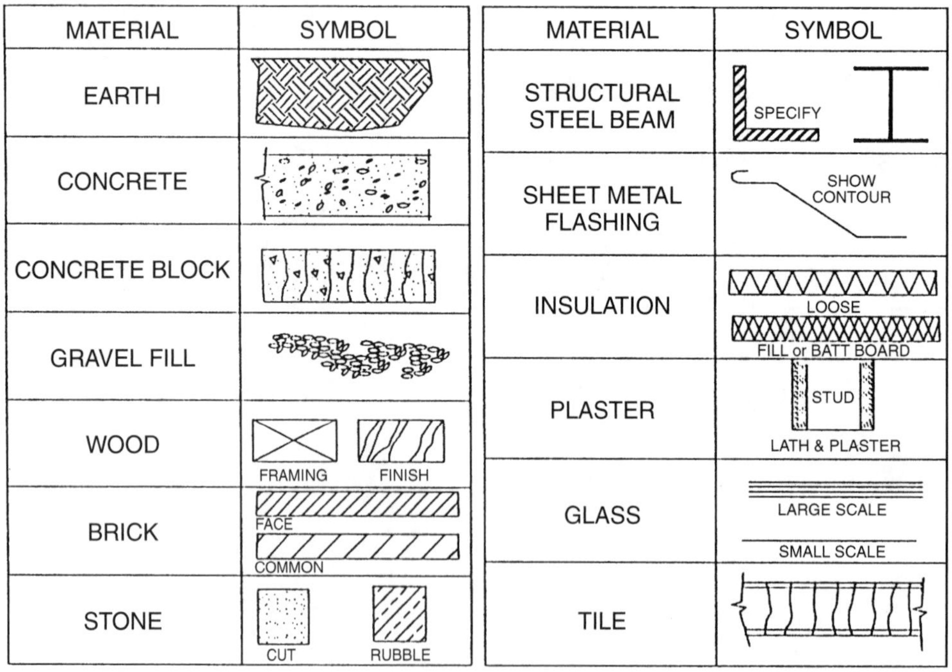

MATERIAL	SYMBOL	MATERIAL	SYMBOL
EARTH		STRUCTURAL STEEL BEAM	SPECIFY / I
CONCRETE		SHEET METAL FLASHING	SHOW CONTOUR
CONCRETE BLOCK		INSULATION	LOOSE / FILL or BATT BOARD
GRAVEL FILL		PLASTER	STUD / LATH & PLASTER
WOOD	FRAMING FINISH	GLASS	LARGE SCALE / SMALL SCALE
BRICK	FACE / COMMON	TILE	
STONE	CUT RUBBLE		

401A03.EPS

DOOR TYPE	SYMBOL	WINDOW TYPE	SYMBOL
SINGLE SWING		AWNING	
SLIDER		FIXED SASH	
BIFOLD		DOUBLE HUNG	
FRENCH		CASEMENT	
ACCORDION		HORIZONTAL SLIDER	

401A04.EPS

ADD.	addition	N	north
AGGR	aggregate	NO.	number
L	angle	OC	on center
B	bathroom	OPP	opposite
BR	bedroom	O.D.	outside diameter
BM	bench mark	PNL	panel
BRKT	bracket	PSI	pounds per square inch
CLK	caulk	PWR	power
CHFR	chamfer	REINF	reinforce
CND	conduit	RH	right-hand
CU FT	cubic foot, feet	RFA	released for approval
DIM.	dimension	RFC	released for construction
DR	drain	RFD	released for design
DWG	drawing	RFI	released for information
ELEV	elevation	SHTHG	sheathing
ESC	escutcheon	SQ	square
FAB	fabricate	STR	structural
FLGE	flange	SYM	symbol
FLR	floor	THERMO	thermostat
GR	grade	TYP	typical
GYP	gypsum	UNFIN	unfinished
HDW	hardware	VEL	velocity
HTR	heater	WV	wall vent
" or IN.	inch, inches	WHSE	warehouse
I.D.	inside diameter	WH	weep hole
LH	left-hand	WDW	window
MEZZ	mezzanine	WP	working pressure
MO	masonry opening		
MECH	mechanical		

401A05.EPS

LIGHTING OUTLETS	CEILING	WALL
Surface or pendant incandescent, mercury-vapor, or similar lamp fixture	◯ ⊕	—◯
Recessed incandescent, mercury-vapor, or similar lamp fixture	Ⓡ	—Ⓡ
Surface or pendant individual fluorescent fixture	▭◯	—▭◯
Recessed individual fluorescent fixture	▭◯ R	—▭◯ R
Surface or pendant continuous-row fluorescent fixture	▭◯▭▭	
Recessed continuous-row fluorescent fixture	▭◯ R ▭▭	
Bare-lamp fluorescent strip	├——┼——┼——┤	
Surface or pendant exit light	Ⓧ	—Ⓧ
Recessed exit light	Ⓡ🇽	—Ⓡ🇽

401A06A.EPS

	CEILING	WALL
Blanked outlet	Ⓑ	—Ⓑ
Fan outlet	Ⓕ	—Ⓕ
Drop cord	Ⓓ	
Junction box	Ⓙ	—Ⓙ
Outlet controlled by low-voltage switching when relay is installed in outlet box	Ⓛ	—Ⓛ

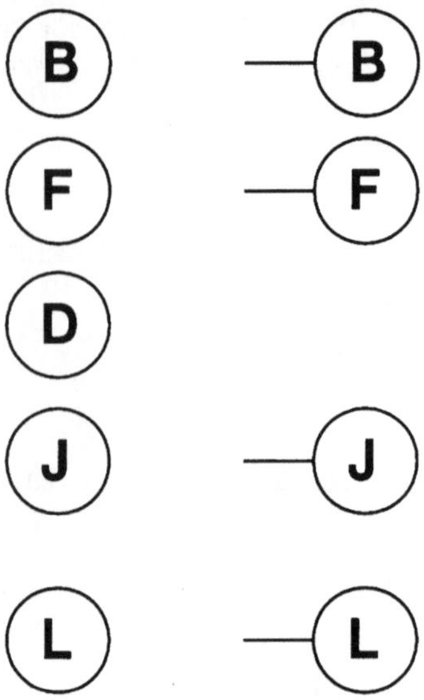

RECEPTACLE OUTLETS

GROUNDED

Single receptacle outlet

Duplex receptacle outlet

Waterproof receptacle outlet   WP

Triplex receptacle outlet

Fourplex (Quadruplex) receptacle outlet

Duplex receptacle outlet, split wired

Triplex receptacle outlet, split wired

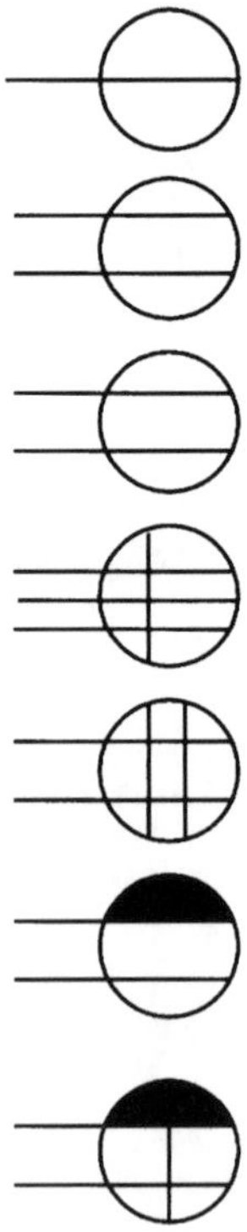

401A06B.EPS

RECEPTACLE OUTLETS

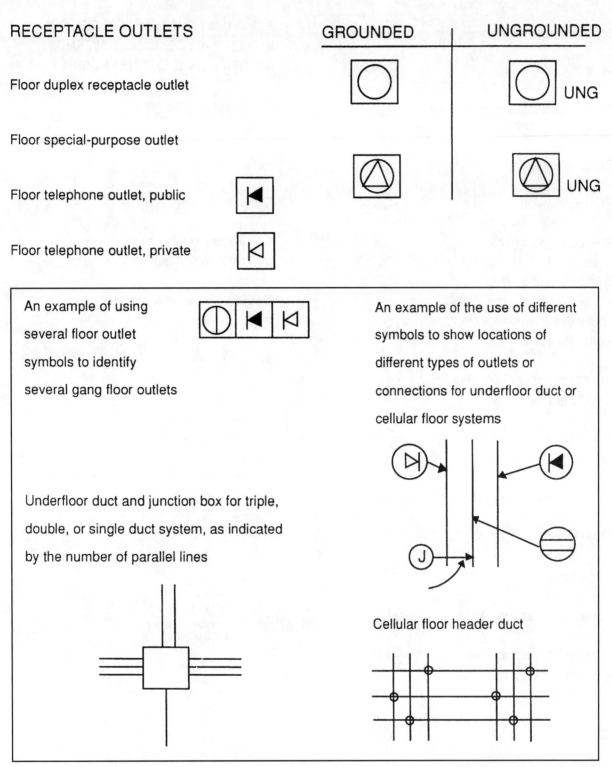

	GROUNDED	UNGROUNDED
Floor duplex receptacle outlet	◯	◯ UNG
Floor special-purpose outlet		
Floor telephone outlet, public	◀	◁ UNG
Floor telephone outlet, private	◁	

An example of using several floor outlet symbols to identify several gang floor outlets

Underfloor duct and junction box for triple, double, or single duct system, as indicated by the number of parallel lines

An example of the use of different symbols to show locations of different types of outlets or connections for underfloor duct or cellular floor systems

Cellular floor header duct

401A06C.EPS

SWITCH OUTLETS

Single-pole switch	S
Double-pole switch	S_2
Three-way switch	S_3
Four-way switch	S_4
Key-operated switch	S_K
Switch and pilot lamp	S_P
Switch for low-voltage switching system	S_L
Master switch for low-voltage switching system	S_{LM}
Switch and single receptacle	$\ominus$ S
Switch and double receptacle	$\ominus$ S
Door switch	S_D
Time switch	S_T
Circuit breaker switch	S_{CB}
Momentary contact switch or pushbutton for other than signaling system	S_{MC}
Ceiling pull switch	Ⓢ

SIGNALING SYSTEM OUTLETS FOR INDUSTRIAL, COMMERCIAL, AND INSTITUTIONAL OCCUPANCIES

Any type of nurse call system device	—○
Nurses' annunciator	—①
Call station, single cord, pilot	—②
Call station, double cord, microphone speaker	—③
Corridor dome light, one lamp	—④
Transformer	—⑤
Any other item on same system (use numbers as required)	—⑥
Any type of paging system device	—◇
Keyboard	—◇1
Flush annunciator	—◇2
Two-face annunciator	—◇3
Any other item on same system (use numbers as required)	—◇4

401A06D.EPS

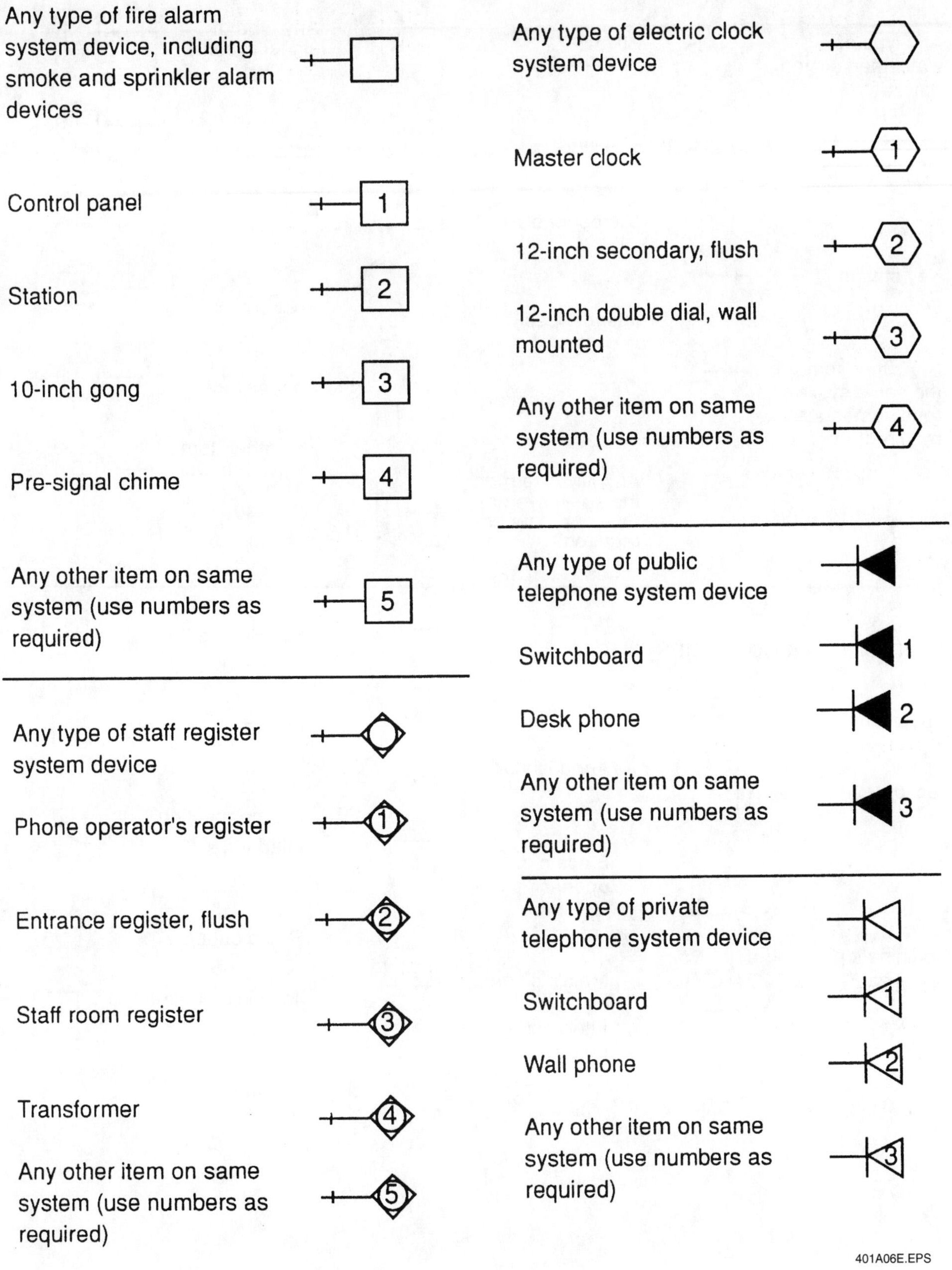

Any type of fire alarm system device, including smoke and sprinkler alarm devices

Control panel   1

Station   2

10-inch gong   3

Pre-signal chime   4

Any other item on same system (use numbers as required)   5

Any type of staff register system device

Phone operator's register   1

Entrance register, flush   2

Staff room register   3

Transformer   4

Any other item on same system (use numbers as required)   5

Any type of electric clock system device

Master clock   1

12-inch secondary, flush   2

12-inch double dial, wall mounted   3

Any other item on same system (use numbers as required)   4

Any type of public telephone system device

Switchboard   1

Desk phone   2

Any other item on same system (use numbers as required)   3

Any type of private telephone system device

Switchboard   1

Wall phone   2

Any other item on same system (use numbers as required)   3

401A06E.EPS

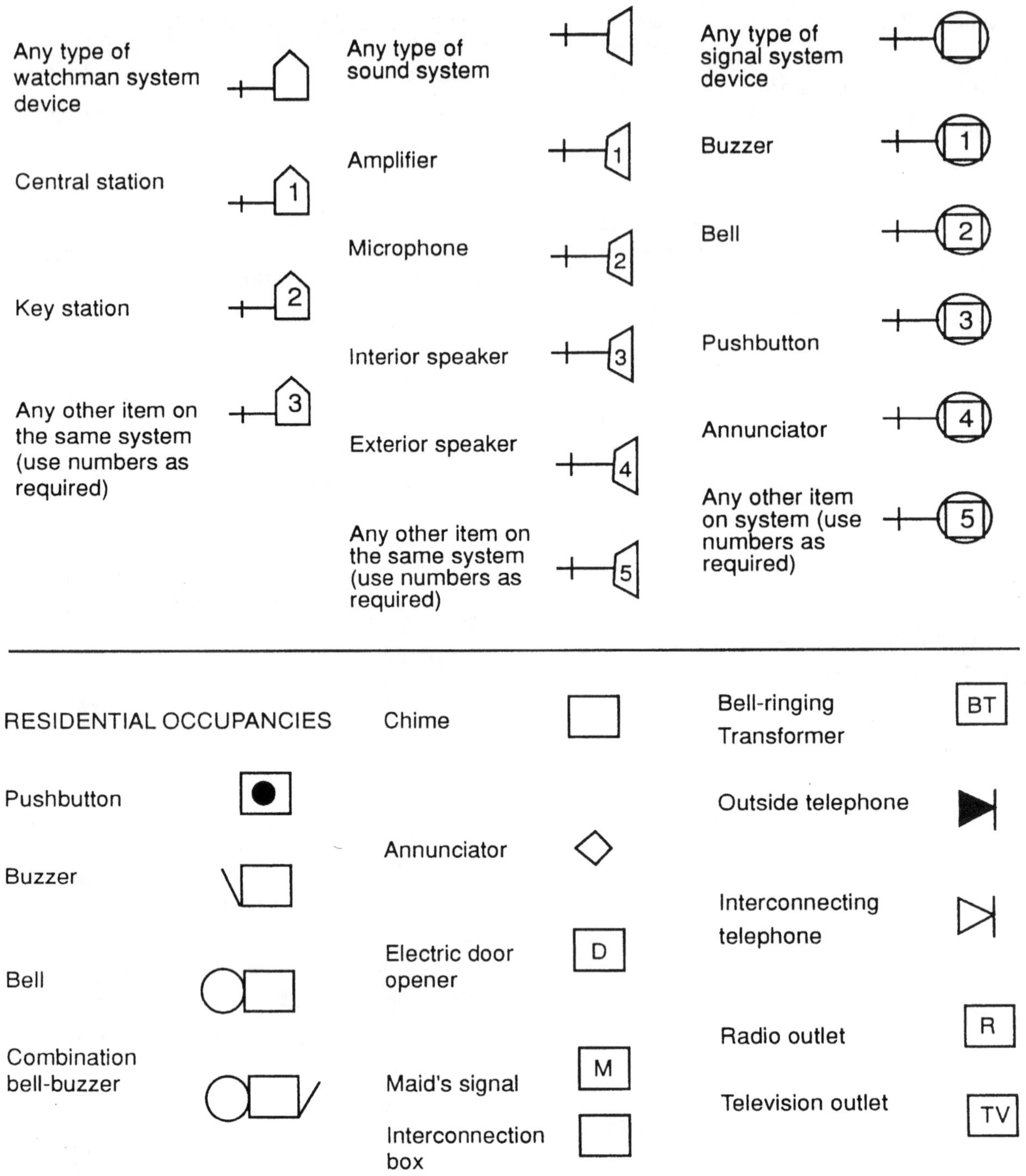

Any type of watchman system device ⌂

Central station ⌂1

Key station ⌂2

Any other item on the same system (use numbers as required) ⌂3

Any type of sound system ◁

Amplifier ◁1

Microphone ◁2

Interior speaker ◁3

Exterior speaker ◁4

Any other item on the same system (use numbers as required) ◁5

Any type of signal system device ⊕

Buzzer ①

Bell ②

Pushbutton ③

Annunciator ④

Any other item on system (use numbers as required) ⑤

RESIDENTIAL OCCUPANCIES

Pushbutton ▣

Buzzer

Bell

Combination bell-buzzer

Chime ▢

Annunciator ◇

Electric door opener ☐D

Maid's signal ☐M

Interconnection box ▢

Bell-ringing Transformer ☐BT

Outside telephone ▶

Interconnecting telephone ▷

Radio outlet ☐R

Television outlet ☐TV

401A06F.EPS

PANELBOARDS, SWITCHBOARDS, AND RELATED EQUIPMENT

Flush-mounted panel board and cabinet *

Surface-mounted panel board and cabinet*

Switchboard, power control center, unit substations (ANSI recommends drawing to scale)*

Flush-mounted terminal cabinet*

Surface-mounted terminal cabinet*

Pull box—identify in relation to wiring system section and size

Motor or other power controller*

MC

Externally operated disconnection switch*

Combination controller and disconnect means*

*Identify by notation or schedule

BUS DUCTS AND WIREWAYS

Trolley duct*

	T			T		T

Busway (service, feeder, or plug-in)*

	B			B		B

Cable trough, ladder, or channel*

	BP			BP		BP

Wireway*

	W			W		W

* Identify by notation or schedule

REMOTE CONTROL STATIONS FOR MOTORS OR OTHER

Pushbutton stations in general

Float switch, mechanical   F

Limit switch, mechanical   L

Pneumatic switch, mechanical   P

Electric eye, beam source

Electric eye, relay

Thermostat   T

401A06G.EPS

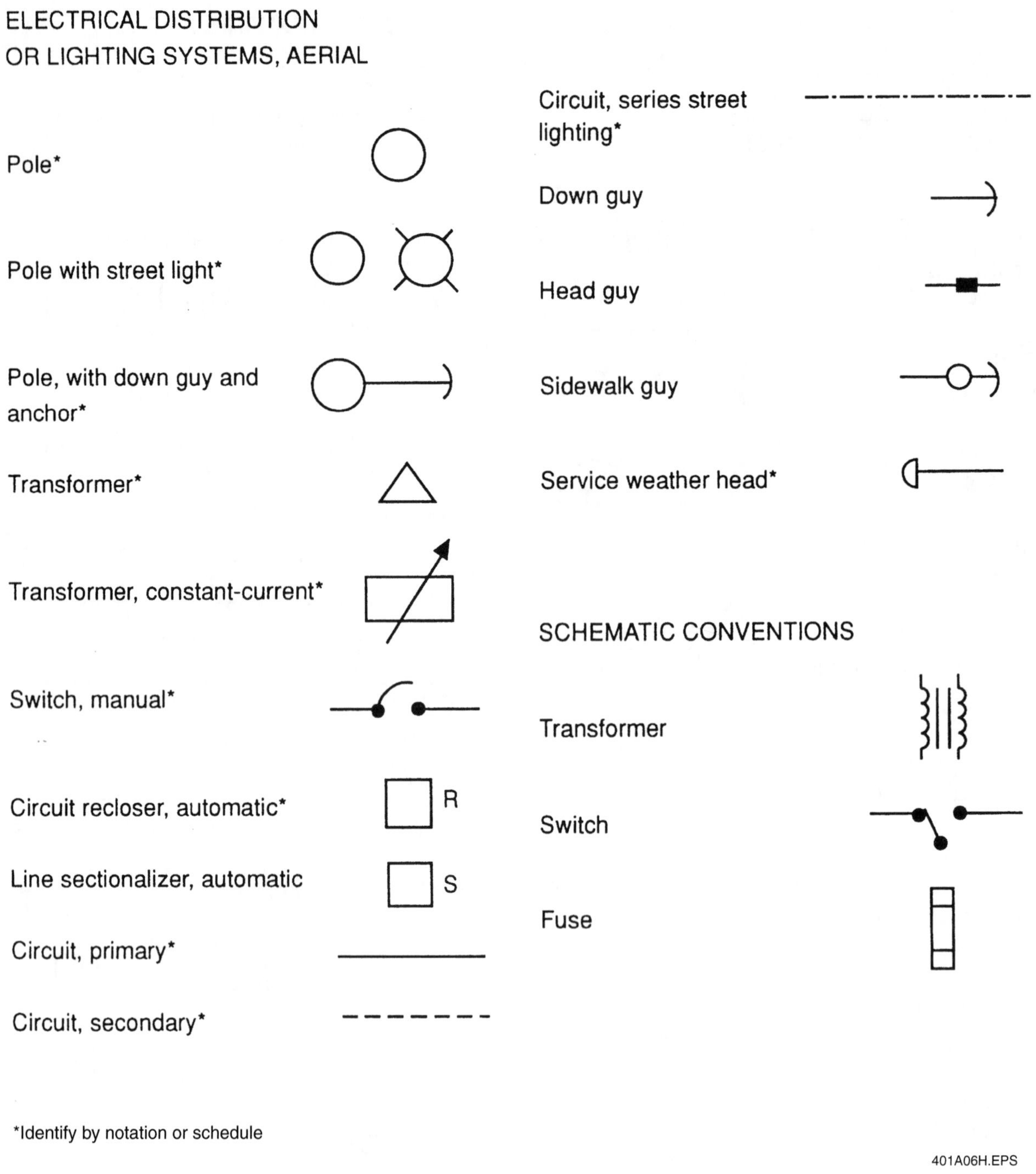

ELECTRICAL DISTRIBUTION
OR LIGHTING SYSTEMS, AERIAL

Pole*

Pole with street light*

Pole, with down guy and anchor*

Transformer*

Transformer, constant-current*

Switch, manual*

Circuit recloser, automatic*   R

Line sectionalizer, automatic   S

Circuit, primary*

Circuit, secondary*

Circuit, series street lighting*

Down guy

Head guy

Sidewalk guy

Service weather head*

SCHEMATIC CONVENTIONS

Transformer

Switch

Fuse

*Identify by notation or schedule

401A06H.EPS

EXPOSED WIRING ——————— E ———————

WIRING CONCEALED IN CEILING OR WALL ——————————

WIRING CONCEALED IN FLOOR – – – – – – – – – – –

WIRING TURNED UP ——————————○

WIRING TURNED DOWN ——————————●

BRANCH-CIRCUIT HOMERUN TO PANELBOARD* ——————▶▶

OR

* Number of arrowheads indicate number of circuits. A number at each arrowhead may be used to identify circuit numbers

** —————————▶ 1 2

** Half arrowheads are sometimes used for homeruns to avoid confusing them with drawing callouts

401A07.EPS

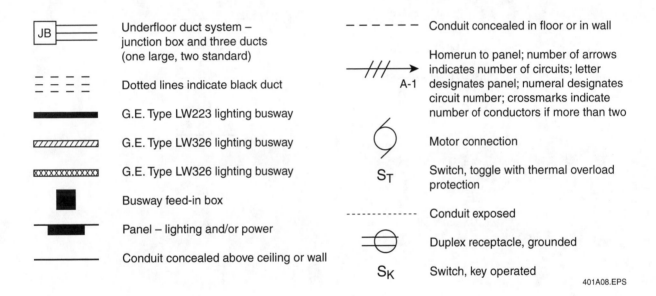

JB	Underfloor duct system – junction box and three ducts (one large, two standard)
	Dotted lines indicate black duct
	G.E. Type LW223 lighting busway
	G.E. Type LW326 lighting busway
	G.E. Type LW326 lighting busway
	Busway feed-in box
	Panel – lighting and/or power
	Conduit concealed above ceiling or wall

– – – – – Conduit concealed in floor or in wall

Homerun to panel; number of arrows indicates number of circuits; letter designates panel; numeral designates circuit number; crossmarks indicate number of conductors if more than two

A-1

Motor connection

S$_T$ Switch, toggle with thermal overload protection

- - - - - - Conduit exposed

Duplex receptacle, grounded

S$_K$ Switch, key operated

401A08.EPS

WASTE WATER

DRAIN OR WASTE – ABOVE GRADE ————————

DRAIN OR WASTE – BELOW GRADE — — —

VENT - - - - - - -

COMBINATION WASTE AND VENT —— CWV ——

ACID WASTE —— AW ——

ACID VENT – – - AV – – -

INDIRECT DRAIN —— D ——

STORM DRAIN —— SD ——

SEWER – CAST IRON _____ S-CI _____

SEWER – CLAY TILE BELL & SPIGOT _____ S-CT _____

DRAIN – CLAY TILE BELL & SPIGOT ————————

OTHER PIPING

GAS – LOW PRESSURE —— G —— G ——

GAS – MEDIUM PRESSURE —— MG ——

GAS – HIGH PRESSURE —— HG ——

COMPRESSED AIR —— A ——

VACUUM —— V ——

VACUUM CLEANING —— VC ——

OXYGEN —— O ——

LIQUID OXYGEN —— LOX ——

401A09.EPS

	FLANGED	SCREWED	BELL AND SPIGOT	WELDED	SOLDERED
Bushing					
Cap					
Cross					
Reducing					
Straight Size					
Crossover					
Elbow 45-Degree					
90-Degree					
Turned Down					
Turned Up					
Base					
Double Branch					
Long Radius					

	FLANGED	SCREWED	BELL AND SPIGOT	WELDED	SOLDERED
Elbow (Cont'd) Reducing					
Side Outlet (Outlet Down)					
Side Outlet (Outlet Up)					
Street					
Joint Connecting Pipe					
Expansion					
Lateral					
Orifice Plate					
Reducing Flange					
Plugs Bull Plug					
Pipe Plug					
Reducer Concentric					
Eccentric					

401A10A.EPS

	FLANGED	SCREWED	BELL AND SPIGOT	WELDED	SOLDERED
Gate, also Angle Gate (Plan)	⊖◁╫	⊖◁├		⊖◁✕	
Globe, also Angle Globe (Elevation)					
Globe (Plan)	⊖◁╫	⊖◁├		⊖◁✕	⊖◁─
Automatic Valve Bypass					
Governor-Operated					
Reducing					
Check Valve (Straight Way)	╫◁◁╫	├◁╲	→◁←	✕◁◁✕	⊖◁◁⊖
Cock					
Diaphragm Valve					
Float Valve					
Gate Valve*	⊣◁▷⊢	⊣◁▷├	→◁▷←	✕◁▷✕	⊖◁▷⊖

*Also used for General Stop Valve Symbol when amplified by specification.

	FLANGED	SCREWED	BELL AND SPIGOT	WELDED	SOLDERED
Motor-Operated					
Globe Valve	⊣◁▷├	⊣◁▷├	→◁▷←	✕◁▷✕	⊖◁▷⊖
Motor-Operated					
Hose Valve, also Hose Globe					
Angle, also Hose Angle					
Gate	⊣◁▷┐	⊣◁▷┐			
Globe	⊣◁▷┐	⊣◁▷┐			
Lockshield Valve					⊖◁▷⊖
Quick-Opening Valve					⊖◁▷⊖
Safety Valve	⊣◁▷├	⊣◁▷├	→◁▷←	✕◁▷✕	⊖◁▷⊖

401A10B.EPS

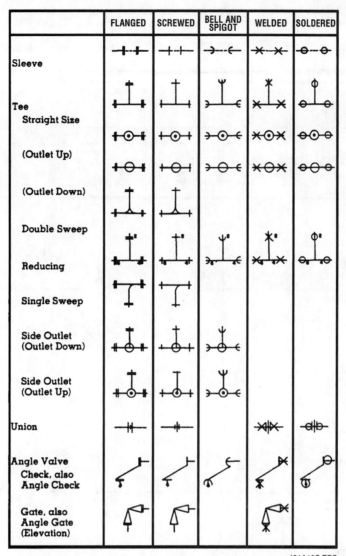

	FLANGED	SCREWED	BELL AND SPIGOT	WELDED	SOLDERED
Sleeve					
Tee Straight Size					
(Outlet Up)					
(Outlet Down)					
Double Sweep					
Reducing					
Single Sweep					
Side Outlet (Outlet Down)					
Side Outlet (Outlet Up)					
Union					
Angle Valve Check, also Angle Check					
Gate, also Angle Gate (Elevation)					

401A10C.EPS

TYPE OF FITTING		SCREWED OR SOCKET WELD	WELDED	FLANGED
		SINGLE LINE	SINGLE LINE	SINGLE LINE
90° ELBOW	TOP			
	SIDE			
	BOTTOM			

401A11.EPS

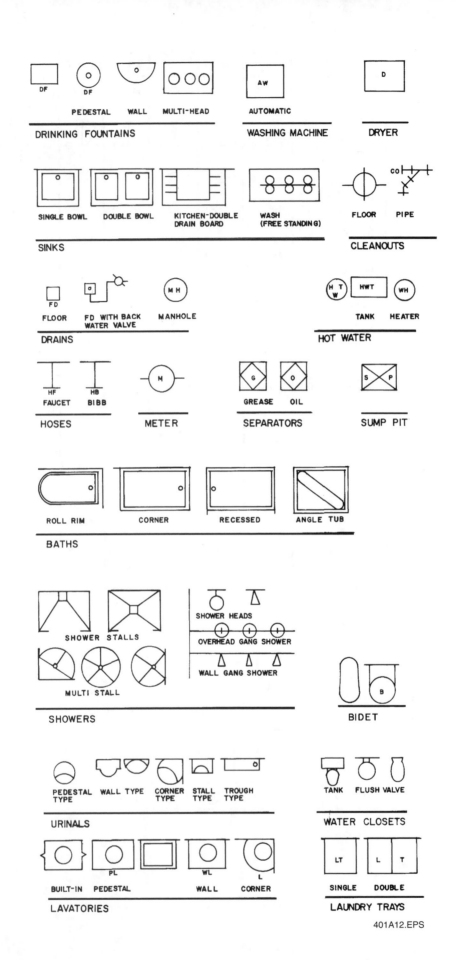

DRINKING FOUNTAINS

PEDESTAL | WALL | MULTI-HEAD

WASHING MACHINE — AUTOMATIC

DRYER

SINKS

SINGLE BOWL | DOUBLE BOWL | KITCHEN-DOUBLE DRAIN BOARD | WASH (FREE STANDING)

CLEANOUTS — FLOOR | PIPE

DRAINS

FLOOR | FD WITH BACK WATER VALVE | MANHOLE

HOT WATER — TANK | HEATER

HOSES — FAUCET | BIBB

METER

SEPARATORS — GREASE | OIL

SUMP PIT

BATHS

ROLL RIM | CORNER | RECESSED | ANGLE TUB

SHOWERS

SHOWER STALLS

MULTI STALL

SHOWER HEADS

OVERHEAD GANG SHOWER

WALL GANG SHOWER

BIDET

URINALS

PEDESTAL TYPE | WALL TYPE | CORNER TYPE | STALL TYPE | TROUGH TYPE

WATER CLOSETS — TANK | FLUSH VALVE

LAVATORIES

BUILT-IN | PEDESTAL | WALL | CORNER

LAUNDRY TRAYS — SINGLE | DOUBLE

401A12.EPS

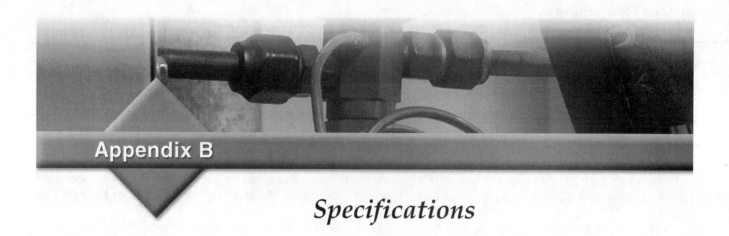

Appendix B

Specifications

CONTENTS:

Special Conditions

XYZ Remodel Project
Bentonville, Arkansas

1. In all instances where contradictions between plans and specifications exist, the plans shall take precedence.

2. The heating and air conditioning equipment noted in Section 15 of the specifications is not to be a part of the General Contract, but will be furnished and installed by XYZ. The General Contractor will be responsible for installing the support frames and platforms and will work with the XYZ equipment installer in providing crane service and assisting with the placement of the rooftop units as noted on the drawings. XYZ's heating and air conditioning installer will direct the placement of the rooftop units.

3. SAFETY — All required exits must be kept usable throughout the construction period. Provide lighted, enclosed walkways through new construction areas as required by governmental authorities having jurisdiction. All work shall comply with OSHA and local requirements.

4. COORDINATION OF WORK — All operations concerning the remodel are to be coordinated with the proper XYZ authority in order to minimize disruption of office areas to remain at the east wall of building. These offices are to remain in operation until the open area offices are set for these people to occupy. Then these offices may be demolished, removed, and completion of construction in this area begun.

5. TEMPORARY PARTITION — When the ceiling and floor covering of the new addition have been constructed, and prior to removing the existing office walls, install a temporary full-height braced 2" x 6" stud wall along the wall of the office area where shown on plan with top of plywood to the roof deck. Distance of temporary wall from existing wall to be coordinated between Owner and Contractor.

6. TEMPORARY INTERIOR ENCLOSURES — The interior remodeling of the building will necessarily require that the work be done in stages and isolated to certain areas. The coordination of the work is to be verified by XYZ Representatives. Areas of the existing work which are undergoing remodeling processes shall be entirely enclosed by the use of temporary partitions and separated from the remainder of the building which is in operation. Partitions shall go from the floor to the ceiling or above the ceiling, as necessary, to control dust and noise.

   Method of temporary partitions shall be as approved by XYZ Representative or the Architect.

7. SHOP DRAWINGS — The following shop drawings and submittals are required. Submit five copies of each to designated XYZ Representative.

   A. Folding Acoustical Partitions
   B. Exterior Canopy System
   C. Glass and Glazing in the Storefront
   D. Electrical Submittals
   E. Plumbing Submittals
   F. Millwork, including Doors
   G. Hardware Schedule
   H. Ceiling Tile Sample
   I. Overhead Doors and Dock Equipment
   J. Toilet Partitions
   K. Sprinkler Drawings Approved by Mutual Insurance

8. SALVAGED MATERIALS — Drawings and specifications call for the demolition and remodeling of certain portions of the existing building. Salvaged materials shall remain the property of the Owner. If the Owner does not wish to retain these materials, they shall become the property of the Contractor, who shall be responsible for their disposal. These items include but are not limited to the following:

   - Unit Heaters
   - Doors and Hardware
   - Electrical Panels
   - Switches, Wiring, and Light Fixtures
   - Water Heater
   - Plumbing Fixtures and Trim

Division 1 – General Conditions (Page 1 of 5)

1-1   GENERAL — The work to be done hereunder includes the furnishing of all labor, materials, and equipment necessary to complete the construction of an existing building renovation as shown and described in the drawings, specifications, and contract documents.

Before submitting his quotation, the Contractor shall satisfy himself as to the nature and location of the work; the confirmation of the ground; materials, tools, and equipment; and other facilities required before and during the work, as well as general and local conditions that can affect the work. The contractor accepts the work site as found, unless otherwise specifically stated in his quotation for the work. No compensation shall be made by the Owner to the Contractors for any Contractors' errors in bids.

The Contractor's quotation shall include the cost of all utility tie-ins, including tap fees for sewage, drainage, water supply, sprinkler system supply, natural gas, electricity, street cuts and replacement, etc., whether or not such work is on Owner's and/or public property in order to insure a complete job. Performance bonds, permits, and/or deposits required are also to be included in the quotation. XYZ will not honor any extras for these items.

1-2   DEFINITIONS — The word "XYZ" shall mean XYZ Properties, Inc., Bentonville, Arkansas, acting through its duly authorized representatives.

The word "OWNER" shall mean XYZ Properties, Inc., Bentonville, Arkansas.

The word "ARCHITECT" shall mean Sam Spearing, Architects and Engineers, who prepared contract documents for the project.

The word "CONTRACTOR" shall mean the person, persons, partnership, company, firm, or corporation entering into the contract for the performance of the work required by it, and the legal representative of said party, or agent appointed to act for said party in the performance of the work.

The word "CONTRACT" shall mean, collectively, all of the covenants, terms, and stipulations contained in the various portions of the contract, to-wit: bid letter, bids, specifications, plans, and shop drawings as well as any addenda, letter of authorization, or instruction, and any change orders which may be originated by XYZ.

1-3   CORRELATION AND INTENT OF DOCUMENTS — The contract documents are complementary to the specifications, and what is called for by any one shall be binding as if called for by all. It is the intent of the contractual documents to have a complete operating facility constructed, and with all services that are within the contract connected and in operating condition.

1-4   DRAWINGS AND SPECIFICATIONS — This specification is for general use and may apply to more than one building system such as masonry buildings, pre-engineered metal buildings, etc. Check drawings for items which are included in the work. Drawings shall govern when variations from this specification are found.

Contractor shall promptly call XYZ and the Architect's attention to any apparent contradictions, ambiguities, errors, discrepancies, or omissions in the plans or specifications. No extras shall be allowed for any such items the Contractor fails to report to XYZ and the Architect prior to the award of the contract.

All drawings, prints, specifications, or other documents furnished by XYZ or prepared by the Contractor or his subcontractors specifically for the work shall become the sole property of XYZ and shall be returned to XYZ at completion of the work.

When necessary, furnish shop drawings in sufficient quantity and, after approval by XYZ, these drawings shall be considered a part of the Contract Documents. Checking and approval by XYZ shall not relieve Contractor of any responsibility for errors, omissions, or discrepancies on such shop drawings.

1-5   INSURANCE — The Contractor shall carry insurance as listed below and furnish a certification of insurance before construction is started. The certification must indicate that the insurance will not be cancelled while the work specified therein is in progress, without ten (10) days prior notice to XYZ.

TYPE		LIMITS OF INSURANCE
Workmen's Compensation		Statutory Amount
CONTRACTOR'S PUBLIC LIABILITY		
Bodily Injury (including death):	Each Person	$250,000
	Each Accident	$500,000
Property Damage:	Each Accident	$100,000
	Aggregate	$100,000
CONTRACTOR'S AUTOMOBILE LIABILITY		
Bodily Injury (including death):	Each Person	$250,000
	Each Accident	$500,000
Property Damage		$100,000

The Contractor shall take out and maintain throughout the course of construction a Builders' All Risks Insurance Policy in the amount of the contract. This policy shall name as insureds the Owner and General Contractor; their subcontractors' loss shall be adjusted with payment to the Owner and General Contractor. Until work is fully completed and accepted by the Owner, the Contractor shall promptly repair any damage to the work. The original of said policy shall be delivered to the Owner prior to commencement of work and returned to the Contractor upon completion of work for cancellation and refund of premium, if any.

1-6   BOND — The Contractor shall furnish, where required by Owner, an approved performance bond in an amount equal to one hundred percent (100%) of the contract price. Verify with Owner prior to bid. The bond shall contain the following paragraph.

"And the said surety, for value received, hereby stipulates and agrees that no change, extension of time, alteration, or addition to the terms of the contract or to the work to be performed thereunder or the specifications accompanying the same shall in any way affect its obligations on this bond, and it does hereby waive notice of any change, extension of time, alteration, or addition to the terms of the contract or to the specifications."

1-7   TAXES — The Contractor shall pay all costs of Social Security payments, unemployment insurance, sales tax, and any other charges imposed by federal, state, and local governments.

1-8   PERMITS AND REGULATIONS — The Contractor shall obtain all permits required for the work, including necessary temporary permits, give all notices, and pay all fees. All equipment, materials, and workmanship shall comply with requirements of federal, state, and municipal codes and ordinances and underwriters' rules and specifications. Proposals shall be based on the plans and specifications with any exceptions required by codes and regulations being noted in writing by the Contractor in his proposal so that an equitable adjustment can be made in the contract price. Otherwise, it shall be construed that the Contractor is willing to comply with such codes, ordinances, rules, and regulations without additional cost to the Owner. Any equipment, materials, and workmanship installed contrary to above regulations shall be removed and replaced at the Contractor's expense.

1-9   SUPERINTENDENCE AND EMPLOYEES — The Contractor shall provide competent superintendence satisfactory to the Owner at all times. The superintendent shall not be changed except with consent of the Owner, unless he proves to be unsatisfactory to the Contractor and ceases to be in his employ. The Contractor shall be totally responsible for the employment, supervision, welfare, and compensation of his employees and shall be responsible for any work performed by his employees or his subcontractors.

1-10   SUBCONTRACTS — The Owner shall have the right to approve or disapprove the subcontractors to be used in the work.

1-11   SEPARATE CONTRACTS — The Owner reserves the right to let other contracts in connection with the work. The Contractor shall give other contractors reasonable opportunity for the storage of their materials and the execution of their work. The Contractor shall properly connect and coordinate all work with other Contractors. When any other Contractor's work is involved in the proper execution and results of the Contractor's work, the Contractor shall inspect and measure the other Contractor's work and report any discrepancy to the Owner.

1-12   DISCHARGE OF LIENS — If at any time there is evidence that a claim which is chargeable to the Contractor may become a lien against the premises, the Owner may retain out of the contract price an amount sufficient to indemnify it against any lien grown out of such claim and against all costs and expenses (including attorney's fees) which Owner may incur in connection with the claim of lien or arising out of any action related thereto. If payment of the contract price has been made to the Contractor, he shall reimburse the Owner for all monies, costs, expenses, and disbursements (including attorney's fees) that the Owner may be compelled to pay to discharge any claim or lien against the premises. When requesting the final payment for completion of contract, the Contractor shall submit to the Owner an executed and notarized Contractor's Affidavit stating that all subcontractors and suppliers have been paid in full. Upon Owner's request, Contractor will supply lien waivers from all subcontractors and material suppliers.

1-13   GUARANTEE — The Contractor shall guarantee materials and workmanship against defects for a minimum period of one (1) year from the date of final acceptance. Upon notification from the Owner, the Contractor agrees to promptly repair or replace any defects and all resulting damage to the satisfaction of the Owner and at no cost to the Owner.

1-14   PAYMENTS TO THE CONTRACTOR — At thirty (30) day intervals, the Contractor may submit to the Owner requests for payment for the work completed to date, less a retainage of ten percent (10%) of the amount due, less the amount of payments previously made.

1-15   CREDITS AND EXTRAS — When additions to, or deletions from, the work covered in the contract are required by the Owner, a change order shall be executed by the owner and the Contractor including the costs of the changes involved.

1-16   INSPECTIONS — XYZ and its representatives shall, at all times, have access to the project, and the Contractor shall give XYZ a sufficient advance notice of when the work will be ready for the following inspection at designated stages of construction:

1. Inspection of soil condition before and after excavation for footings.
2. Inspection of steel reinforcing before placement of concrete.
3. Inspection after structural steel is in place.
4. Inspection of electrical and plumbing rough-in before floor slab is poured.
5. Inspection of roofing application as roofing contractor begins work.
6. Inspection of electrical and plumbing rough-in before rough-in is covered in walls and ceiling.
7. Final electrical, mechanical, and plumbing inspections.
8. Inspection of driveway base material after it has been set up prior to paving.
9. Final inspection of complete job.

1-17 CLIMATIC CONDITIONS — When so directed by XYZ, the Contractor shall suspend work that may be subject to damage by climatic conditions.

1-18 TEMPORARY SERVICES — The Contractor shall pay for all fuel, electrical current, and water required for construction purposes. The Contractor shall also provide temporary heat and temporary toilet, if required, as directed by the Owner.

1-19 SUBSTITUTIONS — The Contractor shall base his bid on the cost of the materials and/or products specified. If the Contractor desires to substitute any equal material of another brand or manufacturer, it shall be requested in writing at least five (5) days prior to the bid due date. Requests from subcontractors will not be considered. Samples and/or technical data on the proposed product or material shall accompany these requests.

1-20 MEASUREMENTS, LINES, AND GRADES — The Contractor shall be responsible for the accuracy of all lines and grades of the work. The Contractor shall do all field work necessary to lay out and maintain the work. No extra charge or compensation will be allowed due to differences between actual dimensions and the measurements indicated on the drawings; any difference which may be found shall be submitted to the Owner for his consideration before proceeding with the work. The Owner will provide a survey of the property including reference points, property corners, and bench marks.

1-21 OCCUPANCY PRIOR TO FINAL ACCEPTANCE — The Owner reserves the right to take possession and use any completed or partially completed portion of the project, providing it does not interfere with the Contractor's work. Such possession or use of the project shall not be considered as final acceptance of the project, or any portion thereof.

1-22 SAFETY — The Contractor shall provide and continuously maintain adequate safeguards, such as railings, temporary walks, lights, etc., to prevent the occurrence of accidents, injuries, or damage to any person or property. The Contractor shall adhere to the requirements of the Federal Occupational Safety and Health Act as it relates to the work covered in the contract.

1-23 USE OF PREMISES — The Contractor shall occupy, use, and permit others to use the premises only for the purpose of completing the work to be performed under his contract with the Owner. Storage and other uses required by the Contractor shall be in areas designated by the Owner.

1-24 USE OF ADJOINING PREMISES — The Contractor shall confine his operations to the area contained within the property lines as shown on the plot plan. He may use public streets and alleys as permitted by the jurisdictional authorities. No equipment, forms, materials, scaffold, or persons shall encroach or trespass on any adjoining property, unless prior written consent of the landowner is obtained by the Contractor.

1-25 PROTECTION OF ADJACENT WORK AND PROPERTY — The Contractor shall protect all adjacent work and property such as structures, fences, trees, hedges, etc., from all damage resulting from his operations. Should he find it necessary to remove or trim, etc., an existing tree, hedge, etc., he shall secure all permits and approvals and pay any and all costs arising therefrom. The Contractor shall check all projects, offsets, footings, etc., and determine that there are no encroachments of the building or appurtenances on adjoining property. Where encroachments occur as a result of the work performed under these plans and specifications, the Contractor shall remove such encroachments at his own cost and at no expense to the Owner.

1-26 CLEANING UP — The Contractor shall keep the project clean at all times. No accumulation of waste material or rubbish shall be permitted, and at the completion of the work, the Contractor shall remove all rubbish, tools, and surplus materials, and shall leave the work "broom clean" and ready for use, unless otherwise specified.

Division 1 – General Conditions (Page 5 of 5)

1-27 CONSTRUCTION-RELATED DOCUMENTS REQUIRED BY XYZ — Prior to the payment by XYZ of the final contract retainage for XYZ-built facilities, or the first month's rent for facilities built by developers other than XYZ, the following documents shall have been placed on record with XYZ:

1. Letter from registered consulting engineering firm attesting to the adequacy of the asphalt paving design.
2. Executed copy of roofing guarantee.
3. Completed Maintenance Data Sheet (forms supplied by XYZ).
4. Copies of structural steel columns, beams, bar joists, and steel decking manufacturer's shippers' bills of lading.
5. Copies of soil density tests.
6. Copies of concrete cylinder tests.
7. Copies of test results of asphalt pavement base thickness and density and of asphalt topping conformance with specifications.
8. Final letter of acceptance of fire protection sprinkler system from Mutual Insurance.
9. Copies of manufacturer's five-year warranties on all heating and air conditioning equipment.

Division 15 – Heating, Ventilating, and Air Conditioning (Page 1 of 3)

15-1 GENERAL — Comply with applicable mechanical codes and standards pertaining to materials, products, and installation of air handling, metal ductwork, hot water systems, and chilled water systems.

15-2 DATA — Submit manufacturer's technical product data, assembly type shop drawings, wiring drawings, and maintenance data for each component of each HVAC system.

15-3 MATERIALS — Provide factory fabricated and tested equipment and materials of sizes, rating, and characteristics indicated. Referenced equipment and materials indicate style and quality desired. Contact engineer for preliminary approval of any other manufacturer's submittal. Provide proper quantity of materials and equipment as required for complete installation of each HVAC system.

Identify each HVAC system's components with materials and designations as directed.

15-4 INSTALLATION — Install each system in accordance with applicable mechanical codes and standards, recognized industry practices, and manufacturer's recommendations.

Test and balance each HVAC system in accordance with applicable mechanical codes and standards; balance air handling system to CFMs shown on drawings. Report findings to engineer.

15-5   KITCHEN — Provide and install all supply, return, and ventilation ductwork shown, together with plenums, casings, dampers, turning vanes, grilles, ceiling outlets, heating coils, etc., including the setting of fans, filters, and air units.

Kitchen exhaust duct shall be 16 gauge steel assembled by welding. It shall drain back to the hood, using a slope as required by code. Access doors shall be grease-tight construction with hinges and latches; locate where required by code and as required to provide access to fire protection devices in the duct. Make rigid connections to the hood and the fan; transition where required for connection to the fan; mastic at connection to the fan inlet. Coordinate required two hour fire-rated drywall enclosure of kitchen exhaust duct provided by others with general contractor.

Spiral conduit and fittings shall be lock-forming quality galvanized prime-grade steel; both conduits and fittings from the same manufacturer; spiral locks seam; fittings constructed with welded seams.

Kitchen dishwasher exhaust shall be fabricated from new prime-grade aluminum or stainless steel per SMACNA guidelines and installed watertight with rigid connections with slope back to dishwasher as required by code.

15-6   SHEET STEEL — All other supply, return, and ventilation ductwork, plenums, dampers, etc. shall be new lock-forming quality galvanized prime-grade steel sheets fabricated, supported, and installed per SMACNA guidelines.

15-7   FLEXIBLE DUCTWORK — Flexible ductwork shall be factory fabricated with vinyl coated spring steel wire helix bonded to a continuous layer of vinyl impregnated and coated fiberglass mesh inner sleeve, a 1-inch thick glass fiber blanket insulation layer, and an outer moisture barrier jacket of Mylar/neoprene or vinyl conductance of 0.23 Btu/hr./sq. ft./°F at 75°F and shall be UL-listed and shall comply with NFPA Standard 90A. All connection points shall be bonded with a mechanical fastener and taped. Maximum length of flexible duct shall be two feet on inlet of all VAV and fan-powered boxes and six feet on all grille connections.

15-8   CEILING — Ceiling supply outlets, grilles, and registers shall be provided in accordance with the schedule on the drawings by specified manufacturers.

Ceiling-mounted devices shall be finished in off-white baked enamel. Sidewall supplies and registers shall be finished in a prime coat or a baked enamel finish and job painted in color required.

15-9   DUCT CONSTRUCTION — Per SMACNA guidelines, except for special duct construction specified hereinbefore, ductwork shall be constructed in accordance with SMACNA "HVAC Duct Construction Standards" rated for two-inch static pressure. This includes all ductwork accessories, including but not limited to dampers, vanes and vane runners, hangers, and supports.

15-10   SUPPLY DUCT — Supply duct between air handling units and VAV and fan-powered boxes shall be constructed in accordance with SMACNA "HVAC Duct Construction Standards" reinforced for 4-inch static pressure.

15-11   FIRE WALLS — Where ducts penetrate two-hour fire walls or floors, fire dampers shall be installed in the construction in a manner directed by the manufacturer of the dampers, with or without attached sleeves as required. Install angles completely framing each opening on both sides and attach to the damper body or sleeve with screws, bolts, or approved fasteners. Leave space between the edges of the construction and the face of the sleeve or damper to accommodate expansion and contraction of the metal.

15-12   FITTINGS — All fittings in primary duct shall be radiused per SMACNA. No vanes, splitters, or dampers are allowed.

15-13 SEAMS AND JOINTS — All ductwork seams and joints shall be sealed according to SMACNA Class B requirements. Fabricate ducts to prevent seams of joints being cut for the installation of grilles, registers, or outlets.

15-14 OPENINGS — Openings through structure required for ductwork will be provided by others unless otherwise shown.

15-15 REINFORCEMENT — Reinforce all ducts to prevent buckling, breathing, vibrations, or unnecessary noises during start-up, shutdown, and continuous operation of air handling system.

15-16 BALANCING DAMPERS — Provide balancing dampers at points of low-pressure supply and exhaust where branches are taken from larger ducts. Use splitter dampers only where indicated.

15-17 SPACE REQUIREMENTS — In adherence to ceiling height schedules indicated; consult with other trades, and in conjunction with them, establish necessary space requirements for each trade so as to maintain required clearances. Where no ceiling height is stated, ductwork shall run as high as possible unless noted otherwise. Penetration of ductwork by pipes, conduits, electrical fixtures, or structural members is not acceptable.

15-18 FIRE DAMPERS — Fire dampers shall be type A with wall or floor sleeves as required for proper protection of penetration; UL labeled and conforming to UL-555. Dampers shall be 90% free area as a minimum. Mechanical contractor must provide and install all fire dampers in accordance with the fire damper system design and required codes.

15-19 DUCT ACCESS DOORS — Duct access doors shall be Air Balance FSA 100 or equivalent, galvanized steel frame and double wall hinged door with 1-inch insulation, gasketing, and latch. It shall be sized to permit servicing of fusible links in fire dampers and/or sized as noted into plenums for access to coils, filters, etc., where installed in lined ducts and/or plenums.

Division 16 – Electrical

16-1 GENERAL — All work shall be in strict compliance with latest edition of the National Electric Code and applicable state and local codes. All work shall be done using IMC, EMT, PVC, ENT, flexible conduit, surface raceway, cable tray, MC cable, AC cable, NM cable, as appropriate for the specific application.

Material and equipment shall be new, of standard manufacturer's construction, installed in accordance with accepted practice by competent workmen.

16-2 WIRE — Wire shall be #12 AWG minimum unless otherwise noted, with THHN or THW insulation and copper conductor.

16-3 OUTLET BOXES — Outlet boxes shall be four inches square, fitted with appropriate device ring or single piece masonry type set flush with finished surface.

16-4 SWITCHES — Switches shall be 15 amp specification grade equal to P&S 501, 120-277 volts, mounted at 18" above floor unless otherwise noted.

16-5 RECEPTACLES — Receptacles shall be 15 amp self-grounding specification grade equal to 5242, mounted at 18" above floor unless otherwise noted.

Receptacles, switch, telephone, and cover plates shall be smooth ivory plastic equal to sierra "P" series.

Additional Resources

This module is intended to be a thorough resource for task training. The following reference works are suggested for further study. These are optional materials for continued education rather than for task training.

Blueprint Reading for Construction, 1998. James A. S. Falzinger, Upper Saddle River, NJ: Prentice Hall.

Construction Blueprint Reading, 1985. Robert Putnam. Upper Saddle River, NJ: Prentice Hall.

Reading Architectural Work Drawings, 1996. Edward J. Muller. Upper Saddle River, NJ: Prentice Hall.

Reading Plans and Elevations, NCCER Carpentry Curriculum.

Figure Credits

Topaz Publications, Inc., 401SA01, 401SA03

John Hoerlein, 401F02

Mark Sanford Group LLC, 401F05

Ivey Mechanical Company LLC, 401F06, 401F09

The Construction Specifications Institute, 401F20

The Numbers and Titles used in this textbook are from MasterFormat™ 2004, published by The Construction Specifications Institute (CSI) and Construction Specifications Canada (CSC), and are used with permission from CSI. For those interested in a more in-dept explanation of MasterFormat™ 2004 and its use in the construction industry visit www.csinet.org/masterformat or contact:

The Construction Specifications Institute (CSI)
99 Canal Center Plaza, Suite 300
Alexandria, VA 22314
800-689-2900; 703-684-0300
http://www.csinet.org

Kogok Corporation, 401F21, 401F22, 401F23

Carnes Company, 401F24

Scalex Corporation, 401SA04

NCCER makes every effort to keep these textbooks up-to-date and free of technical errors. We appreciate your help in this process. If you have an idea for improving this textbook, or if you find an error, a typographical mistake, or an inaccuracy in NCCER's Contren® textbooks, please write us, using this form or a photocopy. Be sure to include the exact module number, page number, a detailed description, and the correction, if applicable. Your input will be brought to the attention of the Technical Review Committee. Thank you for your assistance.

Instructors – If you found that additional materials were necessary in order to teach this module effectively, please let us know so that we may include them in the Equipment/Materials list in the Annotated Instructor's Guide.

Write: Product Development and Revision
National Center for Construction Education and Research
3600 NW 43rd St, Bldg G, Gainesville, FL 32606

Fax: 352-334-0932

E-mail: curriculum@nccer.org

Craft   Module Name
_____

Copyright Date   Module Number   Page Number(s)
_____

Description
_____
_____
_____
_____

(Optional) Correction
_____
_____
_____

(Optional) Your Name and Address
_____
_____
_____

03402-09

System Balancing

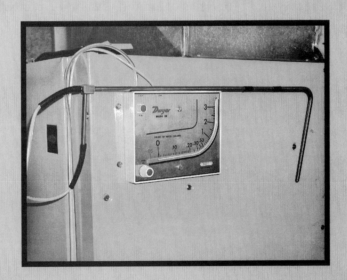

System Balancing

Topics to be presented in this module include:

Overview

Even when a forced-air comfort system is properly installed and operating, the building occupants may not be comfortable. If the air is not being properly distributed to the spaces, some spaces will be too warm, while others may be too cool. Therefore, once the system is installed and running, the distribution of air must be properly balanced. This process requires specialized knowledge and the ability to use specialized test instruments to adjust air flow for optimum performance. Air balancing also requires a knowledge of the physical properties of air and the ability to interpret psychrometric charts.

Objectives

When you have completed this module, you will be able to do the following:

1. Explain the gas laws (Dalton, Boyle, and Charles) used when dealing with air and its properties.
2. Explain the fan and pump laws.
3. Use a psychrometric chart to evaluate air properties and changes in air properties.
4. Explain the principles involved in the balancing of air and water distribution systems.
5. Define common terms used by manufacturers when describing grilles, registers, and diffusers.
6. Identify and use the tools and instruments needed to balance air distribution systems.
7. Balance an air distribution system.
8. Change the speed of an air distribution system supply fan.

Trade Terms

Atmospheric pressure
Boyle's law
Charles' law
Dalton's law
Dew point
Induction unit system
Psychrometrics
Specific density
Specific volume
Total heat
Total pressure
Traverse readings
Velocity
Velocity pressure

Required Trainee Materials

1. Pencil and paper
2. Appropriate personal protective equipment

Prerequisites

Before you begin this module, it is recommended that you successfully complete *Core Curriculum*; *HVAC Level One*; *HVAC Level Two*; *HVAC Level Three*; and *HVAC Level Four*, Module 03401-09.

This course map shows all of the modules in the fourth level of the *HVAC* curriculum. The suggested training order begins at the bottom and proceeds up. Skill levels increase as you advance on the course map. The local Training Program Sponsor may adjust the training order.

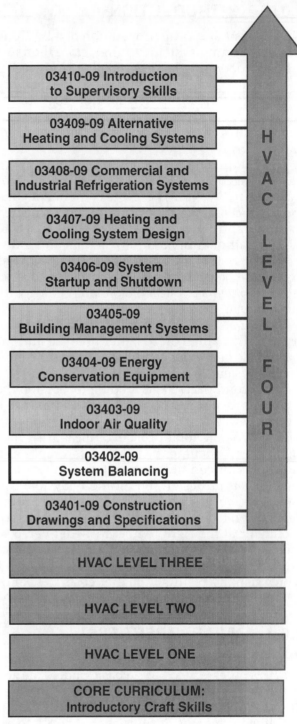

402CMAP.EPS

1.0.0 ◆ INTRODUCTION

The efficient and proper operation of a building air conditioning system requires more than just properly operating heating, cooling, and electrical systems. Of equal importance is the delivery of the correct quantity of conditioned air to the occupied space. This requires that the related air distribution system be properly installed and balanced.

This module builds on the information previously covered in the *HVAC Level One* module, *Air Distribution Systems*. There, the components that form a forced-air distribution system are covered in detail and the various instruments used to measure air quantity and flow within an air distribution system are introduced. Some trainees may find it helpful to review the material covered in *Air Distribution Systems* before proceeding with this module.

Having an understanding of air and its properties will enable you to better analyze and/or predict the operating conditions for an air conditioning system.

The second part of this module focuses on the general procedures used to test and balance a building's air distribution system. Testing and balancing is the total process of measuring, adjusting, and documenting the performance of a building's air conditioning system to make sure that it meets its design specifications. System balancing is necessary because human comfort depends on having the right amount of heated or cooled air at the right temperature delivered to each space in the building. A building that is not properly air balanced is not operating efficiently, which usually results in higher costs to heat or cool the building.

2.0.0 ◆ AIR PROPERTIES

Before you can study how to balance an air system, it is necessary to have a basic understanding of the properties of air and the terms used when describing air. Also, you must become familiar with the basic laws of physics that apply to air systems. This section covers these subjects at an introductory level. For more advanced study, refer to the references and acknowledgments section of this module.

2.1.0 Atmospheric Pressure and Gauge Pressure

The earth is surrounded by a mixture of gases composed of 78 percent nitrogen, 21 percent oxygen, and a 1 percent mix of other gases. Together, they form our atmosphere, which extends about 600 miles above the earth, and is held to the planet by the force of gravity. Being composed of gas, the atmosphere has weight, and that weight is measured in pounds per square inch (psi). At sea level, a square-inch column of air extending 600 miles above the earth exerts a weight and pressure on the earth of 14.7 psi. The pressure exerted on all things on the earth's surface as a result of the weight of our atmosphere is called **atmospheric pressure** (*Figure 1*).

Atmospheric pressure can be measured with a barometer. *Figure 2* shows a basic mercury-tube barometer. It consists of an open dish of mercury, with a mercury-filled tube sealed at one end and inverted vertically into the dish. If the 14.7 psi column of air is applied to the mercury-tube barometer at sea level, it will cause the mercury in the tube to rise to a height of 29.92". This means that every square inch of the earth's surface at sea level has 14.7 pounds of air pressure pushing down on it. These values of 14.7 psi and 29.92 inches of mercury (29.92 in. Hg) for dry air at sea level, at 70°F, are standards that are used frequently in HVAC work. A pressure scale, called the absolute pressure scale, is based on the barometer measurements just described. On this scale, pressures are expressed in pounds per square inch (psi) or pounds per square inch absolute (psia), starting from zero (0) psi.

Another scale, called gauge pressure, is frequently used to define air pressure levels. Gauge pressure scales use atmospheric pressure as their zero starting point. Positive gauge pressures, those above zero (14.7 psi), are expressed in

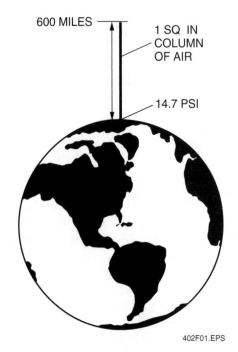

Figure 1 ◆ Atmospheric pressure.

pounds per square inch gauge (psig). Negative pressures, those below 0 psig, are expressed in inches of mercury vacuum (in. Hg vacuum). Gauge pressures can easily be converted to absolute pressures by adding 14.7 to the gauge pressure value. For example, a gauge pressure of 10 psig equals an absolute pressure of 24.7 psia (10 + 14.7). A comparison of the gauge and absolute pressure scales is shown in *Figure 2*. Conversion between the absolute and gauge pressure scales is often necessary when making calculations concerned with air pressure relationships.

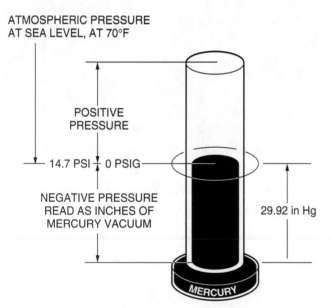

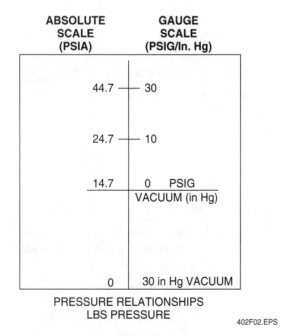

PRESSURE RELATIONSHIPS
LBS PRESSURE

402F02.EPS

Figure 2 ◆ Atmospheric, absolute, and gauge pressure relationships.

2.2.0 Air Temperature, Pressure, and Volume Relationships

Dry air is a gas. As such, air follows the gas laws of Dalton, Boyle, and Charles. These laws pertain to air temperature, pressure, and volume relationships, and are discussed in the following sections.

2.2.1 Dalton's Law

Dalton's law states that the **total pressure** of a mixture of confined gases is equal to the sum of the partial pressures of the individual component gases (*Figure 3*). The partial pressure is the pressure that each gas would exert if it alone occupied the volume of the mixture at the same temperature.

2.2.2 Boyle's Law

Boyle's law states that with a constant temperature, the pressure on a given quantity of confined gas varies inversely with the volume of gas.

DALTON'S LAW

- EACH GAS IN A MIXTURE OF GASES ACTS INDEPENDENTLY.

- TOTAL PRESSURE = SUM OF THE PRESSURES CREATED IN A CYLINDER / OF EACH GAS IN THE MIXTURE

WATER VAPOR

OXYGEN

NITROGEN

COMPRESSED AIR CYLINDER

CARBON DIOXIDE

402F03.EPS

Figure 3 ◆ Dalton's law.

Similarly, at a constant temperature, the volume of a given quantity of gas varies inversely with the applied pressure. Formulas 1 through 3 describe this law mathematically.

Formula 1 $\quad$ Po $\times$ Vo = Pn $\times$ Vn

Formula 2 $\quad$ Vn = $\dfrac{\text{Vo} \times \text{Po}}{\text{Pn}}$

Formula 3 $\quad$ Pn = $\dfrac{\text{Po} \times \text{Vo}}{\text{Vn}}$

Where:
Po $\quad$ = original absolute pressure (psia)

Pn $\quad$ = new pressure (psia)

Vo $\quad$ = original volume (cubic feet)

Vn $\quad$ = new volume (cubic feet)

psia = psig + 14.7

Study Example

What is the new volume of 2 cubic feet of gas at 30 psig if it is compressed to 60 psig, providing the temperature remains constant?

Using *Formula 2:*

Vn = $\dfrac{\text{Vo} \times \text{Po}}{\text{Pn}}$ = $\dfrac{2 \text{ cu. ft.} \times (30 \text{ psig} + 14.7)}{60 \text{ psig} + 14.7}$ =

$\dfrac{2 \times 44.7}{74.7}$ = 1.196 (1.2 rounded off)

2.2.3 Charles' Law

Charles' law states that with a constant pressure, the volume for a given quantity of confined gas varies directly with its absolute temperature. Similarly, with a constant volume of gas, the pressure varies directly with its absolute temperature. Formulas 4 through 6 describe this law when working with a constant pressure. Formulas 7 through 9 describe it when working with a constant volume.

With constant pressure:

Formula 4 $\quad$ Vo $\times$ Tn = Vn $\times$ To

Formula 5 $\quad$ Vn = $\dfrac{\text{Vo} \times \text{Tn}}{\text{To}}$

Formula 6 $\quad$ Tn = $\dfrac{\text{Vn} \times \text{To}}{\text{Vo}}$

Where:
To = original absolute temperature

Tn = new absolute temperature

Vo = original volume (cubic feet)

Vn = new volume (cubic feet)

Absolute temperature = °F + 460

With constant volume:

Formula 7 $\quad$ Po $\times$ Tn = Pn $\times$ To

Formula 8 $\quad$ Tn = $\dfrac{\text{Pn} \times \text{To}}{\text{Po}}$

Formula 9 $\quad$ Pn = $\dfrac{\text{Po} \times \text{Tn}}{\text{To}}$

Where:
To = original absolute temperature

Tn = new absolute temperature

Po = original absolute pressure (psia)

Pn = new absolute pressure (psia)

Absolute temperature = °F + 460

psia = psig + 14.7

Study Examples

1. What is the new volume of 8 cubic feet of gas at 40°F if the temperature is raised to 80°F, at a constant pressure?

Using *Formula 5:*

Vn = $\dfrac{\text{Vo} \times \text{Tn}}{\text{To}}$ = $\dfrac{8 \text{ cu. ft.} \times (80°\text{F} + 460)}{40°\text{F} + 460}$ =

$\dfrac{8 \times 540}{500}$ = 8.64 cu ft

2. What is the new pressure (in psig) of a quantity of gas at 40°F and 35 psig if its temperature is raised to 60°F, and if it is at a constant volume?

Using *Formula 9*:

$$Pn = \frac{Po \times Tn}{To} = \frac{(35 \text{ psig} + 14.7) \times (60°F + 460)}{40°F + 460} =$$

$$\frac{49.7 \times 520}{500} = 51.69 \text{ psia} - 14.7 = 36.99 \text{ psig}$$

3.0.0 ◆ PSYCHROMETRICS

Psychrometrics is the study of dry air and water vapor mixtures. Proper measurement and control of air is necessary so that the human body can feel comfortable in a room environment. It is also critical to many commercial and manufacturing processes. As an HVAC technician, you need to understand air and the relationships that exist among its various properties.

3.1.0 Dry Air

Air has weight, density, temperature, specific heat, and heat conductivity. In motion, it has momentum and inertia. It holds substances in suspension and in solution. Dry air is a mixture of gases composed of about 78 percent nitrogen, 21 percent oxygen, and 1 percent other gases (*Figure 4*).

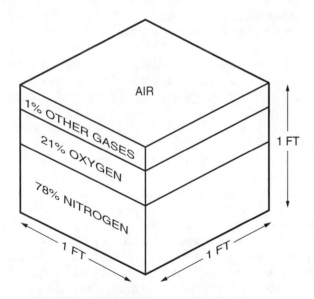

SPECIFIC VOLUME = 13.33 CU FT PER LB
SPECIFIC DENSITY = 0.075 LB PER CU FT
SPECIFIC HEAT = 0.24 BTUS PER LB PER °F

402F04.EPS

Figure 4 ◆ Standard air – dry air at 70°F at sea level.

Specific volume describes how much space one pound of dry air occupies. At 70°F at sea level (29.92 in. Hg), one pound of dry air occupies a volume of 13.33 cubic feet. If air is heated and maintained at a constant pressure, it will expand and weigh less per cubic foot of volume. This property of air is called **specific density**. At 70°F at sea level, one pound of dry air weighs .075 pound per cubic foot. The specific density per pound of dry air is found by dividing the volume of air into one pound.

Specific density of air = 1 lb ÷ 13.33 ft³ = 0.075 lb/ft³

The specific heat of air determines its ability to get hot. Air at sea level has a specific heat of 0.24 Btu/lb/°F. The sensible heat formula can be used to calculate how many Btus are needed to raise the temperature of dry air. It is called the sensible heat formula because sensible heat is the amount of heat which, when added to the air, causes a change in temperature with no change in the amount of moisture present. Sensible heat is measured with a thermometer, and the reading is known as a dry-bulb temperature.

Sensible heat (Btuh) = specific heat × specific density × 60 min./hr. × cfm × ΔT

Btuh = (0.24 × 0.075 × 60) × cfm × ΔT

Btuh = 1.08 × cfm × ΔT

Where:

cfm = **velocity** of airflow in cubic feet per minute

ΔT = change in temperature (°F)

3.2.0 Humidity

Water vapor, called humidity, is almost always suspended in the air. Humidity has many sources. Outdoors, it comes from the evaporation of Earth's oceans and other bodies of water into the atmosphere. Inside buildings, it comes from water vapor added by cooking, showers, and similar activities. The amount of moisture that the air will hold depends on the temperature of the air. Warm air will hold more moisture than cold air.

The amount of moisture contained in the air can be expressed as pounds of moisture per pound of dry air, or as grains of moisture per pound of dry air. Grains of moisture per pound of dry air is usually used, because the numbers are larger and easier to work with. They are also used because the amount of water vapor that is present in each pound of air is normally very small. At 70°F at sea level, one pound of water contains 7,000 grains of moisture (*Figure 5*).

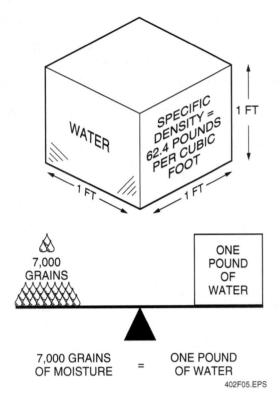

Figure 5 ◆ Relationships of water at 70°F at sea level.

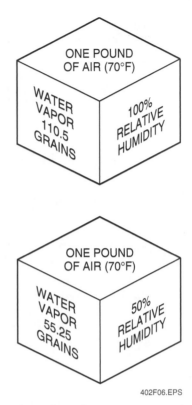

Figure 6 ◆ Relative humidity.

A maximum of only 110.5 grains of moisture are contained in a pound of saturated air at 70°F at sea level. If this same amount of moisture were expressed in pounds of moisture per pound of dry air, it would weigh 0.01579 pound.

3.2.1 Relative Humidity

The amount of humidity in the air affects the rate at which perspiration evaporates. Dry air causes rapid evaporation, which makes the skin feel cooler than the actual temperature. Moist air prevents rapid evaporation, making it feel warmer than the actual temperature. Relative humidity (RH) is the ratio of the amount of moisture present in a given sample of air to the amount it can hold at saturation. Relative humidity is expressed as a percentage. If a volume of air is totally saturated with moisture, its relative humidity is 100 percent. If it contains only half of the moisture it can hold, its relative humidity is 50 percent (see *Figure 6*).

In HVAC work, relative humidity measurements are made to determine the level of indoor environmental comfort that exists in the various rooms of a house or other conditioned space. Proper control of relative humidity is also critical to many commercial and manufacturing processes. Most people feel comfortable when the indoor temperature and humidity conditions fall within certain ranges, called comfort zones. For winter, temperatures between 67°F and 76°F and a relative humidity of about 30 percent are considered comfortable. For summer, the comfort zone is between 72°F and 81°F with a relative humidity of about 40 percent. Properly controlled temperature and humidity conditions are important because they improve personal comfort and health conditions in all seasons and reduce the cost of operating heating and cooling equipment.

Sling psychrometers (*Figure 7*), hygrometers, and some electronic thermometers can be used to measure relative humidity. These instruments contain two identical thermometers, one to measure dry-bulb temperatures and the other to measure wet-bulb temperatures. Dry-bulb temperatures measure the amount of sensible heat in the air. Wet-bulb temperatures are taken with a thermometer that has a wick saturated with distilled water wrapped around the sensing bulb. The reading from a wet-bulb thermometer takes into account the moisture content of the air, thus reflecting the **total heat** content. Evaporation occurs at the wick of the wet-bulb thermometer, giving it a lower temperature reading. The sling psychrometer is fitted with a handle by which it can be whirled with a steady motion through the sur-

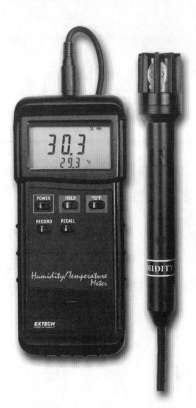

HUMIDITY/TEMPERATURE METER

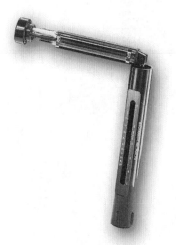

SLING PSYCHROMETER

402F07.EPS

Figure 7 ◆ Air temperature and humidity measuring instruments.

rounding air. The whirling motion is periodically stopped to take readings of the wet- and dry-bulb thermometers (in that order) until consecutive readings become steady.

Wet-bulb and dry-bulb temperatures measured with a sling psychrometer are used with a psychrometric chart to find the percent of relative humidity in the measured air. Electronic thermometers and hygrometers usually give a direct reading of the measured temperature and relative humidity.

3.3.0 Dew Point

The **dew point** is the temperature at which water vapor in the air becomes saturated and starts to condense into water droplets. If the relative humidity of air is 100 percent (saturated), the dew point temperature, wet-bulb temperature, and the dry-bulb temperature are all the same. This is true because the air cannot hold any more moisture, so no water can be evaporated from the wet-bulb thermometer.

3.4.0 Enthalpy

Moisture vapor in the air has its own heat content, called latent heat. This latent heat added to the sensible heat of a quantity of dry air yields the total heat of the air. The total heat content of the air and water vapor mixture as measured from a predetermined base or point is called enthalpy. Enthalpy is measured using a wet-bulb thermometer and is expressed in Btus per pound (Btu/lb).

When dealing with changes in both sensible and latent heat, such as in heating and humidification or cooling, the difference in enthalpy can be used to calculate how many Btus per hour (Btuh) the temperature of the air has been raised or lowered. This is useful when you need to determine the capacity of a typical air conditioning system. Using the measured dry-bulb and wet-bulb temperatures for the air stream entering and leaving the equipment, the values of enthalpy, found with a psychrometric chart, can be used to find the total difference in heat content (change in enthalpy) between the two readings in Btu/lb of air. This value of change in enthalpy, along with the value for the air stream velocity in cubic feet per minute (cfm), is used to calculate the equipment capacity in Btuh using the total heat formula shown below.

Total heat (Btuh) = specific density of air $\times$
60 min/hr $\times$ cfm $\times$ ΔH

Btuh = (0.075 $\times$ 60) $\times$ cfm $\times$ ΔH

Btuh = 4.5 $\times$ cfm $\times$ ΔH

Where:
cfm = velocity of airflow in cubic feet per minute
ΔH = change in enthalpy (Btu/lb)

4.0.0 ◆ PSYCHROMETRIC CHART

The psychrometric chart (*Figure 8*) is a graph of air properties. It is used to determine how air properties vary as the amount of moisture in the air changes.

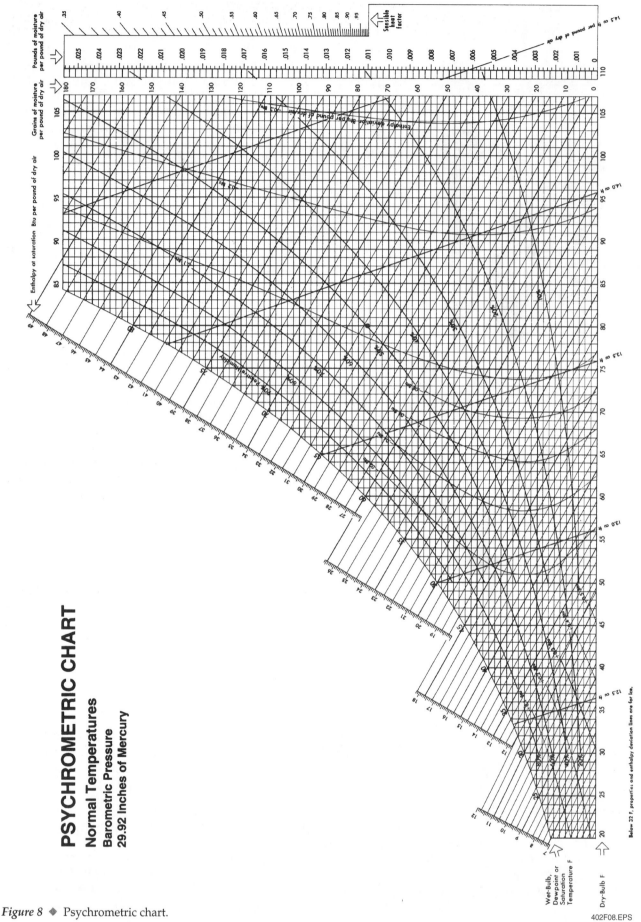

Figure 8 ◆ Psychrometric chart.

402F08.EPS

The chart shown in *Figure 8* is based on normal temperatures at sea level (29.92 in. Hg). Charts are also available that graph the air properties in low temperatures from –20°F to 50°F, high temperatures from 60°F to 250°F, and at various elevations other than sea level, to correct for changes in barometric pressure. They are also available in metric form.

With easily measured values of dry-bulb and wet-bulb temperatures, the psychrometric chart can be used to find the value for one or more of the remaining properties related to a given sample of conditioned air. For HVAC work, the dry-bulb and wet-bulb temperatures are frequently used to find the relative humidity of a conditioned space. One advantage in using a psychrometric chart instead of a direct-reading instrument is that the chart can be used to estimate the changes in value for various properties as a result of making adjustments to an HVAC system.

4.1.0 Psychrometric Chart Layout

The scales shown on the psychrometric chart are:

- Dry-bulb temperature °F
- Grains of moisture per pound of dry air
- Relative humidity
- Dew point temperature °F
- Wet-bulb temperature °F
- Enthalpy
- Specific volume
- Sensible heat factor

4.1.1 Dry-Bulb Temperature Scale

The sensible heat temperature scale, called the dry-bulb temperature scale, is laid out horizontally with increment lines extended vertically, as shown in *Figure 9(A)*. The scale ranges from 20°F to 110°F.

4.1.2 Grains of Moisture (Specific Humidity) Scale

The grains of moisture scale depicts the amount of water vapor mixed with each pound of air. As shown in *Figure 9(B)*, this scale runs vertically with increment lines extended horizontally. The scale ranges from 0 to 180 grains of moisture per pound of dry air at standard atmospheric pressure. Another scale on the chart, located to the right of the grains of moisture scale (*Figure 8*), can be used to determine the pounds of moisture per pound of dry air. This scale ranges from 0 to 0.025 pound.

4.1.3 Relative Humidity Lines

Relative humidity is shown as lines from 0 percent to 100 percent RH in increments of 10 percent. See *Figure 9(C)*. At the 100 percent relative humidity line, the air is fully saturated and can hold no more moisture. Any addition of moisture would cause condensation.

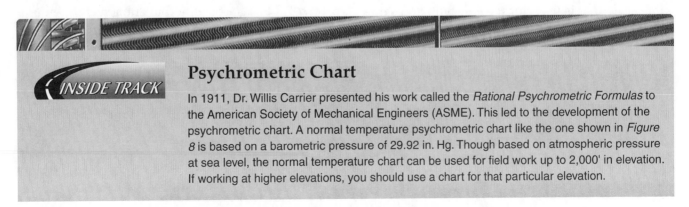

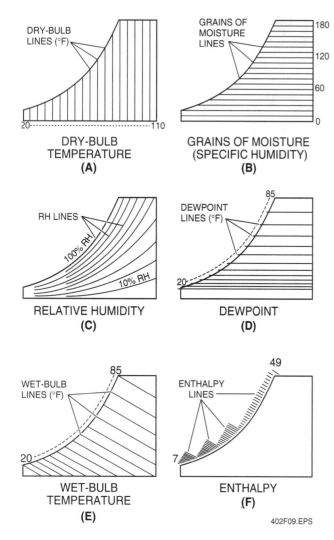

Figure 9 ◆ Psychrometric chart scales.

4.1.5 Wet-Bulb Temperature Lines

The wet-bulb temperature above 32°F is the temperature indicated by a thermometer whose bulb is covered by a wet wick and exposed to a stream of air moving at a velocity between 1,000 and 1,500 fpm. The wet-bulb temperature lines slant to the left at about a 45° angle and the scale ranges from 20°F to 85°F. See *Figure 9(E)*. The scale is the same one that is used with the dew point and saturation (100 percent RH) temperatures.

4.1.6 Enthalpy Scale

Enthalpy is very useful in determining the amount of heat that is added to or removed from air in a given process. Enthalpy is the measurement of the total heat in one pound of air. An approximate value for enthalpy is found by extending the wet-bulb temperature line past the 100 percent saturation line on the chart. The enthalpy scale ranges from 7 to 49 Btus per pound. See *Figure 9(F)*.

Enthalpy is the total heat in the air at 100 percent saturation. If the air is not completely saturated, a slight error is present in the enthalpy reading. When extreme accuracy is needed, the enthalpy reading can be corrected using the enthalpy deviation lines on the chart (*Figure 8*). These lines range from +0.1 to +0.5 and –0.02 to –0.3 Btu. This is the amount of heat that would be added to or subtracted from the enthalpy readings. These lines are used for design work, where precise measurement is required. They are normally not used for field service work.

4.1.4 Dew Point Temperature Lines

The temperature at which the moisture content or relative humidity has reached 100 percent is called the dew point. If the temperature drops below the dew point, the moisture vapor begins to condense. The dew point temperature lines, as shown in *Figure 9(D)*, run horizontally like the grains of moisture lines. The dew point temperature scale ranges from 20°F to 85°F and is the same scale that is used for the wet-bulb and saturation (100 percent RH) temperatures.

4.1.7 Specific Volume Scales

Specific volume is used primarily when checking fan performance and determining fan motor sizes. Specific volume is the number of cubic feet occupied by one pound of dry air at any given temperature and pressure. Specific volume lines ranging from 12.5 to 14.5 cubic feet per pound are shown on the chart as almost vertical lines that slant to the left. See *Figure 10(A)*. These lines represent the space occupied in cubic feet per pound of dry air.

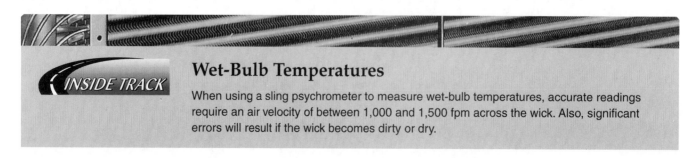

Wet-Bulb Temperatures

INSIDE TRACK

When using a sling psychrometer to measure wet-bulb temperatures, accurate readings require an air velocity of between 1,000 and 1,500 fpm across the wick. Also, significant errors will result if the wick becomes dirty or dry.

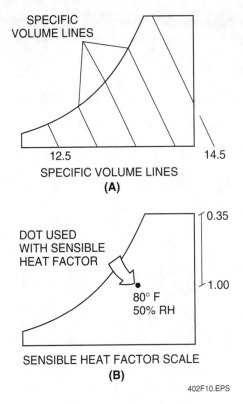

Figure 10 ◆ Psychrometric chart specific volume and sensible heat factor scales.

Referring back to *Figure 8*, you see that the specific volume of one pound of dry air at 75°F dry bulb displaces a volume of 13.5 cubic feet at sea level. If the air is heated to 95°F, it expands and takes up 14 cubic feet. If the air is cooled to 35°F, it occupies only 12.5 cubic feet. As described earlier in this module, the volume of a gas (air) varies directly with its absolute temperature. The higher the temperature, the greater its volume; the lower the temperature, the smaller its volume.

4.1.8 Sensible Heat Factor

A useful method for finding sensible heat involves the sensible heat factor scale. This scale, located on the right side of the chart, provides the sensible heat factor. See *Figure 10(B)*. The scale range is from 0.35 to 1.00 or 35 to 100 percent, which represents the percentage of sensible heat in the air. This scale is used when plotting processes such as cooling or dehumidification. Once the process is plotted, the ratio of sensible heat to total heat (sensible + latent heat = total heat) of the air can be found. The remaining percentage of heat from 100 percent is the percent of latent heat in the air. The dot located at the intersection of the 80°F dry-bulb temperature and the 50 percent

humidity lines on the chart is provided for use in conjunction with the sensible heat factor scale. It is used when finding the dew point of a device when the device's sensible heat factor is known.

4.2.0 Use of the Psychrometric Chart

The psychrometric chart can be used to show what is happening during a specific heating, ventilating, or air conditioning process. It can be used both when designing systems and when servicing them. The remainder of this section focuses on the use of a psychrometric chart for servicing HVAC equipment. For more advanced study, see the references and acknowledgments section at the back of this module.

4.2.1 Using Known Properties of Air to Find Unknown Properties

Figure 11 shows a psychrometric chart with lines plotted for dry-bulb temperature, wet-bulb temperature, dew point temperature, specific humidity, and relative humidity. When any two of these values are known, the exact condition of the air can be located on the chart and the values for all of the other properties can be found from this one point.

For example, assume that the dry-bulb temperature is 75°F and the wet-bulb temperature is 65°F for a given sample of air. Plotting the dry-bulb and wet-bulb temperatures on the psychrometric chart (*Figure 8*), you will find that the relative humidity is about 58 percent, the dew point is 59.5°F, and the specific humidity is 76 grains. A simplified plot of this example is shown in *Figure 12*.

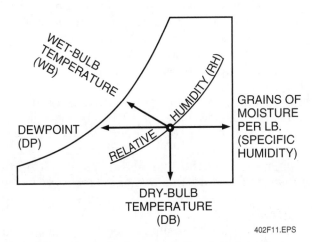

Figure 11 ◆ Example of air properties plotted on a psychrometric chart.

Relative Humidity

Using a psychrometric chart, determine the relative humidity, dew point, and specific humidity of an air sample having a dry-bulb temperature of 90°F and a wet-bulb temperature of 80°F.

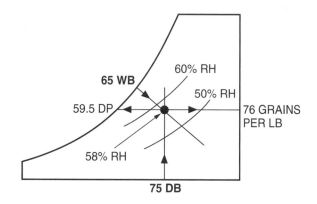

(GIVEN VALUES ARE SHOWN IN BOLD)

402F12.EPS

Figure 12 ◆ Finding unknown air properties when the dry-bulb and wet-bulb temperatures are known.

4.2.2 Working With Changes in Sensible Heat

The psychrometric chart can be used to evaluate air conditioning processes, such as changes in the sensible heat of air. Sensible heat results in a change in temperature and is indicated by a horizontal line on the chart.

An example of a change in sensible heat caused by heating is plotted in *Figure 13*. Assume that air at 30°F dry bulb and 80 percent humidity is heated to 75°F by passing it over a heating coil. If the air is heated with no moisture added, the process results in a horizontal line from point 1 to point 2 as marked on the chart. With the information given, you can determine the values for the unknown wet-bulb temperatures, relative humidity, and the dew points related to each point (*Figure 13*). As shown, the dew point temperature remains the same, because no water vapor has been added or condensed. Also, heating air with a constant moisture content reduces its relative humidity level. This is because the warmer air can hold more moisture; therefore, the ratio of the moisture in the air to the total amount it is capable of holding decreases.

An example of a change in sensible heat caused by cooling is plotted in *Figure 14*. Assume that air at 95°F dry bulb and 75°F wet bulb is cooled to 80°F dry bulb by passing it over a cooling coil.

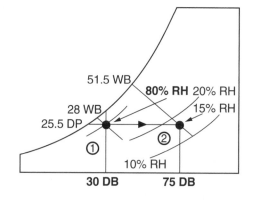

	DRY-BULB (DB) TEMP °F	WET-BULB (WB) TEMP °F	% RELATIVE HUMIDITY (RH)	DEW POINT (DP) °F
AIR AT	**30**	28	**80**	25.5
IS HEATED TO	**75**	51.5	15	25.5

(GIVEN VALUES ARE SHOWN IN BOLD)

402F13.EPS

Figure 13 ◆ Psychrometric chart plot of sensible heat increase (heating).

If the air is cooled without condensing moisture, the process results in a horizontal line from point 1 to point 2 as marked on the chart. With the information given, you can determine the values for the unknown wet-bulb temperatures, relative humidity, and the dew points related to each point (*Figure 14*). As shown, the dew point temperature remains the same because no water vapor has been added or condensed. Also, cooling air with a constant moisture content increases its relative humidity level.

4.2.3 Working With Enthalpy (Total Heat)

Enthalpy is the total heat content of the air and water vapor mixture. It is found on the chart by following along a wet-bulb temperature line, past the saturation line, and out to the enthalpy scale. Enthalpy can be used to determine the total heat that is added to or removed from a volume of air.

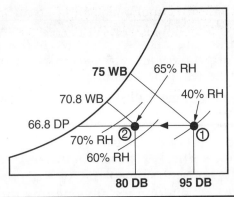

	DRY-BULB (DB) TEMP °F	WET-BULB (WB) TEMP °F	% RELATIVE HUMIDITY (RH)	DEW POINT (DP) °F
AIR AT	**95**	**75**	40	66.8
IS COOLED TO	**80**	70.8	65	66.8

(GIVEN VALUES ARE SHOWN IN BOLD)

402F14.EPS

Figure 14 ◆ Psychrometric chart plot of sensible heat decrease (cooling).

This is done by reading the scale between the two wet-bulb lines.

For example, *Figure 15* shows a plot of the dry-bulb and wet-bulb temperatures measured for the air entering and leaving an evaporator. The measured wet-bulb temperatures can be used to check the capacity of the evaporator. As shown, the wet-bulb temperatures for the air entering and leaving the evaporator are extended past the saturation line, and out to the enthalpy scale. In this example, the measured wet-bulb temperatures of 51°F and 61.5°F correspond to enthalpy readings of 20.8 Btu/lb and 27.5 Btu/lb, respectively. By subtraction, the difference in enthalpy, or the heat removed, is found to be 6.7 Btu/lb. Using the difference value of enthalpy found with the chart, and the value for the measured volume of airflow through the evaporator in cfm, the capacity of the evaporator can be calculated using the following formula:

Capacity (total heat) Btuh = 4.5 × cfm × ΔH

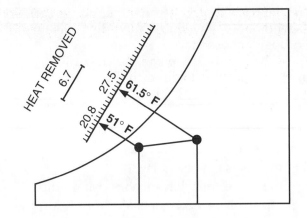

	WET-BULB (WB) TEMP °F	ENTHALPY (BTU/LB)	AIRFLOW VOLUME (CFM)
AIR ENTERING EVAPORATOR	**61.5**	27.5	**1000**
AIR LEAVING EVAPORATOR	**51**	20.8	**1000**

(GIVEN VALUES ARE SHOWN IN BOLD)

402F15.EPS

Figure 15 ◆ Using enthalpy values to find evaporator capacity.

Where:

   4.5 = constant (pounds of air per cu ft = 0.075 × 60 minutes/hour)

   cfm = velocity of airflow in cubic feet per minute

   ΔH = change in enthalpy in Btu/lb (difference in enthalpy between air in and air out)

If 1,000 cfm of air is circulated over the evaporator in this example, then 30,150 Btuh is removed, as follows:

Capacity (total heat) Btuh = 4.5 × 1,000 × 6.7 = 30,150 Btuh

5.0.0 ◆ AIR BALANCE

Air balancing means the proper delivery of conditioned air in the correct amounts to each of the areas in the structure being conditioned. The

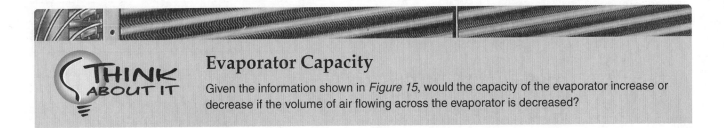

Evaporator Capacity

Given the information shown in *Figure 15*, would the capacity of the evaporator increase or decrease if the volume of air flowing across the evaporator is decreased?

satisfactory distribution of conditioned air depends upon a well-designed duct system and a properly chosen fan.

Air balancing also means that the correct amount of air must be returned to the heating or cooling unit. If the air distribution system is not balanced, several problems could arise:

- The spaces or rooms within a structure will have different temperatures.
- Some ducts may be noisy.
- Some spaces may have incorrect humidity.
- Some areas may contain stale air (low airflow).

Once the conditioning system is installed, it must be adjusted to make sure the right amount of conditioned air is circulated in the required spaces to meet design conditions. Balancing the air distribution system will often require the adjustment of the fan and the volume dampers so they will deliver enough air at the proper velocity to provide satisfactory heating and/or cooling. Balancing is simply a method of adjusting air volume to the designed value.

Information on proportional balancing for commercial units is covered in *Appendix B*. However, an understanding of fan laws and the principles of airflow are prerequisites for performing proportional balancing. Fan laws are covered later in this module.

For detailed information about airflow and the components that form air distribution systems, refer to the *HVAC Level Two* module, *Air Distribution Systems*.

5.1.0 Air Balance Terminology

The following information should assist you in identifying the terms used with the terminal outlets for space conditioning. These terms are generally broad-based and are not specific to any particular unit or system.

- *Aspect ratio* – The ratio of width to height of the grille core. Some manufacturers contend that the aspect ratio of a grille has only a negligible effect on the throw and drop of an airstream, regardless of whether the grille outlet face is square or rectangular in shape.
- *Diffuser* – A device that discharges air in a flat pattern parallel to the face of the diffuser. They may be square or round and are usually located in the ceiling. Ceiling diffusers are usually placed in the exact center of the room or space to be conditioned. If more than one diffuser is required, the ceiling area is usually

divided into equal parts, and one diffuser is placed in each part to supply the necessary conditioning. They are arranged to promote a mix of primary air with secondary room air (entrainment or induction).

- *Drop* – The vertical distance that the lower edge of a horizontally projected airstream falls or rises (*Figure 16*). It is the result of the combined effect of the density of cool or warm air and the vertical expansion of the air path that results from entrainment.
- *Grille* – A covering for any opening through which conditioned air passes. When grilles have a single flap or several adjustable blades affixed to a collar fastened behind the front face, they are called *registers*. When grilles and registers deliver supply air, the outlets discharge the air in an outward pattern.
- *Noise criteria (NC)* – A grille sound rating in pressure level at a given condition of operation. It is based on established criteria and a specific room acoustic absorption value. NC levels have been established for various types of conditioned areas (*Appendix A*).
- *Occupied zone* – The area of a conditioned space which extends to within 6" of all room surfaces and up to a height of 6'-0".
- *Primary air* – The air delivered to the room or conditioned space from the supply duct.
- *Register* – A grille fitted with a damper that is used to control the quantity of air passing through it. Registers are grilles with dampers.
- *Return inlet* – An opening in a wall, ceiling, or floor through which the space air is returned to the conditioning equipment. The opening is usually fitted with a grille.

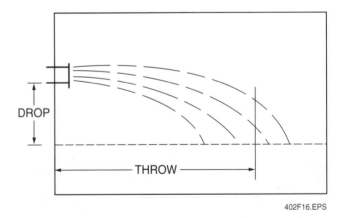

Figure 16 ◆ Air drop.

- *Secondary air or entrainment* – The induced flow of room air by the primary air from an outlet. This creates a mixed air path commonly called secondary air motion. It is sometimes referred to as induced air.
- *Spread* – The horizontal divergence (*Figure 17*) of an airstream after it leaves the outlet.
- *Supply outlet* – An opening in a wall, ceiling, or floor through which conditioned air flows into a conditioned space. The opening is usually fitted with a grille or register.
- *Throw (T)* – The throw (T) of an outlet is measured in feet and is the distance from the center of the outlet to a point in the mixed airstream where the highest sustained velocity (fpm) has been reduced to a specified level, usually 50 fpm (see *Figure 17*).
- *A* – The area of a duct measured in square feet.
- A_k – An area factor of an outlet or an inlet stated in square feet. It is a flow factor determined from the discharge or intake velocity (V_k) and the volume of air (cfm). The manufacturer will specify this value.
- *cfm* – The volume rate of airflow measured in cubic feet per minute.
- *fpm* – The velocity rate of airflow measured in feet per minute.

- P_s – Static pressure measured in inches of water column (in. w.c.). It is the pressure exerted against the sides of the duct by the pressure delivered by the fan or blower. It will be in effect as long as the fan operates. It is potential energy.
- P_t – The total pressure of a duct system measured in in. w.c. It may be obtained from the performance data of a system or component and is the sum of **velocity pressure** (P_v) and the static pressure (P_s). It is the total pressure available in a length of duct.
- P_v – Velocity pressure measured in inches of water column. It is the pressure needed to move the air at the desired speed in the duct. It is kinetic energy.
- *t* – The temperature differential, expressed in °F or °C, between ambient room temperature and the supply air temperature.
- t_a – The ambient temperature expressed in °F or °C.
- *V* – The duct velocity measured in feet per minute (fpm).
- V_k – The discharge or intake velocity of an outlet or an inlet. It is stated in fpm and is measured with a calibrated velometer and a special probe or jet on either the face of the outlet or the inlet. The manufacturer will specify the location.
- V_r – The room velocity of air in fpm. It is determined from velocity measurements in the occupied zone.
- V_t – The terminal velocity of an outlet measured in fpm. It is the highest sustained velocity in the mixed airstream at the end of the throw. Simply stated, it is the flow factor area in square feet. It is determined by the manufacturer for each model and size of outlet and inlet. The manufacturers also determine where the test instrument probe should be positioned on a particular supply outlet and the number of readings required for that particular outlet to determine actual cfm being delivered.

To help explain some of the terms associated with air distribution in a room from a given grille or register, refer to *Figure 18*. For example, the performance table for throw (T) is based on wall mounting heights of 8' to 10', and a 0-degree setting of both single and double deflection blades. The rated throw is based on the full distance the mixed airstream will travel to a terminal velocity (V_t) of 50 fpm.

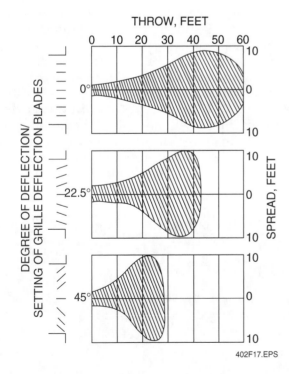

Figure 17 ◆ Air patterns.

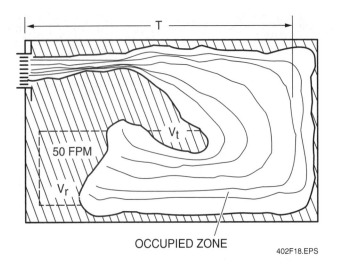

Figure 18 ◆ Air pattern terms – elevation.

Figure 18 graphically shows the T (throw), V_r (room velocity), and V_t (terminal velocity).

In multiple unit applications, grilles of equal capacities should be sized for throw equal to one-half the distance between them. Throw from single outlets should be sized within the range of 75 percent to 100 percent of the distance to the opposite wall. The room velocity (V_r) is affected by changes in the mounting height of the grille on the wall. The throw listed for a particular grille should be decreased by one foot for each foot of mounting height above 10'. This is necessary to maintain the V_r standard of 50 fpm.

To aid in understanding the use of area factor (A_k) in determining the actual cfm being discharged from a given outlet, use the following formula:

$$cfm = V_k \times A_k$$

Where:

cfm = cubic feet per minute

V_k = average outlet velocity in fpm

A_k = area factor from manufacturer

5.2.0 Air System Measuring Instruments

Several instruments and accessories are used to measure the pressures and velocity of airflow in an air distribution system. Common instruments and devices used for this purpose are:

- Manometer
- Differential pressure gauge
- Pitot tube and static pressure tips
- Velometers/anemometers

5.2.1 Manometers

Manometers are used to measure the low-level static, velocity, and total air pressures found in air distribution duct systems. Manometers used for air distribution servicing are calibrated in inches of water column (in. w.c.). Manometers can use water or oil as the measuring fluid. Popular models use an oil which has a specific gravity of 0.826 as the measuring fluid. The manufacturer of the gauge specifies the type of oil to be used; therefore, substitution for the specified oil is not recommended.

Manometers come in many types, including U-tube, inclined, and combined U-inclined. Electronic manometers are also widely used. Pitot tubes or static pressure tips, described later in this section, are almost always used with manometers when measuring pressures in duct systems. *Figure 19* shows some common types of manometers.

Manometers work on the principle that air pressure is indicated by the difference in the level of a column of liquid in the two sides of the instrument. If there is a pressure difference, the column of liquid will move until the liquid level in the low-pressure side is high enough so that its weight and the low air pressure being measured will equal the higher pressure in the other tube.

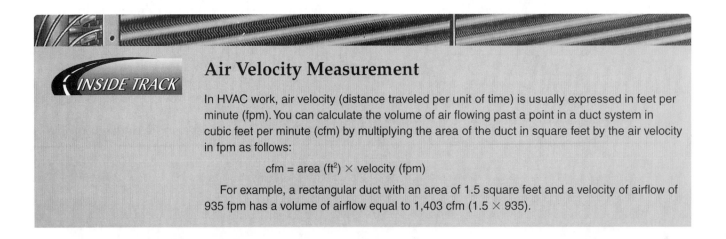

Air Velocity Measurement

INSIDE TRACK

In HVAC work, air velocity (distance traveled per unit of time) is usually expressed in feet per minute (fpm). You can calculate the volume of air flowing past a point in a duct system in cubic feet per minute (cfm) by multiplying the area of the duct in square feet by the air velocity in fpm as follows:

$$cfm = area \ (ft^2) \times velocity \ (fpm)$$

For example, a rectangular duct with an area of 1.5 square feet and a velocity of airflow of 935 fpm has a volume of airflow equal to 1,403 cfm (1.5 × 935).

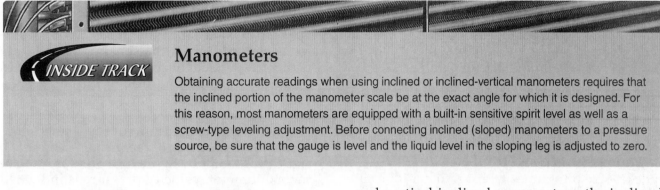

Manometers

Obtaining accurate readings when using inclined or inclined-vertical manometers requires that the inclined portion of the manometer scale be at the exact angle for which it is designed. For this reason, most manometers are equipped with a built-in sensitive spirit level as well as a screw-type leveling adjustment. Before connecting inclined (sloped) manometers to a pressure source, be sure that the gauge is level and the liquid level in the sloping leg is adjusted to zero.

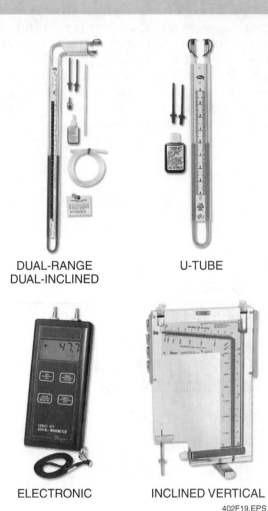

DUAL-RANGE
DUAL-INCLINED

U-TUBE

ELECTRONIC

INCLINED VERTICAL

402F19.EPS

Figure 19 ◆ Manometers.

Individual U-tube and inclined manometers are available in many pressure ranges. Inclined manometers are usually calibrated in the lower pressure ranges and are more sensitive than U-tube manometers. U-inclined manometers combine both the sensitivity of the inclined manometer with the high-range capability of the U-tube manometer in one instrument. Inclined-vertical manometers combine an inclined section for high accuracy and a vertical manometer section for extended range. They also have an additional scale that indicates air velocity in feet per minute. To get accurate readings with inclined

and vertical-inclined manometers, the inclined portion of the scale must be at the exact angle for which it is designed. A built-in spirit level is used for this purpose. Most also have a screw-type leveling adjustment.

Electronic manometers can typically measure differential pressures of –1 to 10 in. w.c. Many can give direct air velocity readings in the range of 300 fpm to 9,990 fpm, eliminating the need for calculations.

5.2.2 Differential Pressure Gauge

The differential pressure gauge provides a direct reading of pressure. These gauges are typically used to measure fan and blower pressures, filter resistance, air velocity, and furnace draft. Some are capable of measuring just pressure or both pressure and air velocity. Single-scale pressure models are calibrated in either in. w.c. or psi. Dual-scale gauges are normally calibrated for pressure in in. w.c. and for air velocity in fpm.

Several models are available covering pressures from 0.0 to 10 in. w.c., and air velocity ranges from 300 fpm to 12,500 fpm. Normally, these gauges are installed in the equipment, but portable models are available.

Pitot tubes and/or static pressure tips are normally used with portable models to make air pressure and velocity measurements in air distribution system ductwork. *Figure 20* shows a portable differential pressure gauge.

402F20.EPS

Figure 20 ◆ Portable differential pressure gauge.

5.2.3 Pitot Tubes and Static Pressure Tips

Pitot tubes (*Figure 21*) and static pressure tips are probes used with manometers and pressure gauges when making measurements inside the ductwork of an air distribution system. The standard pitot tube used for measurements in ducts 8" and larger has a ⁵⁄₁₆" outer tube with eight equally spaced 0.04" diameter holes used to sense static pressure.

For measurements in ducts smaller than 8", pocket-size pitot tubes with a ⅛" outer tube and four equally spaced 0.04" diameter holes are recommended. The pitot tube consists of an impact tube which receives the total pressure input, fastened concentrically inside a larger tube which receives static pressure input from the radial sensing holes around the tip. The air space between the inner and outer tubes permits transfer of pressure from the sensing holes to the static pressure connection at the opposite end of the pitot, and then through the connecting tubing to the low- or negative-pressure side of the manometer.

When the total pressure tube is connected to the high-pressure side of the manometer and the static pressure tube to the low-pressure side of the manometer, velocity pressure is indicated directly. To be sure of accurate velocity pressure readings, the pitot tube tip must be pointed directly into the duct airstream. Pitot tubes come in various lengths ranging from 6" to 60", with graduation marks at every inch to show the depth of insertion in the duct.

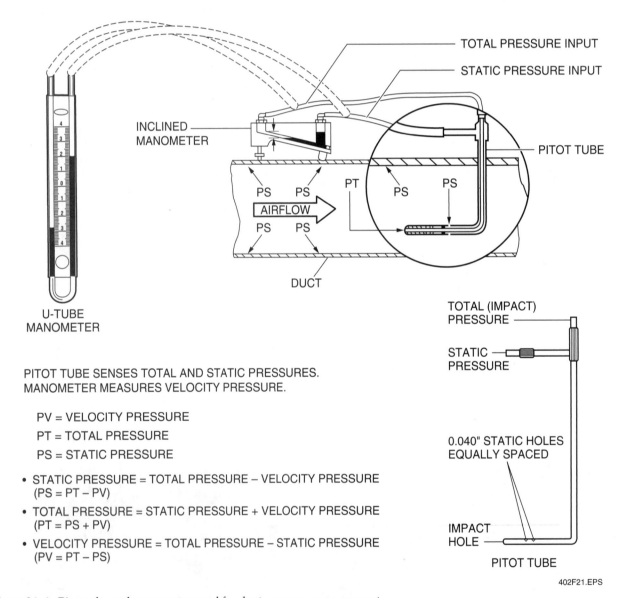

Figure 21 ◆ Pitot tube and manometer used for duct pressure measurement.

Static pressure tips, like pitot tubes, are used with manometers and differential pressure gauges to measure static pressure in a duct system. They are typically L-shaped with four radially drilled 0.04" sensing holes.

5.2.4 Velometers/Anemometers

Velometers (*Figure 22*) are used to measure the velocity of airflow. Measurement of air velocity is done to check the operation of an air distribution system. It is also done when balancing system airflow. Most velometers give direct readings of air velocity in fpm. Some can provide direct readings in cfm. Velometers with analog scales and digital readouts are in common use.

Some velometers use a rotating vane (propeller) or balanced swinging vane to sense air movement. When the rotating vane velometer is positioned to make a measurement, the vane rotates at a rate determined by the velocity of the airstream. This rotation is converted into an equivalent velocity reading for display. In the swinging vane velometer, the airstream causes the balanced vane to tilt at different angles in response to the measured air velocity. The position of the vane is converted into an equivalent velocity reading for display.

Another type of velometer, also called a hot wire anemometer, gives direct readings of air velocity in fpm. This instrument uses a sensing probe containing a small resistance heater element. When the probe is held perpendicular to the airstream being measured, the temperature of

the heater element changes due to changes in the airflow. This causes its resistance to change, which alters the amount of current flow being applied to the meter circuitry. There, it automatically calculates the air velocity for display on the meter.

Some velometers use probes that have a sensitively balanced vane or a small resistance heater element that, when placed in the airstream, produces a measurement of airflow for display on the velometer meter scale.

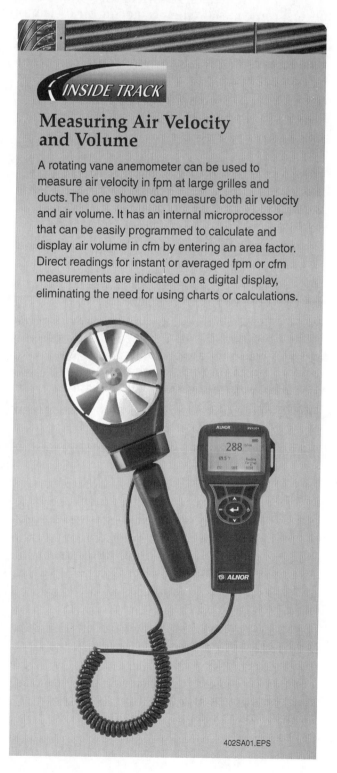

INSIDE TRACK

Measuring Air Velocity and Volume

A rotating vane anemometer can be used to measure air velocity in fpm at large grilles and ducts. The one shown can measure both air velocity and air volume. It has an internal microprocessor that can be easily programmed to calculate and display air volume in cfm by entering an area factor. Direct readings for instant or averaged fpm or cfm measurements are indicated on a digital display, eliminating the need for using charts or calculations.

402SA01.EPS

SWINGING VANE

FLOW HOOD

AIR VOLUME BALANCER

ROTATING VANE

402F22.EPS

Figure 22 ◆ Velometers.

Depending on the sensing probe or attachment used, velometers can measure air velocities in several ranges within the overall range of 0 to 10,000 fpm. Some electronic velometers that use a microprocessor can automatically average up to 250 individual readings taken across a duct area to provide an average air velocity. Certain velometers also include an optional micro-printer to record the readings.

Special velometers, known as air volume balancers, can be used when balancing air distribution systems. This type of velometer is held directly against the diffuser or register to get a direct reading of air velocity. Another type of velometer, called a flow hood, is frequently used to get direct velocity readings in cfm when measuring the output of large air diffusers in commercial systems.

5.2.5 Grille CFM Measurements

To calculate the actual cfm being delivered at a given grille or register, proceed as follows:

Step 1   Use the recommended velometer, in this case an Alnor velometer equipped with a 2220-A jet.

Step 2   Position the jet as recommended by the manufacturer (*Figure 23*), in this case between the blades with the shank of the jet parallel to the face and across the blades of the grille. As indicated, the jet position remains the same for angled blades and for straight blades.

Step 3   Take several readings across the face of the grille (*Figure 24*).

Step 4   Find the average outlet velocity; in this case, 800 fpm.

Step 5   Use the manufacturer's table for the grille chosen; for example, a Carnes Model 150 grille, 24" by 12", with a blade deflection of 45 degrees. The A_k for the grille is 1.20 as determined from the manufacturer's table.

Step 6   Complete the calculation:

cfm = 800 × 1.20

Therefore:

cfm = 960

All air-measuring devices must be used according to the manufacturer's instructions, be periodically calibrated, and be protected from damage.

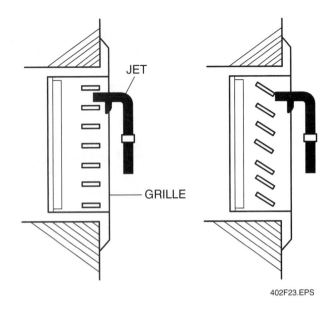

402F23.EPS

Figure 23 ◆ Jet placement.

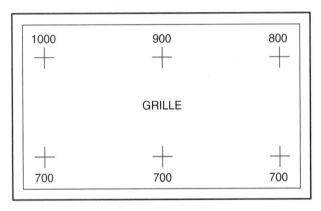

VELOCITY READINGS (fpm)

700 + 700 + 700 + 800 + 900 + 1000 = 4,800

$$\frac{\text{AVERAGE}}{\text{VELOCITY}} = \frac{4,800}{6} = 800 \text{ fpm}$$

402F24.EPS

Figure 24 ◆ Average outlet velocity readings.

Correct airflow measurements depend on the proper positioning of the probe and the use of the recommended anemometer. Each manufacturer will recommend the kind of equipment and the placement of the probe necessary for determining the actual cfm of their device. *Figure 25* illustrates the use of an Alnor velometer for determining the actual air volume delivered through a stamped steel louvered ceiling diffuser. The field balancing information identified in the text surrounding the figure describes the proper placement of the probe and the number of readings to be taken.

STAMPED STEEL LOUVERED DIFFUSERS

$$CFM = V_k \times A_k$$

AREA FACTOR (A_k) TABLE – MODEL SFA

NECK SIZE	NOMINAL LOUVERED AREA		
	12×12	18×18	24×24
	Horizontal Throw		
5	12	–	–
6	14	0.22	0.29
7	17	0.25	0.28
8		0.27	0.28
10		0.38	0.42
12		0.48	0.50
14		–	0.62

AREA FACTOR (A_k) TABLE – MODEL SAA

NECK SIZE	NOMINAL LOUVERED AREA					
	Horizontal Throw			Vertical Throw		
	12×12	18×18	24×24	12×12	18×18	24×24
	0.09	–	–	0.10	–	–
	0.12	0.20	0.35	0.11	0.21	0.30
	0.15	0.23	0.31	0.12	0.23	0.27
	–	0.26	0.33	–	0.24	0.36
	–	0.29	0.32	–	0.23	0.30
	–	0.38	0.39	–	0.30	0.33
	–	–	0.60	–	–	0.48

FIELD BALANCING

The actual volume of air being discharged from an outlet can be determined by measuring the outlet velocity in feet per minute (fpm) and multiplying by an area factor (A_k).

$$CFM = V_k \times A_k$$

The Alnor velometer, with the 2220-A jet is the recommended equipment for balancing Carnes stamped diffusers.

The Alnor Model 6000P with 6070P probe can be used with the same A_k factors.

Place the Alnor jet into the correct louvered space as shown in the sketches below. Point the jet as directly as possible into the air stream and move jet slowly along the lip of the cone to obtain the highest reading. Average the readings from all four sides to obtain V_k. Select the correct A_k from the tables and apply the formula to obtain the CFM.

ALNOR JET POSITION – MODEL SFA

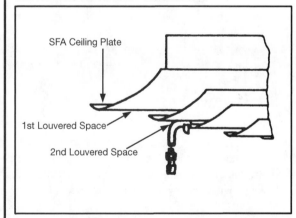

ALNOR JET POSITION – MODEL SAA

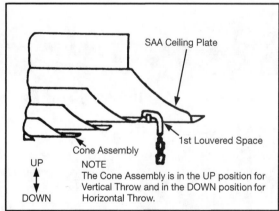

402F25.EPS

Figure 25 ◆ Field balancing data.

5.3.0 Air System Balancing Considerations

Special considerations should be given to properties of air and system components that affect the balance of air distribution systems. Many problems in balancing a duct system can be eliminated if the technician has a basic understanding of how air moves in a duct system and how it is affected by duct run components.

5.3.1 Airflow

In a duct system, air flows from a high-pressure region to a low-pressure region. The indoor or conditioning system blower creates the pressure differential necessary to cause air to move in the duct system. In a typical system (*Figure 26*), the highest air pressure in the system is found at blower outlet C. This pressure gradually and constantly decreases throughout the supply duct until it becomes equal to the atmospheric pressure in the space to be conditioned after it leaves supply grille A.

The lowest pressure in the system is at the blower inlet B. The pressure of the air in the return duct constantly decreases from atmospheric pressure from return grille D in the room to a pressure lower than atmospheric at the blower. Therefore, the total system pressure can be measured across the blower inlet B and outlet C.

Pressure losses in the duct run result from several factors that create a resistance to airflow in the system. The blower fan must overcome these pressure losses as air moves through the system.

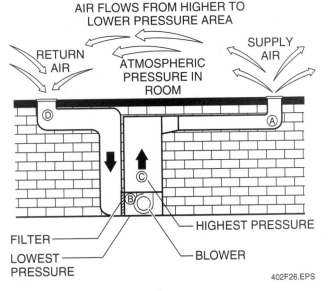

Figure 26 ◆ Airflow principles.

Most of the pressure losses in a duct are due to friction caused by the air molecules rubbing against the walls of the duct. At the walls, they travel at a slower speed than the air molecules moving in the center of the duct.

Losses also occur when air molecules bump into each other as they move along the duct. This happens because the air molecules tend to tumble and mix rather than flow in a straight pattern. Friction losses will also increase with any increase in air speed or change in air direction. The velocity of air moving through a duct is measured in fpm. Doubling air velocity increases the friction losses by almost four times.

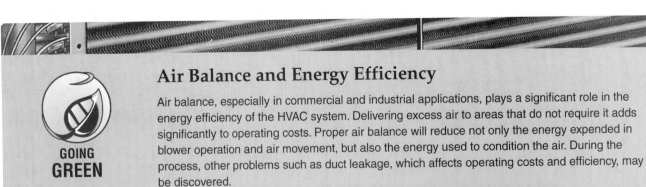

Air Balance and Energy Efficiency

GOING GREEN

Air balance, especially in commercial and industrial applications, plays a significant role in the energy efficiency of the HVAC system. Delivering excess air to areas that do not require it adds significantly to operating costs. Proper air balance will reduce not only the energy expended in blower operation and air movement, but also the energy used to condition the air. During the process, other problems such as duct leakage, which affects operating costs and efficiency, may be discovered.

In systems that require the addition of outside air and exhaust ventilation, proper air balancing helps to ensure that proper pressures are maintained in the building. In many cases, poor air balance can cause one or more areas to operate at negative pressures, causing uncontrolled infiltration from process areas or the outdoors to occur. This further increases energy use while impacting comfort and human health, potentially allowing harmful fumes to enter an area. Although indoor air quality and energy-efficient operation of the HVAC system may appear to be concepts at odds with each other, effective air balancing can substantially reduce energy use and help to cover the operating costs of beneficial indoor air quality systems and accessories.

To overcome additional friction losses within a duct system, the blower must exert greater pressure to keep the air moving through the duct. This greater pressure exerted by the blower results in a higher air pressure within the duct. This pressure, called static pressure, is exerted equally in all directions throughout the system.

The volume of air moving in a duct system is measured in cubic feet per minute (cfm). It can be determined if the velocity (speed of air moving in a duct) and the cross-sectional area of the duct are known. The velocity is measured at several points across the duct (**traverse readings**) and averaged for accuracy (*Figure 27*).

An anemometer (*Figure 28*) measures air velocity and shows the result on a dial or digital display that reads velocity in feet per minute.

Velocity measurements are usually taken at the face of a return or supply air grille or diffuser. The cross-sectional area of a grille is not equal to its actual measured area because the grille offers additional resistance to airflow. Grille manufacturers normally give data pertaining to the equivalent open area rather than the actual grille measurement.

If the open area correction factor is not known, the measured area may be multiplied by 0.65 as a correction factor when calculating velocity.

Total system airflow may be determined by using the manufacturer's performance data pertaining to a particular size or model blower. This method of determining total system airflow requires measurement of total system static pressure drop (*Figure 29*). A manometer and pitot tube are usually used for this purpose. The manufacturer's service manuals give data indicating the cfm air handling capabilities of their units. This data may indicate the capabilities over the normal operating range of static pressure for each blower listed for use in their equipment. It may be presented on a chart or a graph.

5.3.2 Air Weight

Air is compressible and has weight. Its density and weight vary with the elevation above the earth's surface. At higher elevations, it weighs less than at lower elevations. Thus, the layer of air is not as deep or as heavy at Denver, Colorado (elevation 5,500'), as it is at sea level.

Heating and cooling loads and equipment capacities are based on British thermal units per hour (Btuh). Most calculations used in HVAC work are based on the pounds of air to be handled in an hour. Therefore, it is important to know how to convert the volume of air flowing in a system from cfm to pounds per hour.

At sea level, each cfm represents 4.5 pounds of air per hour. For example, if a system is handling

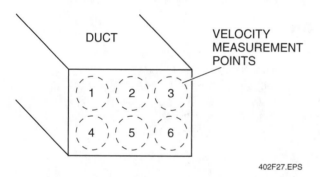

402F27.EPS

Figure 27 ◆ Velocity measurement.

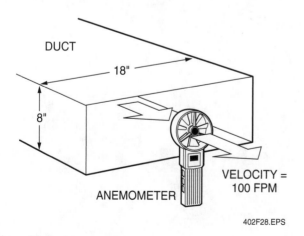

402F28.EPS

Figure 28 ◆ Anemometer.

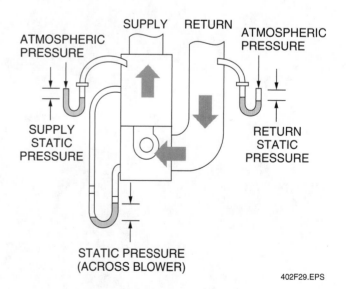

402F29.EPS

Figure 29 ◆ Determining static pressure.

450 cfm per ton of air conditioning, and the cfm per ton is multiplied by 4.5 pounds of air per hour, well over one ton of air per hour is moved for each ton of cooling capacity.

Since air is lighter at higher elevations, the weight per hour of air for each cubic foot of air per minute would be less. At elevations above sea level, air is lighter, and its ability to carry heat is less. At sea level, it takes 1.08 Btuh to raise the temperature of one cfm (4.5 pounds per hour) 1°F; whereas, at 5,000' of elevation, it takes 0.892 Btuh to raise the temperature of this lighter air 1°F for each cfm (3.717 pounds per hour). This indicates a need to move more air at higher elevations to do the same amount of work that can be done at sea level. The higher air quantity can be determined mathematically by using appropriate altitude correction factors.

The cfm required to satisfy a heating or cooling load increases as the elevation increases. One manufacturer uses the following rule of thumb for air conditioning equipment:

- Sea level – 400 cfm per ton of cooling
- 2,500' elevation – 440 cfm/ton
- 3,500' elevation – 460 cfm/ton
- 4,500' elevation – 475 cfm/ton
- 5,500' elevation – 500 cfm/ton

Heat pump installations usually require more precise calculations.

6.0.0 ◆ AIR SYSTEM BALANCING PROCEDURES

Almost every air delivery duct system requires balancing before it can be considered complete and ready for delivering conditioned air. Balancing involves setting the dampers to deliver the necessary amounts of air and making final adjustments to the registers and grilles to obtain the proper air distribution pattern. The person responsible for balancing can do very little about existing duct design and diffuser locations.

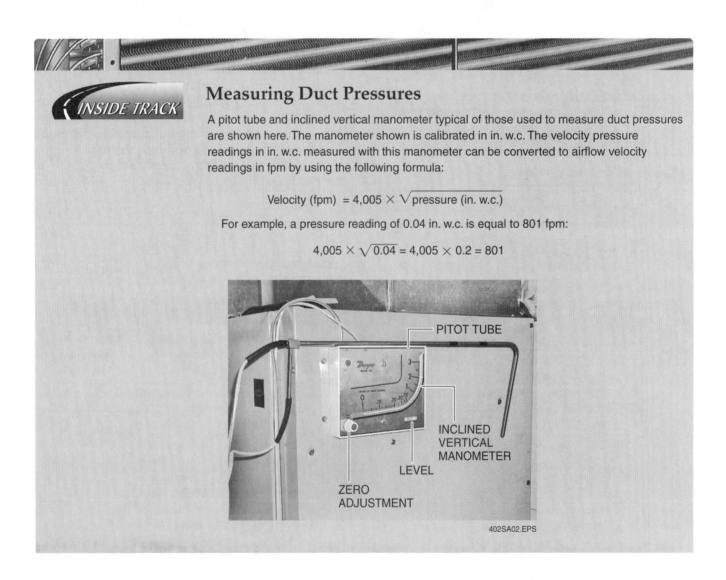

Measuring Duct Pressures

A pitot tube and inclined vertical manometer typical of those used to measure duct pressures are shown here. The manometer shown is calibrated in in. w.c. The velocity pressure readings in in. w.c. measured with this manometer can be converted to airflow velocity readings in fpm by using the following formula:

$$\text{Velocity (fpm)} = 4{,}005 \times \sqrt{\text{pressure (in. w.c.)}}$$

For example, a pressure reading of 0.04 in. w.c. is equal to 801 fpm:

$$4{,}005 \times \sqrt{0.04} = 4{,}005 \times 0.2 = 801$$

INSIDE TRACK

PITOT TUBE

INCLINED VERTICAL MANOMETER

LEVEL

ZERO ADJUSTMENT

402SA02.EPS

When adjustment is needed to obtain a satisfactory compromise between temperature rise in heating or temperature drop in cooling, total airflow, and/or grille velocities, the only choice is to alter blower speed. Most blower systems are equipped with multi-speed, direct-drive motors, or pulleys on belt-drive systems that can be adjusted to alter their speed.

Several factors must be considered when the airflow in a heating mode duct system is altered, including:

- An increase in airflow will increase static loss in the system and increase the percentage of total air leaving the supply grilles and diffusers closest to the blower. This could result in too much heat in areas nearest the blower.
- Higher velocities at return and supply grilles can cause objectionable noise.
- Complaints about drafts can be caused by lower discharge temperatures at higher velocity.
- A decrease in airflow can cause excessive temperature rise at the furnace which can stress the heat exchanger and cycle the limit switch.

- Hot spots near grilles and cold corners can cause complaints when velocities are lowered.
- Stratification can result from low airflow rates.

A satisfactory compromise must be made between temperature rise or temperature drop, total airflow, and diffuser velocities. For fossil fuel heating systems, an ideal compromise may result in a temperature rise between 70°F and 95°F and a sufficient volume of air throughout the room without drafts.

> **CAUTION**
>
> Under no circumstances should the temperature rise exceed the manufacturer's specified rise range.

The ideal compromise can be achieved with a properly designed duct system and with properly sized heating and cooling equipment. However, these conditions do not always exist, and the person balancing the system must do so within the limitations of that system. For example, many duct systems are equipped with balancing dampers in the branch ducts, but some

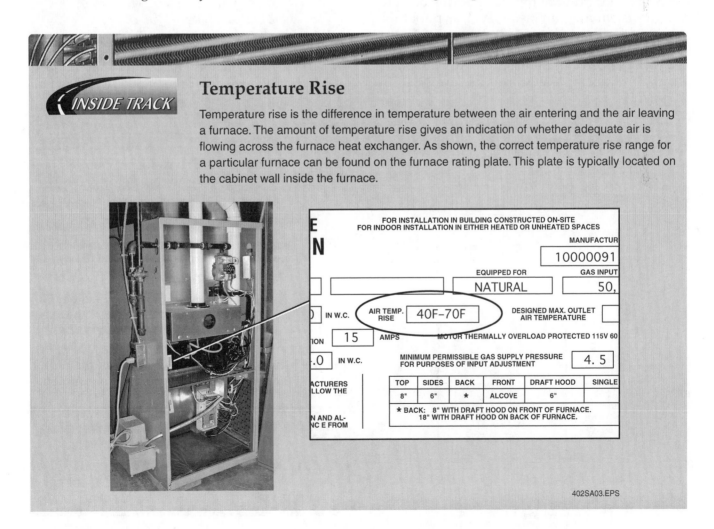

Temperature Rise

Temperature rise is the difference in temperature between the air entering and the air leaving a furnace. The amount of temperature rise gives an indication of whether adequate air is flowing across the furnace heat exchanger. As shown, the correct temperature rise range for a particular furnace can be found on the furnace rating plate. This plate is typically located on the cabinet wall inside the furnace.

402SA03.EPS

economy systems have dampers only in branches serving areas closest to the blower. This is not ideal, but it may be adequate in many instances.

Balancing comfort conditioning systems usually requires time and patience rather than sophisticated equipment. If the individual balancing the system follows a systematic procedure, little difficulty should be encountered in obtaining a properly balanced system.

A systematic balancing procedure begins with preparing a schematic layout of the system. The layout should address the following items:

- Indicate all dampers, regulating devices, terminal units, and outlets and inlets.
- Indicate the sizes and cfm for main ducts and the outlets and inlets, as well as the quantities of fresh air and return air where such occur.
- Check the shop drawings against design drawings so that construction changes can be shown on the schematic layout.
- Number the outlets for reporting purposes.
- Convert all cfm to fpm because velocities are easier to work with.

As needed, prepare reporting forms used to record system measurements, such as pitot tube traverse readings (*Figure 30*). Forms available from the National Environmental Balancing Bureau (NEBB) can be used for this purpose. Examples of completed forms are provided in *Appendix C*.

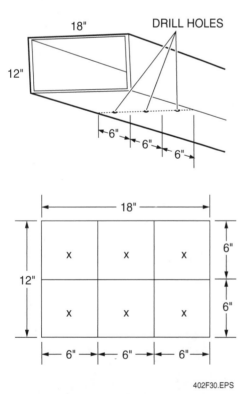

Figure 30 ◆ Traverse readings are taken every 6".

Select the instruments that will do the best job and check them for calibration. Whenever possible, use the same instrument for the entire job to minimize possible calibration errors. If more than one instrument of a similar type is used, make a comparison check. The variation between instrument readings should be limited to ±5 percent.

6.1.0 Pre-Balance Checks

Prior to beginning the balancing procedure, make sure that the equipment and duct system components are clean and operational.

WARNING!

Be sure all electrical power to the equipment is turned off. Open, lock, and tag disconnects. Watch out for pressurized or hot components. Follow all safety instructions labeled on the equipment and given in the manufacturer's service manual.

Balancing air distribution systems normally involves the use of ladders to gain access to measurement points and adjustment dampers. Make sure to use ladders in accordance with the good safety practices you learned in the Core Curriculum module, *Basic Safety*.

Step 1   Check the blower and its components. Make sure that the blower wheel is free and the housing is clear of obstructions.

Step 2   Be sure that the bearings are properly lubricated.

Step 3   Examine the drives for proper belt tension and pulley alignment.

Step 4   Check the vibration eliminators for placement and proper adjustment.

Step 5   Examine the duct run system to make sure that the outlet, inlet, and terminal dampers are in their marked positions.

Step 6   Make sure that all fire dampers are open.

Step 7   Check the filters and coils for cleanliness.

Step 8   Turn the power on and check the motor and fan for correct direction of rotation.

WARNING!

Watch out for rotating, pressurized, or hot components. Follow all safety instructions labeled on the equipment and given in the manufacturer's service manual.

Traverse Readings

The velocity profile of an air stream is not uniform across the cross-section of a duct, as shown here. Friction slows the air moving close to the walls, so that the velocity is greater in the center of the duct. To obtain the average velocity in ducts, a series of velocity pressure readings, called a *traverse*, must be taken at points of equal area. To do this, a pattern of pitot tube sensing points across the duct cross-section must be made.

For the example shown in *Figure 30*, the longest duct dimension (18") is divided into 6" increments and a hole is drilled into the center of each increment. The aim is to take a velocity pressure measurement at the center of each 6" square marked X on the figure. For the duct shown, six measurements are needed. Regardless of the duct size, 6" should be the maximum distance between measurement points. The pitot tube is marked with insertion depth graduations to help in positioning the tube in the duct to the desired depth.

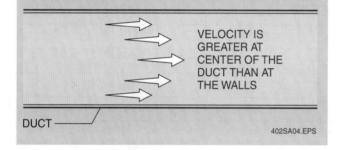

VELOCITY IS GREATER AT CENTER OF THE DUCT THAN AT THE WALLS

DUCT

402SA04.EPS

Step 9 Energize all exhaust devices that affect each other, such as kitchen, bathroom, and general exhaust.

Step 10 Close all windows and outside doors, or open the doors within the building that may affect the procedure.

Step 11 Take amperage and voltage readings at blower motors to check for proper load conditions.

6.2.0 Duct System Balancing: Mains

To balance a duct system, proceed as follows:

 NOTE

Supply air fan adjustment information and procedures are covered in detail later in this module.

Step 1 Check the speed of the fan.

Step 2 Check the manufacturer's performance tables or curves to determine the cfm delivered by the fan.

Step 3 Adjust the speed to meet the desired conditions if the fan is functioning above or below the design cfm. Recheck the fan after each damper adjustment, if needed.

Step 4 Take a pitot traverse of the main branch ducts. The location of the pitot tube in the traverse is not easy to determine on round duct runs. In order to achieve accuracy, proceed as follows:

  a. Drill two holes at right angles to each other, as shown in *Figure 31*.

  b. The number of pitot tube readings necessary for accuracy depends on the duct diameter (*Figure 32*). The larger the duct diameter, the more readings are needed to make sure a good average is obtained. As shown in *Figure 33*, a 4" duct diameter requires six readings, whereas a 32" duct diameter requires ten readings on each diameter.

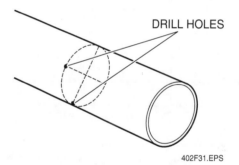

DRILL HOLES

402F31.EPS

Figure 31 ◆ Round duct readings.

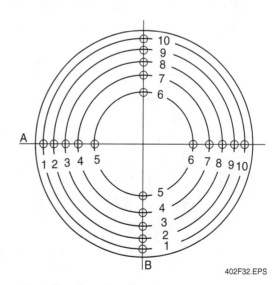

402F32.EPS

Figure 32 ◆ Number of readings.

For each reading, insert the pitot and use the graduated scale to determine how far the pitot is inserted.

c. The length of straight ductwork before and after the pitot tube location should be at least ten times the diameter of the duct for the most accurate readings (*Figure 34*).

Step 5 Adjust the volume dampers to deliver the design cfm. Continue the procedure throughout the system. Read and adjust all branch ducts.

DUCT DIAMETER (inches)	POINTS IN ONE DIAMETER	DISTANCE FROM INSIDE OF DUCT (POINT A AND B) TO PITOT TUBE TIP									
		1	2	3	4	5	6	7	8	9	10
4	6	–	–	¼	⅝	1¼	2⅞	3½	3⅞	–	–
6	6	–	–	¼	⅞	1¾	4¼	5⅛	5¾	–	–
8	6	–	–	⅜	1¼	2⅜	5⅝	6⅞	7¾	–	–
9	6	–	–	⅜	1⅜	2¾	6⅜	7¾	8⅝	–	–
10	8	–	⅜	1⅛	2	3¼	6¾	8⅛	9	9¾	–
12	8	–	⅜	1¼	2	3⅞	8⅛	9	10¾	11⅝	–
14	10	⅜	1⅛	2⅛	3¼	4¾	9¼	10⅞	12	12⅞	13⅝
16	10	½	1⅜	2⅜	3¾	5½	10½	12⅜	13⅝	14¾	15⅝
18	10	½	1½	2⅝	4⅛	6⅛	11⅞	14	15⅜	16½	17⅝
20	10	½	1⅝	3	4½	6⅞	13¼	15½	17⅛	18⅜	19½
24	10	⅝	2	3½	5½	8¼	15⅞	18½	20½	22⅛	22⅜
28	10	¾	2¼	4⅛	6⅜	9⅝	18½	21¾	24	25¾	27¼
32	10	⅞	2⅝	4¾	7¼	11	21⅛	24¾	27¼	29⅜	31⅜

FOR OTHER DUCT DIAMETERS, USE THE FOLLOWING TABLE:

POINTS IN ONE DIAMETER	CONSTANTS TO BE MULTIPLIED BY DUCT DIAMETER FOR DISTANCES OF PITOT TUBE TIP FROM INSIDE OF DUCT (POINT A AND B)									
	1	2	3	4	5	6	7	8	9	10
6	–	–	0.0435	0.1465	0.2959	0.7041	0.8535	0.9564	–	–
8	–	0.0323	0.1047	0.1938	0.3232	0.6768	0.8052	0.8953	0.9677	–
10	0.0257	0.0817	0.1465	0.2262	0.3419	0.6581	0.7738	0.8535	0.9133	0.9743

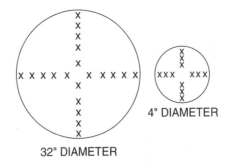

32" DIAMETER   4" DIAMETER

402F33.EPS

Figure 33 ◆ Duct diameters.

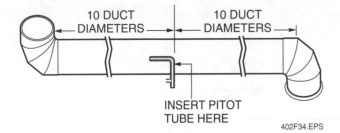

Figure 34 ◆ Pitot tube location.

6.3.0 Terminal Balancing

To balance the terminal and inlet devices, proceed as follows:

Step 1 Check the farthest outlet in each branch first. Use a velometer.

Step 2 If the farthest outlet is below design fpm, leave the damper fully open and go on to next outlet upstream.

Step 3 Proceed, adjusting as necessary, until all the outlets in the branch have been adjusted.

Step 4 If the outlet velocity at the farthest device is above design specifications, throttle it down and then proceed to the next outlet.

> **NOTE**
>
> Follow the manufacturer's recommendations for taking diffuser readings on different models of their devices.

Step 5 Repeat the procedure one or more times (some experts recommend at least three), making finer adjustments each time because the adjustment of one outlet affects the others.

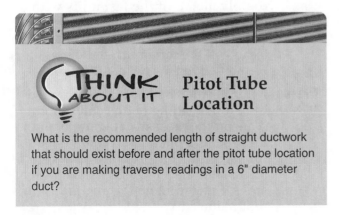

THINK ABOUT IT **Pitot Tube Location**

What is the recommended length of straight ductwork that should exist before and after the pitot tube location if you are making traverse readings in a 6" diameter duct?

Step 6 Readjust the branch duct air volume if the throttle process requires closing the terminal dampers to the extent that they generate noise.

6.4.0 Mixed Air System Adjustment

Where outdoor air is introduced into the return air duct system, it is necessary to determine the percentage of outdoor air needed to meet the installation requirements. Dampers are provided for this purpose. The damper is the primary element in the duct system used for controlling airflow by introducing resistance in the system. Partial closing of the damper increases resistance to airflow, whereas partial opening of the damper decreases resistance to airflow.

However, the reduction in airflow with closure of the damper may or may not be proportional to the amount of adjustment of the damper. In other words, closing the damper halfway does not necessarily mean that the air volume will be reduced to 50 percent of the volume that flows through the damper when it is in the wide-open position. This relationship between the position of the damper and the percentage of air that flows through the damper is called the flow characteristic or mean effective airflow of the damper.

For example, *Figure 35* illustrates the action of an opposed-blade damper. The curve O shows that to obtain 25 percent of the airflow, the damper should be 23 percent open; for 50 percent airflow, the damper is 38 percent open. The curve P demonstrates the characteristics of parallel-blade dampers. Notice that these damper blades need to open only 20 percent to achieve 50 percent airflow. It is important to remember that these percentages are approximate, and vary according to the system characteristics.

To set the fresh-air damper linkage, proceed as follows:

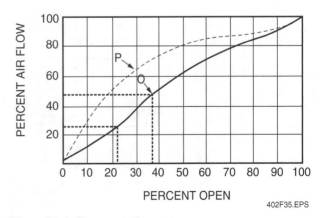

Figure 35 ◆ Damper ratios.

Step 1 Remove the damper drive rod from the damper and motor.

Step 2 Move the damper drive motor to the fully closed position.

Step 3 Move the damper to the fully closed position.

Step 4 Install and tighten the drive rod to the fully closed motor and damper.

Step 5 Find the percent minimum fresh air requirement from the specifications. If the specifications are given in cfm instead of percent, divide the fresh air cfm requirements by the total cfm of the system to obtain percent.

Step 6 Set the fresh-air control potentiometer to the proper specification (for example, 30 percent).

Step 7 With the system operating and stabilized in the correct mode, measure the return air temperature in the return air duct at a location upstream from the fresh air inlet (*Figure 36*).

Step 8 Measure the fresh-air temperature at the fresh air entry to the ductwork.

Step 9 Measure the mixed-air temperature downstream from the fresh-air inlet.

Step 10 Determine the actual fresh air needed by inserting temperature measurements into the following formula:

$$MB = IB + (\%FA)(OB - IB)$$

Where:
FA = fresh air
OB = outdoor air dry-bulb temperature
IB = indoor air dry-bulb temperature
MB = mixed air dry-bulb temperature

We can prove this formula by the following example:

FA = 30 percent
OB = 95°F
IB = 78°F
MB = 78 + (0.30)(95 − 78) = 83°F (approx.)

Step 11 Adjust the damper drive rod until the calculated mixed air temperature is reached.

Step 12 Recheck the outdoor and return air temperatures for change during adjustment.

6.5.0 Balancing Dual-Duct Systems

Most dual-duct systems (*Figure 37*) are designed to handle 100 percent of the total system supply through the cold duct and from 75 to 100 percent of the supply through the hot duct. To balance a dual-duct system, proceed as follows:

Step 1 Set all room control thermostats for maximum cooling. This fully opens the cold valves.

Step 2 Check the speed of the blower.

Step 3 Check the manufacturer's performance tables or curves to determine the cfm delivered by the fan.

Step 4 If the fan is functioning above or below the design cfm, adjust the speed to meet the desired conditions.

Step 5 Take a pitot traverse of the main branch ducts.

Step 6 Adjust the volume dampers to deliver the design cfm. Continue the procedure throughout the system; read and adjust all branch ducts.

Step 7 Proceed to the extreme ends of the system and check the static pressure immediately ahead of the last units with an inclined manometer. The static pressure of these points should be equal to, or in excess of, the minimum static pressures recommended by the manufacturer.

Step 8 For units that are not factory preset, dial the proper setting on the cfm calibrated scale on the mechanical volume regulator type. If a pneumatic volume regulator is used, the pressure differential across the balancing orifice must be read and set according to the manufacturer's calibration curve.

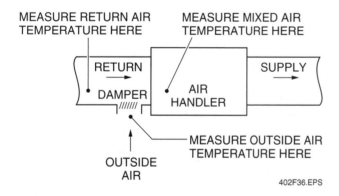

MEASURE RETURN AIR TEMPERATURE HERE

MEASURE MIXED AIR TEMPERATURE HERE

RETURN

SUPPLY

DAMPER

AIR HANDLER

MEASURE OUTSIDE AIR TEMPERATURE HERE

OUTSIDE AIR

402F36.EPS

Figure 36 ◆ Mixed air system adjustment.

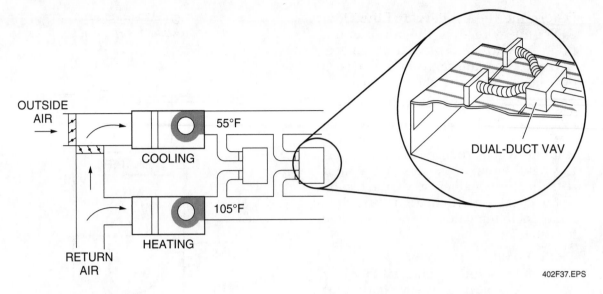

Figure 37 ◆ Dual-duct system.

402F37.EPS

Step 9 Balance the diffusers on the low-pressure side of the box in the usual manner.

Step 10 Change the control setting to full heating and recheck the airflow at one diffuser to make certain the controls and dual-duct box are functioning properly.

6.6.0 Balancing Induction Systems

Induction unit systems have been used extensively for the exterior zones of large, high-rise offices. The induction unit system uses a combination of air and water. The unit shown in *Figure 38* does not have a fan. The fan is replaced by primary air fed to the unit by a high-pressure fan and duct system. The primary air is discharged through a series of nozzles. The streams of primary air induce air from the space to flow over the coil to be either heated or cooled and then discharged into the room. To balance an induction system, proceed as follows:

Step 1 Check the fan and main trunk capacities for cfm.

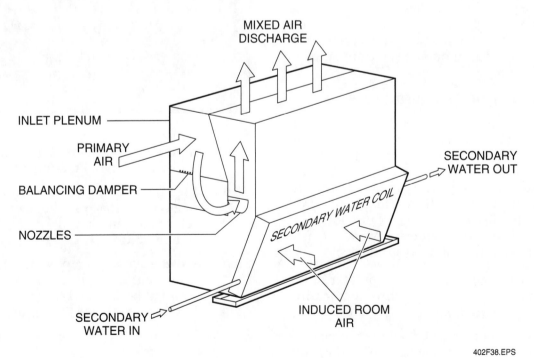

Figure 38 ◆ Induction unit system.

402F38.EPS

Step 2 Determine the primary airflow from each unit by reading the unit plenum pressure with a portable draft gauge. Refer to the manufacturer's charts to determine cfm from static pressure.

Step 3 Make a spot check of the air distribution by reading the first and last unit on each riser. If necessary, the riser with the high readings will be cut back to improve airflow in the rest of the system. Do not reset the units; rather, study the results of the checks and then adjust the riser dampers to regulate the proper volume to each riser.

Step 4 On the first pass around the system, read and adjust as you go. Start on the floor nearest the supply duct. If all unit dampers are open, set the units on the floor next to the supply header about 10 percent under the design specifications.

Step 5 Three complete passes around the entire system are normally necessary for proper adjustment.

Step 6 If unit dampers are throttled back excessively, objectionable noise may occur. Use branch and riser dampers to prevent this.

Step 7 The flow of water is automatically controlled to adjust room temperature. On those systems that use the primary air source to power the controls and move the secondary air dampers for adjusting room temperature, it is extremely important to maintain the manufacturer's recommended minimum static pressure.

6.7.0 Measuring Temperature Rise

Temperature rise is the difference in air temperature entering and leaving heating equipment. The rate of airflow has an appreciable effect on the temperature rise. Low airflow can cause excessive temperature rise and result in insufficient heat output due to the cycling of the limit switch. High rates of airflow can cause low temperature rise and complaints of drafts. To measure temperature rise, proceed as follows:

Step 1 Insert a calibrated thermometer in the main supply duct leaving the heating unit (*Figure 39*).

Step 2 Insert another calibrated thermometer in the return duct entering the heating unit.

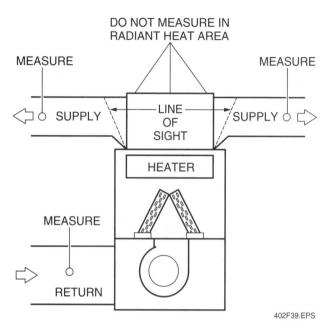

Figure 39 ◆ Measuring temperature rise.

Step 3 The supply thermometer should be out of the line of sight of the heater or heat exchanger where it will not be affected by radiant heat.

Step 4 The difference in readings between the two thermometers is the temperature rise (temperature differential, TD, Delta T, or ΔT).

 CAUTION

In fossil fuel furnaces, the temperature rise should never exceed the manufacturer's recommended value. A low temperature rise is likely to cause complaints about cold drafts. It can also result in condensation that can damage the heat exchanger.

Step 5 It may be necessary to use the same thermometer to measure both supply and return air to avoid thermometer error.

Step 6 Measure within six feet of an air handler or electric furnace. Do not measure at the return and supply grilles; this procedure is inaccurate.

Step 7 Use the average temperature when more than one duct is connected to the plenum. Be sure that the air temperature has stabilized prior to measurement.

Step 8 Measure downstream from any mixed air source.

Step 9 Record the temperature differential in the return air and supply air runs (ΔT).

When temperature rise through an electric furnace or electric heater is known, system airflow can be quickly determined by using the following formula:

$$cfm = \frac{volts \times amps \times 3.414}{(\Delta T)(1.08)}$$

To use the formula for determining airflow, the applied voltage, current draw of the heater, and the temperature rise must be known.

The output in Btuh of an electric resistance heating system may be determined using the following formula if the applied voltage and current draw of the heater are known:

$$Btuh = volts \times amps \times 3.414$$

6.8.0 Measuring Temperature Drop

Temperature drop is used on ducted cooling equipment and can be used for determining total cooling capacity. To measure temperature drop, proceed as follows:

Step 1 Take temperature measurements on ducted cooling equipment in the ducts near the cooling unit. Measurements taken at the grilles are not accurate.

Step 2 Measure the discharge air from the cooling coils at least 6' downstream from the coils.

Step 3 Take the measurements as near to the center of the duct as possible.

Step 4 Use the same thermometer or sensing device for obtaining the entering and leaving air measurements or use calibrated devices.

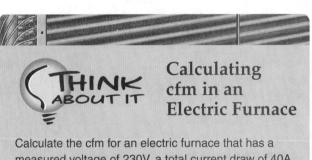

Calculating cfm in an Electric Furnace

Calculate the cfm for an electric furnace that has a measured voltage of 230V, a total current draw of 40A, and a temperature difference between the entering and leaving air of 30°F. What is the capacity of this furnace in Btuh?

NOTE
Accurate wet-bulb measurements cannot be made in high-velocity duct systems. Entering air wet-bulb measurements can be made near the return air grilles with a sling psychrometer if the system does not contain mixed air inlets, humidifiers, or reheat equipment.

Step 5 Determine the total cooling capacity:

a. Use a sling psychrometer to find the entering air and leaving air wet-bulb temperatures.

b. Use a psychrometric chart to convert the wet-bulb temperatures to enthalpy (heat content in Btuh/lb).

c. Find the total heating or cooling capacity in Btuh:

$$Btuh = (cfm)(4.5)(H1 - H2)$$

Where:

4.5 = constant

cfm = evaporator air volume

H1 = entering air enthalpy

H2 = leaving air enthalpy

6.9.0 Balancing by Thermometer

The thermometer method and the thermometer and velocity meter method are two commonly used procedures for balancing residential systems.

For proper balancing of duct systems by the thermometer method, the following pre-checks should be made:

Step 1 Use a thermocouple or thermistor-type thermometer.

Step 2 Make temperature measurements at the center of each room, about 4' above floor level.

Step 3 Locate the temperature-sensing element where air from the supply registers will not affect the temperature reading.

Step 4 Check the air filters and the blower wheel for dirt and obstructions.

Step 5 Check all registers and grilles for blockage by furniture, drapes, or appliances.

Step 6 Check the main ducts for balancing dampers; place them in the fully open position before starting to balance the duct system.

Step 7 If airflow appears to be low in a section of the duct run, check for blockage by a duct liner.

Step 8 Check the ducts for leakage of conditioned air into unconditioned space.

Step 9 Check the return air ducts for unconditioned air leaking into the return air duct run.

Step 10 If the system is equipped with an outside air mixing system, check the makeup air and mixed-air percentage.

Step 11 If the airflow in the system appears to be low, check for dirty filters, blower wheels, or coils, or an iced evaporator.

6.9.1 Thermometer Balancing for Heating

When balancing for winter operation, proceed as follows:

Step 1 Place a thermometer at the center of each room, about 4' above the floor.

Step 2 On supply registers equipped with adjustable vanes, set the vanes for optimum distribution.

Step 3 Fully open the shutoff devices in the register.

Step 4 Set the balancing dampers in the branches farthest from the blower to the fully open position.

Step 5 Check the space thermostat heat anticipator setting. Ensure that the wire entry hole behind the thermostat is plugged to prevent drafts from affecting the thermostat.

Step 6 Adjust the thermostat about 2°F above the room temperature at the thermostat. Balancing should be done when the temperature within the space is within 5°F of normal operating temperature.

Step 7 Operate the system for about 15 minutes before balancing.

Step 8 Read and record the thermometer reading in each area.

Step 9 Adjust the balancing damper in the branch supplying the warmest area to throttle the airflow to that area.

Step 10 Do not reduce the airflow too much because the airflow and temperature in the remaining areas will increase. Several adjustments to each damper may be required.

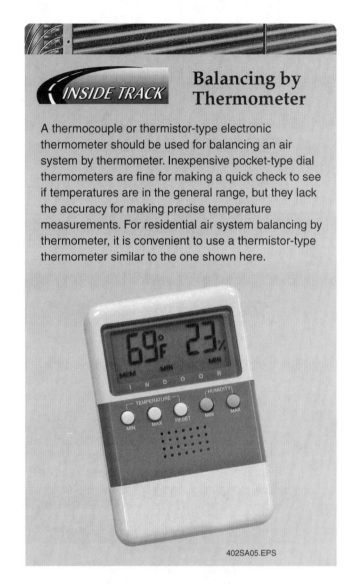

Balancing by Thermometer

INSIDE TRACK

A thermocouple or thermistor-type electronic thermometer should be used for balancing an air system by thermometer. Inexpensive pocket-type dial thermometers are fine for making a quick check to see if temperatures are in the general range, but they lack the accuracy for making precise temperature measurements. For residential air system balancing by thermometer, it is convenient to use a thermistor-type thermometer similar to the one shown here.

402SA05.EPS

Step 11 Adjust the remaining dampers. Work from the warmer areas to the cooler areas.

Step 12 Repeat the balancing process until the temperature in all areas is as near comfort conditions as possible.

Step 13 Check the temperature rise across the heat exchanger. The temperature rise should not exceed the manufacturer's specifications. If a blower speed adjustment is required because the rise is outside the range, the system must be rebalanced.

6.9.2 Thermometer Balancing for Cooling

When balancing for summer operation, proceed as follows:

Step 1 Place a thermometer at the center of each room about 4' above the floor.

Step 2   If the supply grilles are equipped with adjustable vanes, set them for optimum distribution.

Step 3   Position all supply volume dampers in the half-open position. Ensure that all return air grilles are open and not obstructed.

Step 4   Set the thermostat to call for cooling and allow the system to operate until the room temperature stabilizes. Check to make sure that the blower is delivering airflow near that recommended at the existing external static pressure.

NOTE

The air volume should be between 350 and 450 cfm per ton of cooling and the temperature drop across the indoor coil should range from 15°F to 20°F. Check the manufacturer's specifications.

Step 5   When the area thermometers have indicated a stabilized temperature, slightly open the volume damper in those spaces having temperatures higher than the thermostat setting.

Step 6   Slightly close the volume dampers in those spaces having temperatures lower than the design temperature or thermostat setting.

NOTE

The room thermostat and thermometers should be checked for calibration prior to balancing. Thermometers may be checked using a mixture of water and ice for 32°F. Stir the mixture continuously during calibration.

Step 7   Repeat Steps 5 and 6 as often as necessary to provide even temperatures within each of the conditioned spaces (±2°F).

6.9.3   Thermometer and Air Velocity Meter Balancing

To balance an air distribution system by the thermometer and velocity meter method, proceed as follows:

Step 1   Open the room diffusers or grilles.

Step 2   Calibrate the thermometers and locate them in the center of each room, about 4' above the floor.

Step 3   Take several readings of the air velocity from each supply outlet in the building while waiting for the thermometers to stabilize.

Step 4   Make a sketch of the building and locate each supply outlet on the sketch; identify them by a number or symbol.

Step 5   Record the readings for each outlet as they are taken. Record the velocity readings at each diffuser and identify the section of the diffuser that produced this value.

Step 6   Take the readings at the same location on the diffuser during succeeding passes.

Step 7   Take a temperature check after the room temperatures have stabilized within the structure and note which rooms may be overheating and which ones may be underheating.

Step 8   Reduce the volume controls to each space that is overheating. Recheck the velocity at these diffusers. It is easier to know exactly how much the air volume has been reduced at that diffuser with the use of the velocity meter.

Step 9   Make one or more additional passes throughout the structure, rechecking the temperatures and velocities until you are satisfied that the comfort conditioning requirements are being met as closely as possible.

NOTE

The velocity meter in this procedure is used only as an indication of what is happening at each outlet; therefore, little attention is paid to the actual velocity readings.

7.0.0 ◆ SUPPLY AIR FAN ADJUSTMENTS

The blower or supply fan provides the pressure difference necessary to force the air into the supply duct system, through the grilles and registers, and into the conditioned space. It must overcome the pressure loss involved in the return of the air as it flows into the return air grilles and through the return ductwork system back to the air handler. In addition, the supply fan must also overcome any resistance of other components in the system through which the air passes.

Two types of blowers are commonly used in air distribution systems: belt-drive and direct-drive (*Figure 40*). In belt-drive blowers, the blower motor is connected to the blower by a belt and

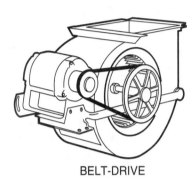

BELT-DRIVE

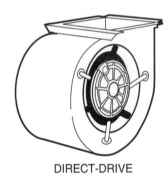

DIRECT-DRIVE

402F40.EPS

Figure 40 ◆ Belt-drive and direct-drive motors.

pulley. The blower speed is changed mechanically by adjusting the diameter of the motor pulley. In direct-drive blowers, the blower wheel is mounted directly on the motor shaft. The blower speed is changed electrically by selecting motor winding speed taps, or changing the settings of motor speed selection switches on a related motor control board. Most newer residential equipment uses multi-speed direct-drive motors. This enables the speed of the motor to be adjusted to match the requirements of the individual heating or cooling air distribution system. It also allows the speed to be changed automatically between heating and cooling modes, since each mode will have different airflow requirements.

When balancing air distribution systems, adjustment of the fan speed may be required. It is important to remember that any change made in fan speed will affect other system parameters, such as the volume of airflow, static pressure, and motor horsepower. Because of these relationships, the fan laws previously studied in the *HVAC Level One* module, *Air Distribution Systems*, are reviewed here. Also covered is other information needed for making adjustments to fan speeds.

7.1.0 Fan Laws

The performance of all fans and blowers is governed by three rules commonly known as the Fan Laws. Cubic feet per minute (cfm), revolutions per minute (rpm), static pressure (s.p.), and horsepower (hp) are all related. For example, when the cfm changes, the rpm, s.p., and hp will

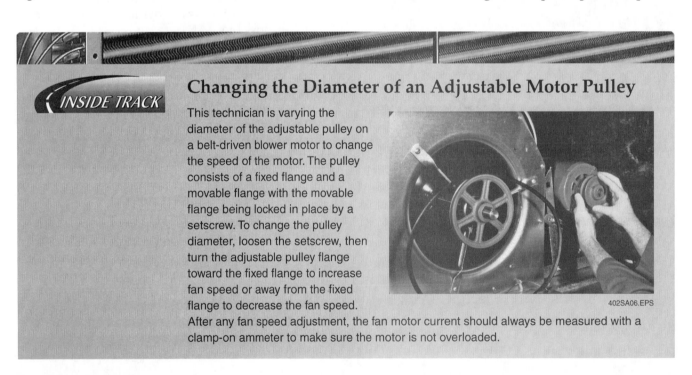

INSIDE TRACK

Changing the Diameter of an Adjustable Motor Pulley

This technician is varying the diameter of the adjustable pulley on a belt-driven blower motor to change the speed of the motor. The pulley consists of a fixed flange and a movable flange with the movable flange being locked in place by a setscrew. To change the pulley diameter, loosen the setscrew, then turn the adjustable pulley flange toward the fixed flange to increase fan speed or away from the fixed flange to decrease the fan speed.

After any fan speed adjustment, the fan motor current should always be measured with a clamp-on ammeter to make sure the motor is not overloaded.

402SA06.EPS

also change. Rpm is the speed at which the shaft of an air-moving device is rotating, measured in revolutions per minute. The easiest way to determine the fan rpm is to measure it directly with a tachometer.

Fan Laws 1, 2, and 3 are as follows:

- Fan Law 1 states that the amount of air delivered by a fan varies directly with the speed of the fan. Stated mathematically:

New cfm = (new rpm × existing cfm) ÷ existing rpm

or

New rpm = (new cfm × existing rpm) ÷ existing cfm

- Fan Law 2 states that the static pressure (resistance) of a system varies directly with the square of the ratio of the fan speeds. Stated mathematically:

New s.p. = existing s.p. × (new rpm ÷ existing rpm)^2

- Fan Law 3 states that the horsepower varies directly with the cube of the ratio of the fan speeds. Stated mathematically:

New hp = existing hp × (new rpm ÷ existing rpm)^3

Study Example

Assume that the existing system conditions are 5,000 cfm, 1,000 rpm, and 0.5 in. w.c. static pressure, with a ½ hp fan. With an increase in airflow to 6,000 cfm, find the new rpm, s.p., and hp.

Use Fan Law 1 to calculate the new rpm:

New rpm = (new cfm × existing rpm) ÷ existing cfm
= (6,000 × 1,000) ÷ 5,000
= 1,200

Use Fan Law 2 to calculate the new s.p.:

New s.p. = existing s.p. × (new rpm ÷ existing rpm)^2
= 0.5 × (1,200 ÷ 1,000)^2
= 0.72 in. w.c.

Use Fan Law 3 to calculate the new hp:

New hp = existing hp × (new rpm ÷ existing rpm)^3
= 0.5 × (1,200 ÷ 1,000)^3
= 0.864

In this example, the system would require a larger fan motor (probably a 1-hp model), and an adjustment of the motor pulley if a belt-driven blower is used.

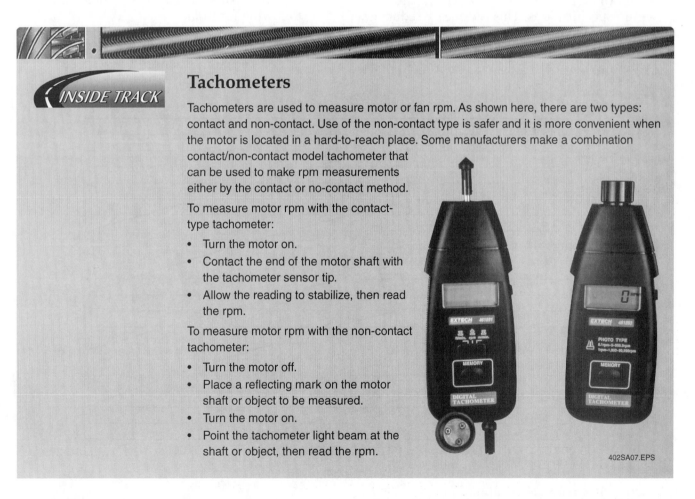

INSIDE TRACK

Tachometers

Tachometers are used to measure motor or fan rpm. As shown here, there are two types: contact and non-contact. Use of the non-contact type is safer and it is more convenient when the motor is located in a hard-to-reach place. Some manufacturers make a combination contact/non-contact model tachometer that can be used to make rpm measurements either by the contact or no-contact method.

To measure motor rpm with the contact-type tachometer:

- Turn the motor on.
- Contact the end of the motor shaft with the tachometer sensor tip.
- Allow the reading to stabilize, then read the rpm.

To measure motor rpm with the non-contact tachometer:

- Turn the motor off.
- Place a reflecting mark on the motor shaft or object to be measured.
- Turn the motor on.
- Point the tachometer light beam at the shaft or object, then read the rpm.

402SA07.EPS

7.2.0 Fan Curve Charts

Manufacturers' fan curve charts can also be used to find the relationships that exist for a set of system conditions involving static pressure, blower/fan rpm, and cfm. *Figure 41* shows an example of a typical fan curve chart. If you know the values for any two of the three characteristics (s.p., rpm, and cfm) shown on the chart, you can easily find the value for the other characteristic. For example, assume that the static pressure is 1.4 in. w.c. and the blower is running at 900 rpm. To find the cfm, locate the intersection point of the 1.4 in. w.c. static pressure line and the 900 rpm curve. From this point, drop down vertically to the cfm scale, then read the value of 7,500 cfm.

7.3.0 Changing the Speed of the Fan

When balancing a system, the fan speed may need to be changed for one or more reasons, including:

- When there is more air being circulated than required and the supply static pressure is higher than specified. The static pressure needs to be reduced while keeping the cfm within specifications. This is done by reducing the supply fan rpm.
- To achieve the airflow needed to balance the system and meet building design requirements.

- When a supply fan motor is not running at the specified rpm.
- When a supply fan motor is overloaded. This is indicated by a motor current draw greater than the full load amps specified on the motor nameplate. Assuming that the high current draw is not the result of some other electrical or mechanical problem, the fan speed must be lowered to reduce the motor load.

If the system supply fan is belt-driven, and the fan motor has a variable-pitch pulley, the speed can be increased by narrowing the width of the pulley V-groove, or decreased by widening the width of the pulley V-groove.

Typically, variable-pitch pulleys allow the speed of the driven fan to be varied by as much as 30 percent.

If the supply fan motor has a fixed-drive pulley, the size of the pulley must be changed to vary the speed. Use of a larger pulley will cause the fan speed to increase; use of a smaller pulley will cause the fan speed to decrease.

The size of the pulley diameter needed to achieve the desired speed can be calculated using the following formula:

$$\text{New pulley diameter} = \frac{\text{existing pulley diameter} \times \text{new speed}}{\text{existing speed}}$$

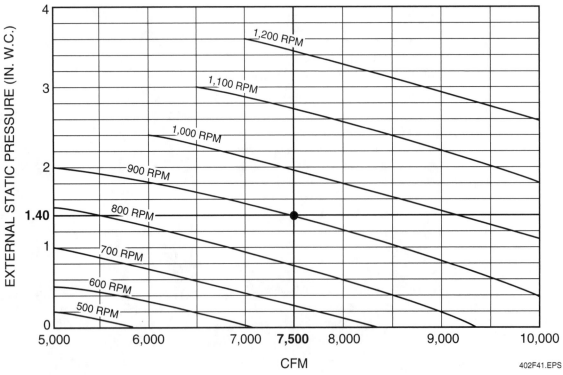

Figure 41 ◆ Typical fan curve chart.

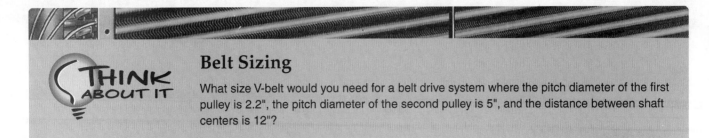

Belt Sizing

What size V-belt would you need for a belt drive system where the pitch diameter of the first pulley is 2.2", the pitch diameter of the second pulley is 5", and the distance between shaft centers is 12"?

Study Example

Assume that an existing supply fan is running at a speed of 900 rpm with a motor drive pulley that is 2" in diameter. What is the size of the drive pulley needed to change the fan speed to 1,000 rpm?

New pulley diameter =

$$\frac{\text{existing pulley diameter} \times \text{new speed}}{\text{existing speed}}$$

$$\frac{2 \times 1,000}{900} = 2.2"$$

The size of the V-belt may need to be changed as a result of changing fan motor pulley sizes. When determining the belt length for most drives, it is not necessary to be exact. This is because of the adjustment built into most drives, and also because the belt selection is limited to the standard lengths available. To determine the belt length for most drives, use the following calculation as a rule of thumb:

Belt length = A + B + C

Where:

A = pitch diameter of first pulley × 1.57

B = pitch diameter of second pulley × 1.57

C = distance between shaft centers × 2

1. An atmospheric pressure of 0 psia is equal to a pressure of _____.
   a. 0 psig
   b. 30 in. Hg vacuum
   c. 30 in. Hg
   d. 0 in. Hg vacuum

2. With a constant temperature, the pressure of a gas in a container will _____.
   a. increase if the volume increases
   b. stay the same if the volume increases or decreases
   c. increase if the volume decreases
   d. decrease if the volume decreases

3. With a constant volume, the pressure of a gas in a container will _____.
   a. decrease if the temperature is decreased
   b. decrease if the temperature is increased
   c. remain the same if the temperature is increased or decreased
   d. increase if the temperature is decreased

4. The condition in which the dry-bulb, wet-bulb, and dew point temperatures are all the same is when the _____.
   a. specific humidity is 50 grains of moisture per pound of dry air
   b. relative humidity is 50 percent
   c. specific humidity is 100 grains of moisture per pound of dry air
   d. relative humidity is 100 percent

5. Plotting a 95°F dry-bulb and a 75°F wet-bulb temperature on a psychrometric chart shows that the relative humidity is _____ percent.
   a. 30
   b. 40
   c. 50
   d. 60

6. Plotting a 95°F dry-bulb and a 75°F wet-bulb temperature on a psychrometric chart shows that the dew point is _____.
   a. 62°F
   b. 65°F
   c. 67°F
   d. 72°F

7. Plotting a 95°F dry-bulb and a 75°F wet-bulb temperature on a psychrometric chart shows that the enthalpy of the air is _____ Btu/lb.
   a. 24.7
   b. 38.0
   c. 38.6
   d. 40.6

8. Primary air is the _____.
   a. air delivered to a room from the supply duct
   b. same as induced air
   c. air in the heat exchanger
   d. induced flow of room air caused by the airflow from an outlet

9. The term that describes the distance from the center of an outlet to a point in the mixed airstream where the highest sustained velocity has been reduced to a specified level is called the _____.
   a. drop
   b. throw
   c. spread
   d. terminal velocity

10. The static, velocity, and total pressures measured in an air duct system are measured in _____.
    a. psi
    b. inches mercury
    c. inches of water column
    d. inches mercury vacuum

11. The noise criteria (NC) for a grille in an urban private home is _____.
    a. 20 – 30
    b. 25 – 35
    c. 30 – 40
    d. 40 – 50

12. When using a manometer and pitot tube to measure velocity pressures in a duct system, the pitot tube tip should be pointed _____.
    a. at right angles to the duct airstream
    b. directly into the duct airstream
    c. away from the duct airstream
    d. in any direction you choose

13. A special velometer that gives a direct reading of air velocity when held directly against the surface of a diffuser or register is called a(n) _____.

    a. rotating vane velometer
    b. hot wire anemometer
    c. swinging vane velometer
    d. air volume balancer

14. If the average velocity measured across a grille is 800 fpm, and the area factor (A_k) for the grille is 1.5, then the actual airflow is _____ cfm.

    a. 960
    b. 980
    c. 1,000
    d. 1,200

15. Within an air distribution system, the highest pressure level is found at the _____.

    a. conditioned space
    b. input to the return duct
    c. output of the blower
    d. input to the blower

16. In Denver, Colorado, elevation 5,500', the approximate cfm/ton needed to satisfy cooling loads is _____ cfm/ton.

    a. 400
    b. 440
    c. 450
    d. 500

17. When making traverse readings on a 16" round duct, _____ readings should be taken on each diameter.

    a. four
    b. six
    c. eight
    d. ten

18. Fan Law 3 applies to the calculation of _____.

    a. rpm
    b. static pressure
    c. cfm
    d. horsepower

19. If a typical air distribution system fan is running at 900 rpm at a static pressure of 1.2 in. w.c., what is the airflow in cfm? Hint: Use the fan curve chart shown in *Figure 41*.

    a. 6,250
    b. 7,500
    c. 8,000
    d. 8,500

20. Assume that an existing supply fan is running at a speed of 1,000 rpm with a motor drive pulley that is 11" in diameter. What is the approximate size of the drive pulley needed to change the fan speed to 900 rpm?

    a. 9"
    b. 10"
    c. 11"
    d. 12"

Summary

Psychrometrics is the study of the thermodynamic properties of moist air and the application of these properties to the environment. Understanding the properties of air enables the HVAC technician to better analyze the operating conditions associated with a selected HVAC system. The gas laws formulated by Dalton, Boyle, and Charles were reviewed in this module. These laws explain the basic relationships between the properties of temperature, pressure, and volume as they relate to a confined quantity of air or other gases.

Dry air is a mixture of gases containing mostly nitrogen. However, air in the atmosphere is not dry but contains small amounts of moisture in the form of water vapor, thus adding another gas to the mixture. The percentage of water vapor in the air varies depending on location and other environmental conditions. This air-water mixture is the moist air referred to in the subject of psychrometrics. The different thermodynamic properties of moist air and how they relate to one another were described in detail in this module. This coverage included how to use and interpret the various scales shown on a psychrometric chart in order to determine and/or evaluate the different air properties, including:

- Dry-bulb temperature
- Wet-bulb temperature
- Dew point temperature
- Grains of moisture per pound of dry air
- Sensible heat factor
- Enthalpy
- Relative humidity
- Specific volume

The second part of this module covered the terminology, tools, instruments, and basic procedural methods used by HVAC technicians when discussing and/or balancing air distribution systems. Air system balancing involves following a systematic procedure for decreasing and/or increasing the air volume at various points within a duct system when there is too much or too little heating or cooling for a given space.

To properly air balance a system, it normally takes several passes through the system during which the main and branch duct dampers are adjusted in order to make sure that all conditioned spaces fall within the comfort zone. Air balancing often requires that the system blower speed also be adjusted in order to achieve the proper volume of system airflow accompanied with acceptable system static pressures. Balancing of an air distribution system should never be attempted if the system installation is not complete, or prior to the pre-balance checks having been performed. Balancing of air distribution systems may be accomplished in various ways depending on the type of equipment or system involved and its application. Regardless of the balancing method used, a system is typically considered balanced when the value of the air quantity of each inlet or outlet is measured and found to be within ±10 percent of the design air quantities.

Atmospheric pressure: The pressure exerted on all things on the surface of the earth as the result of the weight of the atmosphere.

Boyle's law: With a constant temperature, the pressure on a given quantity of confined gas varies inversely with the volume of the gas. Similarly, at a constant temperature, the volume of a given quantity of confined gas varies inversely with the applied pressure.

Charles' law: With a constant pressure, the volume for a given quantity of confined gas varies directly with its absolute temperature. Similarly, with a constant volume of gas, the pressure varies directly with its temperature.

Dalton's law: The total pressure of a mixture of confined gases is equal to the sum of the partial pressures of the individual component gases. The partial pressure is the pressure that each gas would exert if it alone occupied the volume of the mixture at the same temperature.

Dew point: The temperature at which water vapor in the air becomes saturated and starts to condense into water droplets.

Induction unit system: An air conditioning system that uses heating/cooling terminals with circulation provided by a central primary air system that handles part of the load, instead of a blower in each cabinet. High-pressure air (primary air) from the central system flows through nozzles arranged to induce the flow of room air (secondary air) through the unit's coil. The room air is either cooled or heated at the coil, depending on the season. Mixed primary and room air is then discharged from the unit.

Psychrometrics: The study of air and its properties.

Specific density: The weight of one pound of air. At 70°F at sea level, one pound of dry air weighs .075 pound per cubic foot.

Specific volume: The space one pound of dry air occupies. At 70°F at sea level, one pound of dry air occupies a volume of 13.33 cubic feet.

Total heat: Sensible heat plus latent heat.

Total pressure: The sum of the static pressure and the velocity pressure in an air duct. It is the pressure produced by the fan or blower.

Traverse readings: A series of velocity readings taken at several points over the cross-sectional area of a duct or grille.

Velocity: How fast air is moving. The rate of air flow is usually measured in feet per minute.

Velocity pressure: The pressure in a duct due to the movement of the air. It is the difference between the total pressure and the static pressure.

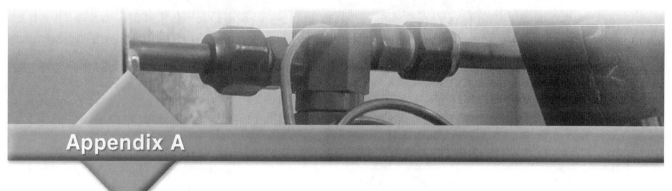

Noise Criteria

TYPE OF AREA	RANGE OF NC CRITERIA CURVES
Residences	
Private homes (urban)	20 – 30
Apartment houses, two- and three-family units	25 – 35
Private homes (rural and suburban)	30 – 40
Hotels	
Individual rooms or suites	30 – 40
Ballrooms, banquet rooms	30 – 40
Halls and corridors, lobbies	35 – 45
Garages	40 – 50
Kitchens and laundries	40 – 50
Hospitals and Clinics	
Private rooms	25 – 35
Operating rooms, wards	30 – 40
Laboratories, halls, and corridors	35 – 45
Lobbies and waiting rooms	35 – 45
Washrooms and toilets	40 – 50
Offices	
Boardrooms	20 – 30
Conference rooms	25 – 35
Executive suite	30 – 40
Supervisor's office, reception room	30 – 45
General open offices, drafting rooms	35 – 50
Corridors	35 – 55
Tabulation and computation	40 – 60
Auditorium and Music Halls	
Concert and opera halls	20 – 25
Studios for sound reproduction	20 – 25
Legitimate theatres, multi-purpose halls	25 – 30
Movie theatres, TV audience studios	30 – 35
Semi-outdoor amphitheaters	30 – 35
Lecture halls, planetariums	30 – 35
Lobbies	35 – 45

TYPE OF AREA	RANGE OF NC CRITERIA CURVES
Churches and Schools	
Sanctuaries	20 – 30
Libraries	30 – 40
Classrooms	30 – 40
Laboratories	35 – 45
Recreation halls	35 – 50
Corridors and halls	35 – 50
Kitchens	40 – 50
Public Buildings	
Public libraries, museums, courtrooms	30 – 40
Post offices, general banking areas, lobbies	35 – 45
Washrooms and toilets	40 – 50
Restaurants, Cafeterias, Lounges	
Restaurants	35 – 45
Cocktail lounges	35 – 50
Nightclubs	35 – 45
Cafeterias	40 – 50
Stores (Retail)	
Clothing stores	35 – 45
Department stores (upper floors)	35 – 45
Department stores (main floor)	40 – 50
Small retail stores	40 – 50
Supermarkets	40 – 50
Sport Activities (Indoor)	
Coliseums	30 – 40
Bowling alleys, gymnasiums	35 – 45
Swimming pools	40 – 55
Transportation (Rail, Bus, Plane)	
Ticket sales offices	30 – 40
Lounges and waiting rooms	35 – 50

Proportional Balancing

Building owners, buildings occupants, facility managers, and maintenance engineers all benefit from proper balancing of HVAC systems. One of the benefits for owners is lower operating cost. For example, the life of the HVAC equipment is extended and there can be large savings in energy usage in instances when flow quantities can be reduced through proper balancing. For the occupants, there's increased comfort resulting from proper air movement, the correct number of air changes, and the proper amount of ventilation air. For facility managers and maintenance engineers, balancing can uncover problems in the operation, installation, and design of the system.

This material explains one method of proportional balancing. This method follows a scientific approach with predictable results. First, it establishes a starting point. Then, it sets a logical sequence for the balancing of the low-pressure system. Generally, with this method, large systems can be balanced more quickly. In addition, with proportional balancing, the least resistance is imposed on the system while still delivering air to the distribution devices at design airflow, or at the highest possible airflow quantities.

This application outlines the procedure for proportionally balancing low-pressure constant volume systems and the low-pressure side of any system—constant or variable air volume systems; low-, medium-, or high-pressure systems; single-zone, multi-zone, induction, or dual-duct systems.

Fans, Fan Performance Curves, and System Curves

This section begins with a general overview of fans, fan performance curves, and system curves. HVAC fans are normally designated by type, class, fan rotation, drive arrangement, motor location, air discharge direction, and width of fan wheel. Fans are further classified into these general categories: axial fans (propeller, tube-axial, and vane-axial), centrifugal fans (forward curved, backward curved, backward inclined, airfoil, and radial), and specific design fans (tubular centrifugal fans and centrifugal or axial power roof ventilators).

A fan performance curve is a graphic representation of the performance of a fan from free delivery to no delivery. The air density, wheel size, and fan speed are usually stated on the curve and are constant for the entire curve. Static pressure (s.p.), static efficiency (SE), total pressure (TP), total efficiency (TE), which is also known as mechanical efficiency (ME), and brake horsepower (bhp) may be plotted against air volume, which is stated in cubic feet per minute (cfm). Fan performance curves are developed from actual tests. The Air Movement and Control Association (AMCA) and other fan manufacturers have established procedures and standards for the testing and rating of fans. According to AMCA publications, their testing procedure normally requires that the entire range of the fan's performance be tested from free delivery to no delivery. Both the discharge pressure and the inlet pressure are measured. These measurements are then mathematically translated to air volume and fan pressure. The test fan is usually driven by a dynamometer which provides values of torque at each of the operating points while readings of fan speed are taken simultaneously. This allows calculation of horsepower input for each of the settings. Dry-bulb and wet-bulb temperature and barometric pressure readings are also taken so that the air density can be calculated. These measured and calculated values are then plotted to develop the fan performance curve.

The system curve for any fan and duct system is a plot of the pressure needed to move the air through the system and overcome the total of all the pressure losses through the system components. For any point on the system curve, there is a given static pressure for a given flow rate. A fixed system is one in which there are no changes in the system resistance resulting from closing or opening of dampers, or changes in the condition of filters or coils. For a fixed system, an increase or decrease in system resistance results only from an increase or decrease in flow rate, and this change in resistance will fall along the system curve.

Each duct system has its own system curve and each fan has its own performance curve for a given flow rate (*Figure B-1*). The intersection of these two curves is the operating point for the fan. On the fan curve, a decrease in static pressure (system resistance) will mean an increase in flow rate and vice versa. On the duct system curve, the system resistance (static pressure) is proportional to the square of the flow rate in accordance with the fan law formula. For example, a point on a given system curve is 1" of static pressure. On this curve, the 1" of static pressure corresponds to an air volume of 600 cfm. Any other point on the curve can be found by using the fan laws. To continue the example, a second point on the curve is 1,200 cfm and 4" static pressure. A third point is 1,800 cfm and 9" static pressure.

Fan law formula:

$$\frac{\text{s.p.}_2}{\text{s.p.}_1} = \left(\frac{\text{cfm}_2}{\text{cfm}_1}\right)^2$$

Where:

s.p.$_1$ = original static pressure in inches of water column

s.p.$_2$ = new static pressure in inches of water column

cfm$_1$ = original volume of airflow in cubic feet per minute

cfm$_2$ = new volume of airflow in cubic feet per minute

Using the fan laws, the fan performance curve, and the system curve, any change to the fan speed or the duct system resistance can be calculated and graphically depicted. For example, in *Figure B-1*, point 1 is the operating point for a fan on fan curve A, with duct system I. To increase or decrease airflow, a physical change must be made to either the duct system or the fan speed or both.

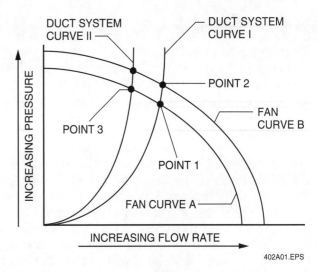

Figure B-1 ◆ Performance curve.

If the change is to the fan speed, the fan will be operating on a new performance curve that runs parallel to the original curve. Since the duct system has remained unchanged, the system curve (I) also remains unchanged.

In *Figure B-1*, an increase in fan speed results in a new fan curve (B). The fan is now operating at a higher cfm and a higher static pressure at point 2.

However, if the increase or decrease in airflow is made by reducing or adding system resistance when opening or closing a main damper, a new system curve is established. The fan performance curve is unchanged. In *Figure B-1*, a fan is operating at point 1 on fan curve A. A main damper is closed and an increase in system resistance results in a new system curve (II). The fan is now operating at a lower cfm and a higher static pressure at point 3.

Field measurements of fan pressures won't always fall into place on the fan performance curve. More often than not, there will be differences. This is due in part to poor locations at the fan at which to take the measurements. Fan inlet or outlet duct connections which aren't properly designed or installed change the aerodynamic characteristics of the fan. These conditions, called system effect, reduce fan performance.

To reduce or eliminate system effect caused by outlet conditions, AMCA Standard 210 specifies an outlet duct that's no greater than 107.5 percent, nor less than 87.5 percent of the fan outlet area. It also calls for the slope of the outlet transition to be no greater than 15 percent for converging transitions nor greater than 7 percent for diverging transitions. Since the normal velocity profile at the fan outlet isn't uniform, a length of straight duct is needed to establish a uniform velocity pattern. The outlet duct, including any transition duct, should extend out at least one duct diameter for each 1,000 fpm of outlet velocity with a minimum length of 2½ duct diameters. For example, a fan with an outlet velocity of 1,500 fpm would need a straight duct of 2.5 duct diameters, while a fan with an outlet velocity of 3,000 fpm would need a straight duct of three duct diameters.

To reduce or eliminate system effect caused by inlet conditions, *AMCA Standard 210* specifies an inlet duct that's no greater than 112.5 percent nor less than 92.5 percent of the fan inlet area. It also calls for the slope of the inlet transition to be no greater than 15 percent for converging transitions nor greater than 7 percent for diverging transitions. To reduce losses at the fan inlet, the inlet duct or fan inlet should have a smooth, rounded entry, a converging taper entry, or a flat flange on the end of the duct. An elbow at the fan inlet will result in turbulent and uneven flow distribution at the fan wheel. Losses can be reduced if straight duct and turning vanes are added. A major cause of reduced fan performance is an inlet duct condition that produces a pre- or counter-rotational spin to the entering air. These inlet conditions can be improved by installing turning vanes, splitters, or airflow straighteners. Fan performance is also reduced if the space between the fan inlet and the fan enclosure is too restrictive. Allow at least one-half the wheel diameter between the enclosure and the fan inlet. Inlets of multiple double-wide centrifugal fans in parallel in a common enclosure should be at least one wheel diameter apart. A belt guard in the plane of the fan inlet will reduce fan performance. Belt guards in the airstream should be made of open construction with as much free air passage as possible and shouldn't be more than one-third of the inlet area.

Proportional Balancing

The principles of proportional balancing require that all the dampers in the distribution system be fully open and that at least one outlet volume damper (the outlet with the lowest percentage of design flow) will remain open. If the system has branch ducts, at least one branch volume damper (the branch with the lowest percent of design flow) will also remain fully open. Because the air outlets are on the low-pressure side of any system, the following proportional balancing procedure can be used on constant or variable air volume systems; low-, medium-, or high-pressure systems; or single-zone, multi-zone, or dual-duct systems.

Proportional Balancing Procedure

Step 1   Determine which outlet has the lowest percentage of design flow (%D). Typically, the outlet with the lowest %D will be on the branch farthest from the fan. This outlet will be called the key outlet.

a. Design flow is either the original design flow per the contract specifications or a new calculated design flow.

b. Percent of design flow is equal to the measured flow divided by the design flow:

$$\%D = \frac{M}{D}$$

The term percent of design flow (%D) is meant for new construction where balancing is done by an agency according to the fluid quantities specified on the contract documents. These numbers are based on expected heating and cooling loads. However, %D can also stand for percent of desired flow quantity. This is used for in-house balancing or balancing to owners' specifications and is based on the actual space operating loads, location of occupants, and equipment and comfort conditions.

c. If anemometers are being used, the measured and design flows will be in feet per minute; whereas, if capture hoods are used, the flows will be in cubic feet per minute. All the system balancing examples in this application will use cfm as the measured flow.

Step 2   Starting with the key outlet, as needed, adjust each outlet on that branch duct in sequence, from the lowest percentage of design flow to the highest percentage of design flow.

a. The ratio of the percentage of design flow between each outlet must be ±10 percent. Ratio of design flow is equal to outlet X %D divided by outlet Y %D.

b. To reduce airflow, volume dampers in the system should be adjusted in the branch ducts and at the takeoffs and not at the outlet since dampering at the outlets results in excessive noise and poor air distribution.

Step 3 Go to the branch that has the outlet with the next lowest percent of design flow as determined from the initial readout. Typically, this key outlet will be on the second farthest branch. Balance all the outlets on this branch to the key outlet to within ±10 percent of design flow.

Step 4 Continue until all the outlets on all the branches have been balanced to within ±10 percent of each other.

Step 5 Starting with the branch with the lowest percent of design flow as the key branch, proportionately balance all branch ducts from the lowest %D flow to the highest %D flow to within ±10 percent of each other.

Step 6 Continue until all branches have been balanced.

Step 7 Adjust the fan speed if needed to bring the system to within ±10 percent of design flow.

Step 8 Reread all the outlets and make any final adjustments.

Example 1

Balance the constant volume, low-pressure system illustrated in *Figure B-2*.

1. The system has been inspected and all dampers are open. The fan is operating correctly. Traverse and static pressure readings have been taken at points A, B, and C. The results are:

Point A: 2,185 cfm (115%D)
Point B: 975 cfm (122%D)
Point C: 1,180 cfm (107%D)

2. All the outlets have been read and the results are:

Outlet	Design	Measured
1	250	235
2	300	280
3	250	265
4	300	340
5	200	230
6	200	210
7	200	260
8	200	225
Total	1,900	2,045

Branch	Design	Measured	%D
B	800	925	116
C	1,100	1,120	102
Total	1,900	2,045	108

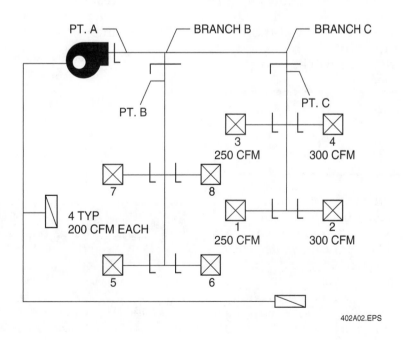

402A02.EPS

Figure B-2 ◆ Constant volume low-pressure system.

Balancing Procedure

Step 1 Determine which outlet has the lowest percentage of design flow.

Outlet	Design	Measured	%D
1	250	235	94
2	300	280	93
3	250	265	106
4	300	340	113
5	200	230	115
6	200	210	105
7	200	260	130
8	200	225	113
Total	1,900	2,045	108

Step 2 Outlet No. 2 is the key outlet. Therefore, the balance will start with outlet No. 2 on branch C. The balancing damper at the takeoff to outlet No. 2 will remain open.

  a. Outlet No. 1 is the next highest %D at 94 percent.

  b. Determine if No. 1 and No. 2 are within ±10 percent of each other. The ratio of No. 1 to No. 2 is 1.01 (94 percent/93 percent). The ratio between No. 1 and No. 2 is within ±10 percent. This ratio will remain in effect for as long as the dampers for the two outlets aren't moved. Therefore, whatever percentage of design flow outlet No. 2 goes to, outlet No. 1 will be 1.01 times No. 2. For example, if the airflow to outlet No. 2 is brought to 100%D, outlet No. 1 will be 101%D.

Outlet	Design	Measured	%D	Ratio
1	250	235	94	1:2 = 1.01
2	300	280	93	

  c. Go to the outlet with the next highest %D, outlet No. 3. Compare outlet No. 3 to outlet No. 2. The ratio of No. 3 to No. 2 is 1.14 (106 percent/93 percent). The ratio between No. 3 and No. 2 is greater than ±10 percent (1.10).

  d. To balance No. 3 to No. 2, the volume damper for outlet No. 3 will be closed. Arbitrarily close No. 3 to 100%D or 250 cfm. Go back and read No. 2. It reads 285 cfm. Determine %D for No. 3 and No. 2 and the ratio between them.

  No. 3 is 100%D.

  No. 2 is 95%D (285/300).

The ratio between No. 3 and No. 2 is 1.05 (100 percent/95 percent).

Since the ratio is within 10 percent, the damper on No. 3 is locked. Once the ratio between two outlets is set between ±10 percent, the damper is locked and the damper is not moved again until the final system balance.

Outlet	Design	Measured	%D	Ratio
1	250	240*	96*	1:2 = 1.01
2	300	285	95	
3	250	250	100	3:2 = 1.05

* Calculated for the purpose of this exercise only to illustrate what is happening at the outlets already set. Once an outlet or branch damper is set, the cfm is normally not reread until the final balance and generally there is no need to calculate the airflow.

  e. Go to outlet No. 4 and read it. In this example, it still reads 340 cfm, 113%D. Outlets No. 4 and No. 2 aren't within ±10 percent of each other (113 percent/95 percent). Arbitrarily cut No. 4 to 100%D or 300 cfm. Read No. 2. It reads 295 cfm. Determine %D for No. 4 and No. 2 and the ratio between them.

  No. 4 is 100%D

  No. 2 is 98%D

The ratio between No. 4 and No. 2 is 1.02 (100 percent/98 percent).

Since the ratio is within ±10 percent, the damper on No. 4 is locked. All the outlets on branch C have been balanced.

Outlet	Design	Measured	%D	Ratio
1	250	248*	99*	1:2 = 1.01
2	300	295	98	
3	250	258*	103*	3:2 = 1.05
4	300	300	100	4:2 = 1.02
* Calculated				

Step 3 Go to the key outlet on branch B. This will be outlet No. 6 at 105%D.

Balance the other outlets to No. 6.

Outlet	Design	Measured	%D
5	200	230	115
6	200	210	105
7	200	260	130
8	200	225	113

    a. No. 8 to No. 6 is 1.08 (113 percent/105 percent).

Outlet	Design	Measured	%D	Ratio
5				
6	200	210	105	
7				
8	200	225	113	8:6 = 1.08

    b. No. 5 to No. 6 is 1.1 (115 percent/105 percent).

Outlet	Design	Measured	%D	Ratio
5	200	230	115	5:6 = 1.1
6	200	210	105	

    c. No. 7 to No. 6 is 1.24 (130 percent/105 percent). The volume damper for outlet No. 7 is arbitrarily cut so that outlet No. 7 reads 115%D or 230 cfm.

Read No. 6. It reads 220 cfm. Determine %D for No. 7 and No. 6 and the ratio between them.

No. 7 is 115%D

No. 6 is 110%D

The ratio between No. 7 and No. 6 is 1.05 (115 percent/110 percent). Since the ratio is within ±10 percent, the damper on No. 7 is locked. All the outlets on branch B have been balanced.

Outlet	Design	Measured	%D	Ratio
5	200	242*	121*	5:6 = 1.1
6	200	220	110	
7	200	230	115	7:6 = 1.05
8	200	238*	119*	8:6 = 1.08
* Calculated				

Branch Balancing

To proportionally balance the branches, start with the lowest percent of design airflow. You can balance branches using a representative outlet on each branch or using static pressures.

Branch Balancing Using Outlets

Example 2

Balance the constant volume low-pressure system illustrated in *Figure B-2* using representative outlets on each branch.

Outlet	Design	Measured	%D	Ratio
1	250	248*	99*	1:2 = 1.01
2	300	295	98	
3	250	258*	103*	3:2 = 1.05
4	300	300	100	4:2 = 1.02
5	200	242*	121*	5:6 = 1.10
6	200	220	110	
7	200	230	115	7:6 = 1.05
8	200	238*	119*	8:6 = 1.08
* Calculated				

1. After proportionally balancing the outlets, the results are:

2. Use the key outlet on each branch to determine which branch has the lowest percent of design flow. In this case, use outlets No. 2 and No. 6. Outlet No. 6 is 110%D and outlet No. 2 is 98%D. The ratio between the outlets is 1.12 (110 percent/98 percent).

The volume damper on branch B is arbitrarily closed to bring outlet No. 6 down to 105%D (210 cfm). Read outlet No. 2. It reads 300 cfm.

Determine %D for No. 6 and No. 2, and the ratio between them.

No. 6 is 105%D

No. 2 is 100%D

The ratio between No. 6 and No. 2 is 1.05 (105 percent/100 percent). Since the ratio is within ±10 percent, the branch volume damper on branch B is locked. The branches are proportionally balanced to each other. As all the outlets have been proportionally balanced to each other by branch, an adjustment at the branch damper will increase or decrease all the outlets on each branch proportionally.

Outlet	Design	Measured	%D	Ratio
1	250	253*	101*	1:2 = 1.01
2	300	300	100	
3	250	263*	105*	3:2 = 1.05
4	300	306*	102*	4:2 = 1.02
5	200	230*	115*	5:6 = 1.10
6	200	210	105	6:2 = 1.05
				(branch B/C)
7	200	220	110*	7:6 = 1.05
8	200	226*	113*	8:6 = 1.08
* Calculated				

3. Reread all the outlets and make any final adjustments. Record final readings on the report forms.

Outlet	Design	Measured	%D
1	250	250	100
2	300	300	100
3	250	265	106
4	300	310	103.3
5	200	225	112.5
6	200	210	105
7	200	220	110
8	200	220	110
Total	1,900	2,000	105.3

Branch	Design	Measured	%D
B	800	875	109.4
C	1,100	1,125	102.2
Total	1,900	2,000	105.3

4. The system is within ±10 percent of total design flow; therefore, it does not require major adjustments. However, outlet 5 is more than 10 percent out of design and should be adjusted down slightly.

Branch Balancing Using Static Pressure

Another method for proportionally balancing branches is performed using static pressure. If a traverse of the branch was made, use this as total cfm. If a traverse was not made, total the outlets. Next, take a static pressure reading at the traverse point, if it hasn't already been done. These readings will be called $s.p._1$ and cfm_1 for total airflow in the branch and its corresponding static pressure. Use the fan laws to calculate the required static pressure ($s.p._2$) that will result in the design airflow (cfm_2).

When using the fan laws, it is absolutely necessary to be sure that no adjustments or changes are made downstream of the branch damper during the branch balancing process. To illustrate what would happen, let's say that after taking the branch traverse and static pressure, a damper is closed in one of the takeoffs. This means that the total cfm at the traverse point will be reduced and the static pressure will be increased. Just the opposite would happen if an outlet damper was found closed and then opened after the traverse and static pressure had been taken; the flow rate would increase and the static pressure would decrease at the traverse point. Under both of these circumstances the fan law couldn't be used until new flow rate and static pressure readings were taken, after the changes to the system had been made. Therefore, it's extremely important to be sure that all the dampers are open before starting the balance and that nothing is disturbed during the balance.

Example 3

Balance the constant volume low-pressure system illustrated in *Figure B-2* using branch static pressures.

Initial conditions:

Location	Design cfm_2	Measured cfm_1	%D	S.P._1	Ratio
Point B	800	930	116	0.70	
Point C	1,100	1,100	100	0.52	B:C 1.16

To bring the branches to within ±10 percent of each other, it will be necessary to close the volume damper on branch B.

Step 1 Arbitrarily elect to close branch B to 108%D.

Step 2 Use the fan law below to determine what new static pressure at point B will correspond to 108%D or 864 cfm.

$$\frac{s.p._2}{s.p._1} = \left(\frac{cfm_2}{cfm_1}\right)^2$$

$$s.p._2 = s.p._1 \left(\frac{cfm_2}{cfm_1}\right)^2$$

$$s.p._2 = 0.70 \left(\frac{864}{930}\right)^2$$

$$s.p._2 = 0.60 \text{ in. w.g.}$$

Step 3   Close the volume damper on branch B until you read 0.60 inches w.g. s.p. on the gauge. Go to point C and read the new static pressure. It reads 0.57 inches w.g. Calculate new cfm, %D flow, and ratio.

$$\left(\frac{cfm_2}{cfm_1}\right)^2 = \frac{s.p._{·2}}{s.p._{·1}}$$

Location	Design cfm₂	Measured cfm₁	%D	S.P.₁	Ratio
Point B	800	864	108	0.60	
Point C	1,100	1,152	105	0.57	B:C 1.03

Step 4   The branches are now balanced to each other within ±10 percent and since all outlets have been proportionally balanced to each other by branch, an adjustment at the branch damper will increase or decrease all the outlets on each branch proportionally. The ratio of the outlets remains the same as when initially set since no outlet volume dampers have been changed. To determine the cfm at each outlet, read outlets No. 2 and No. 6 and calculate new cfm for the other outlets.

Outlet	Ratio
1	1:2 = 1.01
2	Measured cfm
3	3:2 = 1.05
4	4:2 = 1.02
5	5:6 = 1.10
6	Measured cfm
7	7:6 = 1.05
8	8:6 = 1.08

Airflow Instrumentation and Flow Factors

Anemometers

Airflow through the outlets (or inlets) may be measured using a rotating vane anemometer for sidewall grilles and registers, or a deflecting vane anemometer for both sidewall grilles and registers and ceiling diffusers. Anemometers require a correction or flow factor for each outlet to convert velocity readings to cfm. In addition to the manufacturer's flow factor for the outlets/inlets, the anemometer may also have a calibration or correction factor.

Flow Factor

A flow factor, sometimes called a K-factor or A_k factor, is the effective area of the grille or diffuser as determined by the manufacturers' own airflow tests. Check with the manufacturer for a catalog of grilles and diffusers and their corresponding flow factors. These flow factors must be used to calculate cfm (cfm = $A_k \times V_k$). It's also important to understand that these flow factors apply only when using the instrument specified by the outlet manufacturer and in the prescribed method.

If a flow factor is not available or is not producing a satisfactory result, a new flow factor can sometimes be field-determined. To calculate a new flow factor, take a pitot tube traverse in the duct for the outlet or inlet terminal in question. The duct must be free from obstructions and air leaks from the point of traverse to the terminal. Take readings at the traverse point and at the terminal. The cfm calculated at the traverse point divided by the average velocity at the terminal is the flow factor.

Capture Hood

A capture hood or flow hood may also be used to measure sidewalls and ceiling diffusers. The capture hood is the easiest, quickest, and most reliable instrument to take outlet/inlet readings. Airflows are read directly in cfm and flow factors for the air devices aren't needed. However, if a capture hood reading is in question, and a correction factor is indicated because of very high or very low velocities or an unusual use of the capture hood, take a traverse of the duct and determine a correction factor. In addition to a field-measured correction factor for air devices, check the capture hood for a manufacturer's calibration or correction factor. Whether or not a correction factor is needed, the duct system can still be proportionally balanced from the measured airflow. The actual airflow values can be corrected at any time.

Example 4

Readings are taken on several linear air diffusers (LAD) located in a drop ceiling. To determine if a field correction factor is needed, a pitot tube traverse is made and compared with the capture hood readings. The cfm calculated at the outlet divided by the cfm at the traverse point is the correction factor. In this example, the total at the outlet is 450 cfm. The total at the traverse is 500 cfm. There is no noticeable air leakage between the traverse point and the outlet. The correction factor is 0.90. The airflow from any other similar LAD will be measured with the capture hood in the same manner, and the total cfm at the outlet divided by the correction factor to get actual airflow.

VAV Troubleshooting Guidelines

Diffuser dumps cold air:

- Airflow is too low (velocity too slow).
- Check to determine if box is reducing too far.
- Evaluate box minimum setting.
- Diffuser is too large; check installation.

Conditioned space too cold:

- Supply air temperature is too cold.
- There is too much supply air.
- Diffuser pattern or throw is incorrect, causing drafts.
- Temperature sensor is located incorrectly or needs calibration.

Conditioned space too warm:

- Supply air temperature setting is too warm.
- There is not enough supply air.
- Refrigeration system is not operating properly.
- Fan-coil evaporator is iced over due to low airflow.
- Temperature sensor is located incorrectly or needs calibration.
- Low-pressure duct is leaking.
- Low-pressure duct is not insulated.
- Cold air from diffuser isn't mixing properly with room air; increase volume or velocity, change or retrofit diffuser.

Noise:

- There is too much air in low-pressure duct; check box maximum setting.
- Static pressure in the system is too high.
- Diffuser is too small.
- Diffuser is dampered at face; always damper at takeoff.
- Pattern controllers are loose; tighten or remove.

Not enough air:

- Box is not operating properly; check minimum setting and reset as necessary.
- There is not enough static pressure at box inlet for proper operation.

- Damper in VAV box is closed; may be loose on shaft or frozen.
- Low-pressure damper is closed.
- There are restrictions in low-pressure duct.
- Remove pattern controllers in diffusers.
- Low-pressure duct is leaking, disconnected, or twisted.
- Install VAV diffusers where applicable; these diffusers close down at the face, producing higher air velocities as they reduce in size.
- Install fan-powered boxes.

Box not operating properly:

- There is not enough static pressure at the inlet.
- There is too much static pressure at the inlet.
- Static pressure sensor is defective, clogged, or located incorrectly.
- Static pressure setting on controller is incorrect.
- Static pressure controller needs calibrating.
- Fan speed is not correct.
- Inlet duct is leaking or disconnected.
- Main ductwork has been improperly designed.
- There is not enough straight duct on the inlet of the box.
- Diversity is incorrect.
- Box is wrong size or has wrong nameplate; check installation; leak test if necessary.
- Damper is loose on shaft.
- Linkage from actuator to damper is incorrect or binding.
- Actuator is defective.
- Controls are defective, need calibration, or are set incorrectly.
- Volume controller is not set properly for normally open or normally closed operation.
- Damper is linked incorrectly, NO for NC operation, or vice versa.
- Pneumatic tubing to controller is piped incorrectly, leaking, or pinched.
- Restrictor in pneumatic tubing is missing, broken, placed incorrectly, clogged, or wrong size.
- Oil or water is in pneumatic lines.
- There is no power to controls.
- Box is wired incorrectly.
- PC board is defective.

Fan not operating properly:

- Inlet vanes on centrifugal fans are not operating properly.
- Pitch on vane-axial fans is not adjusted properly.

- Fan is rotating backwards.
- Return air fan is not tracking with supply fan; check static pressure sensors, airflow measuring stations and move, clean, or calibrate; consider replacing return fan with a relief fan.

Negative pressure in the building:

- Office buildings should be maintained at +0.03 to +0.05 in. w.g.
- Problem may be caused by stack effect or improper return air control.
- Seal building properly.
- Balance return system and install manual balancing dampers as needed to control outside air (OA), return air (RA), and exhaust air (EA) at the unit.
- Get return fan to track with supply fan; consider replacing return fan with relief fan.
- Check that static pressure sensors are properly located and working.
- Install pressure-controlled return air dampers in return air shafts from ceiling plenums.

- Supply fan is reducing air volume and reduction of outside air for the (1) constant volume exhaust fans and (2) exfiltration. Increase minimum outside air as follows:
  - Open manual volume damper in OA duct and/or increase OA duct size.
  - Control OA from supply fan; as fan slows, outside air damper opens.
  - Control OA damper from flow monitor in OA duct to maintain a constant minimum OA volume.

Inadequate amount of air for proper ventilation:

- Problem may be caused by the supply fan reducing air volume and a reduction of outside air for code or building requirements. Increase minimum outside air as follows:
  - Open manual volume damper in OA duct and/or increase OA duct size.
  - Control OA from supply fan; as fan slows, outside air damper opens.
  - Control OA damper from flow monitor in OA duct to maintain a constant minimum OA volume.

Examples of Completed
Air Balance Forms

Mailing Address
P.O. Box 5782
Greenville, SC 29606-5782

165 South Hammett Road
Greer, SC 29651

Phone (864)877-6832

Fax (864)877-5490

email palmettoab@aol.com

Palmetto Air & Water
B A L A N C E
Raising the level of efficiency and comfort

RECTANGULAR DUCT TRAVERSE TEST REPORT

PROJECT: CHARTER COMMUNICATIONS

Zone: DT-1 SUPPLY

SYSTEM/ UNIT: SCU-1

Correction Factor: 1.00

Width:	30"	Duct Area			Actual		
Height:	22"	4.583 Sq Ft			Vel.	1640 CFM	7516

Position	1	2	3	4	5	6	7	8	9	10	11	12	13	14
1 >>>	1487	1689	1893	1818	1716	1933								
2 >>>	1690	1656	1505	1614	1730	1833								
3 >>>	1619	1589	1539	1416	1472	1573								
4 >>>	1713	1381	1467	1740	1673	1610								
5 >>>														
6 >>>														
7 >>>														
8 >>>														
9 >>>														
10 >>>														
Subtotal	6509	6315	6404	6588	6591	6949	0	0	0	0	0	0	0	0

Total FPM / Number of Readings = Average Velocity X Duct Area = CFM				Final S.P.	
39356	24		1640	4.583	7516	1.31

Remarks: *SUPPLY CFM TOTAL = DT-1 + DT-2
Filename: CCG049
Test Date: Jan–03
Readings By: GF

402A04.EPS

Mailing Address
P.O. Box 5782
Greenville, SC 29606-5782

165 South Hammett Road
Greer, SC 29651

Phone (864)877-6832

Fax (864)877-5490

email palmettoab@aol.com

Palmetto Air & Water

B A L A N C E

Raising the level of efficiency and comfort

RECTANGULAR DUCT TRAVERSE TEST REPORT

PROJECT: CHARTER COMMUNICATIONS   Zone: DT-2 SUPPLY

SYSTEM/ UNIT: SCU-1   Correction Factor: 1.00

Width:	34"	Duct Area										Actual		
Height:	24"	5.667 Sq Ft								Vel.		1892 CFM	10719	

Position	1	2	3	4	5	6	7	8	9	10	11	12	13	14
1 >>>	1314	2351	2436	2231	1348	–412								
2 >>>	2745	2551	2260	2308	1619	335								
3 >>>	2585	2337	2368	2419	1572	461								
4 >>>	2537	2423	2509	2479	1825	796								
5 >>>														
6 >>>														
7 >>>														
8 >>>														
9 >>>														
10 >>>														
Subtotal	9181	9662	9573	9437	6364	1180	0	0	0	0	0	0	0	0

Total FPM /	Number of Readings =	Average Velocity	X Duct Area = CFM	Final S.P.	
45397	24	1892	5.667	10719	1.62

Remarks:
Filename: CCG050
Test Date: Jan–03
Readings By: GF

402A05.EPS

Mailing Address
P.O. Box 5782
Greenville, SC 29606-5782

165 South Hammett Road
Greer, SC 29651

Phone (864)877-6832

Fax (864)877-5490

email palmettoab@aol.com

Palmetto Air&Water
BALANCE
Raising the level of efficiency and comfort

VAV TEST REPORT

PROJECT: CHARTER COMMUNICATIONS
SYSTEM/ UNIT: SCU-1

Box Number	Type	Design CFM	All Boxes Maxed	Damper Position	Flow Corr
FPB-1-1	SUPPLY	1015	1006	67%	0.729
FPB-1-2	SUPPLY	360	*	0%	0.579
FPB-1-3	SUPPLY	1980	1960	66%	0.647
FPB-1-5	SUPPLY	2400	2424	65%	0.753
FPB-1-7	SUPPLY	2400	2156	100%	0.630
FPB-1-8	SUPPLY	1465	1478	70%	0.587
FPB-1-9	SUPPLY	1550	1414	100%	0.597
FPB-1-10	SUPPLY	1550	1630	67%	0.604
VAV-1-2	SUPPLY	1005	1020	55%	0.737
VAV-1-3	SUPPLY	1600	1670	100%	0.755
VAV-1-4	SUPPLY	360	352	54%	0.732
VAV-1-5	SUPPLY	1615	1540	100%	0.462
VAV-1-6	SUPPLY	750	*	0%	0.728
VAV-1-7	SUPPLY	730	710	65%	0.799

TOTAL	
Remarks:	*CLOSED FOR DIVERSITY
Filename:	CCG048
Test Date:	Jan–03
Readings By:	GF

402A06.EPS

Mailing Address
P.O. Box 5782
Greenville, SC 29606-5782

165 South Hammett Road
Greer, SC 29651

Phone (864)877-6832

Fax (864)877-5490

email palmettoab@aol.com

Palmetto Air&Water

B A L A N C E

Raising the level of efficiency and comfort

AIR OUTLET TEST REPORT

PROJECT: CHARTER COMMUNICATIONS
SYSTEM/ UNIT: SCU-1

Area Served	Number	Type	Design CFM	Preliminary CFM	Final CFM	Percent Design
128	1	SUPPLY	800	885	868	109%
131	2	SUPPLY	800	1124	852	107%
134	3	SUPPLY	800	363	720	90%
FPB-1-7 TOTAL			2400	2372	2440	102%
134	1	SUPPLY	265	156	292	110%
133	2	SUPPLY	400	416	379	95%
133	3	SUPPLY	400	411	361	90%
132	4	SUPPLY	200	266	220	110%
132	5	SUPPLY	200	309	195	98%
FPB-1-8 TOTAL			1465	1558	1447	99%
135	1	SUPPLY	280	250	270	96%
135	2	SUPPLY	280	222	252	90%
135	3	SUPPLY	280	272	253	90%
135	4	SUPPLY	355	524	361	102%
135	5	SUPPLY	355	445	356	110%
FPB-1-9 TOTAL			1550	1713	1492	96%
135	1	SUPPLY	280	318	279	100%
135	2	SUPPLY	280	285	272	97%
135	3	SUPPLY	280	268	255	91%
135	4	SUPPLY	100	192	105	105%
135	5	SUPPLY	255	336	237	93%
135	6	SUPPLY	355	601	363	102%
FPB-1-10 TOTAL			1550	2000	1511	97%

TOTAL
Remarks:
Filename: CCG002
Test Date: Jan–03
Readings By: GF

402A07.EPS

Additional Resources and References

Additional Resources

This module is intended to be a thorough resource for task training. The following reference works are suggested for further study. These are optional materials for continued education rather than for task training.

ASHRAE HVAC Applications Handbook, 1999. Atlanta, GA: American Society of Heating and Air Conditioning Engineers, Inc.

HVAC Systems – Testing, Balancing, and Adjusting, 1993. Arlington, VA: Sheet Metal and Air Conditioning Contractor's National Association, Inc. (SMACNA).

Psychrometrics Introduction, 1993. Syracuse, NY: Carrier Corporation.

Testing and Balancing HVAC Air and Water Systems, 2001. Samuel C. Monger. Lilburn, GA: The Fairmont Press, Inc.

Figure Credits

Extech Instruments, 402F07 (top), 402SA07

Bacharach, Inc., 402F07 (bottom), 402F22 (bottom left)

Carrier Corporation, 402F08, 402F30–402F34, 402SA06

Carnes Company, 402F18, 402F24, 402F25

Dwyer Instruments, 402F19, 402F20

TSI Incorporated, 402F22 (top left and right), 402SA01

TIF, SPX Corporation, 402F22 (bottom right)

Topaz Publications, Inc., 402SA02, 402SA03, 402SA05

Vent Products Company, Inc., 402F35

Trane, 402F36, 402F39

Courtesy of Palmetto Air & Water Balance, Inc., Appendix C

NCCER makes every effort to keep these textbooks up-to-date and free of technical errors. We appreciate your help in this process. If you have an idea for improving this textbook, or if you find an error, a typographical mistake, or an inaccuracy in NCCER's Contren® textbooks, please write us, using this form or a photocopy. Be sure to include the exact module number, page number, a detailed description, and the correction, if applicable. Your input will be brought to the attention of the Technical Review Committee. Thank you for your assistance.

Instructors – If you found that additional materials were necessary in order to teach this module effectively, please let us know so that we may include them in the Equipment/Materials list in the Annotated Instructor's Guide.

Write: Product Development and Revision
National Center for Construction Education and Research
3600 NW 43rd St, Bldg G, Gainesville, FL 32606

Fax: 352-334-0932

E-mail: curriculum@nccer.org

Craft   Module Name
_____

Copyright Date   Module Number   Page Number(s)
_____

Description
_____
_____
_____
_____

(Optional) Correction
_____
_____
_____

(Optional) Your Name and Address
_____
_____
_____

03403-09

Indoor Air Quality

03403-09
Indoor Air Quality

Topics to be presented in this module include:

Overview

It is a well-known fact that the air circulating through a building can carry a variety of contaminants that can affect the health and comfort of those breathing the air. These contaminants include molds, bacteria, allergens, fungi, dust, and pollen. Contaminants such as legionella, the bacteria that causes Legionnaire's disease, can cause severe illness and even death. Others can cause respiratory problems, burning eyes, scratchy throat, sneezing, and other symptoms. The terms *sick building syndrome* and *building-related illness* were coined to define a building with poor ventilation and air circulation properties that caused these symptoms among its occupants. A variety of air filtration devices and systems, as well as new standards and methods for ventilating buildings, have been implemented. In addition, local and national government agencies are taking a hard-line approach to ensure that indoor air quality standards are foremost in the minds of HVAC designers, installers, and service personnel.

Objectives

When you have completed this module, you will be able to do the following:

1. Explain the need for good indoor air quality.
2. List the symptoms of poor indoor air quality.
3. Perform an inspection/evaluation of a building's structure and equipment for potential causes of poor indoor air quality.
4. Identify the causes and corrective actions used to remedy common indoor air problems.
5. Identify the HVAC equipment and accessories that are used to sense, control, and/or enhance indoor air quality.
6. Use selected test instruments to measure or monitor the quality of indoor air.
7. Clean HVAC air system ductwork and components.

Trade Terms

Arrestance efficiency
Atmospheric dust spot efficiency (dust spot efficiency)
Biological contaminants
Building-related illness
Desiccant
Environmental tobacco smoke (ETS)
Formaldehyde
Friable
High-efficiency particulate air (HEPA) filter

Microbial contaminants
Microbiological contaminants
Micron
Multiple chemical sensitivity (MCS)
New building syndrome
Off-gassing
Ozone
Pontiac fever
Radon
Sick building syndrome
Volatile organic compounds (VOCs)

Required Trainee Materials

1. Pencil and paper
2. Appropriate personal protective equipment

Prerequisites

Before you begin this module, it is recommended that you successfully complete *Core Curriculum*; *HVAC Level One*; *HVAC Level Two*; *HVAC Level Three*; and *HVAC Level Four*, Modules 03401-09 and 03402-09.

This course map shows all of the modules in the fourth level of the *HVAC* curriculum. The suggested training order begins at the bottom and proceeds up. Skill levels increase as you advance on the course map. The local Training Program Sponsor may adjust the training order.

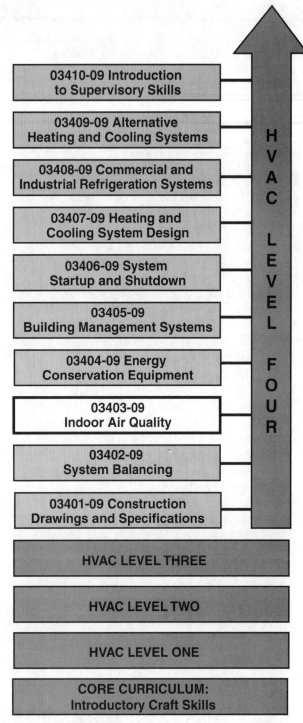

403CMAP.EPS

1.0.0 ◆ INTRODUCTION

Awareness of indoor air quality (IAQ) as a health, economic, and environmental issue is very important. Poor IAQ can cause both long-term and short-term health effects. Excluding overall health issues, the economic impact includes the direct medical cost for people whose health is affected by poor IAQ, lost productivity from absence due to illness, decreased work efficiency, and damage to equipment and materials due to exposure to indoor air pollutants.

One challenge for HVAC technicians, building owners, and occupants is to increase their understanding about IAQ so that a building is used in a manner that does not defeat the capabilities of the building HVAC equipment and/or compromise indoor air quality.

The causes of poor IAQ are varied. Recent energy conservation measures include tighter and better insulated buildings, reduced capacities of HVAC systems, and HVAC system control schemes that minimize air movement in occupied spaces. These changes have often resulted in higher levels of contaminants. Poor IAQ can also be traced to the kinds of materials used to construct, furnish, and maintain buildings. HVAC maintenance is a key factor in the overall health of a building.

This module briefly introduces the subject of indoor air quality. Its focus is on the main causes of poor IAQ and its effect on the health and comfort of building occupants. Also covered are some of the equipment and methods currently being used to test for and achieve good IAQ.

1.1.0 Indoor Environmental Quality Issues

Indoor air quality is a portion of a larger entity known as indoor environmental quality (IEQ). IEQ encompasses all elements of the indoor environment. The U.S. EPA estimates that Americans spend roughly 90 percent of their time indoors. According to the Rocky Mountain Institute, human productivity in the work environment can be increased as much as 16 percent with the correct application of IEQ attributes.

IEQ parameters include thermal comfort and fresh ventilation air, which are also IAQ attributes. Additionally, IEQ includes seating ergonomics, access to day lighting, wall and ceiling colors, sound levels and surface textures. Many of these elements are completely outside of the HVAC realm, but are now being considered part of the overall IEQ landscape of which IAQ is an integral part.

Current building philosophy is strongly leaning toward the understanding that IEQ/IAQ problems are far easier and less expensive to prevent in the building construction phase than they are to resolve after the fact. Many issues such as noise from vibration can be virtually impossible to address in existing buildings.

The principal impacts to the IAQ portion of the IEQ universe include more zoning, more sensors for gases (such as CO, CO_2, NO_x, and VOCs), larger volumes of fresh filtered outside air, and better control over air motion in occupied spaces. These areas of technology will continue to take on a larger role in the design, installation, and maintenance of HVAC systems.

2.0.0 ◆ LONG-TERM AND SHORT-TERM EFFECTS OF POOR IAQ

Health effects resulting from poor IAQ may show up years after exposure. They may also occur after long or repeated periods of exposure. These long-term effects, which can include respiratory diseases and cancer, can be severely crippling or fatal. Long-term health effects are associated with indoor air pollutants such as **radon**, asbestos, and **environmental tobacco smoke (ETS)**. ETS is

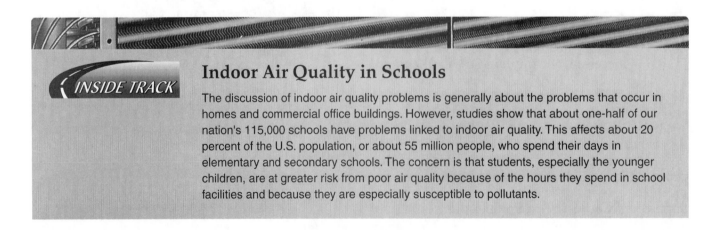

Indoor Air Quality in Schools

INSIDE TRACK

The discussion of indoor air quality problems is generally about the problems that occur in homes and commercial office buildings. However, studies show that about one-half of our nation's 115,000 schools have problems linked to indoor air quality. This affects about 20 percent of the U.S. population, or about 55 million people, who spend their days in elementary and secondary schools. The concern is that students, especially the younger children, are at greater risk from poor air quality because of the hours they spend in school facilities and because they are especially susceptible to pollutants.

GOING GREEN

What is LEED?

LEED stands for Leadership in Energy and Environmental Design. It is an initiative started by the U.S. Green Building Council (USGBC) with the goal of encouraging and accelerating global adoption of sustainable construction standards through a Green Building Rating System™. The rating system addresses six categories. Note that indoor environmental quality, which is covered by this module, is one of the six categories:

- Sustainable sites (SS)
- Water efficiency (WE)
- Energy and atmosphere (EA)
- Material and resources (MR)
- Indoor environmental quality (IEQ)
- Innovation and design process (ID)

LEED is a voluntary program that must be driven by the building owner. While most technicians will not have much, if any, say in whether a building is LEED-certified or not, they will have to maintain the systems to LEED standards in order for the building to retain its LEED rating. In other words, the LEED certification is for the life of the building, not just for the year it gets commissioned.

Additionally, the USGBC was created not as a government entity, but as a private, not-for-profit organization that is attempting to motivate and move market forces to a new place. This initiative is strong enough that ASHRAE has been motivated to create a new STD 189 in code language that is widely expected to be adopted in many localities as code. It is also potentially under consideration for incorporation into the next edition of the International Building Code (IBC).

defined as a combination of sidestream smoke from the burning end of a cigarette, cigar, or pipe and the exhaled mainstream smoke from the smoker.

Short-term or immediate effects of poor IAQ may appear after a single, high-dose exposure or repeated exposures. These effects can include irritation of the eyes, nose, and throat, and headaches, dizziness, and fatigue. The conditions are normally treatable, often by simply eliminating the person's exposure to the source of pollution, if it can be identified. When symptoms of a specific illness can be traced directly to airborne building contaminants, it is referred to as a **building-related illness**.

There are also situations in which occupants experience symptoms that do not fit the pattern of any particular illness and are difficult to trace to any one source. These conditions can be temporary, but some buildings have long-term problems. This situation is referred to by some as **sick building syndrome** or **new building syndrome**.

Sick building syndrome exists when more than 20 percent of a building's occupants complain during a two-week period of a set of symptoms, including headaches, fatigue, nausea, eye irritation, and throat irritation, that are alleviated by leaving the building and are not known to be caused by any specific contaminants. Some causes of these symptoms may include inadequate ventilation, chemical and biological contamination, and other nonpollutant factors such as temperature, humidity, and lighting.

New building syndrome refers to IAQ problems in new buildings. The definition is the same as for sick building syndrome.

3.0.0 ◆ GOOD INDOOR AIR QUALITY

The subject of indoor air quality is a complex one. Some debate exists as to what good IAQ is and what the best methods are of achieving it. For these reasons, you must make an effort to keep current on this subject by reading trade newspapers and journals.

ASHRAE Standard 62.1-2007, Ventilation for Acceptable Indoor Air Quality, specifies minimum ventilation rates and indoor air quality standards acceptable for occupants while minimizing the potential for adverse health effects. In order to

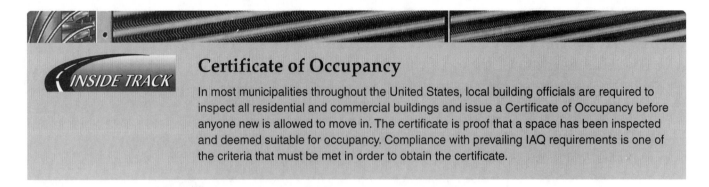

Certificate of Occupancy

In most municipalities throughout the United States, local building officials are required to inspect all residential and commercial buildings and issue a Certificate of Occupancy before anyone new is allowed to move in. The certificate is proof that a space has been inspected and deemed suitable for occupancy. Compliance with prevailing IAQ requirements is one of the criteria that must be met in order to obtain the certificate.

maintain a balance between IAQ and energy consumption, this standard incorporates both a ventilation rate procedure and an air quality procedure for ventilation design. The ASHRAE standard defines acceptable IAQ as follows:

Air in which there are no known contaminants at harmful concentrations as determined by cognizant authorities, and with which a substantial majority (80 percent or more) of the people exposed do not express dissatisfaction.

In order to evaluate a building and its systems with regard to IAQ, it is necessary to know the acceptable levels of contaminants. Again, there is much debate as to what good IAQ is, including the acceptable levels for each contaminant. Current standards can vary widely. For these reasons, specific levels are not given in this module. You should obtain copies of the current federal, state, and local standards for your specific location and use them along with this module.

4.0.0 ◆ SOURCES OF BUILDING CONTAMINANTS

Poor IAQ can result from a building's construction, or it can be caused by pollutants released from sources located in and/or outside of the building. Sources of air pollution can include:

- Building construction
- Human occupancy

- Building materials and furnishings
- HVAC and other building equipment
- Cleaning compounds and pesticides
- Contaminant sources located outside the building

4.1.0 Building Construction

Fresh outdoor air can enter a building by infiltration, natural ventilation, and mechanical ventilation. In infiltration, it enters the building through openings such as joints, cracks in walls, and cracks around windows and doors. In natural ventilation, air enters through open windows and doors. Air movement caused by infiltration and natural ventilation is caused by air temperature differences between indoors and outdoors. Mechanical ventilation involves the use of fans vented to the outdoors that remove air from certain rooms, such as kitchens and bathrooms. It also can be an air handling system that uses fans and dampers to continuously remove indoor air and distribute filtered and conditioned outdoor air to the rooms in the building. The rate at which outdoor air replaces indoor air is called the exchange rate. When there is little infiltration, natural ventilation, or mechanical ventilation, the air exchange rate is low, and pollutant levels can increase.

Older buildings normally allowed more than enough outdoor air to enter by infiltration and

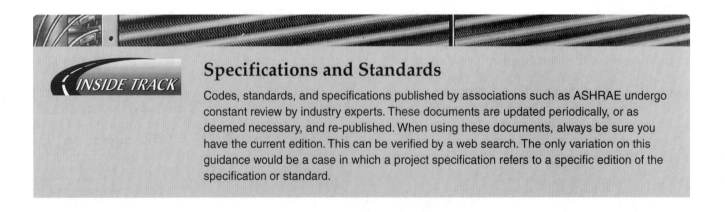

Specifications and Standards

Codes, standards, and specifications published by associations such as ASHRAE undergo constant review by industry experts. These documents are updated periodically, or as deemed necessary, and re-published. When using these documents, always be sure you have the current edition. This can be verified by a web search. The only variation on this guidance would be a case in which a project specification refers to a specific edition of the specification or standard.

natural ventilation to provide good IAQ. Newer buildings are constructed much tighter (*Figure 1*). They are sometimes so tight they create a problem. Without enough inward leakage of outdoor air by natural means or by mechanical ventilation, the indoor air can become unhealthy from internal pollutants. To the building's occupants, this is like living in an airtight box, where the absence of ventilation causes contaminants to accumulate.

4.2.0 Human Occupancy

Most people spend about 90 percent of their time indoors. We consume oxygen and emit carbon dioxide (CO_2). If a person spends one hour in a room without ventilation, there will be a 0.5 percent decrease in the oxygen level and a corresponding 230 percent increase in the CO_2 level. CO_2 concentrations of 400 ppm in an area are generally considered excellent air; 600 ppm, good air; 800 ppm, adequate air; and 1,000 ppm is minimally acceptable air. Experience has shown that CO_2 concentration levels above 1,000 ppm contribute to poor IAQ. At concentrations of about 1,200 ppm, people tend to get drowsy, have headaches, and/or function at lower activity levels. For this reason, the sensing and control of CO_2 concentrations in high occupancy areas of a

403F01.EPS

Figure 1 ◆ Modern tightly constructed residence.

building are important tasks. CO_2 concentrations are widely used as an indicator of IAQ. However, low CO_2 levels do not necessarily mean no IAQ problems exist. There may be other contaminants in the air.

Other kinds of contamination obtained from humans are bioaerosols emitted from the digestive process. In addition, our skin sheds and leaves airborne particles, and our bodies emit

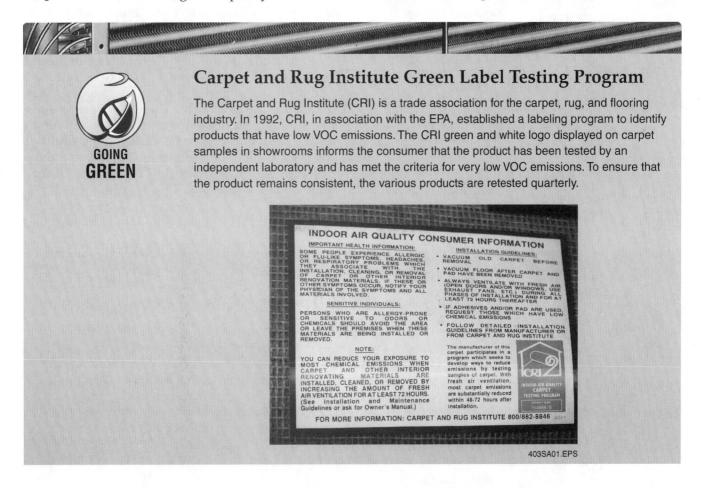

GOING GREEN

Carpet and Rug Institute Green Label Testing Program

The Carpet and Rug Institute (CRI) is a trade association for the carpet, rug, and flooring industry. In 1992, CRI, in association with the EPA, established a labeling program to identify products that have low VOC emissions. The CRI green and white logo displayed on carpet samples in showrooms informs the consumer that the product has been tested by an independent laboratory and has met the criteria for very low VOC emissions. To ensure that the product remains consistent, the various products are retested quarterly.

403SA01.EPS

various odors (*Figure 2*). When we sneeze or cough, we send bacteria and viruses into the air. Personal care products are another source of air pollution. Activities such as smoking and cooking also add to the problem.

4.3.0 Building Materials and Furnishings

Building materials and furnishings are the source of many contaminants. These materials and furnishings can release **volatile organic compounds (VOCs)** and/or have high particulate shed rates. VOCs include a wide variety of compounds and chemicals found in paints, adhesives, sealants, furniture, carpeting, and vinyl wall coverings that vaporize readily at normal air pressure and room temperature. This is called **off-gassing**. The most common VOC is **formaldehyde**, which is heavily used in particle board, plywood, and some foam insulation. It is a colorless, pungent byproduct of hydrocarbons that can cause irritation of the eyes and upper air passages. In high concentrations, it may cause asthma attacks. The rate at which formaldehyde is released is accelerated by heat and sometimes by the humidity level.

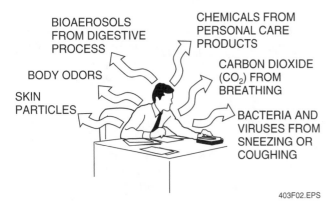

BIOAEROSOLS FROM DIGESTIVE PROCESS

BODY ODORS

SKIN PARTICLES

CHEMICALS FROM PERSONAL CARE PRODUCTS

CARBON DIOXIDE (CO_2) FROM BREATHING

BACTERIA AND VIRUSES FROM SNEEZING OR COUGHING

403F02.EPS

Figure 2 ◆ Human sources of air contamination.

New building materials give off a great deal of moisture and VOCs. The level of VOCs from these materials may remain very high in a new building unless ventilation is used to dilute them during and after construction. New carpet, padding, and adhesives are a major source of VOCs. When practical, new carpeting should be aired out for 24 hours before installation to reduce the amount of VOCs emitted, then for 72 more hours before occupancy of the building or area.

Unlike the materials used in the past to furnish buildings, especially business offices, the fabrics and soft materials used today act as nutrients for microorganisms. Tile floors, painted walls, hard ceilings, and metal furniture used in older buildings have been replaced with carpeting, wallpaper, acoustical tiles, and upholstered furniture in newer buildings. Depending on the temperature and humidity levels maintained in a building, these materials can provide the ideal environment for mold, mildew, fungi, and bacteria.

Another potential source of contamination is asbestos, which is sometimes found in fireproofing materials, thermal insulation, floor tiles, and coverings for structural members in buildings constructed from 1930 to the mid-1970s. Asbestos is known to cause cancer and must therefore be treated with great caution. Where asbestos is used, it is not normally a problem unless the surface of the material is deteriorating or being abraded, allowing asbestos fibers to be released into the air. A material that readily releases fibers is referred to as **friable**. The EPA recommends that undamaged asbestos materials be left alone if they are not likely to be disturbed. Qualified contractors must be used to control any activities that may disturb asbestos and for asbestos removal and cleanup.

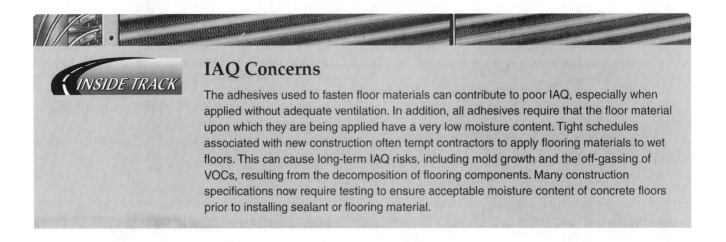

IAQ Concerns

INSIDE TRACK

The adhesives used to fasten floor materials can contribute to poor IAQ, especially when applied without adequate ventilation. In addition, all adhesives require that the floor material upon which they are being applied have a very low moisture content. Tight schedules associated with new construction often tempt contractors to apply flooring materials to wet floors. This can cause long-term IAQ risks, including mold growth and the off-gassing of VOCs, resulting from the decomposition of flooring components. Many construction specifications now require testing to ensure acceptable moisture content of concrete floors prior to installing sealant or flooring material.

4.4.0 HVAC and Other Building Equipment

This section will discuss HVAC and refrigeration equipment, combustion (fuel-burning) equipment, and office equipment and how they can affect IAQ. All of these products can contribute to the pollution of indoor air.

4.4.1 HVAC and Refrigeration Equipment

Widespread use of total comfort heating/cooling systems with forced-air distribution is another cause of increased air pollution. These systems provide a comfortable building environment but, if poorly maintained, can also act to distribute pollutants. HVAC and refrigeration equipment leaks can release toxic refrigerants into the air. Chilled-water cooling coils; evaporator or cooling coil condensate drip pans; humidifiers and dehumidifiers; cooling towers; and evaporative coolers and condensers are all moisture reservoirs that can provide breeding grounds for **biological contaminants**. Biological contaminants, also known as **microbiological contaminants**, include bacteria, fungi, viruses, algae, insect parts, and dust. The sources of these pollutants include the outdoors for pollen, along with human and animal occupants for viruses, bacteria, hair, and skin flakes.

Fiberglass duct board and insulation liners can also trap moisture and contaminants and become a breeding ground for biological contaminants.

Shutting down HVAC systems on weekends to save energy, coupled with water spills, leaks, and dripping plants in the building, can result in the building acting as an incubator. Rust or discoloration inside metal ductwork can be a sign of excessive moisture resulting from faulty humidifier operation.

Some highly publicized cases of Legionnaire's disease and **Pontiac fever** have been attributed to poorly maintained HVAC systems that allowed the incubation and distribution of disease-causing microorganisms.

4.4.2 Combustion (Fuel-Burning) Equipment

Problems can arise in a building from the mixed atmosphere of gases that may exist near, or migrate from, combustion equipment. All products of combustion can be dangerous and may compound preexisting health problems. Furnaces, boilers, space heaters, woodstoves, gas stoves, and fireplaces can produce pollutants such as carbon monoxide (CO), oxides of nitrogen (NO_x), sulfur dioxide (SO_2), and airborne particles.

Carbon monoxide is a highly toxic gas that results from incomplete combustion. It is colorless, tasteless, and nonirritating, yet high levels of CO are extremely harmful. CO can slowly build up in the bloodstream. There, it combines with the hemoglobin in blood and replaces the oxygen until there is too little oxygen to support life. Death from CO poisoning can happen suddenly.

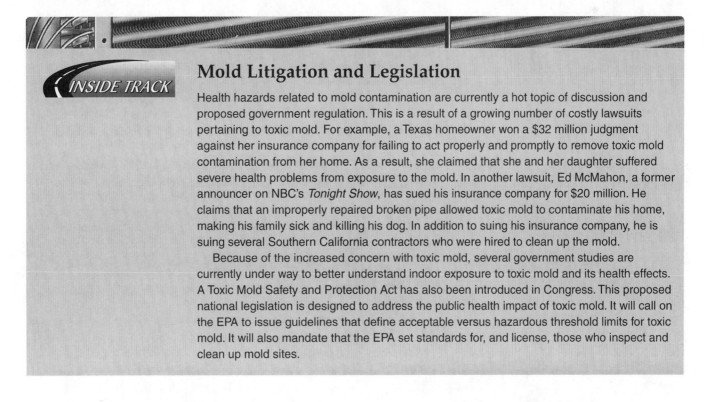

INSIDE TRACK

Mold Litigation and Legislation

Health hazards related to mold contamination are currently a hot topic of discussion and proposed government regulation. This is a result of a growing number of costly lawsuits pertaining to toxic mold. For example, a Texas homeowner won a $32 million judgment against her insurance company for failing to act properly and promptly to remove toxic mold contamination from her home. As a result, she claimed that she and her daughter suffered severe health problems from exposure to the mold. In another lawsuit, Ed McMahon, a former announcer on NBC's *Tonight Show*, has sued his insurance company for $20 million. He claims that an improperly repaired broken pipe allowed toxic mold to contaminate his home, making his family sick and killing his dog. In addition to suing his insurance company, he is suing several Southern California contractors who were hired to clean up the mold.

Because of the increased concern with toxic mold, several government studies are currently under way to better understand indoor exposure to toxic mold and its health effects. A Toxic Mold Safety and Protection Act has also been introduced in Congress. This proposed national legislation is designed to address the public health impact of toxic mold. It will call on the EPA to issue guidelines that define acceptable versus hazardous threshold limits for toxic mold. It will also mandate that the EPA set standards for, and license, those who inspect and clean up mold sites.

Its victims are overcome and helpless before they realize they are in danger. Signs of CO poisoning include headaches, dizziness, fatigue, and nausea. Victims often think they have the flu or a common cold because the symptoms are similar. *Table 1* shows examples of CO levels and related symptoms.

Nitric oxide (NO) and nitrogen dioxide (NO_2) are two oxides of nitrogen (NO_x). All combustion processes can produce NO_x. Oxides of nitrogen form acids in the Earth's lower atmosphere, where they cause acid rain. Also, NO_x and hydrocarbons react with sunlight to produce smog. NO and NO_2 can also displace oxygen in the blood. NO_2 can irritate the skin and the mucous membranes in the eyes, nose, and throat. High levels of NO_2 may result in burning and pain in the chest, coughing, and/or shortness of breath. Oxides of nitrogen released from incomplete combustion can lodge in the lungs and irritate or damage lung tissue. They may also cause cancer.

Chimneys and flues that are poorly installed and maintained, as well as furnaces with cracked heat exchangers, are all sources of pollutants. Negative pressures in tight buildings can cause backdrafting of a combustion appliance and the distribution of combustion byproducts throughout the building. In a warm-air furnace, a cracked heat exchanger can cause a buildup of toxic gases, including CO. These gases would be distributed by the blower into the conditioned space, causing sickness or death.

Visually inspect furnace heat exchangers to make sure that they do not have cracks or pinhole leaks caused by corrosion. However, tiny cracks that may expand as the furnace heats up may not be detected by a visual inspection. A better method is to test and compare combustion gas readings taken before and after the furnace blower has turned on. You should suspect a cracked heat exchanger if there is a change of O_2 concentration in the flue gases of greater than 0.5 percent or a change in the CO level greater than 25 ppm.

Any blockages in a chimney can cause inefficient combustion and produce dangerous levels of gases. Blockages can be caused by birds' nests, chimney deterioration, soot buildup, and other natural causes. High levels of moisture in a chimney can cause the lining materials to decompose and create restrictions.

The routine and proper maintenance of combustion equipment, such as furnaces, flues, and chimneys, is the best way to prevent exposure to CO and NO_x. At a minimum, the equipment should be inspected, cleaned, and adjusted annually. All needed repairs should be made promptly.

4.4.3 Office Equipment

In office buildings, pollutants can be generated by various pieces of office equipment such as photocopiers and copy papers. These devices are sources of irritants such as **ozone**. Ozone is an

Table 1 CO Levels and Related Symptoms

Concentration of CO in the Air	Inhalation Time and Toxic Symptoms Developed
9 ppm (0.0009%)	The maximum allowable concentration for short-term exposure in the living area according to ASHRAE
35 ppm (0.0035%)	The maximum allowable concentration for continuous exposure in any eight-hour period according to federal law
200 ppm (0.02%)	Slight headaches, fatigue, dizziness, nausea after two to three hours
400 ppm (0.04%)	Frontal headaches within one to two hours, life-threatening after three hours; also, the maximum ppm in flue gas (on a free-air basis) according to the EPA
800 ppm (0.08%)	Dizziness, nausea, and convulsions within 45 minutes; unconsciousness within two hours; death within two to three hours
1,600 ppm (0.16%)	Headache, dizziness, and nausea within 20 minutes; death within one hour
3,200 ppm (0.32%)	Headache, dizziness, and nausea within 5 to 10 minutes; death within 30 minutes
6,400 ppm (0.64%)	Headache, dizziness, and nausea within one to two minutes; death within 10 to 15 minutes
12,000 ppm (1.28%)	Death within one to three minutes

unstable, poisonous oxidizing agent that has a strong odor and is irritating to the mucous membranes and the lungs. It is formed in nature when oxygen is subjected to electric discharge or exposure to ultraviolet radiation. It is also generated in devices such as photocopiers, electronic air cleaners, and other equipment that uses high voltage.

4.5.0 Cleaning Compounds and Pesticides

The use of cleaning compounds and pesticides is another source of contaminants in a building. Pesticides are sources of many organic compounds whose effects range from minor irritations to cancer-causing potential. Long-term damage to the liver and the central nervous system is also possible in extreme cases of exposure. The safest way to protect against exposure from chemicals and pesticide pollutants is to always read the MSDS and/or labels so that the specific health hazards related to the product are understood. The product should be used only as directed. If in doubt, contact the EPA for more information.

Chemicals or pesticides used or stored in a building may cause some occupants to suffer from **multiple chemical sensitivity (MCS)**. MCS is a medical condition found in some individuals who are vulnerable to exposure to certain chemicals and combinations of chemicals. There is still debate as to whether or not MCS really exists.

4.6.0 Contaminant Sources Located Outside the Building

Contaminants from outside a building, especially in urban areas, can be a major cause of poor indoor air quality (*Figure 3*). Contaminants may come from aboveground urban or industrial air pollution sources or from belowground sources such as pesticides, fertilizers, and soil gases such as radon.

Contaminant sources include the following:

- Building ventilation air intakes located too close to exhaust gas, loading docks, or kitchen/bathroom exhausts
- Short-circuited HVAC system exhaust
- Exhaust from vehicles
- Exhaust and other airborne discharges from neighboring manufacturing plants
- Urban smog
- Radon
- Pesticides or fertilizers

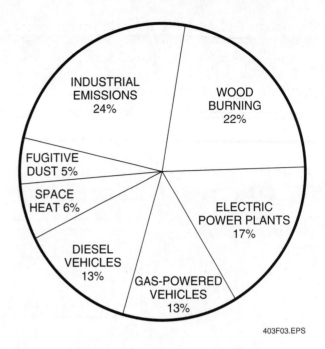

403F03.EPS

Figure 3 ◆ Sources of urban outdoor air pollutants.

- Moisture or standing water that promotes microbial growth
- Neighboring building construction or demolition

4.6.1 Radon Contamination and Testing

Of the sources of outdoor pollution, radon is the least understood. The most common source of radon is uranium in soil and rocks. Another common source of radon is well water. As uranium naturally breaks down, it releases radon, which is a colorless, odorless, radioactive, chemically inert gas. Because it is inert, radon does not combine with other materials in the environment; therefore, it radiates freely from its sources into the atmosphere. It also enters buildings through cracks in concrete, wall and floor joints, hollow concrete block walls, or openings such as those for sewer pipes and sump pumps (*Figure 4*). Radon from well water is released into the air in a home when water is used for cooking, bathing, and other activities. The level of radon can vary depending on the building's construction and location. Generally, the fewer the cracks and openings in a building's foundation, the better the chances of preventing radon entry. However, sealing the various parts of a building foundation is not completely effective by itself. This is because some openings in the building shell may not be accessible, and new openings can develop with time.

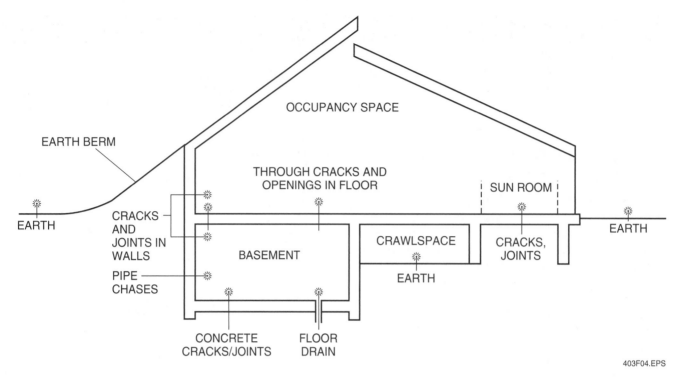

Figure 4 ◆ Passage of radon gas into a building.

Radon produces radioactive decay products, called daughters, that emit high levels of alpha radiation. They are chemically active, which allows them to attach themselves to tobacco smoke and dust particles in the air. When inhaled, these smoke and dust particles can lodge in the respiratory system where they subject the lung tissue to radiation. Radon daughters have relatively short half-lives, so that after being deposited in the lung, they will successively go through their radioactive decay in an hour or less. Currently, there are no reported instances of radon-related problems traced to a short-term exposure period. However, major health organizations agree that extended exposure to radon can increase the risk of lung cancer.

Testing for radon is normally done using electronic radon monitors and/or test kits (*Figure 5*). Inexpensive passive test kits are available for use in residences. Approved test kits must have passed the EPA's testing program or be state-certified. Some of these tests measure radon levels

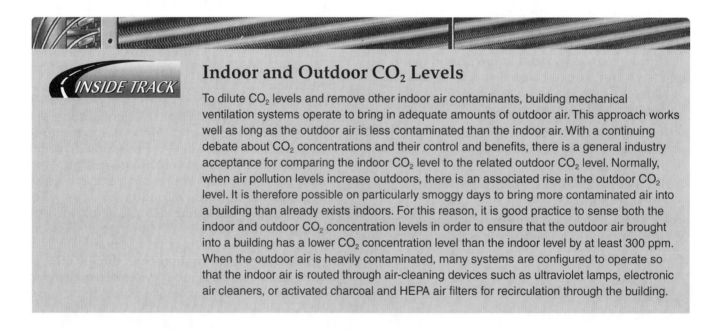

INSIDE TRACK

Indoor and Outdoor CO₂ Levels

To dilute CO_2 levels and remove other indoor air contaminants, building mechanical ventilation systems operate to bring in adequate amounts of outdoor air. This approach works well as long as the outdoor air is less contaminated than the indoor air. With a continuing debate about CO_2 concentrations and their control and benefits, there is a general industry acceptance for comparing the indoor CO_2 level to the related outdoor CO_2 level. Normally, when air pollution levels increase outdoors, there is an associated rise in the outdoor CO_2 level. It is therefore possible on particularly smoggy days to bring more contaminated air into a building than already exists indoors. For this reason, it is good practice to sense both the indoor and outdoor CO_2 concentration levels in order to ensure that the outdoor air brought into a building has a lower CO_2 concentration level than the indoor level by at least 300 ppm. When the outdoor air is heavily contaminated, many systems are configured to operate so that the indoor air is routed through air-cleaning devices such as ultraviolet lamps, electronic air cleaners, or activated charcoal and HEPA air filters for recirculation through the building.

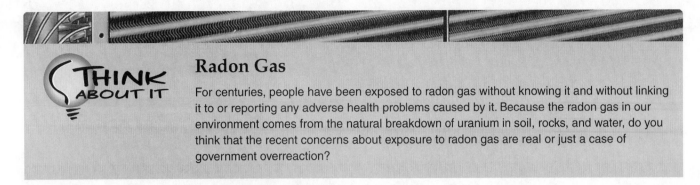
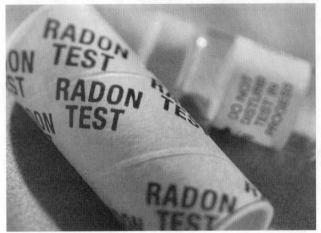

403F05.EPS

Figure 5 ◆ Radon test kit.

over two to three days; others measure it over one to three months. Professional testers may use a method of active sampling that involves the use of a membrane filter and a battery-operated air pump to collect particulate matter to which the radon daughters are attached. After a predetermined time has elapsed, an alpha-particle detector is used to measure the radon level in picocuries per liter (pCi/L). This value is then converted and reported as working levels.

There is some debate regarding the acceptable radon level. The Radon Abatement Act of 1988 established a national goal of achieving indoor radon levels that are no greater than outdoor levels. The EPA suggests a level not to exceed 4 pCi/L. Other countries are using higher threshold levels of about 20 pCi/L.

The most effective measures for reducing radon are the ones that limit soil gas entry into the building. One common method uses subslab depressurization (*Figure 6*). This method uses a radon mitigation exhaust system to reduce the pressure below the floor slab so that the air between the building substructure and the soil tends to flow out of rather than into the building. Positive basement pressurization is a similar method used with some success that keeps the basement indoor air pressure slightly higher than the soil gas pressure. This prevents radon gas, as well as other outdoor contaminants, from entering the building because the radon gas can only flow into an area of negative pressure.

5.0.0 ◆ ELEMENTS OF A BUILDING IAQ INSPECTION/SURVEY

It is not easy to evaluate the performance of a building and its systems for IAQ problems. IAQ problems can be complicated because of personal preferences, the complexity of the building design, and the fact that the evaluation standards and methods may be inconclusive. Ideally, the IAQ inspection team should include, at a minimum, an HVAC engineer and an industrial hygienist. To successfully solve IAQ problems, an organized approach must be followed. The elements of this approach include:

- Problem description
- Site visit and building walk-through
- Building HVAC and ventilation system inspection
- Air sampling and testing for specific contaminants
- Interpretation of test results and corrective actions

5.1.0 Problem Description

To determine the scope of an IAQ problem, perform initial interviews with building staff and/or the individual(s) who requested an investigation. The occupants' symptoms, as well as the location and duration of those symptoms, can be obtained via interviews or the use of questionnaires. Ideally, this information will help to define and determine the complaints. It may also point out if the problem is localized to a particular part of the building and any other relevant circumstance, such as weather, time of day, day of week, building occupancy levels, or activities that improve or worsen the problem.

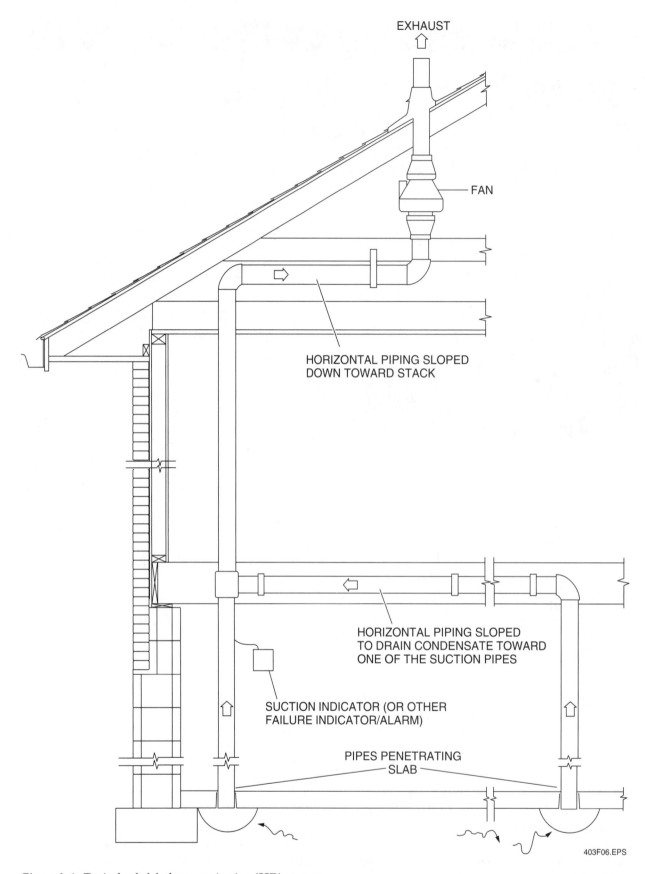

EXHAUST

FAN

HORIZONTAL PIPING SLOPED
DOWN TOWARD STACK

HORIZONTAL PIPING SLOPED
TO DRAIN CONDENSATE TOWARD
ONE OF THE SUCTION PIPES

SUCTION INDICATOR (OR OTHER
FAILURE INDICATOR/ALARM)

PIPES PENETRATING
SLAB

403F06.EPS

Figure 6 ◆ Typical subslab depressurization (SSD) system.

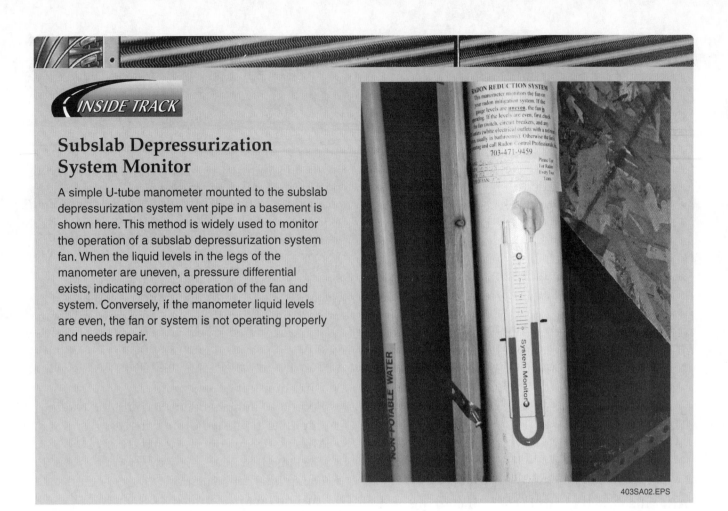

Subslab Depressurization System Monitor

A simple U-tube manometer mounted to the subslab depressurization system vent pipe in a basement is shown here. This method is widely used to monitor the operation of a subslab depressurization system fan. When the liquid levels in the legs of the manometer are uneven, a pressure differential exists, indicating correct operation of the fan and system. Conversely, if the manometer liquid levels are even, the fan or system is not operating properly and needs repair.

403SA02.EPS

5.2.0 Site Visit and Building Walk-Through

A site visit and building walk-through are needed to perform a preliminary evaluation of the overall condition of the building and its operating systems. A set of building plans and specifications is useful during this walk-through, as are any plans or drawings of major upgrades or changes. These drawings can be valuable in answering questions and pinpointing potential problem areas.

To be sure that no potential contamination source or component of the building is missed, use some type of formal checklist to document the results of the evaluation. An official EPA-produced guide entitled *Building Air Quality* (*Figure 7*) is available. It provides the latest information about IAQ problems and how to prevent or correct them. It also contains 15 practical forms and checklists. This guide comes bound in a looseleaf binder so that the forms can be easily removed for on-site use. Because this is a government guide, it can be reproduced without permission.

403F07.EPS

Figure 7 ◆ Checklist for building IAQ evaluation.

Normally, a building's structure should be evaluated relative to the following:

- Tight building construction
- Low ventilation rates
- Rooms properly configured for their intended use
- Adequate separation between activities to control temperature, air movement, odors, liquids, noise, light, or vibration
- Recent addition of partitions or fire walls
- Reduced natural ventilation due to sealed windows or doors
- Windows or doors that have been added, allowing outside pollution to enter
- Asbestos, formaldehyde, and other harmful substances in building materials
- Insulation or recladding added to walls
- Recent use of caulks or sealants
- New openings for doors, ducts, and pipes that allow the transfer of dust, dirt, vapors, and/or odors between occupied spaces
- New furnishings or carpeting

5.3.0 Building HVAC Equipment and Ventilation System Inspection

After the structure has been evaluated, inspect the HVAC equipment, ventilation, and other building systems for potential sources of contamination. Check any building comfort zoning scheme to make sure that system control of any one zone does not affect any other zone. Pay attention to the number of occupants in the various zones so that outside ventilation air is properly distributed without using excess energy. If the building uses a variable air volume (VAV) energy management system, make sure that its layout and control arrangement allow for adequate ventilation to occupied areas when operating at minimum capacities. All systems should supply at least the minimum quantity of outside air required by the current issue of *ASHRAE Standard 62.1* or local codes, whichever is greater.

If a building has an adequate ventilation system, it still may not supply enough air to dilute pollutants. This can happen if the system is shut down too often or for too long, such as during evenings and weekends. Other causes are poor air distribution or mixing, installation flaws, and incorrect system balance. Occupant intervention, such as homemade cardboard diffusers attached to ceiling vents to divert air, boxes or supplies piled near the vents, or office partitions that block airflow may also be the cause.

WARNING!

A full system inspection may expose the inspector to contaminants such as mold and dust. Be sure to wear the company-prescribed PPE, including respiratory protection, when performing this work.

Start examining the ventilation system at the outside air intake. Air intakes are often placed too close to building exhausts or sources of pollution, such as cooling towers, loading docks, and garbage bins. If necessary, a smoke tube may be used to check airflow problems. Make sure that exhaust air is not being short circuited and reintroduced into the building's air supply. Check for areas of negative pressure in the building that allow for migration of ground-level contaminants such as vehicle exhaust. Check the HVAC filtration system for efficiency, fit, condition, and the replacement schedule. Also check for the following:

- Dirty humidifier reservoirs
- Poorly draining condensate pans and trays
- Improperly pitched drain pans
- Missing or improperly designed traps
- Torn insulation
- Damp internal insulation
- Rusting internal surfaces
- Mold on internal surfaces
- Improperly maintained dampers, actuators, or linkages
- Poorly cleaned and maintained pneumatic control systems
- Incorrectly wired or inoperative fans and blowers
- Materials stored within the air handling equipment

5.4.0 Air Sampling and Testing for Specific Contaminants

Testing can be time consuming and expensive, especially when testing for specific sources of contamination. Taking air samples and testing for various contaminants are usually done over an extended period of time (a week, a month, or longer) in order to get enough samples to yield reliable results.

5.4.1 Air Sampling

Basic air sampling typically involves measuring carbon dioxide (CO_2) levels both inside and outside a building. Sampling usually begins in the morning with samples taken at all sample locations, including outdoors. Record the CO_2 levels in ppm by specific location and time of day. The morning measurements will be compared to those taken throughout the day from all the sample locations.

The number of sampling periods is determined by the types and duration of activities that occur in the building, including lunch breaks. Make sure to take a set of samples at the end of the work day just before everyone starts leaving the building. This is important because the increased traffic usually results in more air exchange than during the main part of the day. In order to get a complete picture of the building air pattern, CO_2 measurements should also be made at night when the building is mainly unoccupied.

The building temperature and humidity should be checked at various times and locations throughout the day and night. The airflow at the building air outlets and return grilles should also be checked. This will help determine whether the air outlets are working and whether the airflow is properly directed.

5.4.2 Testing for Specific Contaminants

Testing for specific **microbial contaminants** must be performed by qualified personnel, such as microbiologists or industrial hygienists. Microbial contaminants include bacteria, fungi, viruses, algae, insect parts, pollen, and dust. Their sources include wet or moist walls, duct, duct liner, fiberboard, carpet, and furniture.

Even though sampling for airborne microorganisms is common, the results are often inconclusive. Airborne spores are often only present in large quantities for short periods of time. Spores released into the air depend on the current growing conditions. If growing conditions are intermittent, as is often the case in HVAC systems, then the release of the spores is also intermittent. Significant levels of airborne microbial contamination will be found only if sampling occurs during the time that spores are being released into the air.

A nonairborne method used to test for the presence of microbial contamination is called surface sampling. This method is also referred to as microbial sampling. It provides a historical reference of previous growth and indicates the potential for future growth.

Surface sampling uses sampling strips. Each strip has a pad on one end that contains a growth media for a wide range of organisms. The strip is activated, and one square inch of HVAC duct or other building surface is wiped with the pad. Two strips, each containing a different growth medium, are used at each location. After taking the sample, the strips are placed in sterile envelopes and returned to the manufacturer's laboratory for incubation and analysis. A report is returned with the results stated in colony counts pre square inch, as well as being rated on a multi-step severity index ranging from very low to severe. To reduce or eliminate occupants' complaints resulting from high levels of surface microbial contamination, clean and sanitize building HVAC system components or building surfaces.

5.5.0 Interpreting Test Results and Corrective Actions

Corrective actions are determined by what has been found during the site/building walk-through, HVAC and ventilation system inspections, and/or the air sampling and testing phases of evaluation. Depending on the complexity of the problem, it may or may not be necessary to call in expert help, such as that of an industrial hygienist.

6.0.0 ◆ ACHIEVING ACCEPTABLE INDOOR AIR QUALITY

Acceptable indoor air quality can be achieved through awareness and control of the following related areas:

- Initial building design
- Ventilation control
- Thermal comfort control
- Control of chemical contaminants
- Control of microbial contaminants
- Scheduled building maintenance

6.1.0 Initial Building Design

The methods used to conserve energy in a building can impact indoor air quality. In some instances, they can improve the indoor air and result in better comfort and productivity. In other cases, they can degrade the indoor conditions and result in discomfort, sickness, and lost productivity. As buildings become more energy efficient, they become less forgiving environmentally. Energy conservation is necessary, but it must be done in a manner that also provides for acceptable indoor air quality.

Achieving good indoor air quality begins with the design of a new building, as described here:

- If possible, locate the building on a hill rather than in a valley. Stay away from major highways and parking lots. Locate upwind from a power plant, chemical plant, or other industrial pollution source.

- Incorporate HVAC equipment and mechanical ventilation systems that best meet the needs of the occupants in terms of comfort and performance, while reducing energy consumption and costs.
- Use building materials and furnishings that will keep indoor air pollution to a minimum.
- Consider outside air duct locations, prevailing wind conditions, and neighboring pollutant sources in order to get proper ventilation damper orientation and adequate filtration systems incorporated into the HVAC equipment.
- Ensure that combustion appliances are properly vented and receive enough supply air.
- Provide proper drainage and seal foundations.
- Use radon-resistant construction techniques.
- In landscaping, avoid the use of shrubs and trees, such as olive, acacia, oak, and maple trees, that produce heavy pollen and can aggravate allergies.
- Provide adequate HVAC system access, lighting, and work platforms.

6.2.0 Ventilation Control

Ventilation control can be used to correct or prevent poor indoor air quality. Both outdoor air and recirculated air can be used for this purpose.

6.2.1 Using Outdoor Air

The use of additional outdoor ventilation is currently one of the best methods of correcting and preventing problems related to poor IAQ. Adequate ventilation pertains more to the level of CO_2 in a building than the level of oxygen in the air. When indoor air is stale or stuffy, or when drowsiness sets in, it is not a shortage of oxygen causing the problem; it is an excess of CO_2. If CO_2 or other pollutants are accumulating inside a building, ventilation control is used to bring in more outside air for dilution. Even if a specific contaminant such as formaldehyde is identified as the cause of a problem, dilution can still be the most practical way of reducing exposure.

Outdoor ventilation air should be adequately distributed to all areas of a building during the entire time it is occupied. The ventilation requirements of residential and commercial buildings are constantly changing in response to the latest IAQ standards established by ASHRAE. *ASHRAE Standard 62.1-2007* recommends minimum outdoor air ventilation rates for occupied spaces.

Proper balancing of the air supply and exhaust system may be all that is needed to achieve adequate airflow or air quality in buildings where new walls, partitions, or room dividers have been erected. Proper balancing may also cure problems where insufficient amounts of outdoor makeup air have created negative pressure areas in a building. Negative pressure areas are areas that allow untreated air and/or contaminants to infiltrate from outside. They also allow for the migration of odors or contaminants between areas within a building.

Some other correctable causes of poor airflow or low-quality indoor air related to ventilation problems are as follows:

- Closed or obstructed air outlets or diffusers prevent adequate airflow to the supplied area.
- Outdoor dampers operate improperly. If mechanical ventilation is on, check outdoor air dampers to make sure they are open. If the building is lightly occupied, make sure the outdoor dampers do not close beyond the minimum position; these dampers may have been set to the closed position deliberately to save energy, or they may have been closed automatically by a faulty control device.
- Supply or exhaust fans are inoperative; blowers or fans rotate in the wrong direction.
- Filters in air handling units are dirty.

Proper ventilation does not always guarantee good IAQ. Building CO_2 levels are often used as the basis for controlling building ventilation. However, use of such a system does not necessarily mean that the building is free of IAQ problems. For example, the outside air may have more CO_2 than might be expected. Outdoor concentrations of 400 ppm CO_2 have regularly been recorded in large urban areas. This can create a problem if a building is located in a busy traffic area and outside air is drawn into the building in an attempt to lower the CO_2 level. Another example is in supermarkets, which need extra ventilation during peak customer periods, but cannot lower the ventilation too much at night, even if the CO_2 level is low. This is because chemicals and cleaners used at night can be absorbed into the food if the fumes are not diluted and dispersed.

6.2.2 Using Recirculated Air

There is some controversy as to whether building air should be recirculated to make up part of a building's demand for ventilation. Recirculated air is sometimes considered less healthy than outside air. However, recirculated filtered air is used in most ultraclean environments, such as manufacturing clean rooms. In some cases, recirculated

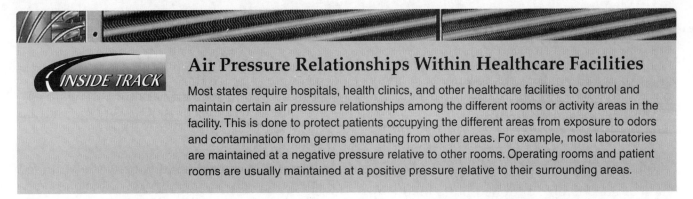

air may actually be better for occupants than outdoor air, especially when a building is located in a highly polluted urban area.

Recirculation substitutes clean recirculated air for some portion of the outdoor air normally used to ventilate a building or space. For instance, assume the level of outdoor air required in a building is 15 cfm per person. To meet this requirement, a reduced outdoor airflow volume of 5 cfm/person might be used along with 10 cfm/person of filtered recirculated air.

Filtration methods can vary widely. Most use multistage filtering that obtains filter efficiencies needed to achieve good IAQ. One advantage of using recirculated air is that the size of building boilers, chillers, and other mechanical equipment can be smaller than the sizes needed to accommodate the use of higher outdoor airflow levels. This is because recirculated air has already been processed through the HVAC system.

6.3.0 Thermal Comfort Control

The human body is not always sensitive enough to tell the difference between slight thermal discomfort and actual IAQ problems. People may complain that the room is stuffy or the air quality is poor, rather than too warm. Generally, IAQ problems are judged to be worse when the ambient room temperature is above 75°F. Stuffiness can be the result of temperature stratification, which occurs when air distribution is inadequate. When air is improperly distributed, poor mixing occurs. Thermostats may not detect the changing room conditions until the temperature in the room becomes uncomfortable. Humidity control is also related to thermal comfort. In summer, the human body can accept higher temperatures if the humidity level is lower than 60 percent relative humidity (RH). In winter, we are comfortable at lower temperatures when the humidity level is greater than 30 percent RH.

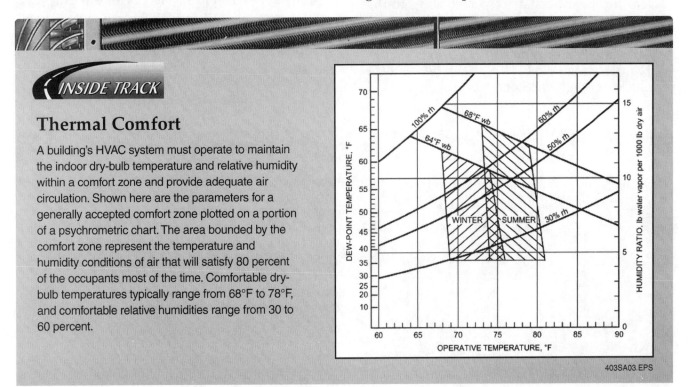

403SA03.EPS

6.4.0 Controlling Chemical Contaminants

IAQ problems related to chemical contaminants can be derived from the organic gases commonly emitted from building materials, as well as cleaning and maintenance products. They can also be derived from pesticides used to kill building pests or those used on lawns and gardens that then drift or are tracked into a building. Exposure to chemical contaminants can usually be eliminated or adequately controlled if the following guidelines are observed:

- Use local exhaust systems where needed to trap and remove contaminants generated by specific processes or equipment such as office machines. Exhaust room air outdoors from areas where solvents are used.
- Areas being remodeled, painted, or recarpeted should be temporarily isolated from other occupied areas in the building. If possible, this includes temporarily isolating any involved HVAC systems. Whenever possible, perform this type of work during evenings and weekends. Also supply the maximum amount of ventilation to the areas on a 24-hour basis to help eliminate any contaminants.
- Apply pesticides and disinfectants only when a building is unoccupied, then thoroughly ventilate the building before it is reoccupied.
- Apply paints, paint strippers, and other solvents, wood preservatives, aerosol sprays, and cleaning products strictly according to the manufacturer's directions and only in the recommended quantities.
- Make sure that outside air intakes or openings are not located close to places where motor vehicle or other emissions collect.

6.5.0 Controlling Microbial Contaminants

Indoor air quality problems related to microbial contaminants are derived from wet or moist sources in building materials, furnishings, and/or equipment. These contaminants can usually be eliminated or controlled if the following guidelines are observed:

- Maintain relative humidity between 30 and 60 percent in all occupied spaces of the building.
- During the summer, make sure cooling coils are operating at low enough temperatures to properly dehumidify the conditioned air.
- Install and use fans in kitchens and bathrooms.
- Regularly check that humidifiers, filters, and sump pumps are clean. Water draining from this equipment can stagnate and promote the growth of microbial contaminants.
- Promptly detect and repair all water leaks. Eliminate or clean areas where water collects.
- Prevent and/or correct any causes of stagnant water accumulation around cooling coils and air handling units.
- If contamination has occurred in the plenum or ductwork downstream from a heat exchanger, add filtering downstream to better filter the air before it is introduced into occupied areas.
- Clean and disinfect surfaces where the accumulation of moisture has caused microbial growth, such as in drain pans and cooling coils. Use only approved biocide agents for this purpose.
- Replace, rather than disinfect, any water-damaged porous furnishings, including carpets, upholstery, and ceiling tiles.

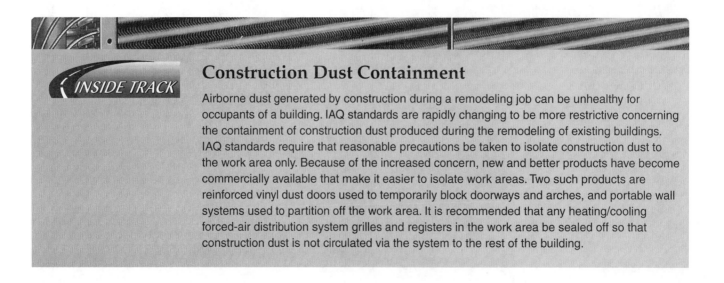

INSIDE TRACK

Construction Dust Containment

Airborne dust generated by construction during a remodeling job can be unhealthy for occupants of a building. IAQ standards are rapidly changing to be more restrictive concerning the containment of construction dust produced during the remodeling of existing buildings. IAQ standards require that reasonable precautions be taken to isolate construction dust to the work area only. Because of the increased concern, new and better products have become commercially available that make it easier to isolate work areas. Two such products are reinforced vinyl dust doors used to temporarily block doorways and arches, and portable wall systems used to partition off the work area. It is recommended that any heating/cooling forced-air distribution system grilles and registers in the work area be sealed off so that construction dust is not circulated via the system to the rest of the building.

7.0.0 ◆ IAQ AND ENERGY-EFFICIENT SYSTEMS AND EQUIPMENT

This section describes IAQ and energy-efficient systems and equipment. Among the methods that can be used to improve indoor air quality are automated building management systems, improved air handling units, unit ventilators, and air filtration equipment.

7.1.0 Automated Building Management Systems

In order to achieve the current IAQ standards, most manufacturers of HVAC equipment and control systems have developed automated systems that control and monitor the ventilation air in a building. These systems communicate directly with the building's air handling and VAV systems (*Figure 8*).

The methods vary, but most involve the use of specialized computer hardware and software that make the air handling units and VAV systems operate more efficiently. These automated systems tie together building heating, cooling, and ventilation equipment in order to provide good IAQ without wasting energy. Through software management of the individual zones, the ventilation, temperature, humidity, and other desired zone parameters are monitored and controlled so that the best operating scheme is selected for each zone in the building. Most of these systems are vote based, in that each zone can communicate its need for heating, cooling, and ventilation. A central controller then allocates resources or establishes priorities to satisfy zone requirements.

Normally, control of the outdoor and exhaust air dampers is provided in these systems so that the dampers are constantly modulated to maintain the fixed ventilation airflow needed to satisfy

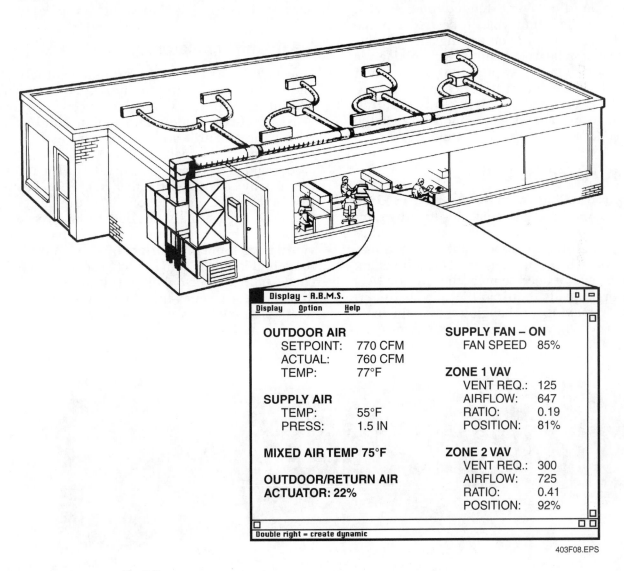

403F08.EPS

Figure 8 ◆ Automated building management system.

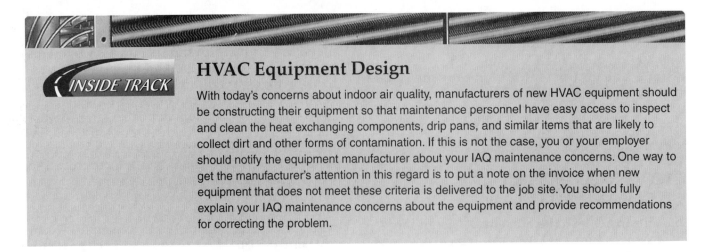

HVAC Equipment Design

With today's concerns about indoor air quality, manufacturers of new HVAC equipment should be constructing their equipment so that maintenance personnel have easy access to inspect and clean the heat exchanging components, drip pans, and similar items that are likely to collect dirt and other forms of contamination. If this is not the case, you or your employer should notify the equipment manufacturer about your IAQ maintenance concerns. One way to get the manufacturer's attention in this regard is to put a note on the invoice when new equipment that does not meet these criteria is delivered to the job site. You should fully explain your IAQ maintenance concerns about the equipment and provide recommendations for correcting the problem.

IAQ requirements for the building. These systems also incorporate special building purge modes to cover temporary IAQ problems. Purge modes can allow for the maximum circulation of ventilation air in the building over extended time periods. The purge mode would typically be used to purge the building air prior to occupying a new building. It may also be used to dilute increased levels of odors or chemical vapors that occur when activities such as painting or carpet cleaning are being performed. You will study building energy management systems in more detail in the *Building Management Systems* module .

7.2.0 Air Handling Units

Newer air handling units respond to the need for improved operation and efficiency by providing more fresh air and better service access, humidity control, and filtration. Many units are modular, like the one shown in *Figure 9*. This allows the unit to be customized to meet the specific IAQ and energy needs required by each customer. Adding, removing, or changing the components in the unit can be accomplished without the need

for major modifications. Modular construction provides a hedge against any modifications that may be needed in the future in response to revisions in air quality and energy conservation standards and codes.

7.3.0 Unit Ventilators

Unit ventilators (*Figure 10*) provide ventilation and temperature control for individual rooms in a building. Unit ventilators have been used for years in offices, schools, and similar buildings. Newer unit ventilators have been vastly improved to provide better indoor air ventilation accompanied by energy conservation. The units can usually be controlled either from a local control panel in the unit or room or by digital command control signals applied from a remote automated building management system. Unit ventilators come in a variety of sizes. A typical room unit is able to provide between 450 and 500 cfm (15 cfm per person) of outside air to the conditioned space. Unit ventilators can be equipped with heat exchangers.

403F09.EPS

Figure 9 ◆ Modular air handler.

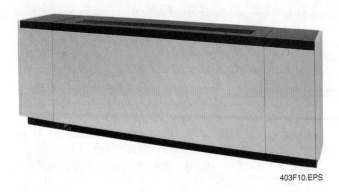

Figure 10 ◆ Unit ventilator.

7.4.0 Air Filtration Equipment

Normal air contains varying amounts of natural and man-made foreign materials. Dirt and pollens contained in outdoor air enter a building as a result of infiltration. Indoor air is recirculated in a building many times, picking up dust, dirt, smoke, and other contaminants. This is especially true in airtight buildings. Airborne bacteria and mold spores are also common both indoors and out. *Figure 11* shows the relative sizes of some common particles that contaminate air. As shown, these particles have diameters that range in size from smaller than 0.001 micron to larger than 10 microns. A **micron** is a unit of length that is one millionth of a meter, or about one 25,400th of an inch. About 99 percent of airborne particles are less than 1 micron in diameter. The remaining 1 percent consists of larger, heavier particles such as dust, lint, and pollen. Several types of air filters can be used to remove contaminants from the air, making the air cleaner and healthier to breath. Both mechanical and electronic air filters are in common use.

Some filters can inadvertently increase the level of microbes. Once trapped in the filter, microbes can grow on the filter material. Unless filters are replaced frequently, or incorporate a safe and effective antimicrobial agent, they can become a major source of IAQ problems.

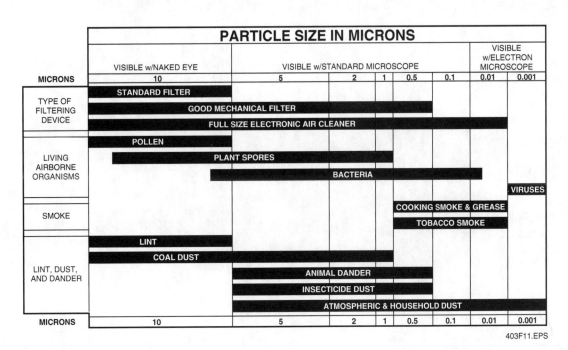

Figure 11 ◆ Particle size in microns.

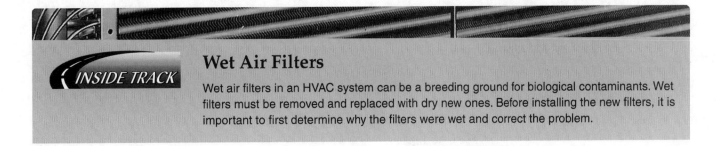

Wet Air Filters

Wet air filters in an HVAC system can be a breeding ground for biological contaminants. Wet filters must be removed and replaced with dry new ones. Before installing the new filters, it is important to first determine why the filters were wet and correct the problem.

7.4.1 Filter Efficiency

ASHRAE Standards 52.1-1992 and *52.2-1999* define different methods for testing and rating filters. In basic terms, *52.1* rates filters on the basis of an overall percentage of **atmospheric dust spot efficiency (dust spot efficiency)** and **arrestance efficiency**. *Standard 52.2* is based on particle size. It uses the minimum efficiency reporting value (MERV) system to rate filters. MERV ratings range from 1 to 20, with 20 being the highest. You will see the MERV rating printed on some filters. *Table 2* shows a comparison of the two standards. Although both standards are currently in use, an effort was being made to combine them at the time of this writing.

7.4.2 Mechanical Air Filters

There are various kinds of mechanical air filters (*Figure 12*). For the purpose of discussion, these mechanical air filters can be divided into four groups based on their performance.

Group 1 filters include fiberglass furnace filters and open-cell foam material commonly found in window air-conditioning units. Also in this group are roll filters, precut pads, filters made of synthetic materials, and metal wire screen-style filters commonly used where high airflow and/or heavy dust loads are encountered. The metal wire screen and open-cell foam filters are the only two permanent filters in the group.

Table 2 Filter Application Guidelines

MERV Std 52.2	Average ASHRAE Dust Spot Efficiency Std 52.1	Average ASHRAE Arrestance Std 52.1	Particle Size Ranges	Typical Applications	Typical Filter Type
1–4	< 20%	60 to 80%	> 10.0 μm	Residential / Minimum Light / Commercial Minimum / Equipment Protection	Permanent / Self Charging (passive) Washable / Metal, Foam / Synthetics Disposable Panels Fiberglass / Synthetics
5–8	< 20 to 35%	80 to 95%	3.0–10.0 μm	Industrial Workplaces Commercial Better / Residential Paint Booth / Finishing	Pleated Filters Extended Surface Filters Media Panel Filters
9–12	40 to 75%	> 95 to 98%	1.0–3.0 μm	Superior / Residential Better / Industrial Workplaces Better / Commercial Buildings	Non-Supported / Bag Rigid Box Rigid Cell / Cartridge
13–16	80–95% +	> 98 to 99%	0.30–1.0 μm	Smoke Removal General Surgery Hospitals & Health Care Superior / Commercial Buildings	Rigid Cell / Cartridge Rigid Box Non-Supported / Bag
17–20[1]	99.97[2] 99.99[2] 99.999[2]	N/A	≤ 0.30 μm	Clean Rooms High Risk Surgery Hazardous Materials	HEPA ULPA

Note: This table is intended to be a general guide to filter use and does not address specific applications or individual filter performance in a given application. Refer to manufacturer test results for additional information:
  (1) Reserved for future classifications
  (2) DOP Efficiency
Source: National Air Filtration Association

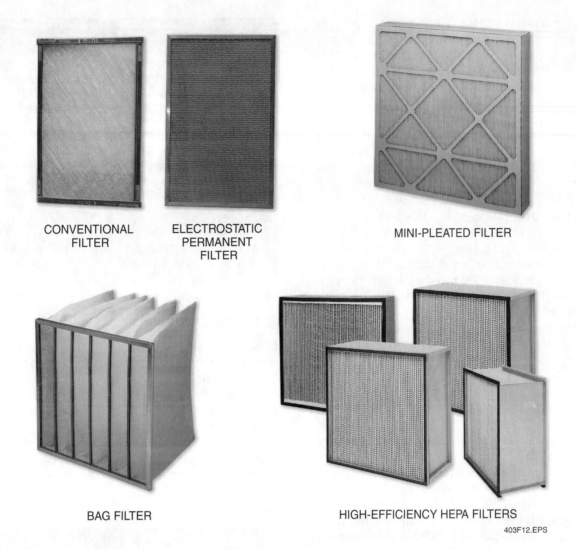

CONVENTIONAL FILTER

ELECTROSTATIC PERMANENT FILTER

MINI-PLEATED FILTER

BAG FILTER

HIGH-EFFICIENCY HEPA FILTERS

403F12.EPS

Figure 12 ◆ Mechanical filters.

These permanent filters can be washed and reused. Group 1 filters are effective at trapping particles that are 10 microns and larger. When coated with a tackifier, some can be effective on particles as small as 5 microns. A tackifier is a material that makes the filter medium sticky, which helps it retain and hold dust particles. Group 1 filters are commonly used as prefilters for higher-efficiency filters. Their low cost and ability to trap dust and dirt make them ideal for helping to extend the life of more expensive final filters.

Group 2 filters consist mainly of pleated panel-type filters. The filter material is usually made of polyester, all natural fiber, or a blend of both. Because the material is more dense, its resistance to airflow is higher. This is overcome by pleating the material to allow for more surface area within a given space. Pleated filters normally have a higher efficiency than most of the flat panel filters described in Group 1. When pleated filters are

used, they are typically the only filter in the system; however, they can be used as a prefilter to protect downstream higher-efficiency filters. Pleated filters are highly effective at removing particles in the 5-micron to 10-micron size range, which means that they can stop all pollen in the 10-micron to 100-micron size range. They also do a good job at trapping mold, spores, and dust, which are 3 to 15 microns in size. Based on current IAQ requirements, Group 2 filters are generally considered to be the minimum standard for new installations.

Group 3 filters include extended-surface supported and nonsupported medium-efficiency to high-efficiency filters, excluding **high-efficiency particulate air (HEPA) filters**. The filtering material is commonly made of ultrafine glass microfibers, electret-type synthetic fibers, or wet-laid paper mat glass fibers. Ultrafine glass material is typically used in non-supported bag-type filters and rigid box-type filters. Electret-type synthetic material is also used in bag-type and box-type filters. Electret material uses an electrostatic charge on the material to improve filter operation. The wet-laid paper mat material is used in box-style filters with pleated paper mat and corrugated separators. It is also used in narrow-pack, close-pleated, rigid-style filters. The box filter typically contains 100 to 140 square feet of material, whereas the narrow-pack filter contains nearly 200 square feet of material. Depending on their construction, Group 3 filters can have dust spot efficiencies ranging between 30 and 95 percent. One thing to keep in mind is that as the filtering material becomes denser to provide higher efficiency, the area of the filter must be increased to maintain an acceptable airflow.

Group 4 filters are the most efficient type of particulate filters. They are called HEPA filters. HEPA filters use a wet-laid ultrafine fiberglass paper filtering material. The material is different from the wet-laid paper material used in Group 3 filters in that it contains fibers with a much smaller diameter. The paper is much more dense, allowing it to remove all particles in the 0.3 micron range, accompanied by a maximum pressure drop of 1.0 inch water gauge for a clean filter when tested at rated airflow capacity. Some newer HEPA filters can remove microscopic particles and microorganisms as small as 0.12 microns. Typically, HEPA filters are used to filter supply air for surgical rooms and in applications where it is necessary to prevent process contamination during critical manufacturing procedures. However, HEPA filters are now being used in some office buildings and residences. They may be prescribed as a means of protecting occupants subject to asthma or allergies to air-borne contaminants.

7.4.3 Adsorption Filters

Adsorption filters remove gaseous vapors. The most common adsorption filter is the activated charcoal filter. This filter blocks materials with high molecular weights and allows materials with low molecular weights to pass through. When this type of filter becomes loaded, it must be replaced or regenerated in order to prevent off-gassing of previously adsorbed materials. Another type of gas filter uses porous pellets impregnated with active chemicals, such as potassium permanganate. These chemicals react with the contaminants and remove them or make them less bothersome or harmful. Maintenance consists of regenerating or replacing the chemicals.

Disposing of Air Filters

Used air filters must be disposed of properly in accordance with the prevailing laws. This can mean disposal in a landfill, by incineration, or by recycling. Some air filters must be handled and disposed of as hazardous waste. Typically, filters used to trap hazardous waste are found in certain areas of hospitals, biomedical facilities, or in industrial processing plants. Do not attempt to dispose of filters used to capture hazardous waste unless you are equipped and licensed to do so.

7.4.4 Air Cleaners

Electronic air cleaners (*Figure 13*) outperform many of the mechanical air filters in trapping air-borne particles and odors. They can be stand-alone units or can be mounted in the A/C system. Most electronic air cleaners contain a pre-filter, ionizer, and collector (*Figure 14*). Charcoal filters are optional. As the air is moved through the filter, larger particles are trapped by the pre-filter section. Smaller particles pass through the

403F13.EPS

Figure 13 ◆ Electronic air cleaner.

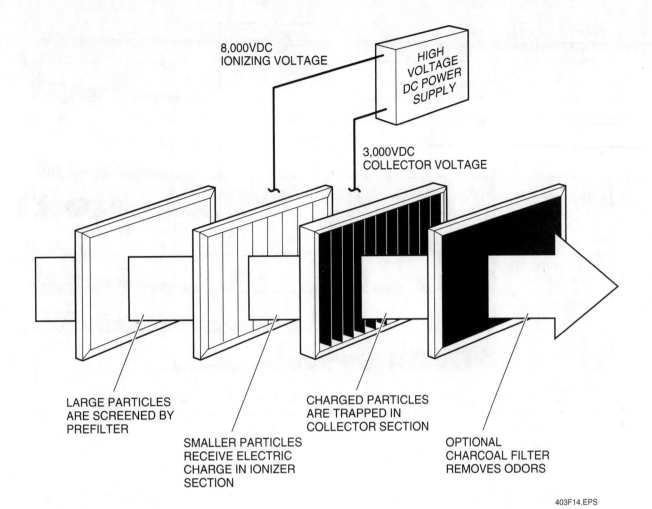

403F14.EPS

Figure 14 ◆ Electronic air cleaner filtration stages.

prefilter to the ionizer section. There, the dirty air particles pass between ionizing wires connected to a high-voltage power supply. Voltage potentials up to 8,000 volts DC strip electrons from the particles, leaving the particles with an intense positive electrical charge. These ionized particles proceed to the collector section, where they encounter closely spaced, oppositely charged collector plates. Particles are both repelled by the positive collector plates and attracted to the negative collector plates where they are collected. The air, cleaned of pollutants, then passes through the charcoal filter where odors are removed.

There are also several types of nonelectronic air cleaners. Most are used as portable room air cleaners. Typically, they consist of a prefilter and multiple stages of mechanical filtering followed by a charcoal afterfilter.

7.5.0 Humidifiers and Dehumidifiers

Improper humidity levels can be a cause of sick building syndrome. Excess humidity promotes the growth of mold and mildew in ductwork,

walls, and other interior spaces. *Figure 15* shows the relationship between humidity and common contaminants. As a rule of thumb, the RH in a building should be maintained at about 30 percent in the winter and 60 percent in the summer. From a system standpoint, RH levels over 30 percent are not practical in very cold weather because condensation on windows and other cold exterior surfaces can cause damage. Similarly, an RH much below 40 percent cannot be achieved with most cooling equipment in the summer. Control of humidity involves the use of humidifiers when humidity levels are too low, and dehumidifiers when the humidity is too high.

7.5.1 Humidifiers

Humidifiers are used to add humidity to a building or conditioned space. This is done by introducing water vapor into a building's conditioned air at a certain rate. Humidifiers can be portable units or mounted in the HVAC system. Operation is controlled by a humidistat located in the conditioned space or unit. Humidifiers were studied in

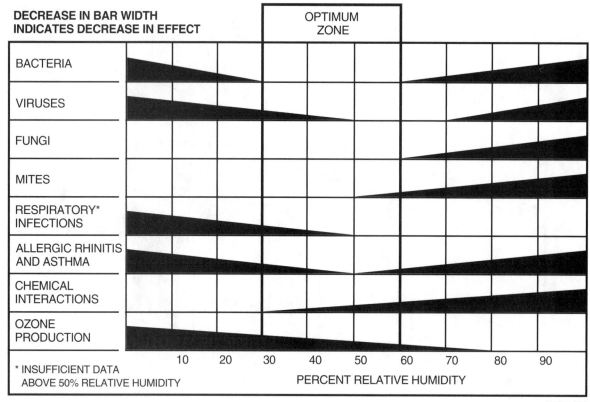

DECREASE IN BAR WIDTH INDICATES DECREASE IN EFFECT

OPTIMUM ZONE

BACTERIA

VIRUSES

FUNGI

MITES

RESPIRATORY* INFECTIONS

ALLERGIC RHINITIS AND ASTHMA

CHEMICAL INTERACTIONS

OZONE PRODUCTION

10   20   30   40   50   60   70   80   90

* INSUFFICIENT DATA ABOVE 50% RELATIVE HUMIDITY

PERCENT RELATIVE HUMIDITY

403F15.EPS

Figure 15 ◆ Relative humidity ranges for health.

detail in the *HVAC Level Two* module *Air Quality Equipment* and the *HVAC Level Three* module *Troubleshooting Accessories*. Review the following humidifier types covered in those modules:

• Wetted element
• Atomizing
• Infrared
• Steam

Excluding a failure of the humidifier or its control circuit, uncomfortable relative humidity (RH) levels in a building can be caused by an incorrect humidistat setting relative to the outdoor temperature. Symptoms of excessive RH are condensation on windows and inside exterior walls. Too low an RH causes dry, itchy skin; static electricity shocks; clothing static cling; sinus problems; a chilly feeling; sickly pets and plants; and loose furniture joints. *Table 3* lists the recommended indoor RH levels for various outdoor winter temperatures.

Another cause of too much or too little humidity can be a poorly sized humidifier. Humidifier capacities are normally rated in gallons of water per day. The capacity depends on the volume of the building or area in square feet (ft²). It also

Table 3 Recommended Indoor Winter Relative Humidity

At Outdoor Temperature (°F)	Recommended Indoor RH (%)
–20	15
–10	20
0	25
10	30
20	35
30	40

Based on an indoor temperature of 72°F

depends on the air tightness of the building's construction. *Figure 16* shows a typical graph used for the selection of residential humidifiers. Similar graphs and/or charts are available for commercial and industrial humidifiers.

7.5.2 Dehumidifiers

Dehumidifiers remove humidity from a building or conditioned space. Dehumidification of air occurs normally in a conventional cooling system. The system cooling coil normally removes

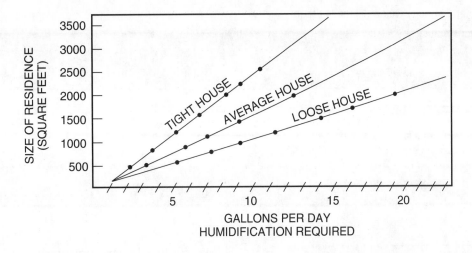

LOOSE	AVERAGE	TIGHT
No weatherstripping	Weatherstripping	Weatherstripping
No infiltration barrier	Vapor barrier	Infiltration barrier
No vapor barrier	Fireplace dampered	Seams/Penetrations
No fireplace damper	Dampered exhausts	sealed
Undampered exhausts	Ductwork taped or in	Fireplace dampered
Ductwork untaped or	conditioned space	Dampered exhausts
in unconditioned space	Indoor combustion air	Ductwork taped or in
Indoor combustion air		conditioned space
		Outdoor combustion air

403F16.EPS

Figure 16 ◆ Humidifier capacity chart.

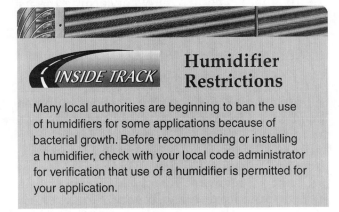

Humidifier Restrictions

Many local authorities are beginning to ban the use of humidifiers for some applications because of bacterial growth. Before recommending or installing a humidifier, check with your local code administrator for verification that use of a humidifier is permitted for your application.

both sensible heat and moisture (latent heat) from the entering air, which is a mixture of water vapor and dry gases. Both lose sensible heat during contact with the first part of the cooling coil, which functions as a dry cooling coil. Moisture is removed only in the part of the coil that is below the dew point of the entering air. When the coil starts to remove moisture, the cooling surfaces carry both the sensible and latent heat loads.

Portable dehumidifiers operate in the same way. These units are controlled by an adjustable humidistat that turns the unit on and off at preselected moisture levels. The capacity of a portable unit is normally rated in pints per 24 hours. Again, the required capacity depends on the volume or area of the building. It also depends on the building's condition (wet, very damp, or moderately damp) without dehumidification during warm and humid outdoor weather conditions. *Table 4* provides sample guidelines for selecting portable dehumidifiers for residential or small commercial use. Similar charts are available for larger commercial and industrial dehumidifiers.

Other dehumidifying equipment uses liquid or solid **desiccants**. These units either collect the water on the surface of the desiccant or chemically combine with the water. One such piece of equipment is an air-to-air heat exchanger wheel (*Figure 17*). During the cooling season, the exhaust airstream recharges (dries out) a desiccant-coated wheel, causing the wheel to cool down. When the wheel rotates into the outside airstream, it absorbs the moisture and cools the outside air before delivering it to the air handler and cooling coils.

Table 4 Dehumidifier Capacity Guide (Pints/24 Hours)

Conditions Without Dehumidification	Area in Square Feet					
	500	1,000	1,500	2,000	2,500	3,000
Moderately damp – Space feels damp and has musty odor in humid weather.	10	14	18	22	26	30
Very damp – Space always feels damp and has musty odor. Damp spots show on walls and floors.	12	17	22	27	32	37
Wet – Space feels and smells wet. Walls and floors sweat, or seepage is present.	14	20	26	32	38	44

AIR-TO-AIR HEAT EXCHANGER WHEEL

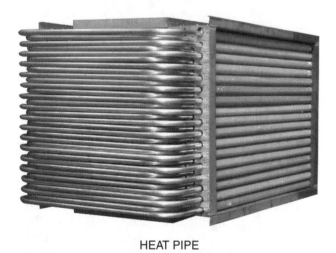

HEAT PIPE

403F17.EPS

Figure 17 ◆ Air-to-air heat exchanger and heat pipe.

Another accessory used to dehumidify an airstream is called a heat pipe. Use of the heat pipe in an HVAC system allows more of the system cooling coil capacity to go towards latent heat cooling by precooling the air before it gets to the cooling coil. Precooled air means less sensible cooling is required at the coil, allowing more capability for latent cooling (dehumidification). You will study heat pipes in more detail in the *Energy Conservation Equipment* module .

7.6.0 Ozone Generators

Ozone generators are sometimes used to break down odor-causing chemicals, such as formaldehyde, into carbon dioxide and water. They are also used to break down odors caused by fire and smoke, garbage wastes, tobacco smoke, sewage gases, and decaying organic matter. Another use is to destroy germs and airborne bacteria. Currently, there is controversy concerning the use and effectiveness of ozone generators.

7.7.0 Ultraviolet Light Air Purification Systems

Ultraviolet (UV) light air purification equipment can be used in HVAC air distribution systems to help prevent the growth of bacteria and other microorganisms known to cause indoor air problems and musty, mold-related odors.

There are many manufacturers and designs of UV air purification equipment. However, the principle of operation is the same. C-band UV light (UVC) energy in the 240- to 280-nanometer wavelength range destroys microorganisms by penetrating the cell wall of the microorganism. High-energy UV photons damage the protein structure of the cell and chemically alter the DNA. Once this occurs, the organism dies or cannot reproduce. Germicidal effectiveness (killing power) is directly related to the UV dose applied, which is a function of time and intensity.

HVAC system air purification by UVC light is done in one of two ways: purification of a fixed object or purification of the moving air stream.

In fixed object purification, the HVAC discharge side evaporator/indoor coil and drain pan are continuously irradiated with light rays generated by stationary quartz UVC lamps or probes (emitters). The UVC rays destroy bacteria and viruses present on the fixed object. The time required to destroy microorganisms on fixed objects depends on a number of things, including the distance the UVC emitter is mounted from the fixed object, the size and intensity or killing power of the UVC emitter, and the temperature of the air and UVC emitter.

In UV purification of the moving air stream, the air in a duct system is irradiated as it moves past a stationary UVC emitter. Achieving air purification using this method is much more difficult because of the short time (dwell time) during which the air moving past the emitter is being irradiated. Typically, the air moves past the UVC emitter at a speed of about 600 feet per minute or faster, spending only about 20 milliseconds in front of the probe/emitter. The intensity or killing power of the UVC emitter, how fast the UV ray intensity decreases with distance as the air moves away from the emitter, and how far into the air stream the UV rays penetrate,

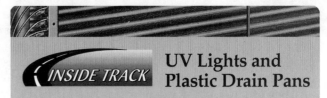
determine the UV purification efficiency on the moving air. Because of the short dwell time, purification of the air stream normally requires the use of multiple UV light sources and reflectors that are capable of producing much stronger UV light rays than needed for a fixed object. For this reason, fixed object air purification systems are more widely used.

Figure 18 shows an example of a typical UVC air purification unit. It is designed to protect coils, drain pans, and humidifiers from mold and bacterial growth while killing some airborne microorganisms. It consists of a housing, power supply, and emitters. The components are incorporated into one assembly that is mounted outside the equipment at the cooling coil ductwork with the emitters protruding into the center of the coil and air stream.

8.0.0 ◆ GAS DETECTORS AND ANALYZERS

Gas detectors and analyzers provide an accurate way to detect and measure the presence of gaseous contaminants in the air. They can be either mechanical or electronic. Electronic instruments simplify testing and are more accurate. They can also perform automatic sampling and calculations. Some models can produce hardcopy reports of the date, time, and test results.

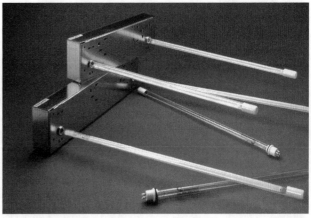

403F18.EPS

Figure 18 ◆ UVC air purification unit.

Others can be connected to computers so that the test results can be transferred to and stored in a remote computer. Electronic detectors and analyzers are available for use as stationary wall-mounted units or portable test instruments. Some units detect and/or measure only one specific kind of gas. Other units can detect and measure several gases. *Figure 19* shows some common gas detectors and analyzers.

Depending on the model, a gas detector may contain one or more sensors. Each sensor can detect a different gas. Typically, a sensing element includes three coated electrodes and a small quantity of an acid solution enclosed in a sealed plastic capsule or body. The three electrodes are related to the sensing, counting, and reference functions of the instrument. In use, the gases being sampled diffuse through a small opening on the sensor instrument's face or probe for application to the electrochemical sensor. There, a small current is generated that is proportional to the level of the gas being measured. Depending on the detector,

this current sets off an alarm and is converted into a digital signal that represents the gas concentration for display on the device. On some detectors, the digital signal can also be used to drive a printer or can be displayed on a computer.

The calibration of an electronic gas detector or analyzer should be done following the manufacturer's schedule and procedure. Most units should be calibrated every six months using certified concentrations of test gases. The average life of a typical sensor is about two years. This type of sensor should be replaced at the interval specified by the manufacturer.

8.1.0 Carbon Dioxide Detectors

CO_2 levels above 1,000 ppm can indicate ventilation problems. The level of CO_2 is widely used as an indicator of suitability for human occupancy. CO_2 sensors are commonly used for monitoring in nonindustrial buildings. Some models can be used as ventilation controllers in demand-based ventilation control systems. When used as a ventilation controller, the CO_2 sensor determines the need for ventilation based on the CO_2 concentration. It then modulates the position of the building dampers to maintain acceptable ventilation. If a space is unoccupied, the CO_2 controller will set the air intake volume at a minimum setting that allows established ventilation rates to be maintained, while reducing overventilation and saving energy.

8.2.0 Carbon Monoxide Detectors

A carbon monoxide (CO) detector is both a safety device and an IAQ device. Early detection of high CO is almost impossible without a CO detector. Stationary CO detectors are made for use in automated systems. They are installed in strategic places throughout a building and will normally activate a contact closure and sound an alarm when a high level of CO is detected. Portable CO detectors are used mainly for testing HVAC combustion equipment.

8.3.0 Volatile Organic Compound Sensors

Volatile organic compound (VOC) sensors are often used to indicate non-occupant-related short-term changes in air contaminant levels. VOC sensors measure and react to a broad range of compounds. VOC sensors use an interactive, chemical-based oxidizing element. When this element is exposed to various compounds in the air, the sensor will vary its electrical resistance and

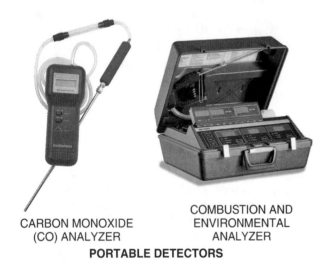

CARBON MONOXIDE
(CO) ANALYZER

COMBUSTION AND
ENVIRONMENTAL
ANALYZER

PORTABLE DETECTORS

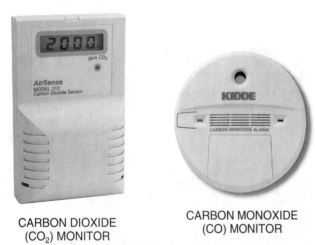

CARBON DIOXIDE
(CO_2) MONITOR

CARBON MONOXIDE
(CO) MONITOR

STATIONARY DETECTORS

403F19.EPS

Figure 19 ◆ Gas detectors and analyzers.

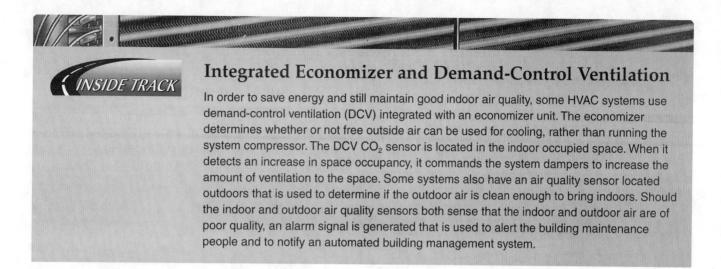

Integrated Economizer and Demand-Control Ventilation

In order to save energy and still maintain good indoor air quality, some HVAC systems use demand-control ventilation (DCV) integrated with an economizer unit. The economizer determines whether or not free outside air can be used for cooling, rather than running the system compressor. The DCV CO_2 sensor is located in the indoor occupied space. When it detects an increase in space occupancy, it commands the system dampers to increase the amount of ventilation to the space. Some systems also have an air quality sensor located outdoors that is used to determine if the outdoor air is clean enough to bring indoors. Should the indoor and outdoor air quality sensors both sense that the indoor and outdoor air are of poor quality, an alarm signal is generated that is used to alert the building maintenance people and to notify an automated building management system.

provide an electrical output. This output is not specific to any one gas, but reflects the total effect of a wide variety of compounds in the air.

One disadvantage is that this type of sensor has no way of telling a harmful gas from a harmless gas. It can only indicate a change in the concentration.

VOC sensors are often tuned to the building space in which they are operating. Each of the individual sensors in the building is adjusted so that it provides a low output signal when the air in the space being monitored is considered to have good air quality. The sensor will then provide a higher signal output when there are more contaminants. The more contaminants, the higher the output signal. Typically, the VOC sensor provides a one-in-five or one-in-ten scale output signal that represents the relative level of contamination. As a control, the sensor output can be used to activate an alarm. It can also be used to regulate building ventilation based on the actual level of pollutants sensed. This may or may not conflict with the established building ventilation scheme.

8.4.0 Other Gas Detectors/Analyzers

Many specialized detectors/analyzers are designed to detect and measure gases other than CO_2, CO, and VOCs. Some of the more common ones are:

- *Oxygen detector* – Monitors the level of O_2 in the area. The normal level is 21 percent.
- *Hydrogen detector* – Monitors the H_2 in the area. Hydrogen is dangerous because of its volatility. It has a lower explosion limit in air of 4 percent. This limit is the level at which the air-gas mixture explodes.
- *Combustible gas detector* – Monitors the LP (propane, butane) and methane (natural) gas in the area. It is used to check for leaks.

- *Air pollution detector* – Monitors various gases that cause air pollution and endanger lives. Typically detects CO, H_2, alcohol, gasoline fumes, cigarette smoke, and exhaust fumes. This type of detector is also referred to as an IAQ detector.
- *Refrigerant gas detector* – Provides area monitoring and early warning of refrigerant leaks.
- *Ammonia detector* – Provides area monitoring and early warning of NH_3. Ammonia is poisonous and is dangerous even at very low levels.

9.0.0 ◆ DUCT CLEANING

Increased emphasis has been placed on the use of duct cleaning as a means of controlling indoor air quality. In the past, there has been some controversy about the effectiveness of duct cleaning and the methods for performing the task. There was also some question as to when duct cleaning should be done and how the job could be validated. Duct cleaning alone does not solve IAQ problems. Dirty ventilation systems are most often the effect, not the cause, of poor indoor air quality. However, when duct cleaning is done along with a program of regular building maintenance, it can help to reduce the threat of indoor air pollution.

In 1989, the National Air Duct Cleaners Association (NADCA) was formed by members of the duct cleaning industry. This organization adopted a standard in 1992 entitled: *NADCA Standard 1992-01, Mechanical Cleaning of Non-Porous Air Conveyance System Components*. This document is now published by NADCA under the title *ACR 2006, Assessment, Cleaning, and Restoration of HVAC Systems*. Always refer to the latest edition of such specifications, as they are often revised.

9.1.0 Duct Cleaning Equipment

Common duct cleaning equipment includes portable and/or truck-mounted HEPA-filtered vacuuming equipment and power brushing equipment to dislodge dirt and debris in the ductwork (*Figure 20*).

Figure 21 shows equipment used to inspect and document the conditions within the ductwork and other components before and after cleaning.

This equipment can include borescopes (tubular devices similar to gun scopes), black and white and/or color video cameras, and VCRs.

9.2.0 Duct Cleaning Methods

When cleaning and accessing air conveyance systems, NADCA Standards and other NADCA published guidelines should be followed. Before beginning cleaning, the operating system must

HEPA VACUUM COLLECTOR

POWER BRUSH

POWER WHIP

403F20.EPS

Figure 20 ◆ Duct cleaning equipment.

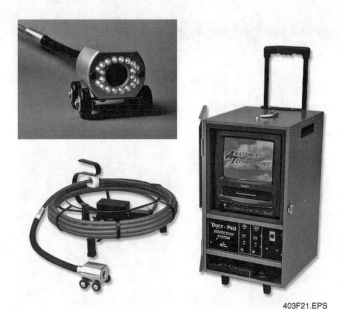

Figure 21 ◆ Duct inspection equipment.

be turned off and locked out using approved lockout/tagout procedures. Drop cloths should also be used to protect furnishings in occupied areas. Common duct cleaning methods include contact vacuuming, air washing, and power brushing.

9.2.1 Contact Vacuuming Method

> **WARNING!**
> Duct cleaning will raise contaminants that may have settled in the ductwork. Be sure to wear the company-prescribed PPE, including respiratory protection, when performing duct cleaning.

Contact vacuuming involves cleaning the interior duct surfaces by way of existing openings and outlets or, when necessary, through openings cut into the ducts (*Figure 22*). The vacuum unit should use HEPA filtering if it is exhausting into an occupied space. Starting at the return side of the system, the vacuum cleaner head is inserted into the section of the duct to be cleaned at the opening furthest upstream, and then the vacuum cleaner is turned on. Vacuuming proceeds downstream slowly enough to allow the vacuum to pick up all dirt and dust particles.

Inspection of each duct section and related components is performed to determine whether the duct is clean. When the section of duct is clean, the vacuum cleaner head is removed from the duct and inserted through the next opening, where the process continues.

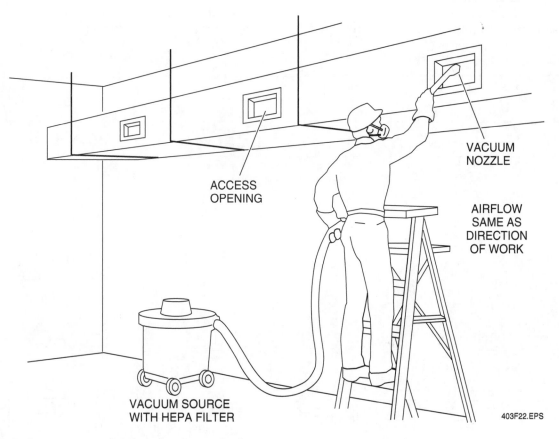

ACCESS OPENING

VACUUM NOZZLE

AIRFLOW SAME AS DIRECTION OF WORK

VACUUM SOURCE WITH HEPA FILTER

403F22.EPS

Figure 22 ◆ Contact vacuuming duct cleaning method.

Duct Cleaning

When cleaning ducts in a large commercial system, it is good practice to obtain a set of air system as-built drawings. These can be used to identify obstructions such as coils, turning vanes, dampers, and similar devices within the duct system. The drawings can also be used to plan the locations for access points in the ductwork and to divide the ductwork system into workable sections for cleaning. Typically, cleaning sections should be no more than about 25' in length. The drawings will also show sections of flexible or lined ductwork, certain types of which cannot be cleaned.

Isolating Cleaning Zones

The ductwork section being cleaned must be sealed off (isolated) from the adjacent sections to prevent loosened debris from contaminating the cleaned section or escaping past the vacuuming device into the downstream section. Isolation of the section being cleaned is typically accomplished using inflatable balloons/bladders inserted into the ductwork at each side of the zone. When these balloons/bladders are inflated, they conform to the interior shape of the ductwork, sealing off the section being cleaned.

9.2.2 Air Washing Method

In the air washing method (*Figure 23*), a vacuum collection unit is connected to the downstream end of the duct section through a suitable opening. The vacuum unit should use HEPA filtering if it is exhausting into an occupied space. The isolated section of duct being cleaned should be subjected to a minimum of 1" negative air pressure to draw loosened materials into the vacuum collection system. Take care not to collapse the duct. Compressed air is then introduced into the duct through a hose equipped with a skipper nozzle. This nozzle is propelled by the compressed air along the inside of the duct. For the air washing method to be effective, the compressed air source should be able to produce between 160 and 200 psi air pressure and should have a 20-gallon receiver tank. This method is most effective in cleaning ductwork interior dimensions no larger than 24" × 24". Inspection of each duct section and related components is performed to determine whether the duct is clean.

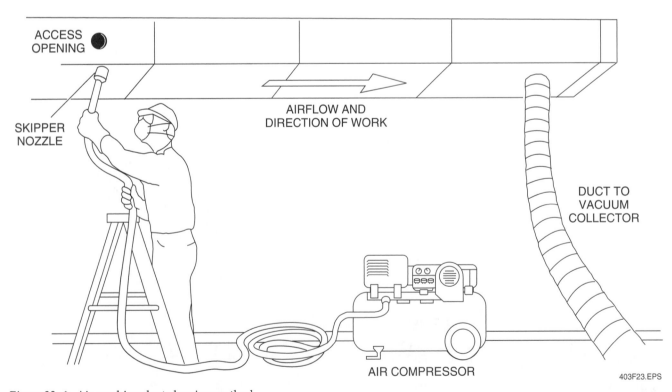

Figure 23 ◆ Air washing duct cleaning method.

9.2.3 Power (Mechanical) Brushing Method

In the power brushing method, a vacuum collection unit is connected to the duct in the same way as with the air washing method. Pneumatic or electric rotary brushes are used to dislodge dirt and dust particles, which become airborne and are then drawn into the vacuum unit (*Figure 24*). Power brushing can be used with all types of ducts and fibrous glass surfaces if the bristles are not too stiff and the brush is not allowed to remain in one place for a long time. Power brushing usually requires larger access openings in the duct in order to allow for manipulating the equipment. The rotary brush is inserted into the duct section at the opening farthest upstream from the vacuum collector. The brush is moved downstream to dislodge dirt and dust particles. Inspection of each duct section and related components is performed to determine if the duct is clean. When the section of duct is clean, the brush is removed from the duct and inserted through the next opening, where the process continues.

10.0.0 ◆ IAQ AND FORCED-AIR DUCT SYSTEMS

In the United States, there are an estimated 60 million homes with forced-air heating and cooling systems. Studies have shown that these

Moisture in Air Ducts

If moisture and dirt are present in air ducts, biological contaminants can grow and disperse throughout the building. Mold contamination in unlined sheet metal ducts can be successfully treated using an EPA-registered biocide. However, if fiberglass-lined sheet metal ducts or ducts made of fiberglass duct board become wet and contaminated with mold, cleaning is not sufficient to prevent regrowth, and there are no EPA-registered biocides for the treatment of porous duct materials. The EPA, National Air Duct Cleaners Association (NADCA), and the North American Insulation Manufacturers Association (NAIMA) all recommend the replacement of wet or moldy fiberglass material.

systems can lose up to 40 percent of the conditioned air through air duct leaks. Translated into wasted energy, it represents an annual fuel usage equivalent to that used by 13 million automobiles. Not only do the leaking ducts waste energy and contribute to poor comfort, they also can adversely affect indoor air quality and create health hazards.

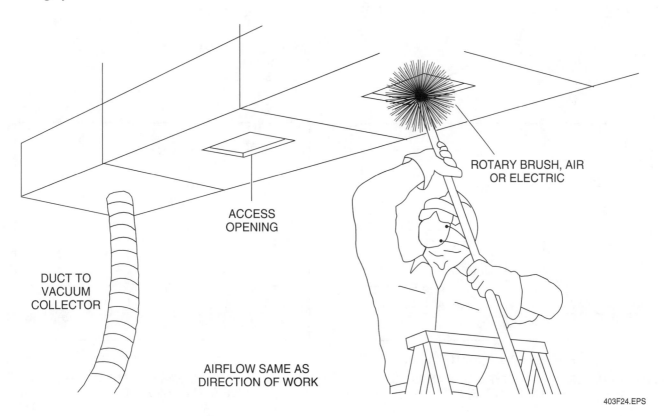

Figure 24 ◆ Power brush duct cleaning method.

10.1.0 Supply and Return Duct Leaks

In an ideal duct system (*Figure 25*), both the supply and return ducts are leak free. During operation, the system fan causes a pressure differential between the supply duct and the return duct. Positive (high) pressure in the supply duct causes the conditioned air to flow into the conditioned space. The negative (low) pressure in the return duct causes the air in the conditioned space to be drawn back into the return duct. The pressure inside the structure itself is essentially neutral. Under these circumstances, the conditioned air inside the structure is circulated with little or no loss.

For the purpose of discussion, assume the return duct is leak free and that the supply duct is leaking into an unconditioned space such as an attic or crawl space (*Figure 26*). Under these conditions, a slightly negative pressure is created in the conditioned space when the system fan runs. The air lost through leaks in the unconditioned space causes this slight negative pressure. To make up this loss, outside air is drawn into the structure by infiltration through cracks and small openings to the outside. Not only does this waste energy, but it can also cause additional airborne contaminants to be drawn into the structure.

Similarly, assume that the supply duct is leak free and the return duct, located in an attic or crawl space, is drawing in unconditioned air

through leaks (*Figure 27*). Under these conditions, a slightly positive pressure is created in the structure. This is because the quantity of air delivered by the supply ducts is greater than the amount of air being drawn from the conditioned space by the return duct. The increased amount of air delivered by the supply duct is a result of the outside air that entered the return ducts through leaks. Leaks in the return duct can bring in all kinds of contaminants from inside and/or outside the structure. In actual practice, most structures have both supply and return duct leaks. In some cases, these two sets of leaks are equal and cancel each other out. In most cases, however, one duct system will leak more than the other, causing the structure to be either positively or negatively pressurized. In either case, contaminants from outside the structure can be brought into the structure.

In addition to the IAQ problems that can be brought about by the introduction of contaminated outside air, leaking ducts can create hazardous conditions within the structure. If leaking return ducts draw in air from a garage or basement where a fuel-burning furnace is located, flue gases can be drawn into and circulated through the structure with the potential for carbon monoxide poisoning of occupants. Leaking return ducts in a basement can also distribute radon gas that has infiltrated the basement throughout the rest of the structure.

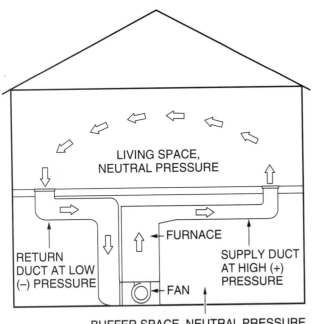

NOTE: BUFFER SPACE IS TYPICALLY AN ATTIC, BASEMENT, OR CRAWL SPACE.

403F25.EPS

Figure 25 ◆ Simplified ideal air duct system.

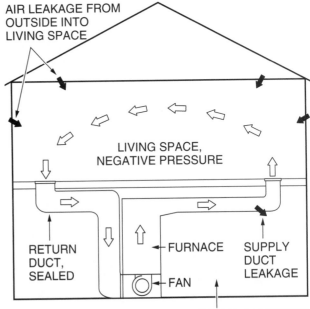

NOTE: SUPPLY DUCT LEAKAGE IS MADE FROM OUTSIDE AIR LEAKING INTO HOUSE.

403F26.EPS

Figure 26 ◆ Simplified air duct system with leaks in the supply duct.

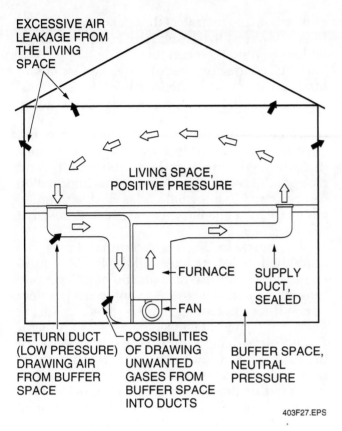

Figure 27 ◆ Simplified air duct system with leaks in the return duct.

10.2.0 Sealing Air Duct Leaks

The solution to air duct leakage problems is to properly seal all ducts during installation. Unfortunately, many commonly accepted installation practices lead to duct systems that leak. Some states and localities are realizing the scope of the problem and are implementing construction practices and modifying building codes to eliminate or greatly reduce this problem. That, however, does not solve the problem of the millions of existing air duct systems that leak. Some duct leaks can be successfully sealed manually using caulks, mastic, or duct tape, assuming the leaks are accessible. However, the use of duct tape is not the preferred method because it tends to lose its adhesiveness after a few years. In such cases, the tape will eventually fall off the ducts or become easy to pull away. Unfortunately, the majority of duct leaks in existing systems are not easily accessible, so manual sealing is not often an option.

New technologies have recently been developed that focus on sealing ducts from the inside. One such technology, called aerosol sealing (*Figure 28*), injects small dry adhesive particles into a pressurized duct system in which the air-conditioning coil, fan, and furnace components are blocked off. This isolates the components from

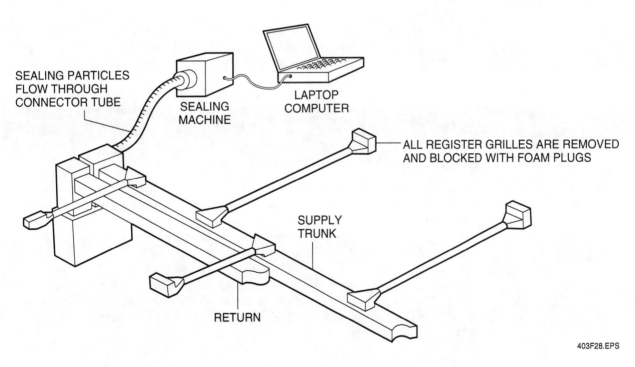

Figure 28 ◆ Aerosol duct sealing system.

the duct system, and all of the registers or grilles are removed and their openings blocked so that they are airtight. A fan (part of a sealing machine) temporarily connected to the supply or return plenum through a plastic connector tube is used to propel dry adhesive particles into and through the duct system. With the duct system plugged and under pressure, the only place air can escape is at the locations of the leaks in the duct. There, the sealant is deposited and collects. Over time, typically two to three hours, enough sealant builds up to stop the leaks. The process and resulting reduction in duct leakage can be monitored in real time by the technician using a laptop computer and related sealing process software. The manufacturer for the process shown claims that leaks up to about ⅝" in diameter can be successfully sealed and that the process can seal 70 to 90 percent of the existing leaks in a duct system.

11.0.0 ◆ HVAC CONTRACTOR LIABILITY

Today, many situations can lead to IAQ lawsuits by building owners, tenants, their employees, or others who use the building. For example, if a tenant discovers an IAQ problem in a building, the tenant may sue the owner, the HVAC contractor, or both, claiming the building or office space was not environmentally safe. If people in a building become ill from an IAQ problem, they may sue the owner and/or HVAC contractor for damages, claiming they suffer from sick building syndrome. For these reasons, HVAC contractors must take steps to protect themselves from unjustifiable lawsuits that result from conditions or situations beyond their control.

HVAC contractors are fully aware of indoor air quality issues associated with HVAC systems, but many of their customers are not, including the owners of commercial buildings. The HVAC contractor should make customers aware and educate them about indoor air quality issues, including all local code requirements and changes. It is recommended that this information be presented in a formal and well-documented manner in case the information becomes relevant in a lawsuit.

The contractor should recommend periodic scheduled maintenance of all the HVAC equipment and should also recommend building walkthrough inspections. In the event that a problem or condition that affects IAQ is detected during such routine maintenance, the HVAC contractor should immediately inform the customer about the problem in writing and give recommendations for correcting it. The customer should be informed of the possible consequences if the repairs are not made and should be encouraged to make the repairs as soon as possible.

Also, when retrofitting existing systems, the HVAC contractor should make sure that the customer is aware of and incorporates all the upgrades necessary to the system so that it will meet all current local IAQ codes and requirements. For example, equipment needed to supply fresh outdoor air may need to be added to a system that originally did not have this capability.

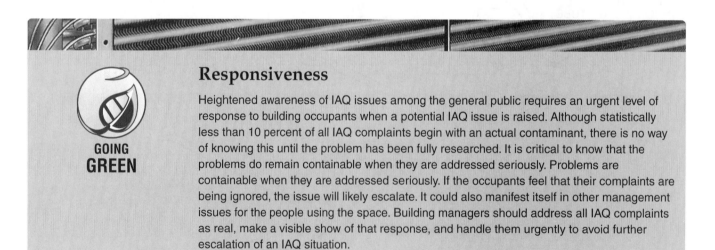

GOING GREEN

Responsiveness

Heightened awareness of IAQ issues among the general public requires an urgent level of response to building occupants when a potential IAQ issue is raised. Although statistically less than 10 percent of all IAQ complaints begin with an actual contaminant, there is no way of knowing this until the problem has been fully researched. It is critical to know that the problems do remain containable when they are addressed seriously. Problems are containable when they are addressed seriously. If the occupants feel that their complaints are being ignored, the issue will likely escalate. It could also manifest itself in other management issues for the people using the space. Building managers should address all IAQ complaints as real, make a visible show of that response, and handle them urgently to avoid further escalation of an IAQ situation.

1. The ASHRAE standard pertaining to acceptable levels of indoor ventilation is _____.
   a. *Standard 69-1990*
   b. *Standard 62-1981*
   c. *Standard 84-1999*
   d. *Standard 62.1-2007*

2. Air quality is considered acceptable when _____ percent or more of the people exposed to the air do not express dissatisfaction.
   a. 60
   b. 70
   c. 80
   d. 90

3. Buildings that have high air exchange rates _____.
   a. are more susceptible to poor indoor air
   b. are less susceptible to poor indoor air
   c. tend to have the same air quality as buildings with low air exchange rates
   d. have high levels of carbon dioxide (CO_2)

4. Personal care products can be a source of _____ pollution.
   a. chemical
   b. carbon dioxide (CO_2)
   c. carbon monoxide (CO)
   d. biological

5. In regard to air level concentrations, at _____, people tend to get drowsy, have headaches, and/or function at lower activity levels.
   a. a 0.5 percent decrease in the O_2 level
   b. a 0.5 percent decrease in the CO_2 level
   c. CO_2 levels of about 400 ppm
   d. CO_2 levels of about 1,200 ppm

6. Building materials and furnishings can be a source of _____.
   a. ozone pollution
   b. radon pollution
   c. volatile organic compounds
   d. nitrogen oxide (NO_2) pollution

7. The best way to prevent exposure to CO and NO_x is to _____.
   a. perform routine and proper maintenance on the source equipment
   b. dilute the indoor air with increased outdoor air ventilation
   c. use furnaces and appliances that burn natural gas
   d. use furnaces and appliances that burn propane gas

8. According to federal law, the maximum allowable concentration of carbon monoxide (CO) that a building's occupants can be continuously exposed to in any eight-hour period is _____ ppm.
   a. 9
   b. 35
   c. 200
   d. 1,600

9. Nitrogen dioxide (NO_2) is an irritant that can be generated by _____.
   a. electronic air cleaners
   b. cleaning compounds
   c. furnaces
   d. building materials

10. Ozone contamination can be caused by _____.
    a. combustion equipment
    b. building furnishings
    c. office copy machines
    d. humidifiers

11. Long-term exposure to radon may cause _____.
    a. lung cancer
    b. headaches and dizziness
    c. irritation of the skin and the mucous membranes in the eyes, nose, and throat
    d. multiple chemical sensitivity (MCS)

12. One method used to help reduce an indoor pollution problem caused by radon is to _____.

    a. eliminate wet or moist building materials and furnishings
    b. dilute the indoor air with increased outdoor air ventilation
    c. reduce pressure under the building's floor slab
    d. increase pressure under the building's floor slab

13. The best way to permanently correct a building indoor pollution problem caused by microbial contaminants is to _____.

    a. close any system dampers to decrease outdoor air ventilation
    b. dilute the indoor air with increased outdoor air ventilation
    c. maintain the building humidity between 30 percent and 60 percent
    d. relocate the source equipment

14. The minimum efficiency filters that should be used in new installations are _____ filters.

    a. HEPA
    b. pleated-panel
    c. fiberglass and open-cell
    d. extended-surface medium to non-HEPA high-efficiency

15. To perform a general test of a building's air for the presence of contamination resulting from building materials, the instrument used is a _____.

    a. volatile organic compound (VOC) sensor
    b. carbon monoxide (CO) detector
    c. carbon dioxide (CO_2) detector
    d. formaldehyde detector

Summary

Most people are aware that outdoor air pollution can damage their health but do not realize that the effects of poor indoor air can be just as harmful. EPA studies show that indoor levels of many pollutants may be two to five times higher than outdoor levels. These pollutants are dangerous because most people spend as much as 90 percent of their time indoors.

Over the past several decades, exposure to indoor air pollutants has increased because of the following factors:

- Construction of tighter buildings
- Reduced ventilation rates to save energy
- Use of synthetic building materials and furnishings
- Increased use of cleaning products, pesticides, and personal care products

Major sources of pollution must be either eliminated or diluted using ventilation. Air systems must circulate reasonable quantities of fresh and filtered air. They must be maintained with proper filtration systems and kept free of accumulations of dust, dirt, and debris. Buildings, HVAC systems, and other systems must not be allowed to become breeding places for microbial contaminants. Indoor air quality is affected by:

- Initial building design
- Ventilation control
- Thermal comfort control
- Control of chemical contaminants
- Control of microbial contaminants

Notes

Arrestance efficiency: The percentage of dust that is removed by an air filter. It is based on a test where a known amount of synthetic dust is passed through the filter at a controlled rate, then the weight of the concentration of dust in the air leaving the filter is measured.

Atmospheric dust spot efficiency (dust spot efficiency): The percentage of dust that is removed by an air filter. It is the number that is normally referenced in the manufacturer's literature, filter labeling, and specifications. The dust spot efficiency of a filter is based on a test where atmospheric dust is passed through a filter, then the discoloration effect of the cleaned air is compared with that of the incoming air.

Biological contaminants: Airborne agents such as bacteria, fungi, viruses, algae, insect parts, pollen, and dust. Sources include wet or moist walls, duct, duct liner, fiberboard, carpet, and furniture. Other sources include poorly maintained humidifiers, dehumidifiers, cooling towers, condensate drain pans, evaporative coolers, showers, and drinking fountains.

Building-related illness: A situation in which the symptoms of a specific illness can be traced directly to airborne building contaminants.

Desiccant: A material that has a high capacity for absorbing moisture; for example, calcium chloride.

Environmental tobacco smoke (ETS): A combination of sidestream smoke from the burning end of a cigarette, cigar, or pipe and the exhaled mainstream smoke from the smoker.

Formaldehyde: A colorless, pungent byproduct of hydrocarbons that can cause irritation of the eyes and upper air passages.

Friable: The condition in which materials can release particulates into the air.

High-efficiency particulate air (HEPA) filter: An extended media, dry-type filter mounted in a rigid frame. It has a minimum efficiency of 99.97 percent for 0.3-micron particles when a clean filter is tested at its rated airflow capacity.

Microbial contaminants: See *biological contaminants*.

Microbiological contaminants: See *biological contaminants*.

Micron: A unit of length that is one millionth of a meter, or about 1/25,400 of an inch.

Multiple chemical sensitivity (MCS): A medical condition found in some individuals who are vulnerable to exposure to certain chemicals and/or combinations of chemicals. Currently, there is some debate as to whether or not MCS really exists.

New building syndrome: A condition that refers to indoor air quality problems in new buildings. The symptoms are the same as those for sick building syndrome.

Off-gassing: The process by which furniture and other materials release chemicals and other volatile organic compounds (VOCs) into the air.

Ozone: An unstable, poisonous oxidizing agent that has a strong odor and is irritating to the mucus membranes and the lungs. It is formed in nature when oxygen is subjected to electric discharge or exposure to ultraviolet radiation. It is also generated by devices such as photocopiers, electronic air cleaners, and other equipment that uses high voltages.

Pontiac fever: A mild form of Legionnaire's disease.

Radon: A colorless, odorless, radioactive, and chemically inert gas that is formed by the natural breakdown of uranium in soil and groundwater. Radon exposure over an extended period of time can increase the risk of lung cancer.

Sick building syndrome: A condition that exists when more than 20 percent of a building's occupants complain during a two-week period of a set of symptoms, including headaches, fatigue, nausea, eye irritation, and throat irritation, that are alleviated by leaving the building and are not known to be caused by any specific contaminants.

Volatile organic compounds (VOCs): A wide variety of compounds and chemicals found in such things as solvents, paints, and adhesives, that are released as gases at room temperature.

Additional Resources and References

Additional Resources

This module is intended to be a thorough resource for task training. The following reference works are suggested for further study. These are optional materials for continued education rather than for task training.

Building Air Quality, a Guide for Building Owners and Facility Managers, Latest Edition. Washington, DC: U.S. Environmental Protection Agency.

Indoor Air Quality, Latest Edition. Chantilly, VA: Sheet Metal and Air Conditioning Contractors National Association (SMACNA).

Indoor Air Quality in the Building Environment. Troy, MI: Business News Publishing Company.

ACR 2006, Assessment, Cleaning, and Restoration of HVAC Systems, Latest Edition. Washington, DC: National Air Duct Cleaners Association.

Figure Credits

Ryan Homes, 403F01

Topaz Publications, Inc., 403SA01, 403SA02, 403F12 (top two left), 403F13

Carrier Corporation, 403F05

ANSI/ASHRAE Standard 55-2004, Thermal Environmental Conditions for Human Occupancy. © American Society of Heating, Refrigerating and Air Conditioning Engineers, Inc., www.ashrae.org, 403SA03

Trane, 403F08

McQuay International, 403F09, 403F10

National Air Filtration Association, Table 2

CLARCOR Air Filtration Products, 403F12 (top right and bottom)

Airxchange, Inc., 403F17 (heat exchanger wheel)

Munters Corporation, 403F17 (heat pipe heat exchanger)

Steril-Aire, Inc., 403F18

Bacharach, Inc., 403F19 (top)

Digital Control Systems, Inc., 403F19 (bottom left)

Brooks Equipment Co., Inc., 403F19 (bottom right)

Abatement Technologies, 403F20, 403F21

NCCER makes every effort to keep these textbooks up-to-date and free of technical errors. We appreciate your help in this process. If you have an idea for improving this textbook, or if you find an error, a typographical mistake, or an inaccuracy in NCCER's Contren® textbooks, please write us, using this form or a photocopy. Be sure to include the exact module number, page number, a detailed description, and the correction, if applicable. Your input will be brought to the attention of the Technical Review Committee. Thank you for your assistance.

Instructors – If you found that additional materials were necessary in order to teach this module effectively, please let us know so that we may include them in the Equipment/Materials list in the Annotated Instructor's Guide.

Write: Product Development and Revision
National Center for Construction Education and Research
3600 NW 43rd St, Bldg G, Gainesville, FL 32606

Fax: 352-334-0932

E-mail: curriculum@nccer.org

Craft _____ Module Name _____

Copyright Date _____ Module Number _____ Page Number(s) _____

Description _____

_____

_____

_____

(Optional) Correction _____

_____

_____

(Optional) Your Name and Address _____

_____

_____

03404-09

Energy Conservation Equipment

03404-09
Energy Conservation Equipment

Topics to be presented in this module include:

Overview

In these times of shrinking energy resources, environmental concerns, and high energy cost, the world is focusing on ways to reduce energy consumption. Because HVAC systems consume a great deal of energy, these systems are primary targets in the energy conservation battle. In response to the need to conserve energy, HVAC manufacturers have developed innovative devices that make use of heat that might otherwise be wasted. They have also come up with novel ways to minimize the use of mechanical cooling when outdoor air can be substituted. Because of the increased use of energy conservation devices in residential and commercial applications, it is important that installers and service technicians understand them.

Objectives

When you have completed this module, you will be able to do the following:

1. Identify selected air-to-air heat exchangers and describe how they operate.
2. Identify selected condenser heat recovery systems and explain how they operate.
3. Identify a coil energy recovery loop and explain how it operates.
4. Identify a heat pipe heat exchanger and explain how it operates.
5. Identify a thermosiphon heat exchanger and explain how it operates.
6. Identify a twin tower enthalpy recovery loop system and explain how it operates.
7. Identify airside and waterside economizers and explain how each type operates.
8. Identify selected steam system heat recovery systems and explain how they operate.
9. Identify an ice bank-type off-peak hours energy reduction system.
10. Operate selected energy conversion equipment.

Trade Terms

Monel®
Retort
Runaround loop
Sensible heat recovery
  device

Thermosiphon
Total heat recovery
  device

Required Trainee Materials

1. Pencil and paper
2. Appropriate personal protective equipment

Prerequisites

Before you begin this module, it is recommended that you successfully complete *Core Curriculum*; *HVAC Level One*; *HVAC Level Two*; *HVAC Level Three*; and *HVAC Level Four*, Modules 03401-09 through 03403-09.

This course map shows all of the modules in the fourth level of the *HVAC* curriculum. The suggested training order begins at the bottom and proceeds up. Skill levels increase as you advance on the course map. The local Training Program Sponsor may adjust the training order.

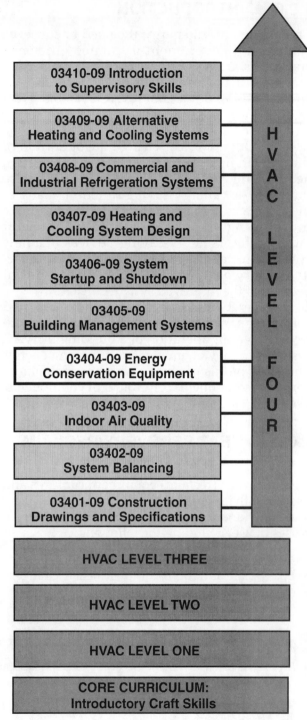

03410-09 Introduction to Supervisory Skills

03409-09 Alternative Heating and Cooling Systems

03408-09 Commercial and Industrial Refrigeration Systems

03407-09 Heating and Cooling System Design

03406-09 System Startup and Shutdown

03405-09 Building Management Systems

03404-09 Energy Conservation Equipment

03403-09 Indoor Air Quality

03402-09 System Balancing

03401-09 Construction Drawings and Specifications

HVAC LEVEL THREE

HVAC LEVEL TWO

HVAC LEVEL ONE

CORE CURRICULUM: Introductory Craft Skills

HVAC LEVEL FOUR

404CMAP.EPS

1.0.0 ◆ INTRODUCTION

The higher cost of energy, the need to conserve energy, and government-mandated efficiency standards are all factors that have caused an increase in the use of heat recovery and/or energy-saving devices in HVAC systems. Heat recovery devices save energy through the capture and reuse of heat that would otherwise be wasted. Other devices change the operation of the system in a way that increases the system heating and cooling efficiencies. In addition to their heat- or energy-saving function, many of these devices are also designed to help improve the building indoor air quality. Use of one or more of these energy-saving devices in a new system often allows the selection of lower capacity primary heating and/or cooling equipment because of the improved system efficiency.

There are both nonautomated and automated energy management systems. These systems control the overall operation of a building's HVAC systems, so they operate without wasting energy. The focus of this module is on some of the more common components, or groups of components, used in HVAC systems to help conserve energy.

2.0.0 ◆ HEAT RECOVERY/RECLAIM METHODS AND EQUIPMENT

Heat recovery (reclaim) equipment captures and uses heat that would otherwise be wasted. There are many kinds of heat recovery devices and processes in use, including the following:

- Energy and heat recovery ventilators
- Fixed-plate and rotary air-to-air heat exchangers
- Condenser heat recovery

- Coil energy recovery loops
- Heat pipe heat exchangers
- Thermosiphon heat exchangers
- Twin tower enthalpy recovery loops

2.1.0 Energy and Heat Recovery Ventilators

Energy-efficient homes and buildings do a good job of keeping conditioned heated or cooled air in, but they also seal in air that has been recirculated within the building many times. This causes the air to become stale and contaminated with airborne particles. ASHRAE standards recommend that a building's indoor air be exchanged for fresh outdoor air at a rate of 0.35 air changes per hour. An alternate method recommended by ASHRAE calls for an exchange rate of 15 cfm per person, 20 cfm per bathroom, and 25 cfm per kitchen. Ventilators are one type of HVAC equipment that can be used to help solve poor indoor air quality problems within a building by bringing a controlled amount of outside air into the building. In addition to helping maintain good indoor air quality, ventilators also help to conserve energy.

There are two types of ventilators: energy recovery ventilators (ERVs) and heat recovery ventilators (HRVs). ERVs are used to supply fresh air and recover from both heating and cooling operations. ERVs are used in most localities in the United States. HRVs are used to supply fresh air and recover heat energy during the heating season. They typically are installed in homes in colder climates that have longer heating seasons, such as those in the northern part of the United States and Canada.

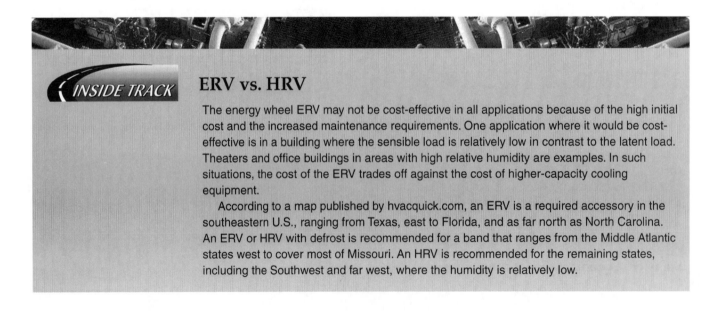

INSIDE TRACK

ERV vs. HRV

The energy wheel ERV may not be cost-effective in all applications because of the high initial cost and the increased maintenance requirements. One application where it would be cost-effective is in a building where the sensible load is relatively low in contrast to the latent load. Theaters and office buildings in areas with high relative humidity are examples. In such situations, the cost of the ERV trades off against the cost of higher-capacity cooling equipment.

According to a map published by hvacquick.com, an ERV is a required accessory in the southeastern U.S., ranging from Texas, east to Florida, and as far north as North Carolina. An ERV or HRV with defrost is recommended for a band that ranges from the Middle Atlantic states west to cover most of Missouri. An HRV is recommended for the remaining states, including the Southwest and far west, where the humidity is relatively low.

According to the U.S. Department of Energy, most models are capable of recovering about 60 to 80 percent of the energy from the exiting air and delivering the energy to the incoming fresh air. Typically, an ERV/HRV improves the indoor air by changing the air about every three hours.

Air from the living space is passed through the ERV or HRV and exhausted outside (*Figure 1*). At the same time, fresh air is brought in from the outside and sent through the unit. When the two airstreams pass through the heat exchanger core, most of the heat or cooling from the outgoing indoor air is transferred to the incoming fresh outdoor air. The core design allows this transfer of heat and cooling between the entering and leaving airstreams to occur without mixing the two airstreams. The result is a constant stream of fresh air being delivered to the living space.

The main difference between an ERV and HRV is the way the heat exchanger core works. In the HRV, only sensible heat is transferred. That's why HRVs are used mainly in colder climates. In the ERV, the core has the capability of transferring both sensible and latent heat, allowing it to transfer heat in the winter and remove moisture from the air during the summer cooling season. This makes the use of ERVs popular in humid climates, such as in the Southeast. Upon installation of an ERV or HRV, balancing of the air distribution system is critical to make sure that the amounts of incoming and outgoing air are equal.

Some commercial building air-conditioning systems use an ERV unit in conjunction with a rooftop-package air conditioning unit. As shown in *Figure 2*, this can be done using either a standalone ERV unit or one that is fastened directly over the outdoor intake of the rooftop unit. With an ERV being used, the outdoor air first enters and is preconditioned by the ERV, rather than entering the rooftop unit directly. Use of a standalone

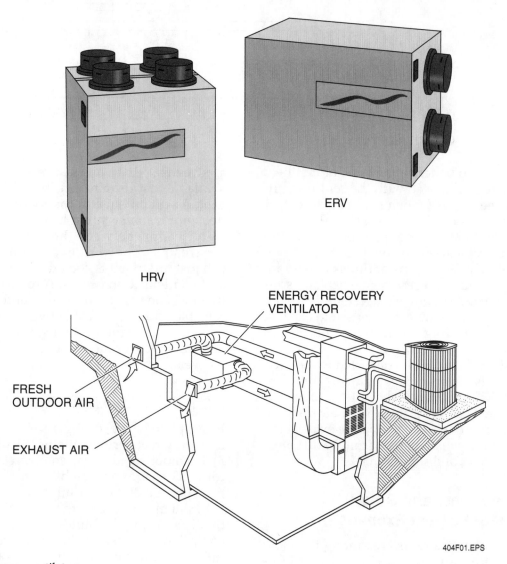

ERV

HRV

ENERGY RECOVERY VENTILATOR

FRESH OUTDOOR AIR

EXHAUST AIR

404F01.EPS

Figure 1 ◆ Recovery ventilators.

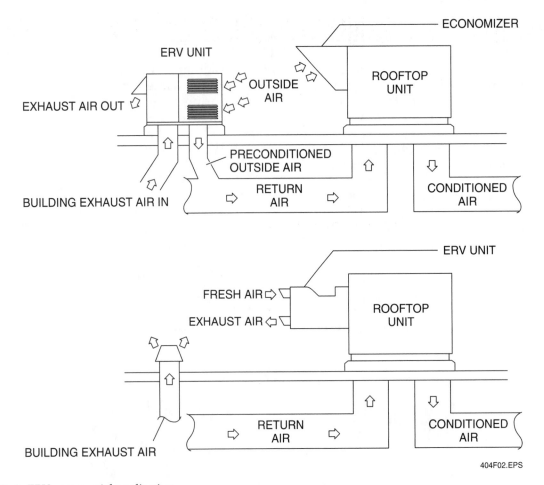

Figure 2 ◆ ERV commercial application.

ERV unit has two benefits. One is that it allows an economizer to be used with the rooftop unit. This is because the ERV mounts on a separate roof curb rather than on the outdoor intake of the rooftop unit. During economizer operation, the rooftop unit typically controls the stand-alone ERV so that its exhaust fan continues to operate but the supply fan and recovery wheel are shut down. The second benefit is that it eliminates the need for an exhaust fan normally used to exhaust air from the building bathrooms, conference rooms, and similar areas. This is because the building exhaust air ductwork is connected to the ERV, and the ERV is used in place of an exhaust fan. Because an ERV fastened to the outdoor intake of the rooftop unit can draw its exhaust air from the return duct only, the use of a separate exhaust fan is still required to exhaust air from bathrooms, conference rooms, and similar areas.

2.2.0 Fixed-Plate and Rotary Air-to-Air Heat Exchangers

Air-to-air heat exchangers are among the most common devices used to recover heat by transferring the heat between the supply and exhaust airstreams. Air-to-air heat recovery devices are available that reclaim sensible heat only or total heat. A **sensible heat recovery device** is one that does not transfer latent heat (heat contained in water vapor) between the supply and exhaust airstreams. The one exception is when the exhaust airstream is cooled below its dew point and condensation occurs. A **total heat recovery device** is one that can recover and transfer both sensible and latent heat between the supply and exhaust airstreams. Total heat recovery devices normally recover more energy than sensible heat recovery devices.

2.2.1 Fixed-Plate Heat Exchangers

Fixed-plate heat exchangers are commonly used in energy recovery ventilator (ERV) units (*Figure 3*). In colder weather, the energy recovery ventilator (ERV) saves energy by transferring from 70 to 80 percent of the warmth contained in the heating system exhaust air to the cold ventilation air that is entering the building. In the summer, air conditioned indoor air flowing through the ERV unit is used to cool the warmer incoming fresh air as it passes through the ERV unit. During both

Heat Recovery Ventilator (HRV) Units

In addition to making ERV units like the one shown here, some manufacturers also make heat recovery ventilator (HRV) units. The HRV units are similar in construction and operation to the ERV unit. However, the fixed-plate heat exchanger in the HRV is designed to transfer only sensible heat between the fresh incoming air and the stale exhaust air. For this reason, the HRV unit is intended for use mainly in colder climates that have longer heating seasons and mild summers. Fixed-plate heat exchangers used in HRVs can accumulate frost as part of their normal operation when outdoor temperatures drop below freezing. For this reason, some fixed-plate HRVs are designed to periodically defrost the heat exchanger for more efficient operation. Different manufacturers use different methods to achieve this defrost.

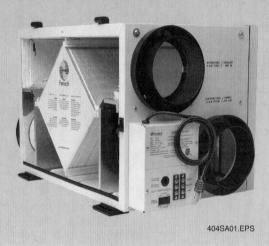

404SA01.EPS

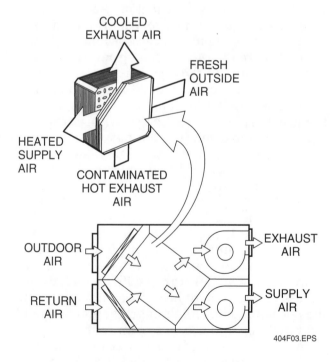

404F03.EPS

Figure 3 ◆ Cross-flow ERV unit with a fixed-plate heat exchanger.

heating and cooling modes of ERV operation, the unit acts to improve the quality of the building air by allowing the exchange of stale, polluted indoor air for fresh outdoor air.

Outdoor and indoor (return) airstreams are drawn by the ERV unit fan(s) through the ERV unit heat exchanger core, and then are discharged from the ERV exhaust and supply air ducts.

The heat exchanger core is a fixed-plate air-to-air heat exchanger that contains no moving parts. Typically, it consists of alternately layered aluminum plates, separated and sealed, that form exhaust and supply airstream passages. Heat is transferred directly from the warmer air through the separating plates into the cooler air. This is done without mixing the two airstreams. The direction of the supply and exhaust airflow through the exchanger can be parallel, counter-flow, or crossflow (*Figure 4*).

Fixed-plate heat exchangers used in ERVs achieve both sensible heat recovery and latent heat recovery. In the winter, the result is that less heat energy is needed to heat the preheated fresh ventilation air than would be needed to heat

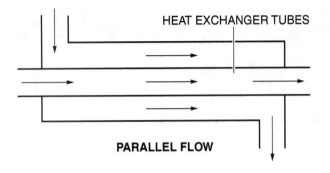

PARALLEL FLOW

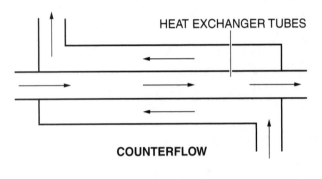

COUNTERFLOW

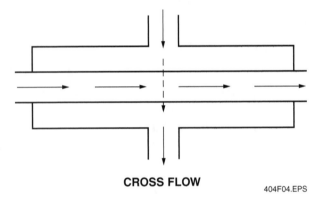

CROSS FLOW

404F04.EPS

Figure 4 ◆ Types of fixed-plate heat exchangers.

cooler air that entered the building solely through natural infiltration or ventilation. Similarly, in the summer, less energy is needed to cool the precooled ventilation air than would be needed to cool warmer air that entered the building through natural infiltration or ventilation. Because the fixed-plate heat exchanger in an ERV transfers some of the moisture from the exhaust air to the usually less humid incoming winter air, the humidity of the building air during the winter tends to remain more constant. In the summer, the ERV fixed-plate heat exchanger transfers some of the water vapor contained in the more

humid incoming air to the drier exhaust air leaving the building, thus providing for dehumidification of the incoming outside air. Because of the ERV's capability to transfer both sensible and latent heat, they are recommended for use in almost all of the United States and some parts of Canada.

Some manufacturers offer an ERV with a fixed-plate heat exchanger that is coated with a desiccant so it can absorb moisture. These heat exchangers are not able to handle extreme moisture, but provide effective energy and enthalpy transfer within reasonable limits.

2.2.2 Rotary (Wheel) Heat Exchangers

A rotary air-to-air heat exchanger, or heat wheel, consists of a motor-driven revolving cylinder containing heat transfer media through which the airstreams pass (*Figure 5*). The supply and exhaust airstreams flow through half of the heat exchanger in a counterflow pattern. To minimize the mixing of the two airstreams (cross-contamination), the sections of the wheel are separated by a partition or purge section. The barriers prevent exhaust air trapped within the heat transfer media from being carried over to the supply side. Depending on the kind of heat transfer media used in the wheel, rotary heat exchangers are available that can recover either sensible heat only or total heat. Types of wheel media commonly used in sensible heat recovery units include aluminum, copper, stainless steel, or **Monel**®. Monel® is an alloy made of nickel, copper, iron, manganese, silicon, and carbon that is very resistant to corrosion. Media used in total heat recovery units include several kinds of metal, mineral, or synthetic materials that are treated with a desiccant such as lithium chloride or alumina.

As shown in *Figure 6*, sensible heat is transferred as the media picks up and stores heat from the warmer airstream and releases it to the cooler one. Latent heat is transferred as the media condenses moisture from the airstream with the higher humidity. It does this either because the media temperature is below its dew point or by means of absorption (liquid desiccants) or adsorption (solid desiccants) with a simultaneous release of heat. Latent heat is also transferred by the release of moisture through evaporation (and heat pickup) into the airstream with the lower humidity ratio. Thus, moist air is dehumidified while the drier air is humidified. Transfer of sensible and latent heat occurs simultaneously.

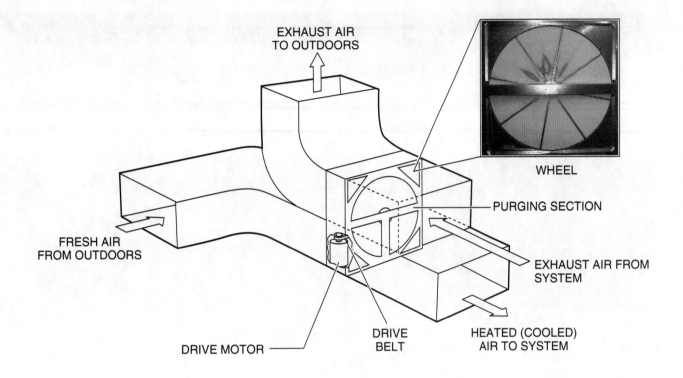

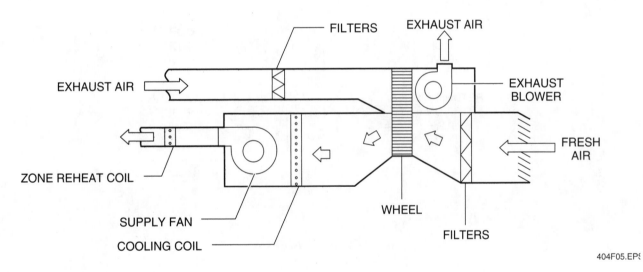

Figure 5 ◆ Rotary air-to-air heat exchanger (heat wheel).

The capacity of rotary heat recovery heat exchangers can be controlled by varying the speed of the wheel rotation via its drive motor. Another method commonly used is a supply air bypass control. This method uses an air bypass damper, controlled by a supply air discharge sensor. The bypass control determines the proportion of supply air allowed to flow through and bypass the heat exchanger wheel.

Rotary heat exchangers should be maintained as directed by the manufacturer. The wheel media should be cleaned when lint, dust, or other foreign materials accumulate. Media with liquid desiccants for total heat recovery must not be wetted.

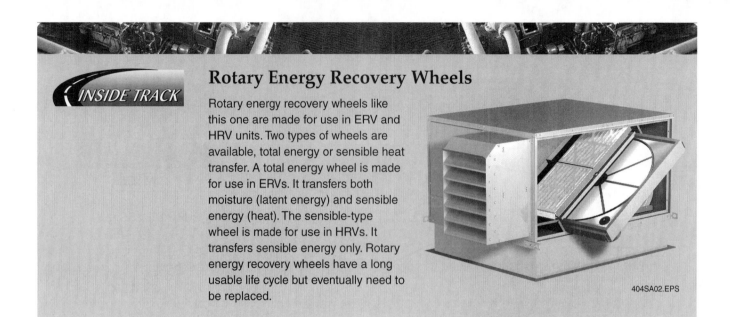

Rotary Energy Recovery Wheels

Rotary energy recovery wheels like this one are made for use in ERV and HRV units. Two types of wheels are available, total energy or sensible heat transfer. A total energy wheel is made for use in ERVs. It transfers both moisture (latent energy) and sensible energy (heat). The sensible-type wheel is made for use in HRVs. It transfers sensible energy only. Rotary energy recovery wheels have a long usable life cycle but eventually need to be replaced.

404SA02.EPS

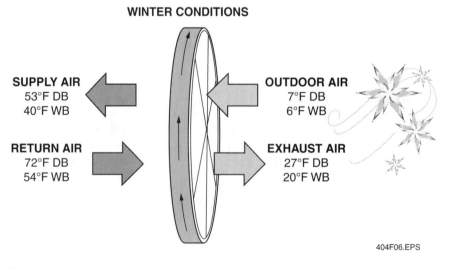

SUMMER CONDITIONS

SUPPLY AIR
81°F DB
68°F WB

OUTDOOR AIR
95°F DB
78°F WB

RETURN AIR
75°F DB
61°F WB

EXHAUST AIR
89°F DB
73°F WB

WINTER CONDITIONS

SUPPLY AIR
53°F DB
40°F WB

OUTDOOR AIR
7°F DB
6°F WB

RETURN AIR
72°F DB
54°F WB

EXHAUST AIR
27°F DB
20°F WB

404F06.EPS

Figure 6 ◆ Heat wheel operation.

2.3.0 Condenser Heat Recovery Systems

Use of rejected condenser heat in an HVAC system is a common heat recovery method. It uses a second (heat recovery) condenser to extract heat from the hot refrigerant gas. This recovered heat is then transferred to air or water that is used to condition the occupied space.

2.3.1 Air-Conditioning/Refrigeration System Condenser Heat Recovery

A dual-condenser refrigeration system typical of those used in commercial buildings is shown in *Figure 7*. When the building thermostat calls for heat, a refrigerant hot gas diverting valve actuates and routes the refrigeration system compressor discharge gas through the heat recovery condenser. This air-cooled condenser is located in the building's air handling unit ductwork. As building air flows through the recovery condenser, the heat in the refrigerant gas is rejected to the cooler air. In turn, the heated air is circulated by the air distribution system blower(s) for subsequent dispersal through the building. When the heating demand is greater than can be supplied by the recovery condenser, the difference is often made up by either gas or electric heaters. When no building heat is required, the hot gas diverting valve causes the compressor hot discharge gas to circulate through the refrigeration system's primary air-cooled condenser, where the heat is rejected into the atmosphere.

A practical example of the system just described is often used in a factory where refrigeration is used as part of an industrial process or to chill edible products in a food processing plant. The condenser heat from the industrial process would be used to heat office areas in the factory or to provide heat for another industrial process.

On a smaller scale, refrigerant-to-water heat exchangers (*Figure 8*) are sometimes used in residences and small businesses to heat domestic water. In these instances, the heat exchanger is installed in the compressor's discharge line.

The compressor discharge gas heats the water flowing through the heat exchanger, which then flows to another water-to-water heat exchanger located in the domestic hot water heater tank.

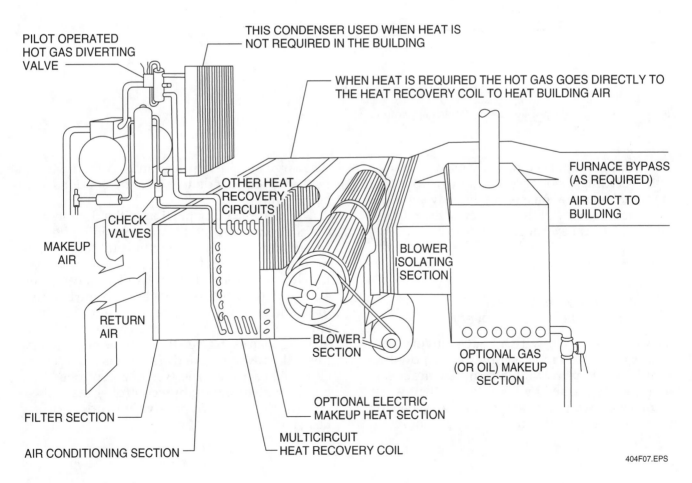

Figure 7 ◆ Dual-condenser system.

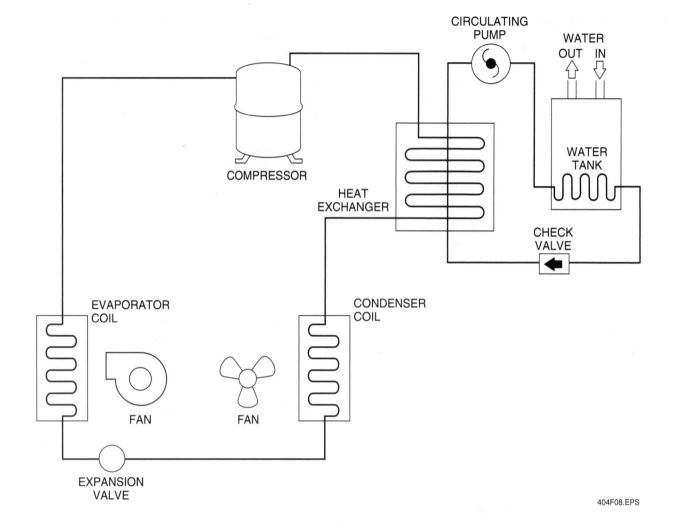

Figure 8 ◆ Refrigerant-to-water heat exchanger application.

This arrangement is commonly used in restaurants where high cooling loads and year-round air conditioning operation ensures a steady supply of hot water. In many of these systems, the refrigerant to water heat exchanger is able to provide sufficient heat, eliminating the need for an additional heat exchanger in the domestic water tank.

2.3.2 Chilled-Water System Condenser Heat Recovery

A chiller equipped with a heat recovery condenser is referred to as a chiller with a double-bundle condenser. Double-bundle condensers are typically formed by two independent water circuits enclosed in the same condenser shell with a common refrigerant chamber. The compressor

discharge hot gas output can be directed to the heat recovery condenser, cooling condenser, or both (*Figure 9*).

The heat recovery portion of the double-bundle condenser is piped into the building heating circuit and can supply all of the building heating needs up to the total heating rejection capacity of the chiller bundle. When the available heat generated from the chiller system exceeds the building load, the surplus heat is rejected into the atmosphere by the cooling tower. When the building heating load is greater than can be supplied by the chiller system, the difference is made up by an auxiliary heater. The system is controlled by a heated water temperature controller that acts to control the tower bypass valve and auxiliary heater.

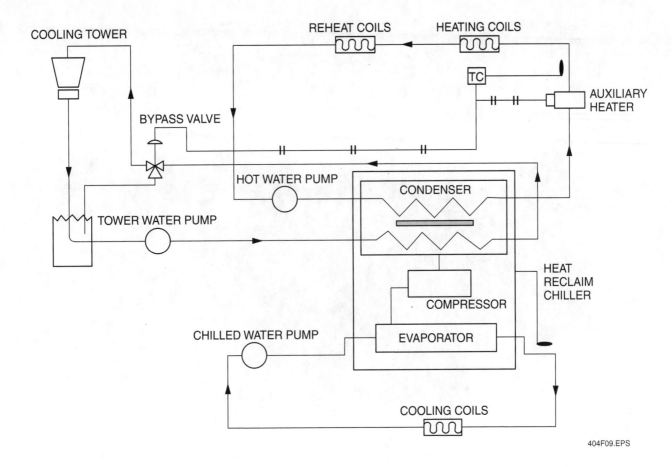

404F09.EPS

Figure 9 ◆ Double-bundle heat reclaim system.

2.3.3 Swimming Pool Heat Recovery Systems

Indoor swimming pools provide an excellent opportunity for heat recovery. Specialized high-capacity dehumidification systems, such as the one shown in *Figure 10*, are specifically designed for the high-moisture environments found in swimming pool structures, as well as in some commercial and industrial environments. This system uses heat pump technology to cool and dehumidify air from the pool enclosure. At the same time, heat from the warm, humid air in the pool enclosure is recovered for re-use. The recovered heat can be used to heat the structure or the pool water.

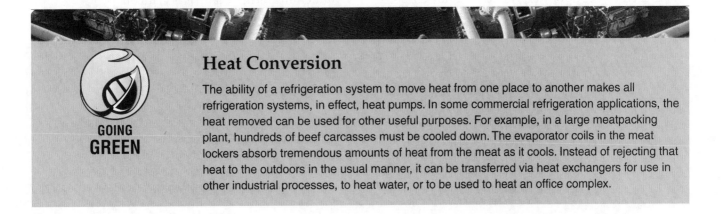

GOING GREEN

Heat Conversion

The ability of a refrigeration system to move heat from one place to another makes all refrigeration systems, in effect, heat pumps. In some commercial refrigeration applications, the heat removed can be used for other useful purposes. For example, in a large meatpacking plant, hundreds of beef carcasses must be cooled down. The evaporator coils in the meat lockers absorb tremendous amounts of heat from the meat as it cools. Instead of rejecting that heat to the outdoors in the usual manner, it can be transferred via heat exchangers for use in other industrial processes, to heat water, or to be used to heat an office complex.

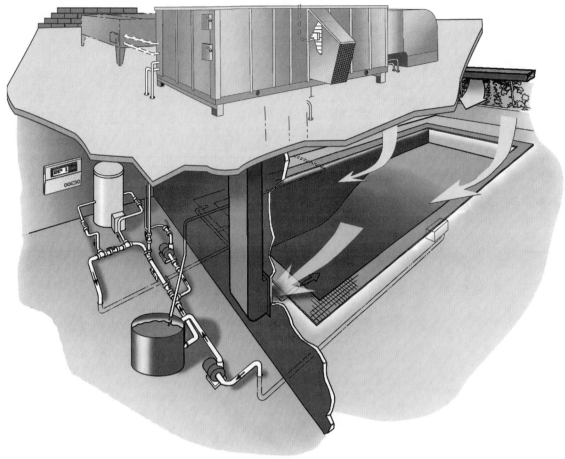

404F10.EPS

Figure 10 ◆ Swimming pool heat recovery system installation.

As shown in *Figure 11*, the hot, high-pressure gas leaving the compressor can be routed to the condenser/reheat coil, pool water condenser, or auxiliary condenser, as needed. A microprocessor control activates the solenoid valves based on demand. Hot liquid refrigerant leaving the condensers is stored in the receiver. As this refrigerant passes through the expansion valve, it is expanded to the operating pressure and tempera-

ture of the evaporator so that it can absorb heat from the pool return air. *Figure 12* shows the physical arrangement of the unit.

Other environments in which this dehumidification and reheat technology can be used include museums, printing facilities, warehouses, plywood manufacturing facilities, and water treatment plants. All of these are applications in which humidity control is critical.

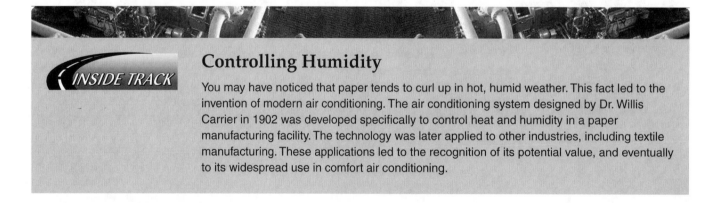

Controlling Humidity

INSIDE TRACK

You may have noticed that paper tends to curl up in hot, humid weather. This fact led to the invention of modern air conditioning. The air conditioning system designed by Dr. Willis Carrier in 1902 was developed specifically to control heat and humidity in a paper manufacturing facility. The technology was later applied to other industries, including textile manufacturing. These applications led to the recognition of its potential value, and eventually to its widespread use in comfort air conditioning.

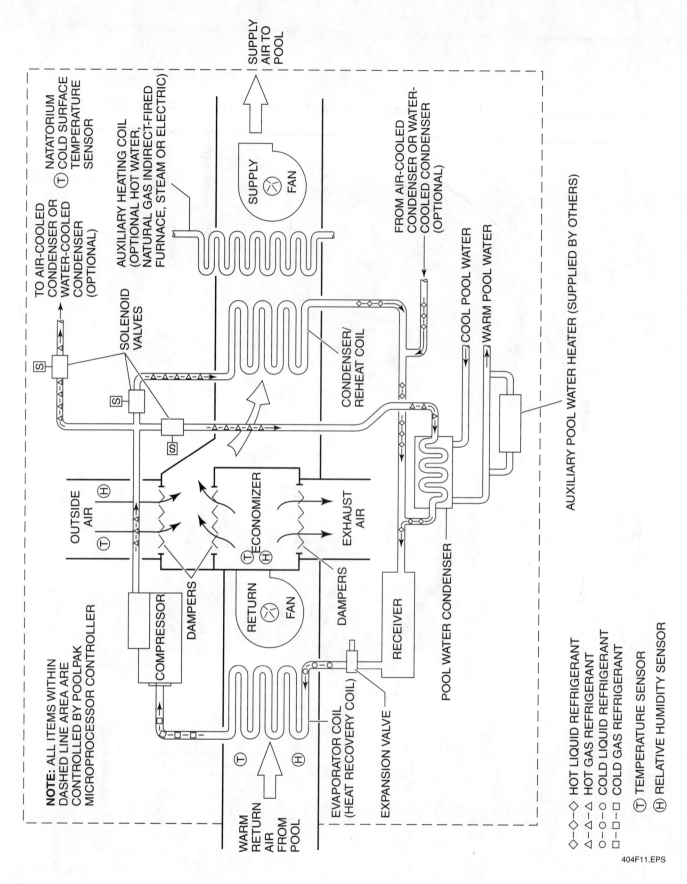

Figure 11 ◆ Swimming pool heat recovery system schematic.

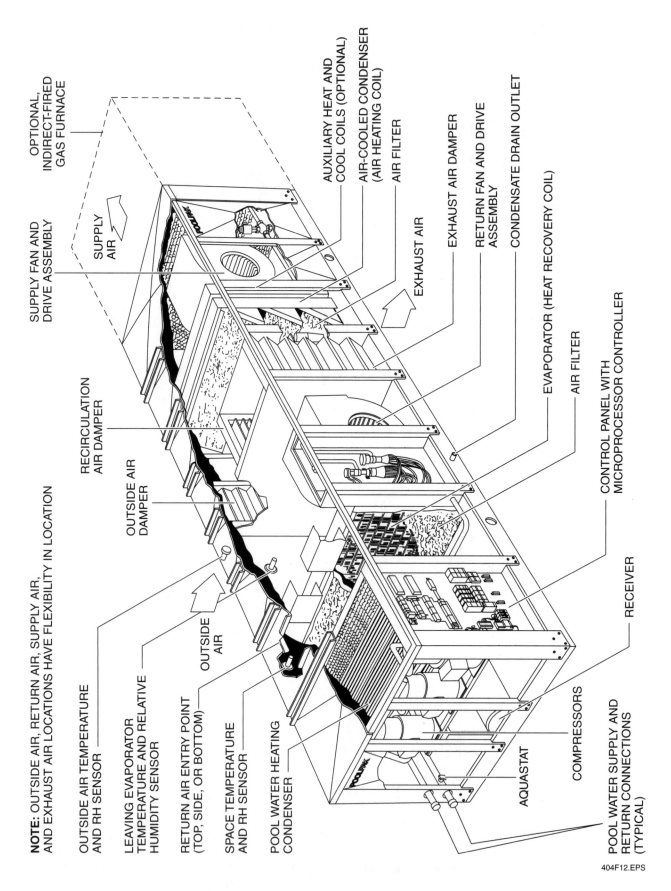

OPTIONAL, INDIRECT-FIRED GAS FURNACE

SUPPLY AIR

SUPPLY FAN AND DRIVE ASSEMBLY

RECIRCULATION AIR DAMPER

OUTSIDE AIR DAMPER

OUTSIDE AIR

AUXILIARY HEAT AND COOL COILS (OPTIONAL)

AIR-COOLED CONDENSER (AIR HEATING COIL)

AIR FILTER

EXHAUST AIR

EXHAUST AIR DAMPER

RETURN FAN AND DRIVE ASSEMBLY

CONDENSATE DRAIN OUTLET

EVAPORATOR (HEAT RECOVERY COIL)

AIR FILTER

CONTROL PANEL WITH MICROPROCESSOR CONTROLLER

RECEIVER

COMPRESSORS

AQUASTAT

POOL WATER SUPPLY AND RETURN CONNECTIONS (TYPICAL)

POOL WATER HEATING CONDENSER

SPACE TEMPERATURE AND RH SENSOR

RETURN AIR ENTRY POINT (TOP, SIDE, OR BOTTOM)

LEAVING EVAPORATOR TEMPERATURE AND RELATIVE HUMIDITY SENSOR

OUTSIDE AIR TEMPERATURE AND RH SENSOR

NOTE: OUTSIDE AIR, RETURN AIR, SUPPLY AIR, AND EXHAUST AIR LOCATIONS HAVE FLEXIBILITY IN LOCATION

404F12.EPS

Figure 12 ◆ Swimming pool heat recovery unit.

2.4.0 Coil Energy Recovery Loops

In some commercial buildings or factories, stale or contaminated air must be exhausted and fresh air brought in. To prevent energy from being lost in the exhaust air and to condition the incoming air, a coil energy recovery loop can be used. The typical coil energy recovery loop (**runaround loop**) consists of two finned-tube water coils, a pump, a thermostatically controlled three-way valve, and related system piping (*Figure 13*). The coils are connected in a closed loop by the piping through which water, or another heat transfer fluid, such as glycol, is pumped. One coil is installed in the exhaust duct. The other coil is installed in the incoming air (supply) duct so that the incoming air that flows through the coil is preheated (or precooled). Installation of the coils in these locations gives the greatest temperature difference between the outside air supply and exhaust airstreams; therefore, the maximum energy recovery occurs. Sensible heat is transferred between the exhaust and incoming supply airstreams

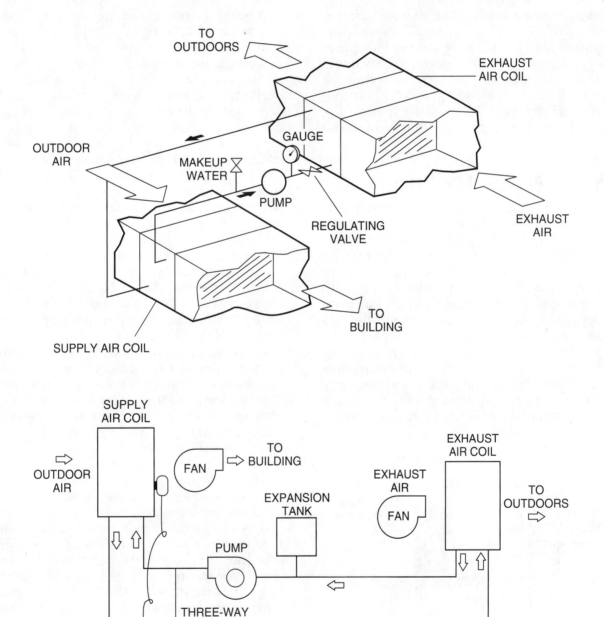

404F13.EPS

Figure 13 ◆ Coil loop heat recovery system.

without any cross-contamination. In comfort air-conditioning systems, the heat transfer can be reversed. In winter, the supply air is preheated when it is cooler than the exhaust air; in summer, the supply air is precooled when it is warmer than the exhaust air. The recovery efficiency for runaround loops averages between 40 and 65 percent; the recovery efficiency of the cooling cycle is somewhat less than that for heating. This is because in the cooling cycle, the temperature difference between the airstreams is not as great.

When a glycol solution is used as the intermediate heat transfer fluid in a runaround loop system, there is some protection against freezing. However, moisture must not be allowed to freeze in the exhaust coil air passages. The dual-purpose, three-way temperature control valve prevents the exhaust coil from freezing. This valve is controlled to maintain the water entering the exhaust coil at a temperature over 30°F. This condition is maintained by bypassing some of the warmer water (or glycol solution) around the supply air coil.

2.5.0 Heat Pipe Heat Exchangers

In a conventional air-conditioning system, dehumidification of the air occurs at the system's cooling coil (*Figure 14*). The coil normally removes both sensible and latent heat from the entering air, which is a mixture of water vapor and dry gases. Both lose sensible heat during contact with the first part of the cooling coil, which functions as a dry cooling coil. Latent heat (heat contained in water vapor) is removed only in the part of the coil that is below the dew point of the entering air. When the coil starts to remove moisture, thus dehumidifying the air, the cooling surfaces carry both the sensible and latent heat load. To remove large amounts of moisture in a hot, humid environment, an air conditioner needs to operate longer; therefore, it consumes more energy. To remove more moisture from the air, the thermostat setpoint is usually lowered. This allows the system to run longer, which removes more moisture. This results in the conditioned air being overcooled and too cool for human comfort. To remedy this condition, the air is usually reheated before it is delivered to the conditioned space. Reheating may also be needed to decrease the relative humidity of the overcooled air. This reheating, which is often done using an electric reheat coil installed in the system, consumes extra energy.

A heat pipe heat exchanger is used to increase the dehumidification capacity of a system and reduce its energy consumption. It does this by precooling the incoming air before it gets to the system cooling coil (*Figure 15*). A heat pipe heat exchanger transfers heat from one end of the exchanger to the other. It is a passive device that does not need an energy input. A heat pipe heat exchanger is an assembly formed by a bank of individual closed copper tubes with aluminum fins that are not interconnected. Each of these tubes is lined with a capillary wick, sealed at both ends, evacuated to a vacuum level, and charged with a refrigerant. The individual heat pipes are assembled into a heat exchanger unit. This type of heat exchanger can only be installed where the supply and exhaust ducts are mounted next to each other. When the exchanger is mounted in the system ductwork, the ends of the heat pipes in the hot duct act as an evaporator, while the ends of the pipes in the cold duct act as a condenser. Heat pipe heat exchangers are sensible heat transfer devices, but condensation on the fins does allow for some transfer of latent heat.

When the evaporator side of the exchanger is exposed to the incoming warm airstream, the refrigerant in each of the pipes absorbs heat and evaporates (*Figure 16*). This precools the incoming air before it contacts the cooling coil.

The refrigerant vapor in each pipe then flows to the cooler condenser end of the pipe. Because the condenser end of the pipe is exposed to the cooler supply airstream, the refrigerant vapor inside the tube transfers its heat to the cooler airstream and condenses. Transfer of this heat warms the cooled supply airstream to a more comfortable temperature. After condensing, the liquid refrigerant in the condenser end of each heat pipe is returned to its evaporator end by gravity and/or the capillary wick. This closed loop evaporation/condensation process in the heat pipe heat exchanger continues as long as there is enough temperature difference between the two airstreams to drive the process.

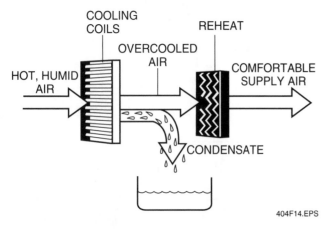

Figure 14 ◆ Dehumidification in a conventional cooling system.

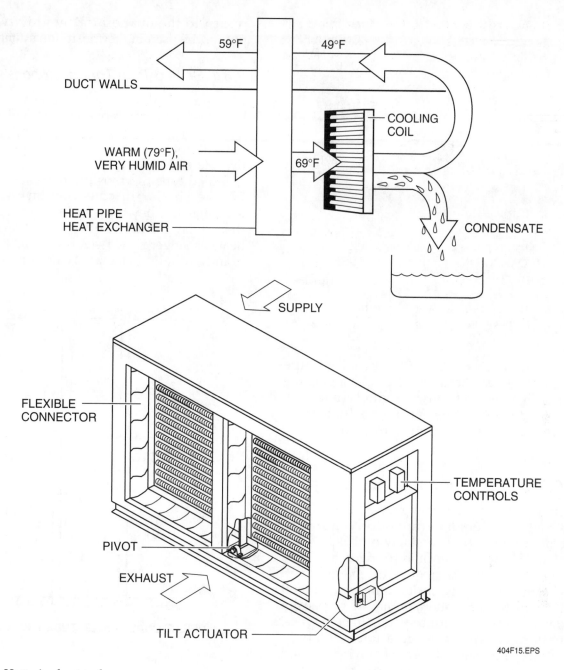

59°F 49°F

DUCT WALLS

COOLING COIL

WARM (79°F), VERY HUMID AIR

69°F

HEAT PIPE HEAT EXCHANGER

CONDENSATE

SUPPLY

FLEXIBLE CONNECTOR

TEMPERATURE CONTROLS

PIVOT

EXHAUST

TILT ACTUATOR

404F15.EPS

Figure 15 ◆ Heat pipe heat exchanger.

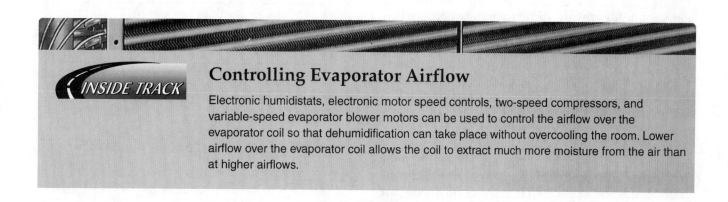

Controlling Evaporator Airflow

Electronic humidistats, electronic motor speed controls, two-speed compressors, and variable-speed evaporator blower motors can be used to control the airflow over the evaporator coil so that dehumidification can take place without overcooling the room. Lower airflow over the evaporator coil allows the coil to extract much more moisture from the air than at higher airflows.

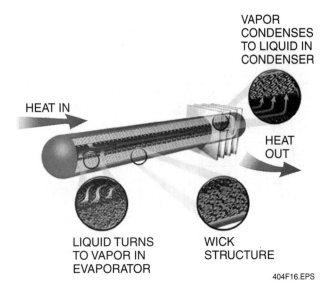

Figure 16 ◆ Heat pipe closed loop evaporation/condensation process.

The amount of heat transferred by a heat pipe heat exchanger can be controlled by changing the slope or tilt of the unit. This tilt control is normally done by a temperature-controlled actuator that rotates the exchanger around the center of its base.

2.6.0 Thermosiphon Heat Exchangers

Thermosiphon heat exchangers are closed systems that consist of an evaporator, condenser, interconnecting piping, and a two-phase (liquid and vapor) heat transfer fluid (refrigerant). They are passive devices that require no energy input. These elements may be enclosed in a single shell (sealed tube thermosiphon) or may be physically separated (coil loop thermosiphon). In both types, the natural convection circulation of the two-phase refrigerant and the force of gravity are used to transfer energy between the two airstreams.

Because part of the system contains vapor and part contains liquid, the pressure in a thermosiphon is determined by the liquid temperature at the liquid-vapor interface. If the surrounding air causes a temperature difference between the liquid and vapor regions, the resulting pressure difference causes the vapor to flow from the warmer region (evaporator) to the cooler region (condenser). This flow is maintained by condensation in the cooler region and evaporation in the warmer region. Depending on the mounting orientation of thermosiphon exchangers, the transfer of heat can be in both directions (bidirectional) or in one direction (unidirectional). When the heat transfer is unidirectional, the evaporator and condenser must be located so that the condensate can

return to the evaporator by gravity, since no pumps are used in thermosiphon systems.

2.6.1 Sealed Tube Thermosiphons

Unlike the heat pipe, sealed tube thermosiphons have no wick; they rely only on gravity to return the condensate to the evaporator end. Heat transfer will not take place if all the liquid resides at the cold end of the tube.

Sealed tube thermosiphon exchangers (*Figure 17*) are similar in construction to heat pipe exchangers. They transfer heat from one end of the exchanger to the other. The thermosiphon heat exchanger is an assembly formed by a group of individual tubes that are not interconnected.

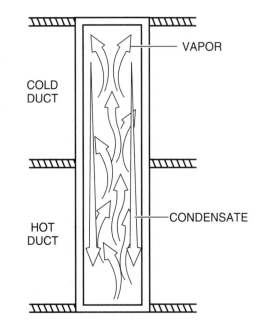

UNIDIRECTIONAL SEALED TUBE THERMOSIPHON

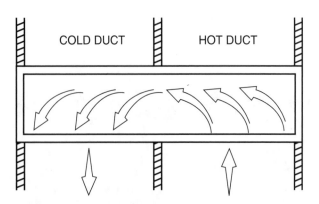

BIDIRECTIONAL SEALED TUBE THERMOSIPHON

404F17.EPS

Figure 17 ◆ Sealed-tube thermosiphon recovery system.

Each of these tubes is sealed at both ends, evacuated to a vacuum level, and charged with a refrigerant. The individual tubes are assembled into an exchanger unit. Like the heat pipe exchanger, the sealed tube thermosiphon exchanger is used only when the supply and exhaust ducts are mounted next to each other. The evaporator and condenser regions are at opposite ends of a bundle of thermosiphon tubes. When the exchanger is mounted in the system ductwork, the ends of the tubes in the hot duct act as an evaporator, while the ends of the tubes in the cold duct act as a condenser.

2.6.2 Coil-Loop Thermosiphons

Thermosiphon loops (*Figure 18*) are used when the supply and exhaust air ducts are not mounted next to each other. A single closed loop consists of two coils interconnected by vapor and condensate return piping. The loop is charged with refrigerant in its saturation state, so that part is filled with liquid and part with vapor. The pressure in the loop depends on the type of refrigerant used in the loop and the fluid temperature at the liquid-vapor interface. Loops may be installed for unidirectional and bidirectional flow. Unidirectional loops are normally more

efficient because the coil and loop charge can be selected to best satisfy only one function, rather than two functions (evaporation and condensation). When necessary, several coil loop thermosiphons can be mounted in the supply and exhaust ducts to achieve a recovery effectiveness greater than that obtained with a single loop.

> **NOTE**
>
> The routing of the interconnecting tubing must be considered, because ambient conditions surrounding this piping can interfere with successful operation. This is not an issue with heat pipe, because heat pipes have adjacent ducts.

2.7.0 Twin Tower Enthalpy Recovery Loops

A twin tower enthalpy system is an air-to-liquid, liquid-to-air recovery system. It can consist of one or more towers (contactor towers) used to process the outdoor supply air and one or more towers used to process the building exhaust air (*Figure 19*). An absorbent solution, typically lithium chloride and water, is continuously circulated by pumps between these supply and exhaust towers. In the towers, the circulated solution is sprayed over the tower contact surfaces where it comes in contact with the related supply or exhaust airstream. Spraying the absorbent solution into the airstream enhances this contact. Because the absorbent solution transfers latent as well as sensible heat, there is a total heat recovery or enthalpy transfer. Recovery efficiencies in the 60 to 70 percent range are typical.

Twin tower enthalpy systems are used mainly for comfort air conditioning. The absorbent solution is an effective antifreeze, allowing the system to operate in winter air temperatures as low as –40°F. In the summer, they can operate with supply temperatures as high as 115°F. When using the twin tower system in colder climates, overdilution of the absorbent solution can occur as the solution becomes saturated. This results in uneven supply air temperatures and humidity levels. To remedy this condition, a thermostatically controlled heater is often used to maintain constant-temperature supply air, regardless of the outdoor air temperature. The heater's control thermostat senses the air temperature leaving the supply tower and turns the heater on and off as needed.

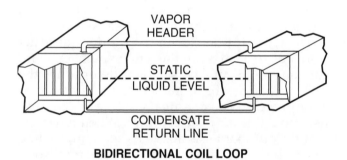

BIDIRECTIONAL COIL LOOP

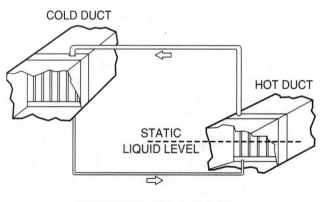

UNIDIRECTIONAL COIL LOOP

404F18.EPS

Figure 18 ◆ Coil-loop thermosiphon system.

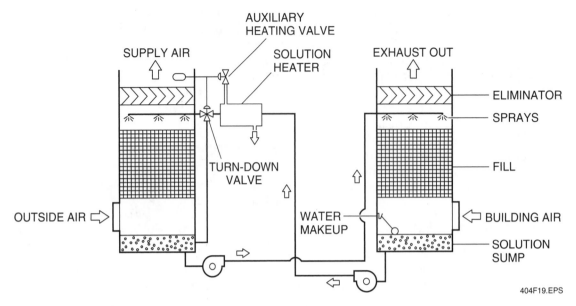

Figure 19 ◆ Twin tower enthalpy recovery loop.

3.0.0 ◆ ECONOMIZERS

An economizer is an accessory typically used in self-contained heating/cooling systems. The benefit of using an economizer is related mainly to the cooling mode of system operation. Economizers can use outdoor airflow (air-side economizer) or cooled water flow (water-side economizer) as the medium to accomplish lower cost cooling.

3.1.0 Air-Side Economizers

An air-side economizer provides control of building cooling and ventilation. It does this by controlling the amount of outside air brought into a conditioned space. *Figure 20* shows a basic economizer system. It consists of a damper actuator assembly and a related economizer control module.

INSIDE TRACK

Restaurant Indoor Air Quality

Many restaurants vent the kitchen exhaust through the roof. The restaurant's HVAC equipment may also be on the roof. To prevent the recirculation of smoke or kitchen odors into the restaurant through the rooftop unit economizer, the HVAC equipment should be located away from the kitchen exhaust and oriented such that the prevailing winds will not carry kitchen-exhausted smoke or odors toward it. Similar concerns exist for plumbing and other building exhausts in the vicinity of the rooftop unit. Also, consider any local code requirements.

Four conditions are used to control operation of an economizer: outside air, return air, mixed air temperature, and ventilation air. Control signals applied to the economizer control module come from the thermostat located in the conditioned space, an enthalpy sensor located in the outdoor air duct, and a discharge air sensor located on the discharge side of the system evaporator coil. There are many kinds of economizers, but most operate in basically the same way.

The enthalpy sensor in the outdoor air duct responds to changes in the air dry-bulb temperature and humidity. Its setpoint determines the system changeover from cooling using compressor operation (mechanical cooling) to cooling using outside air (free cooling). When it detects that the outdoor air is above its setpoint, cooling for the building is provided by mechanical cooling. When the outdoor air falls below the sensor setpoint, the building is cooled with free cooling.

When the space thermostat calls for cooling and the outdoor air sensed by the enthalpy sensor is below its setpoint, the economizer control module initiates the free cooling mode. In this mode, the compressor is turned off, and the indoor fan is used to bring outside air into the building through motor-actuated dampers. Also, the discharge air sensor monitors the temperature of the air being discharged from the face of the system's indoor coil. This air is a mixture of return air from the conditioned space and fresh outdoor air. The sensor compares this temperature to a predetermined setpoint. It sends a voltage level having a magnitude based on this comparison to the economizer control module. In response to the voltage input from the sensor, the

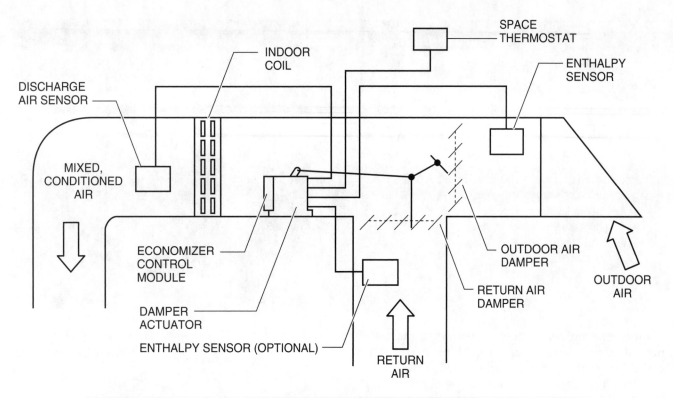

MODE OF OPERATION	OUTDOOR AIR DAMPER	RETURN AIR DAMPER
OFF	CLOSED	WIDE OPEN
FAN ONLY & MECHANICAL COOLING	OPENS TO MINIMUM POSITION FOR VENTILATION	MODULATES TO COMPLEMENT OUTDOOR AIR DAMPER
FREE COOLING	MODULATES TO PROVIDE THE PROPER MIXED AIR TEMPERATURE	MODULATES TO COMPLEMENT OUTDOOR AIR DAMPER
HEATING	OPENS TO MINIMUM POSITION FOR VENTILATION	MODULATES TO COMPLEMENT OUTDOOR AIR DAMPER

404F20.EPS

Figure 20 ◆ Basic air-side economizer.

control module then causes the damper actuator to position (modulate) the outdoor and return air dampers. The outdoor air damper is opened to provide the proper mixed air temperature, while the return damper is closed to complement the outdoor damper. For example, if the outdoor damper is set to the 60 percent open position, then the return damper would be closed to the 40 percent open position. Typically, the economizer works to maintain the temperature of the mixed air between 50°F and 56°F.

When the space thermostat calls for cooling, and the outdoor air sensed by the enthalpy sensor is above its setpoint, the economizer control module turns on the system compressor to provide mechanical cooling. It also causes the damper actuator to modulate the outdoor air damper to its minimum open position to provide ventilation. The return air damper is then opened to complement the outdoor damper.

During system operation, the economizer works to position the outdoor air and return air dampers to achieve the best system performance. The damper positions for the various modes of system operation are summarized in *Figure 20.*

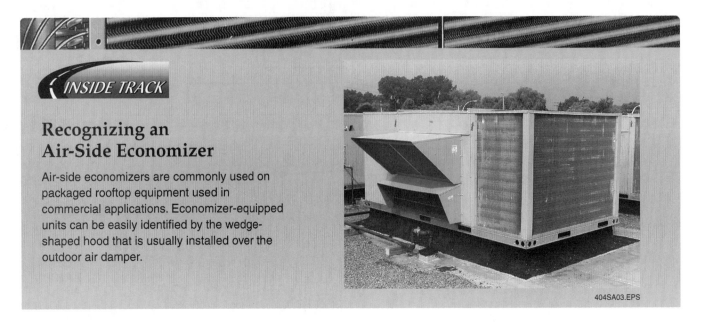

Some economizer systems use an optional second enthalpy sensor located in the return duct. Use of two enthalpy sensors is called differential enthalpy. The differential enthalpy economizer permits the use of outdoor air for cooling, but generally monitors and uses the air with the least enthalpy, regardless of whether it is suitable for total free cooling. It will use mechanical cooling unless outdoor air (OA) enthalpy is sufficiently low to manage the load.

3.2.0 Water-Side Economizers

A water-side economizer uses low-temperature cooling tower water to either precool the entering supply air or to supplement mechanical cooling. If the cooling water is cold enough, it can be used to provide all the system cooling. The water-side economizer consists of a water coil installed upstream of the cooling coil (*Figure 21*). Cooling water flow is controlled by two valves, one at the input to the economizer coil and the other in a bypass loop to the condenser. One method of flow control keeps a constant water flow through the unit. In this mode, the two valves are controlled for complementary operation, where one valve is driven open while the other is driven closed.

Another method provides for variable system water flow through the unit. In this mode, the valve in the bypass loop is an on-off valve and is closed when the economizer is operating. Water flow through the economizer coil is modulated by the valve in its input line. This varies the amount of water flow through the economizer coil in response to the system cooling load. As the cooling load increases, the valve opens more, increasing water flow through the coil. If the economizer valve is fully open and the economizer is unable

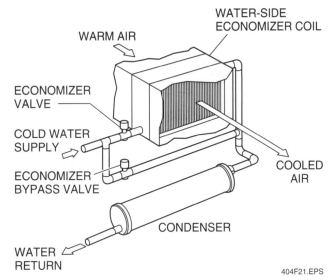

Figure 21 ◆ Water-side economizer.

to satisfy the system cooling load, the system controller turns on the system compressor to supplement the economizer cooling. When the unit is in the heating mode, both the economizer and bypass valves are closed.

4.0.0 ◆ HEAT RECOVERY IN STEAM SYSTEMS

Heat recovery in steam systems is commonly done using direct heat recovery devices, such as heat exchangers/converters to heat air, fluid, or a process. In medium-pressure and high-pressure process steam systems, heat recovery can also be done using the heat in the system's liquid condensate to vaporize or flash some of the liquid to steam at a lower pressure. This lower pressure steam can then be used for comfort heating.

4.1.0 Flash Steam (Flash Tank) Heat Recovery

When hot condensate, under pressure, is released to a lower pressure, part of it is re-evaporated and becomes flash steam. This normally happens when hot condensate is discharged into the condensate return line at a lower pressure than its saturation pressure. Some of the condensate flashes into steam and flows along with the liquid condensate through the return line back to the boiler. This tends to cause an undesirable pressure increase in the condensate return line. In medium-pressure and high-pressure steam systems, flash tanks can be used to remove flash steam from the condensate lines by venting it to either the atmosphere or into a low-pressure steam main for reuse. The heat content of flash steam is the same as that of live steam at the same pressure. Use of the flash steam in a low-pressure steam main allows this heat to be used, rather than wasted. It is commonly used for space heating and heating or preheating water, oil, and other liquids.

When flash steam is used as an energy source, high-pressure steam returns are usually piped to a flash tank (*Figure 22*). Flash tanks can be mounted vertically or horizontally. However, vertical mounting provides better separation of steam and water and, therefore, better quality steam. When the condensate and any flash steam in the return line reach the tank, the high-pressure condensate is released into the lower pressure of the tank, causing some of it to flash into steam. This flash steam is discharged into the low-pressure steam main. The remaining condensate is pumped back to the boiler or discharged to a waste drain.

For proper flash tank operation, the condensate lines should pitch towards the tank. If more than one condensate line feeds the tank, a check valve should be installed in each line to prevent backflow. The top of the tank should have a thermostatic air vent to vent any air that accumulates in the tank. The bottom of the tank should have an inverted bucket or float and thermostatic-type steam trap. The demand steam load on the flash tank should always be greater than the amount of flash steam available from the tank. If it is not, the low-pressure system can be overpressurized. A safety relief valve must be installed at the top of the flash tank to protect the low-pressure line from overpressurization. Because the flash steam produced in the flash tank is less than the amount

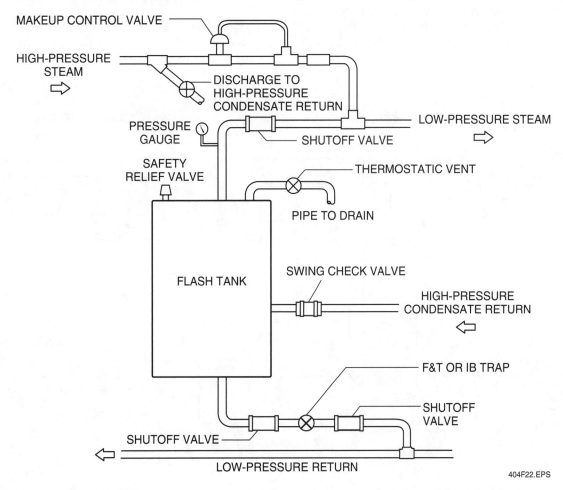

Figure 22 ◆ Steam system flash tank.

404F22.EPS

of low-pressure steam that is needed, a makeup valve in the high-pressure steam line is used to supply any additional steam needed to maintain the correct pressure in the low-pressure line.

4.2.0 Flue Gas Heat Recovery System

The sensible heat available in the flue products of steam/hot water boilers can be recovered and put to use instead of wasted. This can be done by inserting a heat reclaimer (heat exchanger) between the boiler flue output and the stack. *Figure 23* shows a steam boiler with a stack heat exchanger that is used to preheat the boiler's feedwater supply. In this system, a portion of the heat traveling through the exhaust stack is absorbed by the water circulating through the heat exchanger coil. The heated water is then returned back into the reheat tank in the boiler's feedwater system. Once the water circulated in the feedwater system has been initially heated, it is then reheated without using additional fuel. This reduces the boiler burn time and can result in a typical increase in system efficiency of 10 percent or more. An automatic thermostat is normally used to control the feedwater circulating pump. A similar use involving the heat exchanger is to heat water used for the building's domestic hot water supply.

A major consideration when using a stack heat exchanger is that it must offer negligible resistance to the flow of the flue gases in the stack. A heat exchanger that offers too much flow resistance can adversely affect the system by excessively reducing both the flow and the temperature of the flue gases. This can cause moisture condensation in the vent, which can cause corrosion. A lower flue gas temperature can also adversely affect the flow of the flue gases out of the stack, posing a possible safety hazard.

> **NOTE**
>
> Consistent cleaning and maintenance of the heat exchanger are required in order to ensure that the heat exchanger does not become clogged with combustion particulate byproducts. These particulates would block the flue and degrade heat transfer. Because of this requirement, it is very important that the heat exchanger be accessible to maintenance personnel.

4.3.0 Blowdown and Heat Recovery System

Blowdown and heat recovery systems are used to recover heat from the boiler blowdown water and use it to preheat the boiler makeup water.

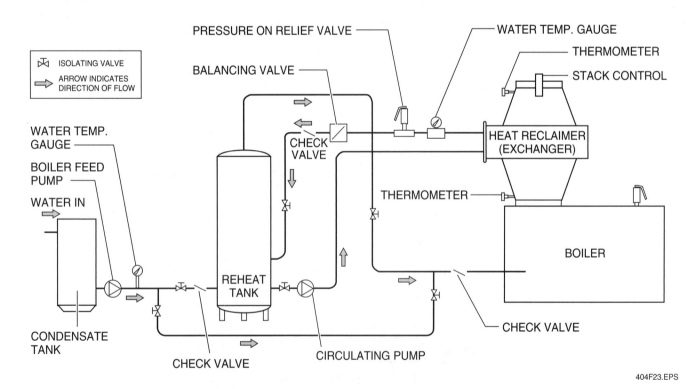

Figure 23 ◆ Flue gas heat recovery system.

Continuous boiler surface blowdown used to purge the solids from a steam boiler system results in constant heat loss. Use of the blowdown and heat recovery system automatically controls the surface blowdown to maintain the desired level of total dissolved solids (TDS) in the boiler. A control valve in the unit senses the flow of makeup water, and positions itself to maintain the desired ratio of blowdown and makeup water. As a result, the concentration of dissolved solids within the boiler is maintained automatically. The control valve also provides for efficient heat recovery (about 90 percent) because the hot blowdown water flows only when there is a corresponding flow of cold makeup water. *Figure 24* shows a typical blowdown and heat recovery system.

> **NOTE**
>
> Codes in many jurisdictions require the blowdown water to be cooled before it is dumped into the municipal sewer system. Note the cold water makeup connection to the heat exchanger in *Figure 24.* This connection provides the means of cooling the blowdown water.

5.0.0 ◆ ELECTRIC UTILITY ENERGY DEMAND REDUCTION SYSTEMS

The electric power consumption demands on an electric utility cycle vary with alternating periods of peak and low demand. For this reason, it is desirable for the utility to try to level-load the demand whenever possible.

With the customer's agreement, some HVAC systems are equipped with power reduction features. One way power reduction can be accomplished is for the utility to cycle equipment off during peak demand periods. The utility accomplishes this by sending a demand reduction signal to a device attached to the customer's HVAC equipment that interrupts power for a short time period. The demand reduction signal can be sent by radio control (*Figure 25*), over phone lines, or through a modem connected to a building's computer-controlled energy management system. The duration of the off cycle is long enough to reduce the utility's peak load but not long enough to noticeably affect indoor comfort.

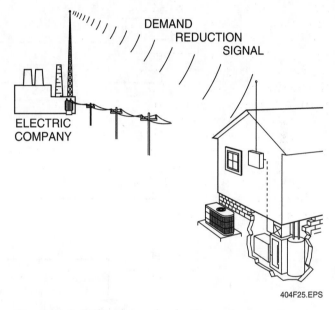

404F25.EPS

Figure 25 ◆ Utility demand reduction system.

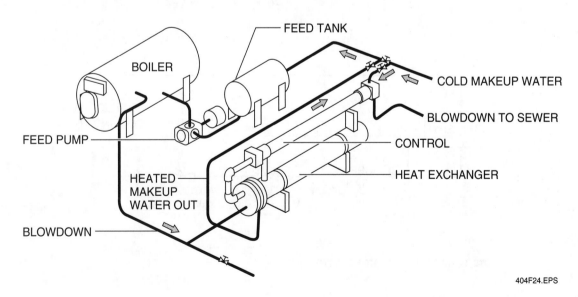

404F24.EPS

Figure 24 ◆ Blowdown and heat recovery system.

If the utility has a large enough customer base participating in the demand reduction program, the duration of the equipment off-time cycle for each customer is reduced to the point that little individual comfort is lost while significantly lowering the peak demand.

5.1.0 Off-Peak Utility Usage

Many electric utilities have a pricing structure that allows commercial and industrial customers to purchase their power at a lower rate during off-peak times. Typically, off-peak times are from midnight to 5:00 AM. At these times, demand is low. By encouraging large customers to use power during these times, the demand during peak usage hours is reduced.

Unfortunately, for many HVAC applications, peak cooling loads and peak utility demand tend to coincide. To reduce HVAC energy use during peak periods, some facilities employ a system involving the use of ice storage tanks. One manufacturer's system, referred to as IceBank® energy storage or off-peak cooling, is described here (*Figure 26*).

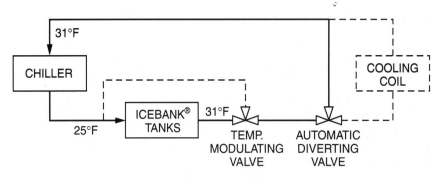

CHARGE CYCLE

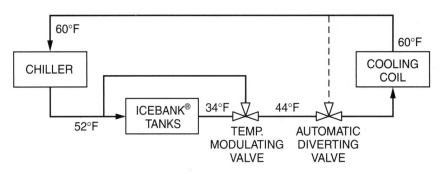

DISCHARGE CYCLE

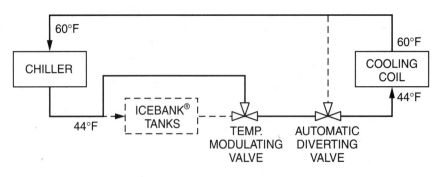

BYPASS CYCLE

NOTE: DASHED LINES INDICATE NO FLOW.

404F26.EPS

Figure 26 ◆ Off-peak cooling system.

During off-peak hours, the cooling system chiller is used to make ice in large tanks called IceBank® tanks (*Figure 27*). These are insulated, water-filled, polyethylene tanks, each containing a spiral-wound plastic tube heat exchanger surrounded by water. These tanks may be installed indoors, outdoors, or even underground. During the night-time off-peak hours, the cooling system chiller is used to produce a chilled glycol solution output at a below-freezing temperature (typically 25°F). This solution is circulated through the ice tank heat exchangers instead of the normal path through the building air handler coil in the conditioned space. The below-freezing solution circulating through the tank heat exchangers causes the water in the tanks to freeze. This cycle of operation is called the charge cycle. It enables the chiller to be operated under the increased load

needed to make ice during the less expensive off-peak power hours.

As the day progresses and the building cooling load increases, the chiller is operated to where chilled glycol solution output is circulated serially, first through the heat exchangers in the ice storage tanks, then through the air handler coil in the conditioned space. The temperature of the solution produced by the chiller for input to the tank heat exchangers is higher than normal, typically 52°F. As the warm solution is circulated through the ice tank heat exchangers, it is cooled by the ice in the tanks to a temperature of about 34°F. The cold 34°F solution output from the tanks is then mixed with some of the warm 52°F solution to produce a solution of about 44°F. This is the normal design temperature range for the cooling solution input to the building air handler coil. This cycle of operation is called the discharge cycle. It reduces the amount of energy needed to run the chiller during the building's peak cooling load interval that is coincident with the more expensive peak demand power daytime interval.

At times when the building's actual cooling load is equal to or lower than the chiller's capacity, the IceBank® tanks are bypassed, and the chiller glycol solution output is circulated at 44°F through the air handler cooling coil in the normal fashion. This cycle of operation is called the bypass cycle.

During the charge, discharge, and bypass cycles of operation, the automatic diverting valve controls the routing of the glycol solution in the system while the temperature modulating valve controls the temperature of the solution.

6.0.0 ◆ FOOD PROCESSING COOLING WATER RECOVERY SYSTEM

Water costs, coupled with reduced availability of water, have resulted in the need to save and recycle both hot and cold water. This section describes a system used in a food processing plant to recover both cold and hot water.

The cooling water recovery system (*Figure 28*) begins with the filling of a water storage tank using city-supplied water. The water is treated with chlorine (about 1 ppm) and is then pumped into the cooking **retorts** where it is mixed with cold recycled process water. The retorts are filled with just enough water to displace the volume of the food containers to be used. Steam is introduced into the retorts to heat the water between 170°F and 180°F to prevent thermal shock and

404F27.EPS

Figure 27 ◆ IceBank® energy storage tanks.

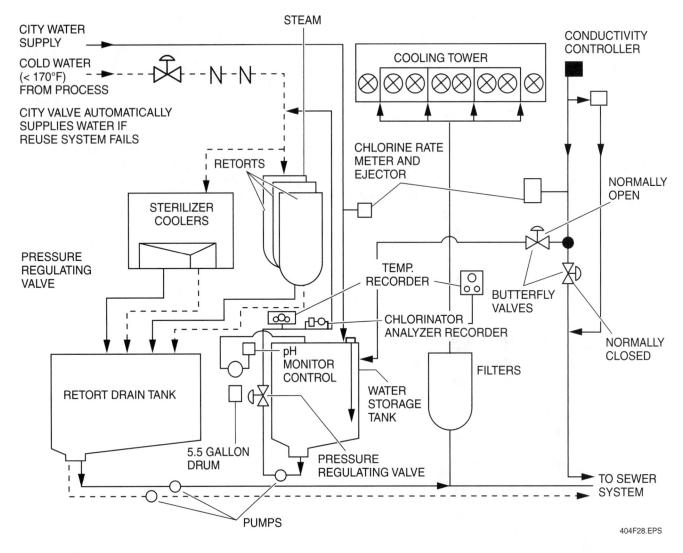

Figure 28 ◆ Food processing cooling water recovery equipment.

breakage of the glass containers. Next, the jars of food are placed in the retort and the water is further heated by steam to the proper cooking temperature. When the cooking process is completed, the water is allowed to cool enough to bring the temperature of the jars down and then is drained from the retort into a drain tank. A water level monitor energizes a pump that forces the water through filters that remove suspended solids (such as glass) and then to a cooling tower. The cooling tower cools the water further to about 70°F before it is returned to the storage tank to be held until the next cooking process begins.

The hot water rejected from the cooking cycle is diverted to the hot water reuse system by a heat-sensitive temperature controller valve (*Figure 29*). This flow control valve will divert any hot water (from 250°F down to 170°F) discharged from the cooking cycle to the hot water reuse system. Any water below 170°F is sent to the cold water reuse system. This allows the hot water from the hot water storage tank to be sent to the precook hot water portion of the system. Therefore, the hot water required for preheating and the cool water required for cooling require less energy to reach their respective process temperatures.

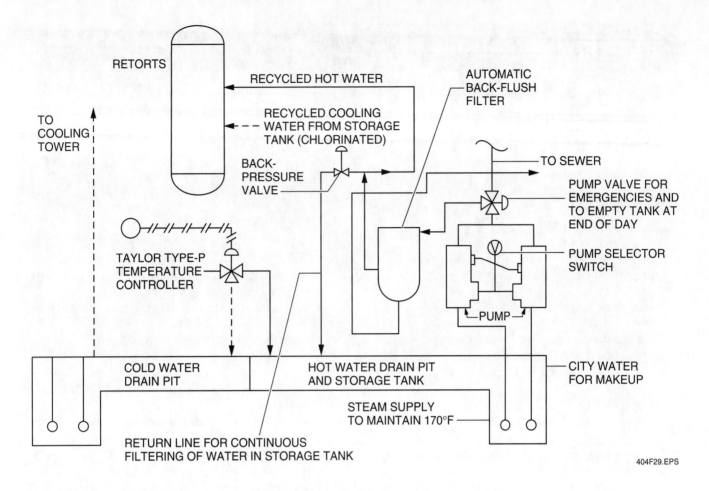

RETORTS

RECYCLED HOT WATER

AUTOMATIC
BACK-FLUSH
FILTER

TO SEWER

RECYCLED COOLING
WATER FROM STORAGE
TANK (CHLORINATED)

TO
COOLING
TOWER

BACK-
PRESSURE
VALVE

PUMP VALVE FOR
EMERGENCIES AND
TO EMPTY TANK AT
END OF DAY

PUMP SELECTOR
SWITCH

TAYLOR TYPE-P
TEMPERATURE
CONTROLLER

PUMP

COLD WATER
DRAIN PIT

HOT WATER DRAIN PIT
AND STORAGE TANK

CITY WATER
FOR MAKEUP

STEAM SUPPLY
TO MAINTAIN 170°F

RETURN LINE FOR CONTINUOUS
FILTERING OF WATER IN STORAGE TANK

404F29.EPS

Figure 29 ◆ Food processing hot water recovery equipment.

1. A fixed-plate air-to-air heat exchanger in an ERV can recover _____.
   a. total heat
   b. sensible heat only
   c. both sensible and latent heat
   d. latent heat only

2. The capacity of a rotary heat exchanger can be controlled by varying the _____.
   a. width of the purge section
   b. flow of air through the cooling coil
   c. flow of air through a return air bypass damper
   d. speed of the wheel rotation

3. One type of heat recovery device in which the hot discharge gas from the system compressor is used to heat domestic water is called a _____.
   a. refrigerant-to-water heat exchanger
   b. coil energy recovery loop
   c. double-bundle condenser system
   d. dual-condenser system

4. When the available heat generated from the chiller system with a double-bundle condenser exceeds the building load, the surplus heat is _____.
   a. routed to and stored in the auxiliary heating system
   b. used to prevent the exhaust coil from freezing
   c. rejected into the atmosphere by the cooling tower
   d. directed to the refrigeration system to be cooled down

5. In a runaround loop system, _____.
   a. the supply air is precooled when it is warmer than the exhaust air
   b. the supply air is preheated when it is warmer than the exhaust air
   c. the supply air is precooled when it is cooler than the exhaust air
   d. latent heat is transferred between the exhaust and incoming airstreams

6. A heat pipe heat exchanger _____.
   a. transfers only latent heat
   b. is used to help increase the dehumidification capacity of a system
   c. operates only when there is no temperature difference between the supply and exhaust airstreams
   d. requires the use of an electric reheat coil

7. In a unidirectional coil loop thermosiphon system, the _____.
   a. coil in the cold duct must be higher than the coil in the hot duct
   b. coil in the cold duct must be lower than the coil in the hot duct
   c. coil in the cold duct must be at the same level as the coil in the hot duct
   d. position of the coil in the hot duct relative to the cold duct is not important

8. A twin tower enthalpy recovery loop system is a(n) _____.
   a. air-to-liquid system
   b. air-to-liquid and liquid-to-air system
   c. liquid-to-air system
   d. air-to-air system

9. In a water-side economizer, variable water flow operation is obtained by _____.
   a. controlling the economizer coil and bypass loop valves for complementary operation
   b. modulating the bypass loop valve and closing the economizer coil valve
   c. closing the bypass loop valve and modulating the economizer coil valve
   d. closing both the bypass loop and economizer coil valves

10. A method used to reduce peak utility electrical usage in HVAC equipment is _____.
    a. IceBank® storage
    b. heat pump defrost
    c. flash tank storage
    d. chiller barrel bypass

Summary

Heat recovery devices save considerable amounts of energy through the capture and reuse of heat that would otherwise be wasted. Heat recovery systems commonly use air-to-air, water-to-air, water-to-water, and steam-to-water heat exchangers and/or coils to transfer heat from one part of an HVAC system for use in another part. This heat recovery can be achieved in process-to-process, process-to-comfort, or comfort-to-comfort applications.

Devices called economizers are used in HVAC systems to save energy by altering the operation cycle of an HVAC system in a way that increases the system cooling efficiency. In the cooling mode, the economizer substitutes outdoor ventilation air (air-side economizer) or cooling tower water flow (water-side economizer) as the medium used to cool a building. This free cooling is used in place of mechanical cooling (compressor generated cooling) whenever the prevailing temperature and humidity conditions permit.

In addition to the heat recovery and energy conservation methods commonly used in HVAC systems, many methods that save energy and/or resources, such as water, are unique to specific commercial or manufacturing processes. The safety and long-term ownership costs must be considered.

Notes

Trade Terms Introduced in This Module

Monel®: An alloy made of nickel, copper, iron, manganese, silicon, and carbon that is very resistant to corrosion.

Retort: A container in which substances are cooked, distilled, or decomposed by heat.

Runaround loop: A closed-loop energy recovery system in which finned-tube water coils are installed in the supply and exhaust airstreams and connected by counterflow piping.

Sensible heat recovery device: An air-to-air recovery device that transfers only sensible heat between the supply and exhaust airstreams. It does not exchange latent heat (heat contained in water vapor) between the supply and exhaust airstreams.

Thermosiphon: A passive heat exchange process in which liquid is circulated by means of natural convection.

Total heat recovery device: An air-to-air recovery device that can transfer both sensible and latent heat (heat contained in water vapor) between supply and exhaust airstreams.

Additional Resources and References

Additional Resources

This module is intended to be a thorough resource for task training. The following reference works are suggested for further study. These are optional materials for continued education rather than for task training.

ASHRAE Handbook – HVAC Systems and Equipment. Atlanta, GA: American Society of Heating, Refrigerating, and Air Conditioning Engineers, Inc. HVAC

Systems Design Handbook. Blue Ridge Summit, PA: TAB Books Inc.

Figure Credits

Fantech, Inc., 404SA01

Greenheck Fan Corporation, 404SA02

Airxchange, Inc., 404F05 (photo), 404F06

Tyler Refrigeration Division, 404F07

Poolpak International, 404F10–404F12

Colmac Coil Manufacturing, Inc., 404F15 (bottom)

Topaz Publications, Inc., 404SA03

CALMAC Manufacturing Corporation, 404F27

NCCER makes every effort to keep these textbooks up-to-date and free of technical errors. We appreciate your help in this process. If you have an idea for improving this textbook, or if you find an error, a typographical mistake, or an inaccuracy in NCCER's Contren® textbooks, please write us, using this form or a photocopy. Be sure to include the exact module number, page number, a detailed description, and the correction, if applicable. Your input will be brought to the attention of the Technical Review Committee. Thank you for your assistance.

Instructors – If you found that additional materials were necessary in order to teach this module effectively, please let us know so that we may include them in the Equipment/Materials list in the Annotated Instructor's Guide.

Write: Product Development and Revision
National Center for Construction Education and Research
3600 NW 43rd St, Bldg G, Gainesville, FL 32606

Fax: 352-334-0932

E-mail: curriculum@nccer.org

Craft _____ Module Name _____

Copyright Date _____ Module Number _____ Page Number(s) _____

Description _____

_____

_____

_____

(Optional) Correction _____

_____

_____

(Optional) Your Name and Address _____

_____

_____

03405-09

Building
Management Systems

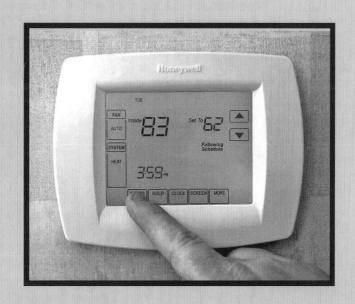

03405-09
Building Management Systems

Topics to be presented in this module include:

Overview

In previous modules, you learned how electronic controls are used to manage multi-zone HVAC systems in both residential and commercial applications. Building management systems take this control to a higher level by using special software to integrate the control of all building systems, including HVAC, lighting, security, and fire control. All the building systems can be monitored and controlled from a central computer located either on site or off site. These types of systems are common in large office buildings and medical facilities. A technician who will be called upon to install and service these systems must be able to interact with the systems in order to monitor performance, diagnose problems, and establish or change operating parameters. This ability requires knowledge of computers, networking principles, and the building management system software.

Objectives

When you have completed this module, you will be able to do the following:

1. Identify the major components of a building management system and describe how they fit together.
2. Operate a basic direct digital controller.
3. List the types of information available on a typical front-end computer screen for a building management system.
4. List the typical steps required to install a building management system.
5. Demonstrate how to install typical sensors, actuators, power wiring, and communication wiring.

Trade Terms

Algorithm	Internet protocol (IP)
Application specific controller	Internet protocol address (IP address)
Baud	Interoperability
Bit	Load-shedding
Building management system (BMS)	Local area network (LAN)
Bus	Local interface device
Control point	Modem
Data collection	Network
Digital controller	Personal digital assistant (PDA)
Direct digital control (DDC)	Product-integrated controller (PIC)
Ethernet	Protocol
Firmware	Software
Gateway	Sustainable
Green building	Tenant billing
Hypertext transfer protocol (http)	Transients
Indoor air quality (IAQ)	User interface module
Integrated building design	Web browser
Internet	Web page
	Wide area network (WAN)
	World wide web

Required Trainee Materials

1. Pencil and paper
2. Appropriate personal protective equipment

Prerequisites

Before you begin this module, it is recommended that you successfully complete *Core Curriculum*; *HVAC Level One*; *HVAC Level Two*; *HVAC Level Three*; and *HVAC Level Four*, Modules 03401-09 through 03404-09.

This course map shows all of the modules in the fourth level of the *HVAC* curriculum. The suggested training order begins at the bottom and proceeds up. Skill levels increase as you advance on the course map. The local Training Program Sponsor may adjust the training order.

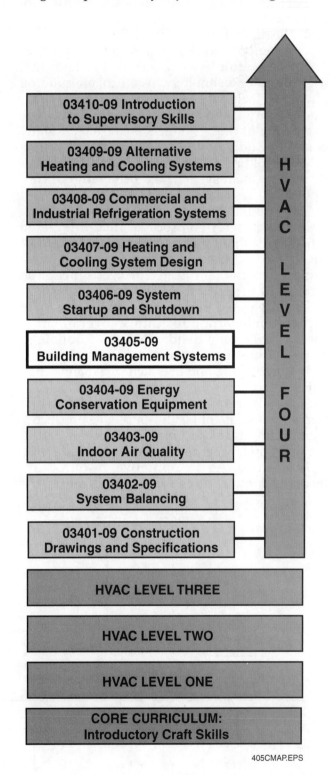

405CMAP.EPS

1.0.0 ◆ INTRODUCTION

Digital control systems were briefly discussed in the Level Two module, *Basic Electronics*. However, most of the previous coverage of electronic controls has involved individual electronic control devices. Examples of such devices are programmable electronic thermostats and electronic modules that control such individual comfort system components as furnaces and heat pumps.

In this module, we will focus on systems that use computer technology to control all building functions, including the HVAC system, lighting, smoke control, and building security access. The composite of these systems is known by many names, including **building management system (BMS)**, building automation system (BAS), facility management system, and **direct digital control (DDC)** system.

Many building management systems are being designed as part of a process called **integrated building design**. Integrated building design is a collaborative decision-making process that uses a project design team from a project's inception through its design and construction phases.

The integrated building design process is, in turn, often incorporated into an overall larger commercial design trend called green buildings. A **green building** is a **sustainable** structure that is designed, built, and operated in a manner that efficiently uses resources while being ecologically friendly. The resources used include, but are not limited to, occupant health, employee productivity, building materials, building systems, building operation, water use, and energy use.

Thus, in order to minimize resource use in many modern green building designs, the building management system must be able to integrate control strategies across multiple building systems.

2.0.0 ◆ BASIC DIGITAL CONTROLLER

The heart of any building management system is a series of **digital controllers** (*Figure 1*) used to monitor and control various building functions. The use of a digital controller is usually referred to as direct digital control or DDC. The term *direct* refers to the direct use of the digital controller to perform control logic normally accomplished with electronic or pneumatic controllers. The digital controller consists of three major components: an input module, a microprocessor, and an output module.

Digital controllers fall into two basic groups: product-specific controllers (also known as product-integrated controllers) and general-purpose controllers.

Product-specific controllers are digital controllers designed to control specific equipment and may be installed by the equipment manufacturer. The controllers for chillers or boilers are examples of product-specific controllers. General purpose controllers, on the other hand, are designed to control a variety of equipment and are typically field installed. Such controllers would be used to control generic, non-complex devices with simple operating schemes, such as exhaust fans, pumps, and similar equipment.

Figure 1 shows a typical digital controller. A sensor located in a hot-air duct is wired to the input module. The sensor is part of a low-voltage circuit in the input module. The input module converts resistance changes in the sensor into

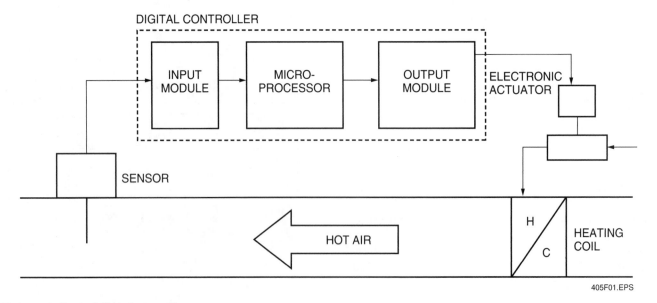

405F01.EPS

Figure 1 ◆ Typical digital controller.

Application of a Building Management System

This is an example of a modern office building in which the building HVAC systems, lighting, smoke control, and other building systems are all under the control of a building management system.

405SA01.EPS

temperature value and transmits this value to the microprocessor. A control routine, called an **algorithm**, compares the duct temperature against a setpoint and sends instructions through the output module to the hot-water valve actuator. The valve opens or closes accordingly. Each input and output device wired to a controller is called a **control point,** or simply, a point.

2.1.0 Control Point Classification

Control points are either analog or binary. Binary points are also referred to as digital or discrete points. An explanation of each point type follows:

- *Analog input (AI)* – A sensor used to measure variables that have a range of values, such as temperature or pressure. The sensor signal varies over a predefined range of values. A duct temperature sensor is an example.
- *Discrete input (DI)* – A sensor used to measure the status of binary devices, such as the position of a pressure switch. The sensor is discrete because the switch can be either open or closed.

- *Analog output (AO)* – An output signal used to modulate the position of a device, such as a valve or damper. The output signal from the controller to the AO point varies typically from 4 to 20mA or from 0 to 10VDC.
- *Discrete output (DO)* – An output signal used to change the position of a device from one state to another. Examples of discrete outputs are the starting and stopping of pump or fan motors.

2.2.0 Input and Output Devices

Analog or discrete devices are wired to the input or output (I/O) modules in accordance with the types of points involved (inputs or outputs). The points on each module are grouped differently by manufacturer; the most common hardware components are available in blocks of 4, 8, or 16 points. These points may be predetermined to be of one type, such as DO, or they may be universal, with their type being established during the installation and programming of the digital controller. The following paragraphs describe typical examples of each type of input and output device.

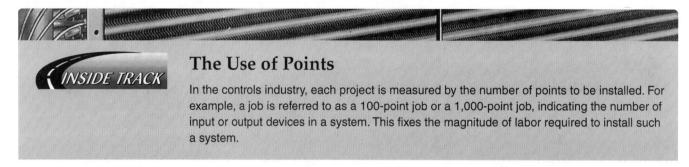
2.2.1 Analog Input Devices

Temperature sensors (*Figure 2*) are the most common analog input (AI) devices used in HVAC control systems. Temperature sensors fall into one of three basic types: thermistor, resistance temperature detector (RTD), or transmitter.

A thermistor is a small bead of material that changes resistance in proportion to changes in temperature. The thermistor senses temperature at a single point. Thermistors can be used to sense the temperature of fluid within a duct or pipe or the temperature of air in a room or outdoors.

An RTD uses a long resistance-sensing element as opposed to a single point. Therefore, RTDs are commonly used to sense an average temperature. Measuring mixed air temperature in an air handler is one use of an RTD.

When the medium requires a sophisticated sensing element, or when there is a long distance between the sensor and the input module, a

405F02.EPS

Figure 2 ◆ Examples of thermistors.

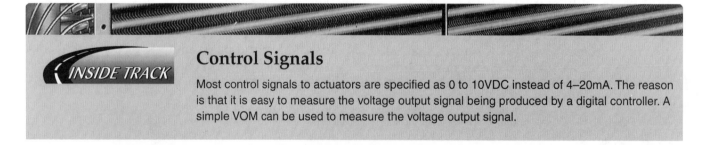

transmitter is used. The transmitter is a device that senses the specific range of a variable (like temperature) and sends (transmits) a linear current (4–20mA) or voltage (0–10VDC) signal back to the controller.

2.2.2 Discrete Input Devices

Discrete input devices are typically dry-contact (no voltage) closures from switches or relays used for feedback information from equipment. Such devices include auxiliary starter contacts,

405F03.EPS

Figure 3 ◆ Differential pressure switch.

differential pressure switches (*Figure 3*), current sensing switches, and flow switches.

Another type of discrete input device is a meter that can send a pulsed signal to the input module. The frequency and duration of the pulses represents power consumption or flow rates. Each meter has a pulse conversion rate that is programmed into the processor to interpret the pulse signal correctly.

2.2.3 Analog Output Devices

Analog output devices are actuators that modulate valves or dampers. Most actuators accept industry standard output control signals of 4–20mA or 0–10VDC (*Figure 4A*).

The transducer is another common analog output device. Transducers, such as the one mounted on the pneumatically-actuated valve shown in *Figure 4B*, convert a digital controller output signal to a pressure range. The use of transducers is common in large or retrofit applications where valves or dampers are equipped with pneumatic actuators. To control a pneumatic actuator, a current-to-pressure (I/P) transducer converts a 4–20mA output signal into a 3–15 psig air signal.

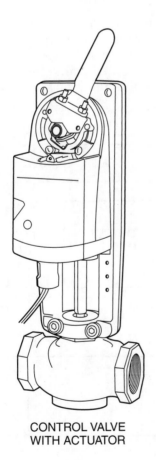

CONTROL VALVE
WITH ACTUATOR

VALVE WITH I/P
TRANSDUCER

405F04.EPS

Figure 4 ◆ Modulating valves.

2.2.4 Discrete Output Devices

Discrete output (DO) devices are typically low-voltage (24VDC) control relays with contacts wired into motor control circuits to start or stop the motors. Pulsed or stepped actuators can also be used as discrete output devices if the devices are designed to receive a pulse width modulating signal. This is called floating-point control. One DO point is used to pulse the device open and another is used to pulse the device closed.

2.3.0 Closed Control Loop

The basis for almost all digital control is a closed control loop. *Figure 5* shows such a loop being used to control the duct temperature in a warm-air heating application. A sensor located in the warm air measures duct temperature. The input module monitors the resistance change of the sensor and transmits duct temperature to the microprocessor.

The microprocessor compares the duct temperature, for example 95°F, to a programmed setpoint, perhaps 110°F, and calculates a milliamp signal to send to the hot-water valve actuator. The output module sends the milliamp signal to the valve actuator, and the valve opens to allow more hot water through the valve. The coil transmits the extra heat to the air, raising its temperature for this example to 100°F. The cycle repeats as the duct sensor reads the new temperature of 100°F.

The automatic feedback created by this loop makes the process a closed control loop. The sensor automatically measures the impact of any change ordered by the microprocessor.

2.4.0 Control Algorithms

Closed control loops are executed by running multiple algorithms, or **software** instructions, simultaneously within the microprocessor. Each algorithm controls one closed control loop. The algorithms interpret input conditions and decide what output signals to send for each control loop. Algorithms consist of a series of logic statements, continuously repeated at high speed. For example, part of the heating coil algorithm for *Figure 5* might consist of something like this:

Step 1   Read the duct temperature.

Step 2   Compare it to the setpoint. If it's too low, calculate what output signal to send to the hot-water valve actuator.

Step 3   Send X milliamps to the hot-water valve actuator.

Step 4   Wait for Y seconds and then go back to Step 1.

Many algorithms also require time schedules to determine when specific equipment should be turned on or off. Setpoint schedule data are also needed to determine the temperatures at which valves and dampers should be opened or closed.

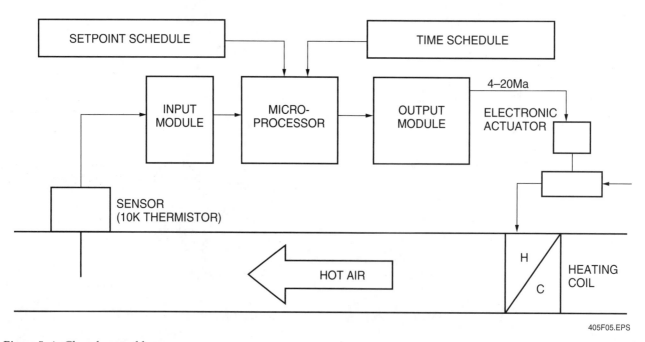

Figure 5 ◆ Closed control loop.

Using setpoints, time schedules, and point data, a microprocessor can use algorithms to perform a wide range of control functions. Some of the most typical HVAC algorithms are listed here:

- Heating/cooling coil control
- Humidification/dehumidification
- Mixed air damper optimization
- VAV fan supply control
- **Indoor air quality (IAQ)**
- Time of day scheduling
- Electric heat staging
- Primary/secondary pump control

The following three strategies are used in operating DDC control loops:

- Proportional (P)
- Proportional-integral (PI)
- Proportional-integral-derivative (PID)

2.4.1 Proportional Control Algorithms

Proportional control is much like the early cruise control used on automobiles. Setting the cruise at 55 miles per hour (mph) meant the car stayed at 55 on a flat road, but when the car started up a hill, it would slow to 50. Missing the setpoint on the low side is called undershoot. When the car traveled down a hill, it would increase to 60 mph. Missing the setpoint on the high side is called overshoot. Thus, due to control inefficiencies as well as overshoot and undershoot, the car's speed fluctuated within a wide range.

Even though the setpoint was 55 mph, in proportional control the speed of the car did not stay constant. The difference between the actual speed of the car when it settled at a fixed speed (the control point) and the desired speed of the car (setpoint) is called offset. Offset is a measure of the inefficiency of proportional control.

In order to respond, proportional control must have a change in the controlled variable. After responding, proportional control will always arrive at a control point that is different from the setpoint.

2.4.2 Proportional-Integral Control Algorithms

An integral control function is added to the control loop of a modern automobile's cruise control to eliminate offset over time. When the speed of the automobile falls below 55 mph while climbing a hill, the integral function moves the accelerator pedal down and causes the car to increase speed until it returns to 55 mph again. The reverse is true when the automobile travels down

a hill. Thus, the integral term eliminates offset over time, and keeps the car at 55 mph regardless of the terrain being traveled.

One problem remains, however. Going up a very steep hill will cause the car to decelerate at a faster rate than going up a mild grade hill. The reverse is true for a steep downhill. Utilizing the PI function, the car's cruise control will take longer to bring the car back to 55 mph on a steep hill than it will for a moderately inclined hill. PI control does not respond to the rate at which the speed change occurs (the acceleration of the error), it simply eliminates the error over time.

PI control is used in most HVAC applications where the controlled variable changes slowly over time. The control of room temperature is one such example because room temperature changes gradually with time.

2.4.3 Proportional-Integral-Derivative Algorithms

Adding a derivative term to the cruise control function allows the control to respond to the car's acceleration or deceleration when it encounters hills. The derivative function measures the rate of change of the variable (speed) and causes the gas pedal to overcompensate beyond that required by the PI function. The net result is that the cruise control now rapidly brings the car back to 55 mph, regardless of the incline of the hill.

PID control is used in HVAC systems when the controlled variable is changing rapidly over time. The control of duct pressure in a VAV system is one such example. *Figure 6* summarizes the three functions involved in PID control and graphically illustrates the benefits of each function.

3.0.0 ◆ DDC NETWORK TYPES

The simplest DDC control system is a single, stand-alone digital controller wired to the input and output devices of a single HVAC system.

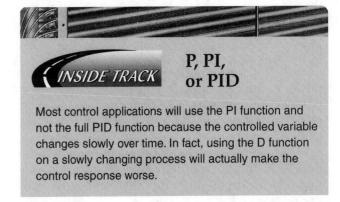

P, PI, or PID

Most control applications will use the PI function and not the full PID function because the controlled variable changes slowly over time. In fact, using the D function on a slowly changing process will actually make the control response worse.

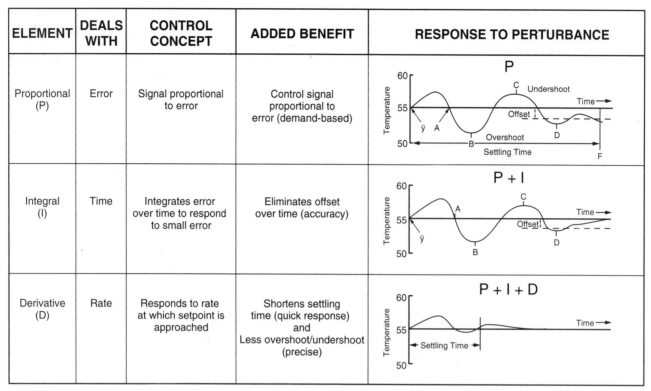

ELEMENT	DEALS WITH	CONTROL CONCEPT	ADDED BENEFIT	RESPONSE TO PERTURBANCE
Proportional (P)	Error	Signal proportional to error	Control signal proportional to error (demand-based)	
Integral (I)	Time	Integrates error over time to respond to small error	Eliminates offset over time (accuracy)	
Derivative (D)	Rate	Responds to rate at which setpoint is approached	Shortens settling time (quick response) and Less overshoot/undershoot (precise)	

405F06.EPS

Figure 6 ◆ PID control summary.

By itself, however, a single digital controller can only control a finite number of input and output points. The number of points is a function of the microprocessor design and varies with each manufacturer. Thus, a stand-alone general-purpose digital controller can only provide comfort and energy management strategies for individual HVAC units within the point limitation of the controller. In many small buildings, this is sufficient.

Multiple digital controllers are required for applications involving multiple pieces of equipment and a large number of points. The ability to share information between controllers becomes important, so a DDC **network** is required.

A DDC network is a system that allows its individual controllers to communicate with each other and share information. Information sharing can take place between two controllers, or it may be broadcast by one controller to all other controllers. For example, multiple controllers in a building may need to know the outdoor air temperature in order to run their control algorithms. Only one outside air temperature sensor needs to be wired to one controller. This controller then broadcasts the sensor value along the communication **bus** to all other controllers in the network. In this manner, all controllers on the network share the outside air temperature, saving sensor and wiring expense.

Information sharing can also be used to achieve HVAC system coordination between water-side and air-side equipment controllers. Air terminals can share information with air handlers that in turn can share information with water chillers.

In such a system, the air handlers continually vary the air quantity and temperature to meet changing air terminal needs. Likewise, chillers continually vary the water quantity and temperature to meet varying air handler needs. Thus, sharing information allows the HVAC system to continually provide optimal comfort while minimizing system operating costs.

There are typically two types of DDC networks: polling and peer-to-peer.

3.1.0 Polling Networks

A polling network (*Figure 7*) consists of several digital controllers linked together, with one controller designated as the polling device. The polling device requests or polls information from each of the other controllers, one at a time, and in a certain order. The polling process is executed in the same order continuously.

In such a network, response time may be slow due to the time required to complete the polling sequence. This is particularly true when considering alarms. For example, consider that controller number 5 is being polled in a network consisting

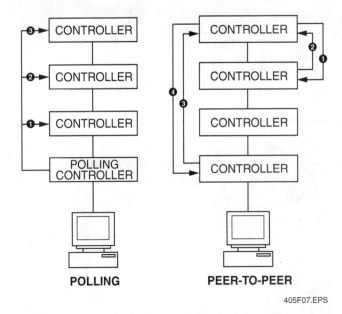

Figure 7 ◆ Polling and peer-to-peer network.

of 40 controllers. If controller 3 goes into alarm, the alarm cannot be communicated until the polling device has communicated with the remaining 35 controllers and returns to polling controller 3.

In a polling network, the user interface can only view information for controllers wired to each polling control panel. For example, a building might have four separate polling networks, one per floor. The user interface would have to be physically moved and connected to each polling panel individually to view the entire building's operating information.

3.2.0 Peer-to-Peer Networks

Peer-to-peer networks (*Figure 7*) get their name from the way in which communication occurs between controllers. Each controller can communicate with any other controller connected on the network and all controllers are peers, having equal priority to use the communication bus at any time. In such a network, any controller can communicate at any time by going through a four-step process:

Step 1 Wait (in milliseconds) to see that the communication bus is quiet.

Step 2 When sufficient quiet time has occurred, send a packet of information to another controller(s).

Step 3 Wait to receive a confirmation that the other controller(s) received the message.

Step 4 Repeat the transmission if confirmation has not been received in a timely manner.

All controller information may also be viewed or modified from a single user interface (computer). In this way, all controllers on all floors of a building can be interfaced from a single computer located anywhere within the building.

3.2.1 Communication Bus

Information sharing between controllers in a peer-to-peer network takes place over a communication bus. Multiple buses (*Figure 8*) can exist in a single network, with one bus designated as the primary bus. Secondary buses extend from the primary bus in order to isolate controllers programmed for polling-type communication or to extend the physical length of the bus system. A communication module, sometimes called a bridge, is used to connect a secondary bus to a primary bus.

The bus cables and cable connectors are similar to those used in hooking up telephones and video equipment. The bus is a simple three-wire cable with a shield that is physically wired to each controller in the network. Fiber-optic cable is also used.

The controllers connected in a network can share information because they use a common language or set of rules known as a **protocol**. In a protocol, the computer words traveling along the bus must be configured in a specific way in order to be recognized. The protocol is designed to check the data for errors and make corrections, provide access for all devices on the network, and make sure that only one device transmits at a time. Each DDC control system manufacturer has a unique protocol that governs communication along the communication bus.

The rate at which information moves on the network is known as the **baud** rate. The baud rate is stated in **bits** per second (BPS). The higher the baud rate, the faster the data transfers. In DDC networks, data can be transferred at baud rates upwards of 100 million bits (megabits) per second. Higher transfer rates are available with fiber-optic cable.

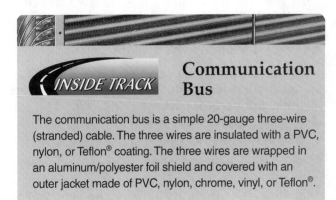

Communication Bus

The communication bus is a simple 20-gauge three-wire (stranded) cable. The three wires are insulated with a PVC, nylon, or Teflon® coating. The three wires are wrapped in an aluminum/polyester foil shield and covered with an outer jacket made of PVC, nylon, chrome, vinyl, or Teflon®.

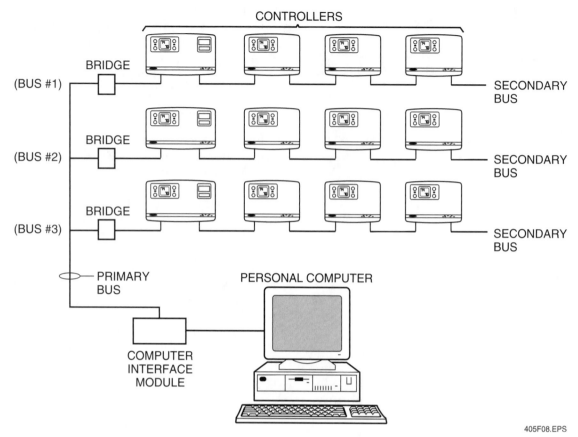

Figure 8 ◆ Communication bus.

4.0.0 ◆ BUILDING MANAGEMENT SYSTEM ARCHITECTURE

Building management systems (BMS) usually start with the computer-controlled automation of a building's HVAC system, and then expand as a building grows in size to include the building's lighting, security access, computer room air conditioning, and smoke control systems. The objective of the BMS is to use the power of the computer to provide energy management routines to save energy; provide monitoring and reporting capability for planned maintenance; and permit human access and intervention from a central point in the building or from a remote location. *Figure 9* shows the architecture of a building management system.

The automation of the building's HVAC system is divided into two broad categories:

• Systems using packaged equipment, such as residential split systems or commercial rooftop units and vertical packaged units
• Applied systems that use air terminals, air handlers, chillers, boilers, and cooling towers

The top portion of *Figure 9* shows the next level of building control, the interface of the building's management functions. Finally, the very top of *Figure 9* shows the final level of building con-

trol—human access capability through the use of local or remote computers, and/or the use of the internet. The key to all building management system functions is the use of digital controllers to control equipment in all the various building systems and the use of a communication network to link all controllers in the building to a centralized computer for human access and integrated building system control strategies. BMS functions can be applied to installations ranging in size from a residence to a multi-building campus.

4.1.0 Packaged HVAC Systems

Packaged HVAC systems are used in residential and commercial buildings. More sophisticated building management functions are added as the building grows in size. Packaged units have a factory-installed control panel that is wired to a remote programmable thermostat. In a large building, the programmable thermostat may be wired to a building-wide communication network. These options are shown on the bottom of *Figure 9* in the categories of non-communicating programmable thermostats, communicating programmable thermostats, and packaged unit digital controllers. The following paragraphs expand on these categories.

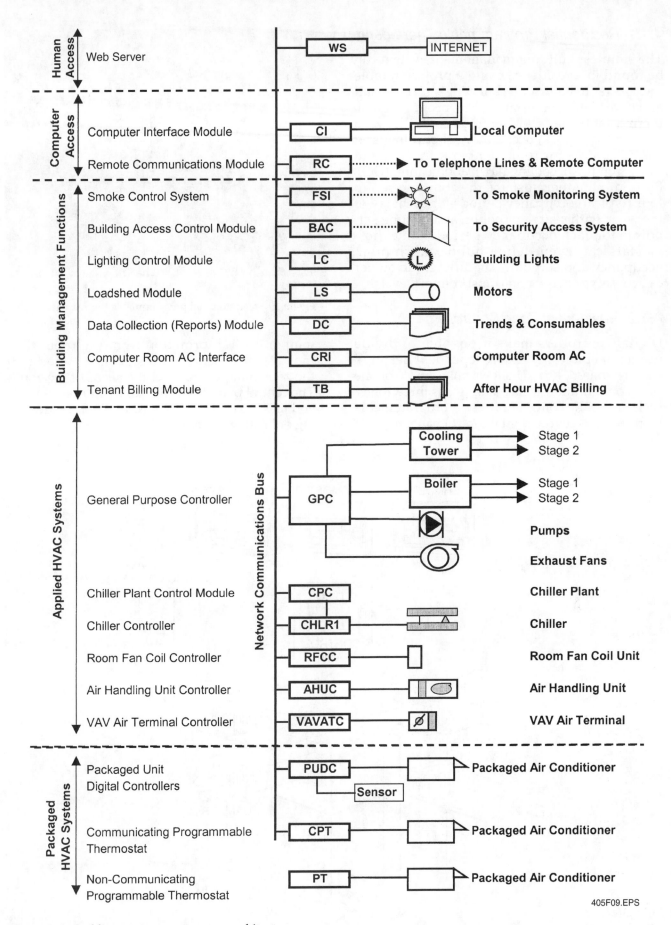

Figure 9 ◆ Building management system architecture.

405F09.EPS

4.1.1 Residential Programmable Thermostat

The simplest building management system can be found in a residence using a programmable thermostat to control the building's furnace and air conditioning equipment. The programmable thermostat (*Figure 10*) is a digital controller with its own time clock, occupancy scheduling, and control points for room temperature (AI), fan start/stop (DO), cooling (DO), and heating (DO). Built into the thermostat is the ability to set up four (typical) occupancy periods for each day of the week with separate cooling and heating setpoints for each occupancy period. Some thermostats can even define holidays. Through occupancy and setpoint scheduling, the owner can conserve energy and adjust comfort levels.

4.1.2 Residential Zone Controllers

Digital controllers make it possible to divide residential buildings into independent heating/cooling zones. *Figure 11* shows an example of one such configuration. This building has three zones, each under the control of its own programmable thermostat. A central digital control panel receives

405F10.EPS

Figure 10 ◆ Residential programmable thermostat.

temperature information from the room thermostats and sends out signals to modulate the zone dampers to meet the needs of the zones. The central panel reads the zone controller's needs and tells the central packaged unit to provide cooling or heating as necessary.

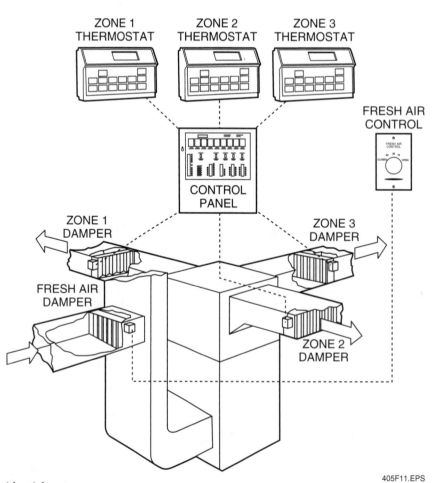

405F11.EPS

Figure 11 ◆ Zoned residential system.

An obvious advantage of using zoning in a residence is that unoccupied rooms can be shut off. If a child is away at college, for example, the thermostat in that bedroom can be turned down until he or she returns for vacations or holidays. In addition, each zone can have individual setpoint and occupancy schedules to further save energy and assure individual comfort. In many of today's high-end homes, systems, such as lighting, entertainment, cooking, and comfort, can all be controlled from a central computer that the homeowner can access from anywhere in the world.

4.1.3 Commercial Programmable Thermostats

The simplest form of a commercial BMS uses a programmable thermostat to control a rooftop or vertical packaged unit (*Figure 12*). The commercial programmable thermostat, however, has more sophisticated occupancy and holiday programming. It controls multiple stages of cooling and heating typically available with commercial packaged units. One such stage of cooling

will use an outdoor economizer to cool the building with cool outdoor air when the outdoor air conditions are appropriate. Thus, the small commercial building owner can save energy and adjust comfort through occupancy, setpoint, and holiday scheduling, and have additional energy saving benefits from using an economizer and multiple stages of cooling and heating.

4.1.4 Communicating Programmable Thermostats

In larger commercial buildings, the cooling and heating loads are likely to vary significantly from one part of the building to another. In buildings with interior zones, it is not unusual for some interior zones to request cooling while some or all exterior zones are demanding heat.

One approach to this type of building is to use multiple constant volume packaged units to zone the building. Each packaged unit is equipped with a communicating programmable zone thermostat (*Figure 13*), and all thermostats are accessible from a central computer. This type of BMS

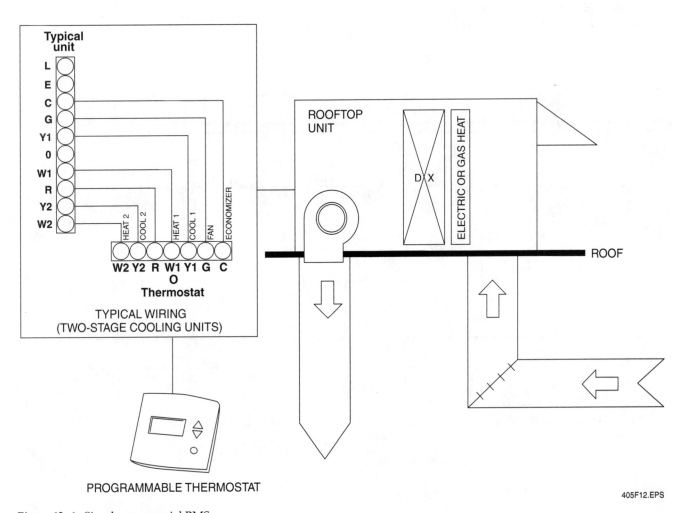

405F12.EPS

Figure 12 ◆ Simple commercial BMS.

provides the owner with all of the previously mentioned energy and comfort benefits of programmable thermostats, and also adds the ability to monitor, adjust, and troubleshoot the systems from a central location in the building.

4.1.5 Packaged Unit Digital Controllers

A further level of BMS sophistication exists when the manufacturer of packaged equipment installs a communicating digital controller on the unit at the factory (*Figure 14*). This type of controller is also referred to as a **product-integrated controller (PIC),** or as an **application-specific controller**. With this controller, all information regarding the packaged unit operating and safety systems is available in the controller. Such a controller is typically provided with a sophisticated 365-day time clock with up to eight occupancy periods and setpoints for occupied and unoccupied periods. Holiday and daylight savings time programming is also provided. In addition, a sensor, not a thermostat, is located in the zone. The sensor is provided with a setpoint adjustment slide bar and an override button, so the occupant can adjust the comfort level and manually start the packaged unit during unoccupied hours.

Thus, connecting the packaged unit controller to the building's network bus makes all this internal unit information and scheduling available from a remote central computer. This is very attractive to owners who are interested in servicing and troubleshooting.

405F14.EPS

Figure 14 ◆ Packaged unit digital controller.

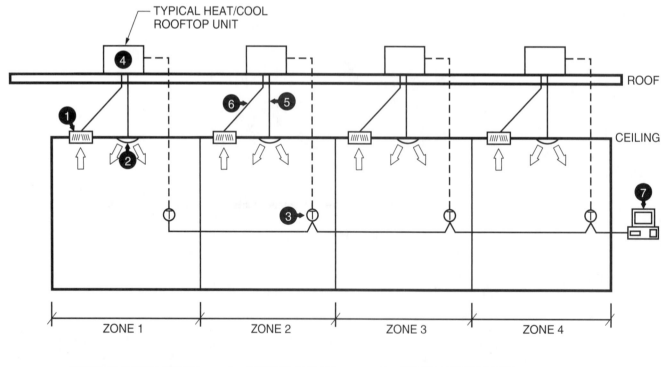

SYSTEM COMPONENTS:

1 – RETURN GRILLE	5 – SUPPLY DUCTWORK
2 – SUPPLY DIFFUSER	6 – RETURN DUCTWORK
3 – ZONE THERMOSTAT	7 – CENTRAL COMPUTER
4 – PACKAGED UNIT	

405F13.EPS

Figure 13 ◆ Application using a communicating programmable zone thermostat.

4.2.0 Applied HVAC Systems

The second level of control sophistication in a BMS is encountered in applied HVAC zoning systems. These systems use either water or air to deliver cooling to the building and have the ability to provide the building zones with cooling and heating. Such systems use application-specific controllers, usually installed at the factory, to control air terminals, fan coil units, air handlers, chillers, cooling towers, and boilers. General-purpose controllers are typically used to integrate the boiler, cooling tower, pumps, and exhaust fans into the HVAC system controls. Sometimes general-purpose controllers are also installed on air handlers in the field by the automatic temperature controls contractor.

The following sections discuss the use of digital controllers in variable volume/temperature (VVT), variable air volume (VAV), and chilled-water systems.

4.2.1 *Variable Volume/Temperature*

Figure 15 shows a zoning approach used on small commercial buildings (25 tons or less in size). In this arrangement, sensor-equipped zone damper controllers make it possible through bus communication to use a single PIC-equipped constant volume packaged unit to meet individual zone cooling and heating needs. Each room is a control zone with its own communicating zone damper controller and room sensor.

In this manufacturer's system, called variable volume/temperature (VVT), there are primary and secondary zone controllers. The primary zone controller (linkage coordinator) sends time of day, system mode of operation, and outdoor temperature information to the secondary controllers (zone controllers). It also monitors the status of the zones on its bus and selects the mode of operation for the central packaged unit. When a sufficient number of zones call for heat, the

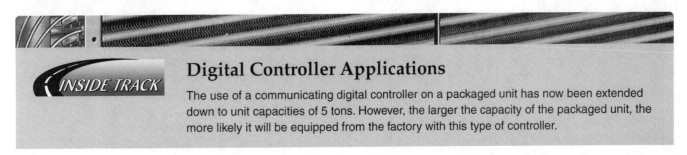

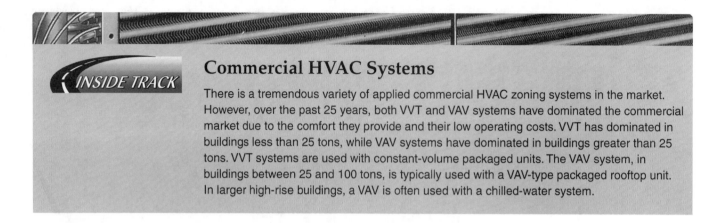

INSIDE TRACK

Commercial HVAC Systems

There is a tremendous variety of applied commercial HVAC zoning systems in the market. However, over the past 25 years, both VVT and VAV systems have dominated the commercial market due to the comfort they provide and their low operating costs. VVT has dominated in buildings less than 25 tons, while VAV systems have dominated in buildings greater than 25 tons. VVT systems are used with constant-volume packaged units. The VAV system, in buildings between 25 and 100 tons, is typically used with a VAV-type packaged rooftop unit. In larger high-rise buildings, a VAV is often used with a chilled-water system.

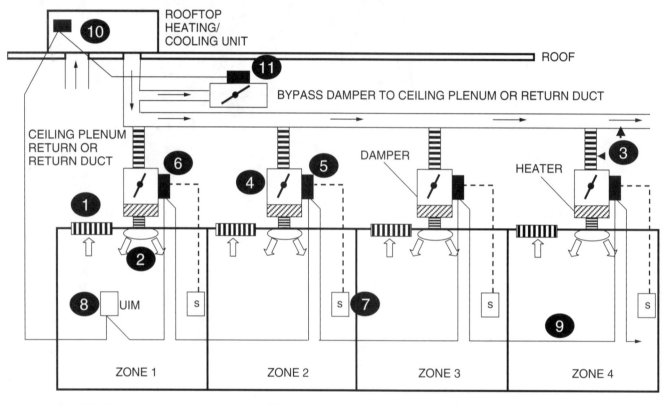

1. RETURN GRILLE
2. SUPPLY DIFFUSER
3. SUPPLY DUCT
4. ZONE DAMPER
5. ZONE CONTROLLER
6. MASTER ZONE CONTROLLER
7. ROOM SENSOR
8. USER INTERFACE MODULE
9. COMMUNICATIONS BUS
10. ROOFTOP UNIT DIGITAL CONTROLLER
11. BYPASS DAMPER & CONTROLLER

405F15.EPS

Figure 15 ◆ Variable volume/temperature system.

linkage coordinator switches the central packaged unit into heat. When all heating zones have been satisfied, the linkage coordinator switches the central packaged unit back into cooling. In this manner, the system provides cooling and heating to a building that has some zones needing cooling while others need heating. Because the central packaged unit is a constant-volume machine, a bypass damper must be provided. As zone dampers throttle, the bypass damper is opened to keep the airflow through the packaged

unit constant. The damper bypasses supply air directly to the return duct or to a ceiling plenum.

Because each system is equipped with a **user interface module**, the owner can program each zone controller with independent occupancy, setpoint, and holiday schedules. Each room sensor is provided with a setpoint adjustment slide bar and an override button, so the occupant can adjust the comfort level and manually start the air source during unoccupied hours.

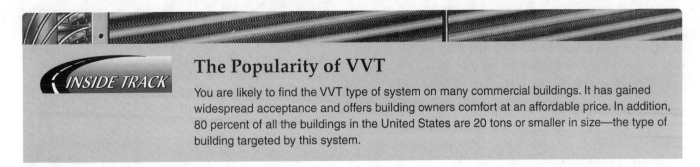

Today's VVT zone controllers, working through bus communication with the central air source, are also equipped with the capability of controlling either a zone's carbon dioxide level or its relative humidity level. More details of how this is accomplished are provided in the Sequence of Operation section of this module.

Finally, control of multiple stages of cooling and heating, including the use of an outdoor economizer, is accomplished by the packaged unit's PIC controller as it responds to the system mode established by the master zone controller.

The VVT system approach is intended primarily for use with smaller tonnage buildings (up to 25 tons) where no zones require year-round cooling.

4.2.2 VAV Systems

VAV systems (*Figure 16*) are used in buildings with loads in excess of 25 tons that require simultaneous cooling and heating at the zone level. A rooftop unit or air handling unit provides a variable volume of air to individual VAV boxes, one per zone. If an air handler is used, a chilled-water

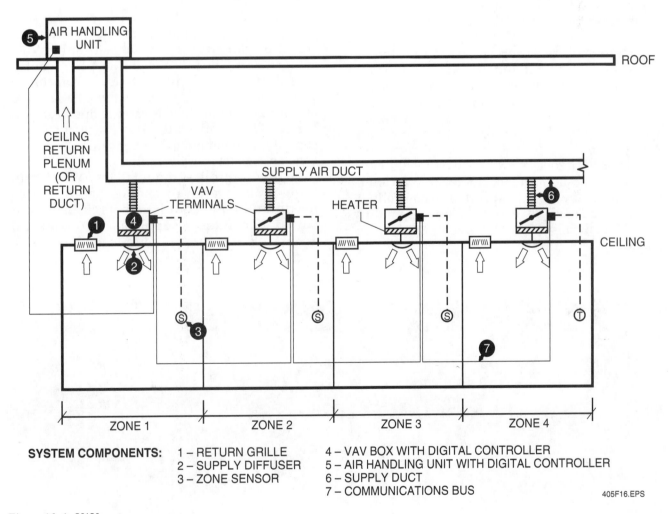

SYSTEM COMPONENTS:
1 – RETURN GRILLE
2 – SUPPLY DIFFUSER
3 – ZONE SENSOR
4 – VAV BOX WITH DIGITAL CONTROLLER
5 – AIR HANDLING UNIT WITH DIGITAL CONTROLLER
6 – SUPPLY DUCT
7 – COMMUNICATIONS BUS

405F16.EPS

Figure 16 ◆ VAV system.

coil is typically provided in the air handler, thus requiring a chiller in the building. The air source is equipped with a means of fan volume control, either variable frequency drive or inlet guide vanes. A digital controller mounted on the air source controls all air source functions. Each VAV box is installed with a zone heater, a digital controller, and a zone sensor. Finally, a communication bus connects all VAV box controllers with the air source controller.

During occupied periods, the air source supplies cool air in the supply duct. Individual zone VAV boxes vary the amount of air necessary to satisfy zone needs, throttling down to a minimum quantity for ventilation. If any zone requires heat, the VAV box controller activates the heater located on the discharge of the VAV box.

Individual zone setpoints and occupancy schedules are programmed into the VAV box controllers. One zone controller is designated as the master. The master controller polls the other zone controllers and tells the air source when to start and stop. It provides zone setpoint and temperature information, so the air source can choose its mode of operation and its supply air temperature.

Each room sensor is provided with a setpoint adjustment slide bar and an override button, so the occupant can adjust the comfort level and manually start the air source during unoccupied hours. Such a system provides optimal zoning comfort while minimizing the use of fan energy.

Today's VVT zone controllers, working through bus communication with the central air source, are also equipped with the capability of controlling either a zone's CO_2 level or its relative humidity level. More details of how this is accomplished are provided later in the module.

4.2.3 Chilled-Water Systems

Figure 17 shows a typical chilled-water system. Two parallel water-cooled chillers supply chilled water to three air handling units (loads) equipped with two-way control valves. Each chiller has its own chilled-water pump and leaving chilled-water sensor (LCHW). Sensors are also installed in the chilled-water plant leaving pipe (CHWST) and in the chilled-water return pipe (CHWRT). Both chillers are typically sized for 50 percent of the building load.

Chiller condensers are piped in parallel, with each chiller having its own condenser pump. Each chiller will also have its own cooling tower or tower cell. Sensors will be installed in the piping entering (CWST) and leaving (CWRT) the tower.

Digital controllers are mounted on each chiller and on each air handler. In addition, a general-purpose controller (GPC) is used to control the cooling tower. Finally, a chiller plant controller (CPC) is provided on the job to manage the two chillers into a single chilled-water plant. All digital controllers are connected together with a communication bus.

Each chiller controller contains its own operating system, safety circuits, time clock, occupancy schedule, and chilled-water setpoint. When requested to start by the chiller plant control module (CPC), each chiller starts its own chilled-water and condenser water pumps. Each chiller also reads its own chilled-water temperature sensor (LCHW) and modulates its refrigeration capacity to maintain its leaving chilled water at a constant level, typically 44°F.

Digital controllers mounted on each air handling unit start the air handler based on zone needs, and they modulate their chilled-water valves to maintain a supply air temperature of about 55°F. This air is supplied to the VAV system as mentioned earlier.

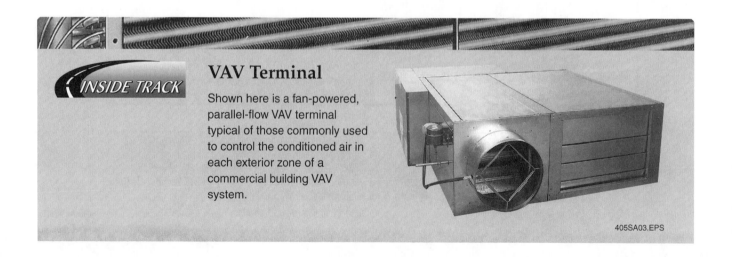

INSIDE TRACK

VAV Terminal

Shown here is a fan-powered, parallel-flow VAV terminal typical of those commonly used to control the conditioned air in each exterior zone of a commercial building VAV system.

405SA03.EPS

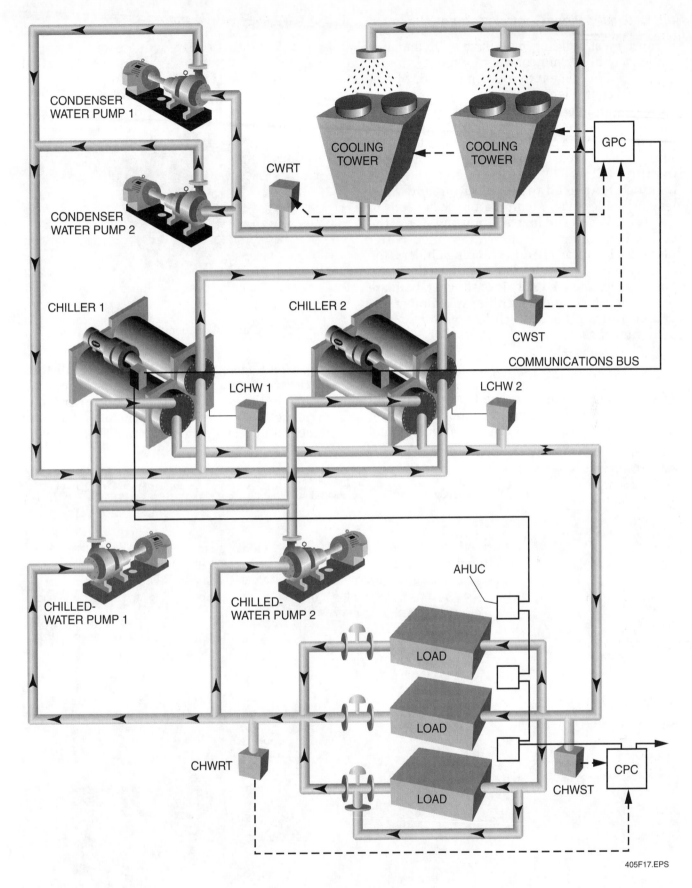

Figure 17 ◆ Chilled-water system.

405F17.EPS

4.2.4 Chiller Plant Control Module

Because most buildings will have more than one chiller, a chiller plant control module (*Figure 18*) is required on the job to coordinate their operation. This controller monitors the needs of the air handlers and starts the chiller plant based on when the first air handler starts. The controller then uses both the plant's leaving water temperature (CHWST) sensor and the return water temperature (CHWRT) sensor to determine how many chillers need to be operating and what leaving chilled-water temperature each needs to produce. When the last air handler stops, the chiller control module stops the chilled-water plant. Besides starting and stopping chillers, the chiller control module also equalizes their run hours, keeps them equally loaded when both are operating, starts another chiller when one fails, and performs other such chiller plant management functions.

405F18.EPS

Figure 18 ◆ Chiller plant control module.

4.2.5 General-Purpose Controller Functions

The last components associated with an applied HVAC system are the building boiler, cooling tower, and exhaust fans. A general-purpose digital controller (*Figure 19*) is installed to integrate these pieces of equipment into the HVAC system. Algorithms in the general-purpose controller (GPC) cycle the cooling tower fans to maintain the water temperature leaving the tower (CWRT).

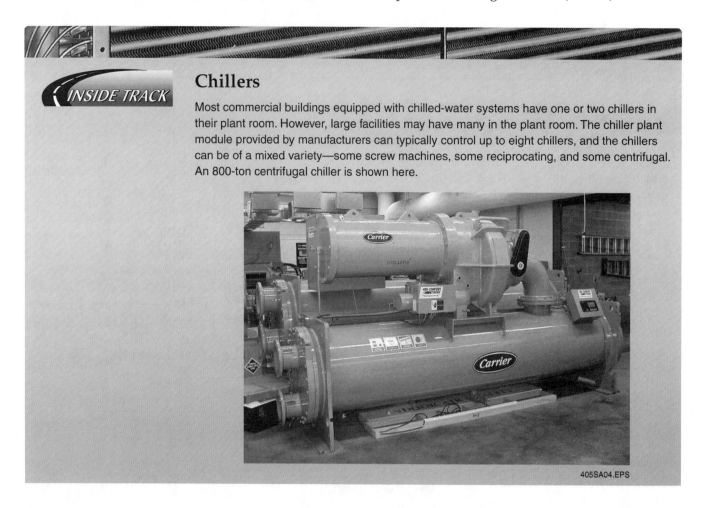

Chillers

Most commercial buildings equipped with chilled-water systems have one or two chillers in their plant room. However, large facilities may have many in the plant room. The chiller plant module provided by manufacturers can typically control up to eight chillers, and the chillers can be of a mixed variety—some screw machines, some reciprocating, and some centrifugal. An 800-ton centrifugal chiller is shown here.

405SA04.EPS

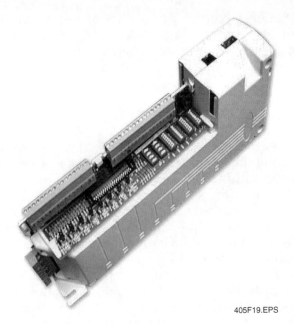

Figure 19 ◆ General-purpose controller.

the boiler with the HVAC system heating needs and to send the boiler's controller a hot-water temperature reset signal as necessary.

Similarly, the GPC is used to cycle building exhaust and pressurization fans as required by the building systems. Finally, the GPC is used to control chilled-water pumps when they are ganged and not started by the individual chiller controls.

4.3.0 Building Management Functions

Digital control modules or software programs running on a central computer are used to achieve a variety of building subsystem management functions. The following paragraphs describe each function and the roles they play in a commercial building.

4.3.1 Tenant Billing

In large commercial buildings, central HVAC systems like VAV or chilled-water systems centralize the production of chilled water and hot water to improve efficiency and reduce operating costs. In addition, each building has an occupancy schedule, such as becoming occupied at 6:00 AM and

Typically, the boiler for such a building comes with its own digital controller mounted at the factory. The boiler controller contains the operating system and safety circuits for the boiler. The GPC is used to integrate the starting and stopping of

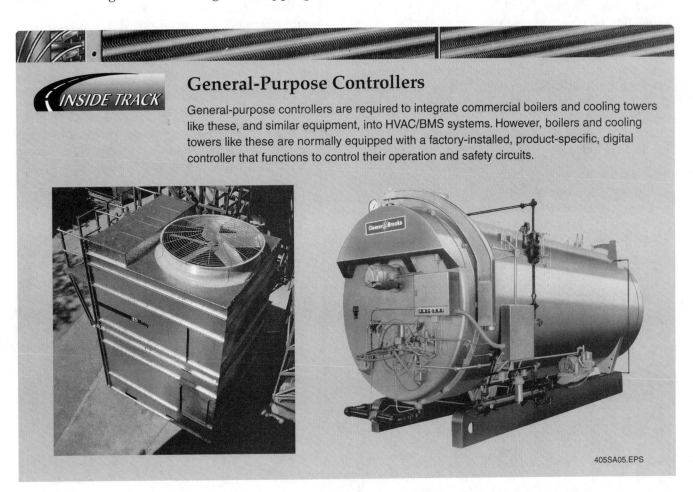

INSIDE TRACK

General-Purpose Controllers

General-purpose controllers are required to integrate commercial boilers and cooling towers like these, and similar equipment, into HVAC/BMS systems. However, boilers and cooling towers like these are normally equipped with a factory-installed, product-specific, digital controller that functions to control their operation and safety circuits.

405SA05.EPS

unoccupied at 6:00 PM. Tenants of the building are charged for their occupied heating and cooling usage by the square footage of the building they occupy. However, when tenants activate the HVAC system during unoccupied periods, building owners want to charge the individual tenants for this usage.

Room sensors in each tenant zone have an override button. When pushed during unoccupied hours, the override button causes the associated air handler and chiller or boiler to activate and provide cooling or heating as needed. The **tenant billing** module, located on the communication bus, registers this after-hours demand and stores the zone that requested the extra HVAC usage and how many hours were used. At the end of the month, the tenant billing module works with the central computer software to produce a bill for each tenant's use.

4.3.2 Load-Shed

One of the most important BMS functions is **load-shedding**, which is a key element of energy cost control. Most utilities charge their commercial and industrial customers higher rates during peak use periods and for consumption above a pre-established demand limit. Therefore, it is in the building owner's best interest to minimize energy consumption during these periods and to keep track of the energy consumption in relation to the demand limit. A load-shed control module uses one or more wattmeters to track energy consumption. As the energy consumption approaches the demand limit, the system is programmed to exercise one or more strategies to shed loads. The following are examples of these strategies:

- Override the zone thermostat setpoints so that the equipment is off for longer periods.
- Deactivate auxiliary heating devices, which use a lot of energy.
- Shut off electrical devices in a prearranged order of priority.

If energy consumption exceeds the demand limit, an alarm may sound. At that time, building managers may implement even more severe measures to curtail the use of energy in order to avoid cost penalties.

4.3.3 Data Collection

Access to historical data is very important to building managers. In nationwide retail chains, hotels, and industries, energy managers constantly need to look at operating data from all of their locations to determine trends, identify problems, and develop energy management strategies. Installing a **data collection** module on the building network and programming it to collect desired information satisfies these needs. The data collection module is capable of collecting trend information on temperatures and pressures, and recording the use of consumables, such as gallons of water and kW power usage. This information is collected on the network and does not require the central computer to be operational.

The central computer is programmed to use the building's communication network to automatically upload stored data from the data collection module on a daily basis. This information is then archived on the central computer into weekly, monthly, and yearly historical files for generating reports needed to manage the building.

4.3.4 Lighting Control

Among the major energy users in a commercial building are the lights, both internal and external. Algorithms are used to schedule their use. For example, programmed daily scheduling can minimize the use of parking lot lights. These algorithms, depending on the control system manufacturer, may be run on the central computer, in a separate control module, or in a general-purpose controller.

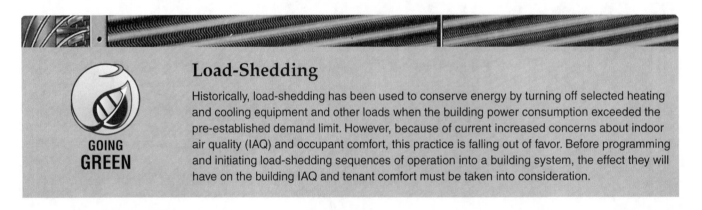

GOING GREEN

Load-Shedding

Historically, load-shedding has been used to conserve energy by turning off selected heating and cooling equipment and other loads when the building power consumption exceeded the pre-established demand limit. However, because of current increased concerns about indoor air quality (IAQ) and occupant comfort, this practice is falling out of favor. Before programming and initiating load-shedding sequences of operation into a building system, the effect they will have on the building IAQ and tenant comfort must be taken into consideration.

Commercial building lights are controlled by their own digital controller system and use one of several unique protocols like digital addressable lighting interface (DALI), ZigBee Wireless, or building automation and control network (BACnet). A protocol is a convention that governs the format and timing of communication between devices in a computer network. In order for the BMS to access information from this system, a lighting interface module is used. The interface module translates between the protocol of the BMS and the protocol of the lighting system.

4.3.5 Computer Room Air Conditioning

Air conditioning the building's main computer center where payroll, business communication, and accounting functions are performed is typically done by specialized equipment. This equipment is controlled by its own digital controllers. Today's communication data centers are most often use networks running with the internet protocol (IP). In order for the BMS to access information from this system, a computer room interface module is used. The interface module translates between the protocol of the BMS and the protocol of the computer room air-conditioning units.

4.3.6 Building Access

For security reasons, many buildings have limited access. Only individuals with access keys (programmed plastic cards) may gain entry to the building, and their access can be tracked. The building access system software has its own protocol. In order for the BMS to gather information from the building access system, an interface module is used. The interface module translates between the protocol of the BMS and the protocol of the building access system.

4.3.7 Smoke Control System

In larger commercial buildings, the ability to sense smoke and control its spread (and therefore the spread of a fire) throughout the building is significant. Typically, this function is controlled by a specialized software system separate from the building's HVAC control system. *Figure 20* shows a typical smoke control system with smoke detectors, detection and alarm panels, and remote control panels located appropriately in the building. The smoke control system has its own unique communication protocol, typically Modbus or BACnet.

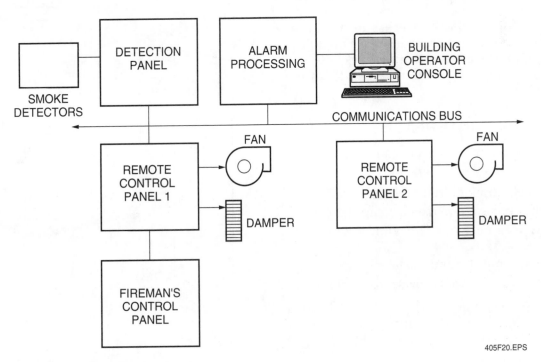

405F20.EPS

Figure 20 ◆ Smoke control system.

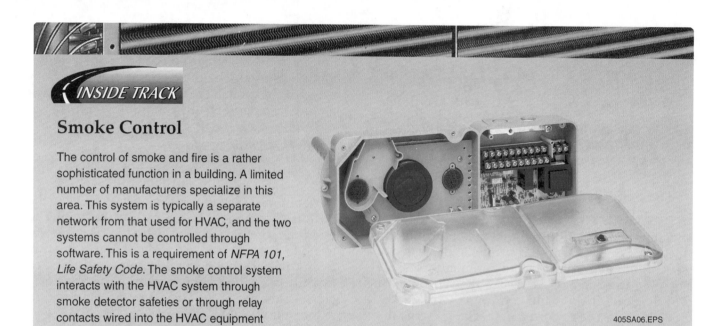
The remote control panels position ventilation and exhaust dampers and operate fans to exhaust smoke and pressurize floors on either side of the floors involved in a fire. An operator's control panel allows building personnel to interface with the system, and a fireman's control panel allows the fire marshal to view equipment status and override the smoke control system.

In order for the BMS to access information from the fire alarm system, an interface module is used. The interface module translates between the protocol of the BMS and the protocol of the fire alarm system.

5.0.0 ◆ USER INTERFACES

Building management system controllers can be interfaced by using the following:

- A handheld local interface device
- A locally connected personal computer
- A **local area network (LAN)** or a **wide area network (WAN)**
- A remote computer and the telephone lines
- The internet with web browser technology

A **local interface device** (*Figure 21*) is a keypad with an alphanumeric data display and keys for data entry. It is connected to the digital controller with a phone-type cable that provides power and communication. Handheld local interface devices may be moved from one controller to another as needed for viewing, diagnosing, and changing controller information.

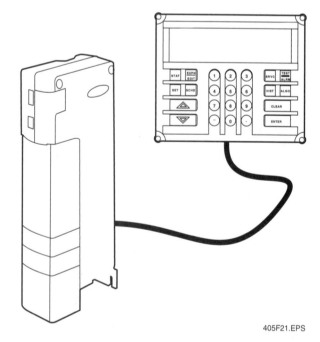

405F21.EPS

Figure 21 ◆ Local interface device.

A popular method for interfacing with the network controllers is to use a single computer as shown in *Figure 22*. Through the single computer, loaded with appropriate control system software, the operator can interact with all controllers on the connected network.

In larger applications, where multiple computer users need simultaneous access to the DDC network, a computer network (*Figure 23*) may be used as the user interface. The use of a local area network (LAN) links computers into a server/client format, where the server computer provides central data storage for all network controller data. Client computers access the network through the server and change the information stored on the server. Thus, all changes made to the network are current and located in one place, making multiple-PC user database management easy.

A LAN is typically used with a single building or building complex, such as a college campus. All computers in the LAN can access any building BMS through the server database.

Computers may also be arranged in a wide area network (WAN). A WAN is a server/client computer network spread over a large geographical area. A WAN is unlimited in geographical size and may connect several LANs together though telephone lines or satellite links.

The telephone lines may be used to connect remote computers to local BMS networks. *Figure 24* shows the use of a remote access interface module to accomplish this. Two modules are required, one on each end of the telephone lines. The module contains a **modem** (modulator-demodulator) and security software to limit access to qualified personnel. The modem is the interface between the computer and the phone lines. Modem transmission speeds are rated in bits per second (bps) and range from 28,000 to 56,000 bps. With this configuration, the local BMS network can automatically call a remote site or beeper service and send alarm messages.

In today's market, web browser **internet** technology may also be used to access the local BMS network from a remote site or from any computer located on a building's **Ethernet** LAN (*Figure 25*). A small industrial server computer (web page server) is wired to any unique protocol bus as well as the building's Ethernet LAN. The **web page** server is programmed to scan the unique protocol bus and create web pages for each controller on the bus. These web pages may be tabular or graphical in nature, and are dynamically updated regularly. In addition, the open **hypertext transfer protocol (http)/internet protocol (IP)** may be used to interface with the web pages.

Thus, an operator may sign in at any computer on the LAN, and using a normal **web browser** program like Internet Explorer, may then perform functions that could have been accomplished through a local computer loaded with a BMS-specific software program.

Further, using the power of the internet, an operator may use a remote computer's web browser, a **personal digital assistant (PDA)**, or a cell phone to interact with the web page server computer through a phone line or a wireless hub.

This system, because of its versatility and use of web browser technology, is rapidly replacing the use of locally connected personal computers loaded with BMS-specific software programs.

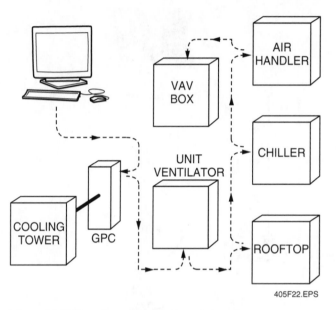

405F22.EPS

Figure 22 ◆ Local computer connection.

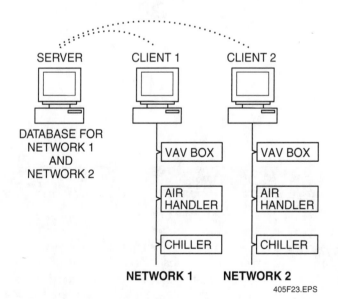

405F23.EPS

Figure 23 ◆ Local area network.

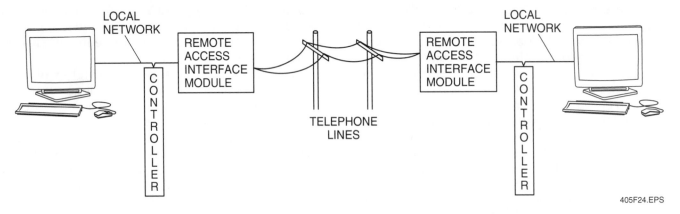

Figure 24 ◆ Remote access—no PC modem.

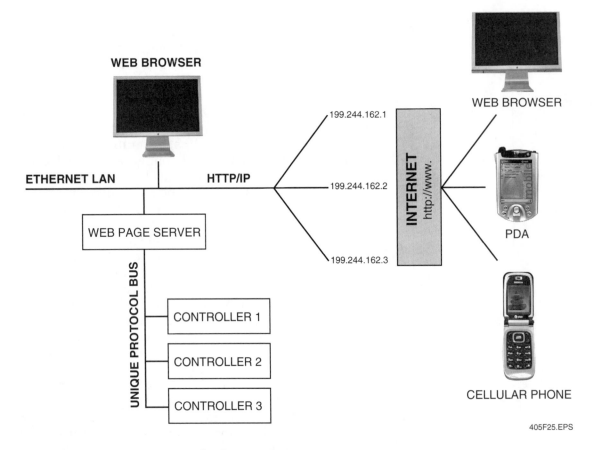

Figure 25 ◆ Web browser internet technology.

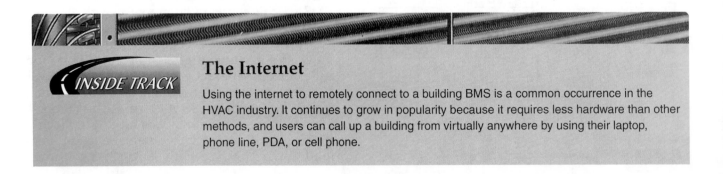

The Internet

Using the internet to remotely connect to a building BMS is a common occurrence in the HVAC industry. It continues to grow in popularity because it requires less hardware than other methods, and users can call up a building from virtually anywhere by using their laptop, phone line, PDA, or cell phone.

Any user equipped with a PC, PDA, or cell phone, and knows the proper access codes and **internet protocol addresses (IP addresses)**, can communicate with the building web page server computer locally, or from any remote location. This system is especially attractive to companies that have a number of technicians on the road. The technician, with no special software, can have access to any number of buildings from home, or wherever he or she is presently located.

6.0.0 ◆ INTEROPERABILITY

Facility managers would like to have a complete real-time model (*Figure 26*) of all the buildings under their care, providing every important detail at their fingertips. They want to know what building space each tenant occupies, when it is being used, what everyone's phone number is, what phones are or are not operational, what furniture (including color and fabric) is assigned to each building space, and what kind of special assistance might be required in case of an emergency. They want easy access to the actual temperature in each building space, the ability to adjust each space's control temperature at will, and the ability to match the operational schedule of each HVAC system in their facilities to its real-time occupant use. They also want to be able to optimize their building's utility costs to match real-time utility rate structures.

Building managers like to have status information about video teleconferencing sites throughout the facilities, and have security video feeds at their fingertips from cameras located throughout the sites. Additionally, they would like to have data exchanged between all computers and voice over IP (VOIP) phone connections.

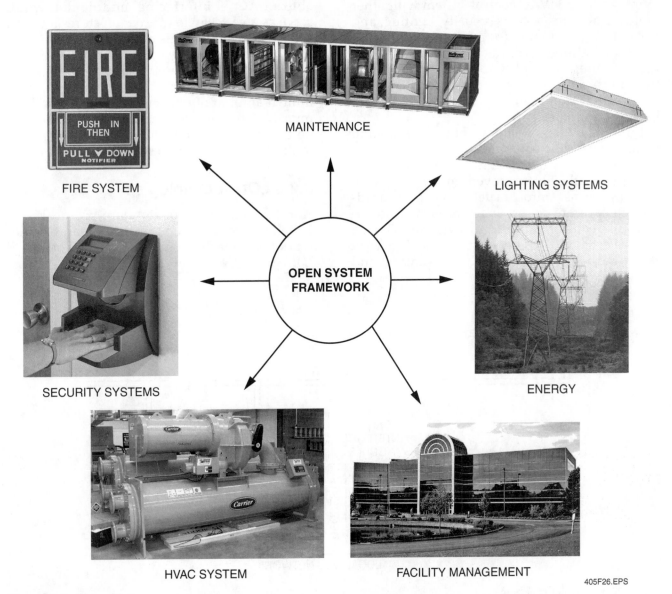

MAINTENANCE

FIRE SYSTEM

LIGHTING SYSTEMS

OPEN SYSTEM FRAMEWORK

SECURITY SYSTEMS

ENERGY

HVAC SYSTEM

FACILITY MANAGEMENT

405F26.EPS

Figure 26 ◆ Connectivity.

They want to know who manufactured each piece of equipment, who installed it, when its warranty period ends, what preventive maintenance has been performed, and when its next maintenance is scheduled. They also want to know where all the equipment is located, where all conduits and piping are, and where all communication cables are located.

The accurate transfer of information and commands required to match the facility manager's desires across building systems with different protocols is called **interoperability**. These desires are becoming more and more possible today with proper planning by a qualified system integration company and utilization of manufacturer's equipment that is becoming more and more interoperable.

With the evolution of open communication protocols in the market place, it is now possible to integrate an HVAC control system with other building systems such as security, lighting, and fire/life safety. Open protocols are defined as communication standards that allow different manufacturer's control systems to share information. Each protocol has a unique set of rules to ensure that good communication take place between controllers.

Building owners do not want to be limited to using any specific equipment vendor they have. Thus, through various industry organizations and groups, they have moved toward the use of open protocols. Unfortunately, none of the leading protocols are perfect and none are able to satisfy the needs of all building systems. Thus, a unique protocol does not exist in today's market that is the best. Each protocol has advantages and disadvantages. With these limitations in mind, however, there are several open protocols that are tending to become most used by commercial building owners.

The following sections describe the most popular protocols being used in building systems today. It should be understood that the list is not exhaustive, and does not cover other industries such as factories and automotive industries.

6.1.0 BACnet

Developed by ASHRAE and approved as a standard in 1995, the building automation and control network (BACnet) protocol provides a comprehensive set of message rules which, if implemented, provide interoperability between the various building systems. As *Figure 27* shows, each manufacturer would design and make available a BACnet interfacing module that would serve as a **gateway** to the manufacturer's proprietary protocol bus. This module would have the capability of two-way communication between the manufacturer's network and the BACnet Ethernet. Each BACnet module would provide a limited amount of information transfer as determined by the module's manufacturer. Depending on the amount of information to be exchanged, multiple BACnet modules would be required.

6.2.0 LON Technologies

Another contender for standard protocol is the Echelon company, with its local operating network (LON) technologies, including LonTalk™, LonWorks™, and LonMark™.

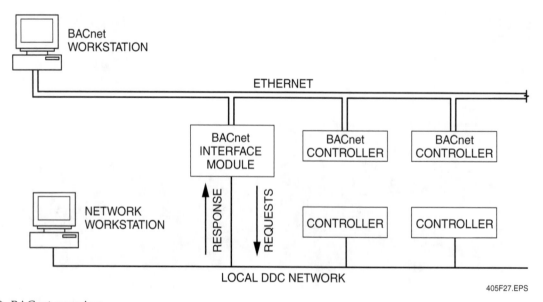

Figure 27 ◆ BACnet overview.

Echelon has developed LON technology based on a neuron chip microprocessor. The neuron chip contains hardware and **firmware** (instructions embedded in the chip) that operate on a seven-layer communication protocol. LonTalk™ is the seven-layer protocol that allows controllers (nodes) to operate using an efficient, reliable communication structure. Each node in the network contains a chip with embedded intelligence that implements the protocol and performs control functions.

LonWorks™ is Echelon's DDC network that uses LonTalk™ protocols, and is part of the ANSI-approved *ASHRAE 135-1995* BACnet standard for building automation.

Echelon's LON technology products were designed to provide solutions for a general manufacturing environment; however, the technology can be applied equally to the HVAC industry. Because each node sold by Echelon has a LON chip in it, any product with this chip has the built-in protocols to communicate with any another product. Likewise, any manufacturer who uses LON chips in their products ensures interoperability with any other manufacturer's devices that include LON chips.

6.3.0 Modbus

The Modbus standard was developed by the Modicon Company in 1979 and later became the trademark of Schneider Electric. Schneider made Modbus available for free and gave the Modbus TCP/IP protocol specifications to the Internet Engineering Task Force. Thus Modbus TCP/IP has become widely used and is popular in the industrial, metering, power generation, and electrical equipment industries. Modbus is typically used as a translation protocol between two other proprietary protocol networks.

The Modbus standard defines both hardware and software standards. Modbus Remote Terminal Unit (Modbus RTU) communicates serially over Electrical Industry of America (EIA) RS232 and RS485 hardware connections. Modbus Transmission Control Protocol (Modbus TCP) communicates over Ethernet LAN type hardware connections.

6.4.0 Hypertext Transfer Protocol (http)

Hypertext Transfer Protocol (http) is the base protocol used by the **world wide web**. This standard is published and updated by The World Wide Web Consortium (W3C), and uses TCP/IP as the means of transporting information over the internet. The http protocol allows a computer operator to use a web browser program, such as Internet Explorer, to view transmitted data in the form of web pages. The http protocol defines how messages are formatted and transmitted. The http protocol also defines what actions web server computers, browsers, and network controllers take in response to various commands.

Thus, control systems can take advantage of this well-defined protocol to share data with end users as well as other controllers. Since the use of this standard extends beyond the HVAC industry, BMS control systems can use http to easily share information, such as corporate financial data, energy data, maintenance management data, and HVAC system operational data across building systems.

6.5.0 Web Browser System Integration

Approximately 80 percent of all commercial building projects will be satisfied with an integration system like that shown in *Figure 25*, but the remaining 20 percent require more complex interoperability to achieve their operational needs. To meet this greater complexity, many manufacturers now provide a low-cost web-based connection to their control systems. Thus, building owners and operators can now use the common http protocol and a standard web browser to interface with their multiple building systems.

Figure 28 shows the network architecture of such a system. In this system, a unique protocol can be used where its qualities best suit the control needs involved. Thus, separate control buses (and protocols) for lighting, HVAC, building security, and other systems, may still exist, but they are seamless to the building operator. In addition, the integrators are capable of implementing control strategies across building systems.

System integration is achieved by the use of multiple industrial integrator computers, known as gateways, located on the building's Ethernet LAN. These integrators do not have keyboards or monitors. They are powerful central processing units with multiple cable connections that match the transport medium requirements of the common open protocols (BACnet, LON, and Modbus). In addition, the integrators have a bus connection that matches the manufacturer's own unique proprietary protocol. The integrators are then programmed to regularly extract data from controllers located in the building's lighting, HVAC, security, and other systems. This extracted data is then complied and temporarily stored in two-way information tables recognized

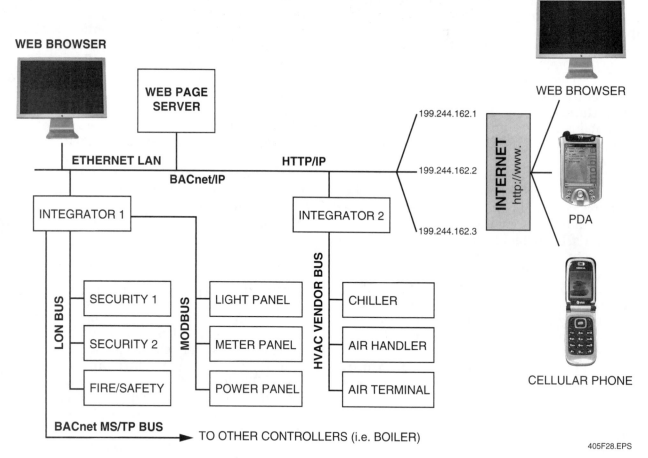

WEB BROWSER

WEB PAGE SERVER

ETHERNET LAN

BACnet/IP

HTTP/IP

199.244.162.1

199.244.162.2

199.244.162.3

INTERNET http://www.

WEB BROWSER

PDA

CELLULAR PHONE

INTEGRATOR 1

INTEGRATOR 2

LON BUS

SECURITY 1

SECURITY 2

FIRE/SAFETY

MODBUS

LIGHT PANEL

METER PANEL

POWER PANEL

HVAC VENDOR BUS

CHILLER

AIR HANDLER

AIR TERMINAL

BACnet MS/TP BUS

TO OTHER CONTROLLERS (i.e. BOILER)

405F28.EPS

Figure 28 ◆ Web browser integration technology.

by a common protocol normally used on the Ethernet LAN. BACnet/IP is becoming the common protocol used on the Ethernet LAN.

Also located on the LAN is a web page server computer. The web page server computer uses the common protocol (typically BACnet/IP) to regularly extract table information from the integrator computers and to create web pages of information.

These web pages are then viewable in tabular or graphical format by any operator through a web browser located on any computer on the LAN. The communication are also two-way, so that any operator may manually intervene at the controller level in any building system. Additionally, the operator may communicate with the web page server over the internet via a computer, a PDA, or a cell phone.

Finally, integrated control strategies across building systems may be achieved through special routines written in the web page server computer. For example, data from a card reader can be sent to the lighting and HVAC system controllers so that they are activated when an employee enters the building, and shut down when he or she leaves.

6.6.0 The Future of Interoperability

In the future, most manufacturers will build their equipment with controllers that have bus connections to support four typical protocols: their own, BACnet, LON, and Modbus. A simple switch setting on the controller will make it ready to communicate with the protocol of choice. This will make it easy to integrate equipment of multiple vendors into a single, seamless system.

In addition, within each building system industry, equipment manufacturers are all moving toward the elimination of their own unique communication bus protocol and tending to standardize around a common open protocol that meets the needs of that industry.

Using the HVAC industry as an example, in a few years a building owner should be able to buy a chiller from Manufacturer A, an air handler from Manufacturer B, and a set of air terminals from Manufacturer C that can talk together on a common HVAC communication bus. The emerging use of the common open protocol (BACnet) in the HVAC industry will make it easy to integrate the chiller, air handler, and air terminals into a cohesive HVAC system.

Because of these trends, the building owner will no longer be locked into a single vendor because of the control system. In addition, during later building expansions or renovations, the owner will enjoy the ability to freely choose between equipment manufacturers within a type of building system without having to worry about interoperability issues.

When it comes to interoperability across building systems, however, it is highly unlikely that any one protocol will be developed to meet the needs of all systems. However, as shown above in *Figure 28*, the building owner will be able to use the power of http internet protocol and web browsers to more easily manage the building through a single seamless interface. The use of web-based technology is already growing rapidly today, and will continue even more rapidly in the future.

7.0.0 ◆ INTERPRETING FRONT-END SOFTWARE

In the past, each building system used a separate computer loaded with software specific to that systems controllers. Thus, separate computers were needed for the building operator to interface with the building security system, lighting system, and HVAC system.

Although using a single computer to interface building systems may not yet be feasible, it is becoming feasible through the movement of integration companies toward the use of internet web-based technology, particularly the use of simple graphical web page interfaces that present key data from multiple building systems to the operator.

Regardless of the means used to gather the data, or the number of computers used to do so, the primary window into a building's systems is a front-

end computer loaded with the appropriate control software, or one that uses a web browser. The front-end information presented to the operator is typically divided into functional areas as follows:

- Sign on/off
- Operator management
- Database management
- Graphical interface
- Alarm management
- Report generation

7.1.0 Sign On/Off and Operator Management

In the sign on/off portion of the software, security controls, including passwords, are used to limit access. In the operator management portion, different operators are given different levels of authority. Some can only look at data, some can run diagnostic tests, and others can change the operating parameters. The higher the level of authority, the more control the individual will have over the system.

7.2.0 Database Management

In the database management portion of front-end software systems, files from all the controllers in the local building are uploaded and stored according to the networks in which they exist. If the computer is used to dial other remote buildings, then the files from those networks are also stored in the database. These files are updated whenever an operator changes any parameter in any controller. In web-based systems, each controller database file exists at the controller level and at the web page server level.

The operator can view and modify the algorithms and points for any controller (*Figure 29*). This information is displayed on dynamic screens that interrogate the highlighted controller through the network bus. Any changes made are automatically stored at the controller level and at the computer that stores a mirror image of the controller file.

7.3.0 Graphical Interface

In the graphical interface portion of the software, graphic images or photographs are presented for each piece of equipment in the building. Dynamic, live information about the operation of the unit is superimposed on these images.

Figure 30 shows a graphical image of the control system for a typical air handling unit.

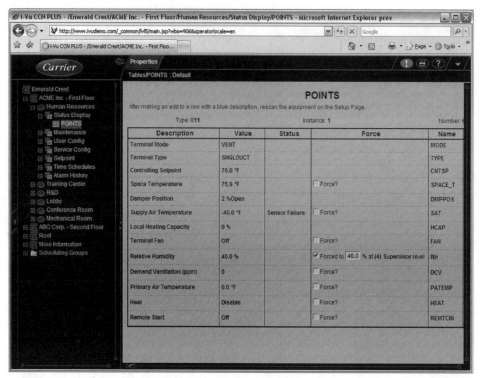

Figure 29 ◆ Controller and point list screens.

Figure 30 ◆ AHU control system.

The diagram shows all the controllable devices in the air handler, including the cooling and heating coil, supply and return fan, mixed air dampers, and spray humidifier. In addition, all the input and output points are represented by symbols and abbreviated descriptions. Next to each point symbol is dynamic data that the software uploads from the associated controller in the network about every five seconds.

This information is very useful for monitoring and troubleshooting. Also, the setpoint and time schedules associated with the air handler can be displayed and modified by using the mouse and clicking on the value of any point on the screen.

Screens similar in format are available for all the building systems. In fact, the trend is to provide simple graphics that provide key information from multiple building systems on a single graphic. A click of the mouse shows the operator greater detail of any of the building systems he or she chooses to look at.

Figure 31 shows a typical time schedule. The schedule has eight periods, with each period defining when the building is considered occupied and when it is unoccupied. Different periods can apply for different days of the week. The right side of the image is a graphical display of the occupied hours for each period. Once the operator makes a change, the change is sent down to the actual controller containing the schedule and then saved in the controller's database on the central computer.

These time schedules can be specific to one controller, grouped to a number of controllers, or it can be global in nature and apply to all controllers in a network or networks. Thus, the operator may impact controller operation across building systems through a simple graphical interaction.

7.4.0 Alarm Management

Whenever an alarm condition occurs in a controller, the alarm is transmitted to the central computer and stored in an alarm file (*Figure 32*). The operator is notified of the incoming alarm visually and audibly and can then deal with the alarm condition by reviewing the alarm file and acting accordingly.

7.5.0 Report Management

Many types of reports are available to the operator of the system. They include operator activity, how alarms were handled, trends of temperatures and pressures, equipment runtimes, tenant billing, and consumable information. *Figure 33* shows a typical daily space temperature report.

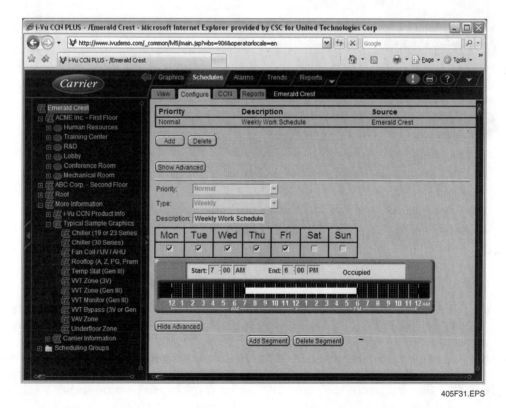

405F31.EPS

Figure 31 ◆ Time schedule.

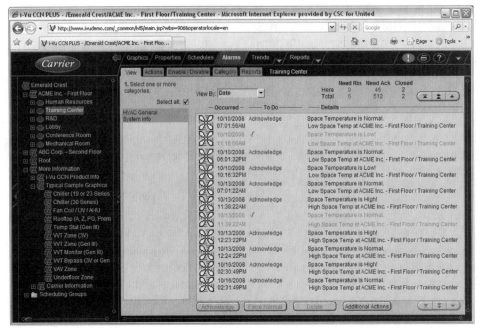

405F32.EPS

Figure 32 ◆ Alarm file.

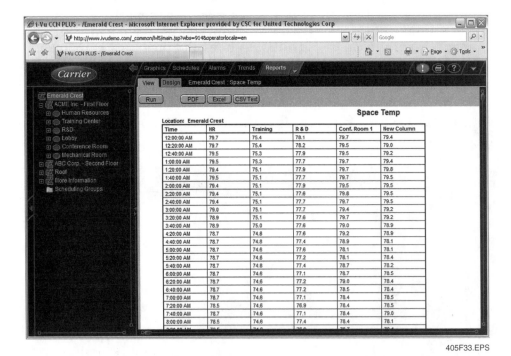

405F33.EPS

Figure 33 ◆ Space temperature report.

Data from the network can be stored on the central computer in daily, weekly, monthly, and yearly files. Reports can then be set up and run by the operator to cover any particular period of time.

With information like that shown in the previous examples, the building operator is well-equipped to monitor the building performance,

adjust operating parameters, and troubleshoot problems when they arise. In addition, building management is well-equipped with historical information to invest in building management strategies that, when implemented across building systems, reduce overall building energy use and operating costs.

8.0.0 ◆ INSTALLATION

Like all systems, the building control system is installed based on design drawings. The drawings indicate the location and type of every sensor, actuator, controller, and computer in the building. In addition, the drawings show where to install pneumatic tubing, electrical conduit, sensor wiring, and communication bus wiring. The installation contractor is usually responsible for sizing and installing all valves and dampers.

NOTE

Details for the types of wires are provided to help the trainee; however, the actual job specifications must be followed for each project.

System installation usually takes place in the following order:

Step 1 Install all sensors, valves, dampers, and meters.

Step 2 Install all control panels that house general-purpose controllers or controllers specific to each building system.

Step 3 Run all sensor, power, and communication wiring.

Step 4 Program each controller.

Step 5 Install all general-purpose controllers and terminate wiring.

Step 6 Install and program the central computer or multiple computers in a LAN or WAN architecture.

Step 7 Program modems for dial-out capability.

Step 8 Set up and program internet web browser capability.

8.1.0 Sensor Installation

Each sensor and actuator comes with mounting and wiring instructions. *Figure 34* is representative of the type of information provided by the sensor manufacturer. In this diagram, the installation of an RTD averaging sensor is shown. All sensors must be mounted according to the locations shown on the design drawings. The location of each sensor is critical for accurate measurement of the fluid or air temperature being sensed.

Likewise, valves, dampers, meters, and other control and monitoring devices must be installed as indicated on the drawings. The control con-

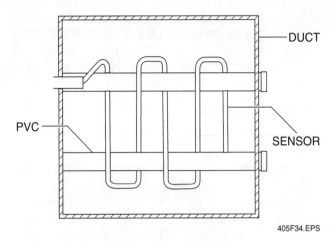

405F34.EPS

Figure 34 ◆ Averaging temperature sensor.

tractor will have sized and ordered these devices early in the job because of the length of time it takes for manufacturers to supply them.

8.2.0 Control Panel Installation

All general-purpose controllers and BMS-specific controllers must be field-installed in control panels. The panels are lockable for the protection of the controllers. The drawings should show where the panels are located throughout the building. Typically, they will be located close to the equipment they control to minimize the length of sensor wiring. Product-specific controllers are usually installed by the manufacturer at the factory in a control panel mounted on the equipment.

8.3.0 Running Sensor, Power, and Communication Wiring

The following rules should be followed when running sensor, power, and communication wiring:

- Power (24V) and sensor wire should be general-purpose, CM-, CL3-, or CL2-rated, two-conductor, 20-gauge, foil-shielded cable.
- Bus communication wire requirements will vary depending upon the bus protocol being used. For example, general-purpose, CMG-, CM-, CL3-, or CL2-rated, three-conductor, 20-gauge, foil-shielded RG-45 cable would be used.
- When wires are run in a ceiling plenum or other environmental air spaces, appropriate cable must be used. RS485, for example, would require CMP-, CL3P-, or CL2P-rated, 20-gauge, foil-shielded cable.

- Similarly, when wires are run in risers, RS485 cable, for example, would require CMR-, CL3R-, or CL2R-rated, 20-gauge, foil-shielded cable. Risers are defined as vertical runs in a shaft or from floor to floor.
- When wires are run in high-temperature areas, RS485 cable would use general-purpose, 105°C CM-rated, 20-gauge, foil-shielded cable.
- It is recommended that stranded copper wires be used because solid wire tends to break over time.
- Never run sensor or communication wires in the same conduit as power wiring. This is a *National Electrical Code*® (*NEC*®) requirement.
- Avoid running wires in close proximity to lights, motors, and other devices that have changing electrical fields. Such dynamic fields will cause interference and result in communication problems.
- Run and tag all sensor wires between each sensor and the appropriate control panel.
- Run the communication bus between all appropriate controllers on the job and the central computer, if applicable.

- Only install the wire specified by the job specifications. All wire is not the same.
- Always use daisy-chain wire terminations (*Figure 35*) at each connection (three wires in and three wires out). Avoid using T-tap connections, unless the manufacturer allows it.
- When making daisy-chain terminal connections, make sure to connect the foil shield wire across the connection. This provides a continuous shield along the wire.
- Ground each communication bus and sensor wire at one end only, as shown in *Figure 35*. Ground each controller to chassis ground. If grounding is not done correctly, **transients** can cause system problems.

All digital controllers require 24VAC power to run. This means that each controller must be provided with a transformer of sufficient VA capacity. Check the drawings for transformer specifications. Avoid installing one large transformer on a floor and connecting all controllers in parallel. This is a poor approach for two reasons: the voltage at the controller will vary with distance, and a failure of the transformer will take the entire floor down.

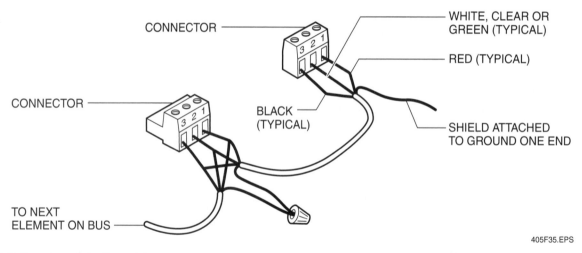

CONNECTOR

WHITE, CLEAR OR GREEN (TYPICAL)

RED (TYPICAL)

CONNECTOR

BLACK (TYPICAL)

SHIELD ATTACHED TO GROUND ONE END

TO NEXT ELEMENT ON BUS

405F35.EPS

Figure 35 ◆ Bus communication wiring.

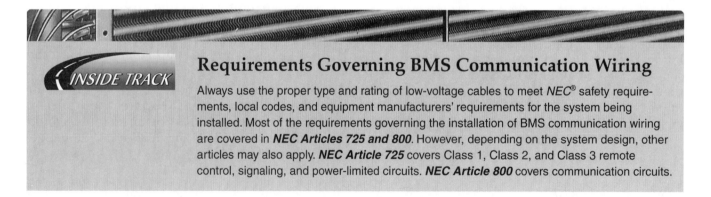

Requirements Governing BMS Communication Wiring

INSIDE TRACK

Always use the proper type and rating of low-voltage cables to meet *NEC*® safety requirements, local codes, and equipment manufacturers' requirements for the system being installed. Most of the requirements governing the installation of BMS communication wiring are covered in *NEC Articles 725 and 800*. However, depending on the system design, other articles may also apply. *NEC Article 725* covers Class 1, Class 2, and Class 3 remote control, signaling, and power-limited circuits. *NEC Article 800* covers communication circuits.

8.4.0 Controller Programming

Product-specific controllers are programmed at the factory. All they need is to be configured for their setpoints, alarm limits, time schedule functions, and optional features such as temperature reset parameters.

General-purpose controllers (GPCs) require much more programming to activate their algorithms. This programming is typically done back at the office in a clean environment, not at the job site.

Programming GPCs usually follows one of two paths. Some manufacturers require that actual code (programming language) be written for each application and downloaded into the controller. *Figure 36* shows a typical coded program that starts one of two pumps based on the pump with the lowest runtime hours.

Other manufacturers provide coded algorithms already in the controller and require the programmer to simply fill in the blanks to tell the algorithm the points from which to gather information, the time schedule and setpoint schedule to use, and the point to which the output signal should be sent. This second process is called configuration and is much easier than learning a programming language. *Figure 37* shows a typical configuration screen for an algorithm that starts and stops the fan (FAN1) of an air handler based on a time schedule stored in the controller. Programming is simply a matter of filling in the blanks with names from the controller database,

such as output point (FAN1), input point (SPT1), time schedule (OCCPC01), setpoint schedule (SETPT01), and so on.

Regardless of the programming method used, the programmer should attend factory training classes to learn how to program the specific product before attempting the first project.

	Value	Units
Discrete Output Point	FAN1	
Sensor Group/SPT Sensor	SPT1	
Time Schedule	OCCPC02	
Setpoint Schedule	SETPT01	
Loadshed	LDSHD01	
NTFC Algorithm	NTFC_01	
Hysteresis	0.5	°F
Duty Cycle		
Duty Cycle Enable	Disable	
First Minute of Hour	0	
Second Minute of Hour	0	
Occupied Off Duration	0	min
Unoccupied Off Duration	60	min
Minimum Off Time	0	min
Redline Bias Time	0	min
Power On Delay	0	sec

405F37.EPS

Figure 37 ◆ Programming configuration.

```
PROGRAM  RUNSTART

~THIS PROGRAM STARTS ONE OF TWO PUMPS BASED ON THEIR RUNTIME
~HOURS.  THE PUMP WITH THE LOWEST RUNTIME HOURS IS STARTED
~EACH DAY.

STEP ONE
IF OCC.OCCPC.MODE=0 THEN PUMP1=0; PUMP2=0; REPEAT ENDIF
IF PUMP1=1 AND PUMP2=1 THEN GOTO TWO
IF PUMP1=1 OR PUMP2=1 THEN REPEAT ENDIF
:TWO
IF RUN1HR>RUN2H2 THEN PUMP2=1; PUMP1=0; REPEAT
ELSE PUMP1=1; PUMP2=0; REPEAT
ENDIF
ENDPROGRAM
```

405F36.EPS

Figure 36 ◆ Programming code.

8.5.0 Controller Installation and Wiring Termination

Each general-purpose controller must be field-installed in a control panel. After the processor is installed, tagged sensor wires must be connected to the controller's I/O modules according to the wiring diagram for each control panel. Each sensor and actuator must be tested to ensure the proper sensor wire terminations have been made. The 240V power must be connected to each controller, and the operation of each algorithm in the controller tested, one at a time. Finally, attach the communication bus to each controller and verify bus communication.

The communication bus must also be attached to each factory-installed product-specific controller and communication must be verified.

8.6.0 Central Computer Installation and Programming

After the physical components of the central computer have been installed, the proper PC operating system must be installed. Next, the front-end control system software is installed and the controller database is created. With the database installed, printers are assigned, operator profiles are created, alarms are configured, and then graphical interfaces and standard reports are created. At that point, the front-end software is active and useable.

8.7.0 Modem Programming

In order to provide remote communication with the local network, remote communication modules must be programmed with passwords and dial-out numbers, and each modem must be loaded with the proper string configuration for communication. Once programmed, a final check of remote communication makes the entire system ready for use.

8.8.0 Web Page Server Programming

If a web page server computer is used to connect to a building system bus (HVAC system bus for this example), it must be installed and programmed. After Ethernet and equipment bus hardware connections have been established, the computer is loaded with its web-based software program. Software installation is completed by obtaining and programming an IP address for this computer. The IP address is obtained from the building's information technology department.

After being loaded with software and assigned an IP address, many of these computers, when prompted, automatically poll their equipment bus for controllers, and automatically establish a series of web pages for each controller, including:

• Dynamic tabular data
• Dynamic graphical interface
• Dynamic alarms
• Standard reports
• Standard point trends

After the polling process, the web page server is ready to communicate with any web browser in a two-way communication mode. It will provide the same information that could be obtained from a front-end manufacturer-specific software program. Some installations may have both the front-end software program on a dedicated computer, as well as the web page server computer. Newer installations are more likely to have only the web page server installed.

9.0.0 ◆ SYSTEM CONTROL STRATEGIES

The *HVAC Level Two* module *Commercial Airside Systems* described the control sequences for constant volume systems. This section expands that discussion to cover systems that are likely to be installed in buildings managed with BMS

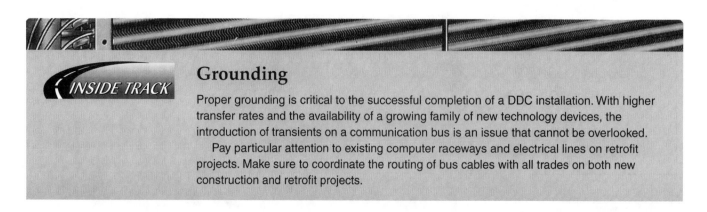

INSIDE TRACK   Grounding

Proper grounding is critical to the successful completion of a DDC installation. With higher transfer rates and the availability of a growing family of new technology devices, the introduction of transients on a communication bus is an issue that cannot be overlooked.

Pay particular attention to existing computer raceways and electrical lines on retrofit projects. Make sure to coordinate the routing of bus cables with all trades on both new construction and retrofit projects.

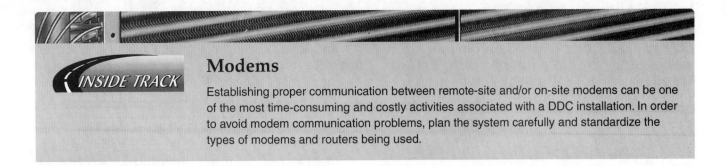

software. The system types include VVT and VAV. Understanding these control sequences will give you insight into how an HVAC system can be integrated with other building systems using BMS software.

9.1.0 VVT Control Sequence

Figure 38 shows a VVT system control schematic. The key to VVT system operation and control is the use of zone controllers that have the ability to communicate with each other across a communication bus. Each zone has its own zone controller mounted on a VVT air terminal. A room sensor monitors zone temperature and allows the zone occupant to modify his own cooling and heating setpoints. Each zone controller also contains a time clock that allows the occupant to define occupied and unoccupied times for each day of the week, weekend days, and holidays.

One zone controller acts as a linkage coordinator and polls the other zone controllers at regular intervals. The linkage coordinator then, through communication, provides zone information to the central air source unit controller. The central air source controller decides the system mode of operation based on zone needs.

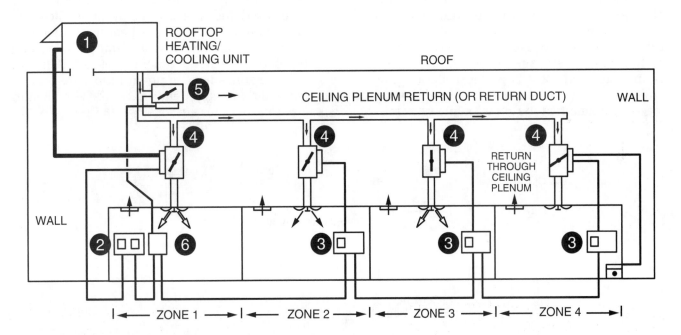

1. PACKAGED AIR CONDITIONER OR SPLIT SYSTEM
2. MONITOR THERMOSTAT WITH TIME CLOCK
3. ZONE CONTROLLER
4. ZONE DAMPER WITH ACTUATOR
5. BYPASS DAMPER WITH ACTUATOR AND STATIC PRESSURE SENSOR
6. BYPASS CONTROLLER

405F38.EPS

Figure 38 ◆ VVT system.

When any zone becomes occupied, the linkage coordinator determines that the building is occupied and tells the packaged unit indoor fan to start. The fan remains on as long as the building remains occupied. At the same time the packaged unit's outside air dampers are set to the minimum ventilation position.

9.1.1 Occupied Cooling

Each zone controller establishes how far away from setpoint it is. This is called the demand. The linkage coordinator regularly polls the individual zone controllers and calculates the total average cooling demand and total average heating demand. These demands are displayed at a user interface module. The average demand is weighted based on the airflow delivered to a zone. An 800 cfm zone will count twice as much in the demand average as a 400 cfm zone.

If the average cooling demand exceeds the average heating demand and is greater than the minimum demand required for the cooling mode, the air source controller positions the air source in the cooling mode of operation and communicates the system cooling mode to each zone controller. A mode timer is also started. The outside air damper is opened to provide ventilation air and compression stages are added as necessary. If the system is equipped with an economizer, it is activated as the first stage of cooling. With cool air being supplied, each zone controller (*Figure 39*) modulates its individual air terminal damper as needed to maintain the zone cooling setpoint.

As the air terminals throttle airflow, the supply duct pressure rises, and the airflow through the central packaged unit decreases. A bypass controller monitors the rise in duct pressure and opens the bypass damper accordingly. This allows supply air to be recirculated back to the return side of the central unit and maintains a constant airflow across the DX cooling coil.

If the average cooling demand continues to rise, the central unit controller adds capacity stages as necessary. Capacity stages are then cycled on and off to satisfy the average demand setpoint for cooling. The cooling mode ends when the average cooling demand has dropped below the field-configured minimum average cooling demand, or a greater average heating demand is encountered.

9.1.2 Occupied Heating

Should any zone require heating while the central unit is in cooling, the zone controller closes that air terminal's damper to the minimum position and sends the zone heating demand to the linkage coordinator. If the average demand for heating becomes greater than the average demand for cooling, and the cooling mode minimum timer has elapsed, the air source controller switches the central unit from cooling to heating. It then communicates the heating system mode to the zone controllers. The mode timer is also restarted.

Those zones needing heat modulate their air terminal dampers open to meet the zone heating setpoint. All other air terminal dampers are set to

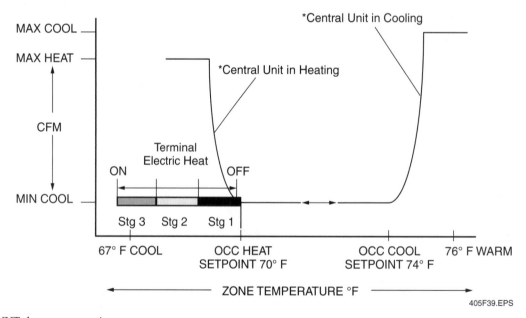

Figure 39 ◆ VVT damper operation.

their minimum ventilation position, even if any zone starts to need cooling. The linkage coordinator continues to record all zone needs.

If the average heating demand continues to rise, the central unit controller adds heating stages as necessary. Capacity stages are then cycled on and off to satisfy the average demand setpoint for heating. The heating mode ends when the average heating demand has dropped below the field-configured minimum average heating demand, or when a greater average cooling demand is encountered.

In this manner, the linkage coordinator cycles the central unit from one mode to another to meet average zone cooling and heating demand. During heating, the bypass controller continues to function, ensuring a constant cfm across the heating elements.

9.1.3 Unoccupied Period

When the linkage coordinator determines that all zones are unoccupied, it requests the central unit to stop and all air terminals to be returned to their minimum positions. The central unit compressors and supply air fan stop, and the outside air damper closes. If the zone controllers are equipped with setback temperatures, the central unit operates just as it did during the occupied period, except that the outside air dampers remain closed.

9.1.4 Simultaneous Cooling and Heating

When the central unit is in cooling, simultaneous heating at the zone level can be provided if zone air terminals are equipped with zone heaters. If the system is so configured, a zone heater can be activated by its zone controller, with the primary cool air set to minimum ventilation position. When the central unit is in heating, however, simultaneous cooling cannot be provided by a VVT system.

Normal operation of the VVT system does not involve the use of zone heaters and comes close to satisfying a building's simultaneous need for cooling and heating by time-sharing central equipment heating and cooling modes. Simultaneous need for cooling and heating in small buildings tends to follow the weather and occurs in the spring and fall when the building loads are significantly less than the peak design loads. Because VVT uses a central unit designed for the peak cooling and heating loads, the unit can quickly switch from one mode to another and satisfy the building's small part-load conditions.

VVT, with its time-cycling of capacity, works well in building sizes up to 25 tons. As buildings become larger, VVT will not be able to meet the simultaneous need for cooling and heating, particularly when buildings have internal zones that require cooling year-round.

9.2.0 VAV Control Sequence

This section covers the sequence of operation of a single duct box VAV system (*Figure 40*). Slight variations in this sequence occur when fan-powered boxes are used. A central air source unit provides air through ductwork to single-duct VAV boxes equipped with zone re-heaters. System control modes are established through communication between digital controllers that are mounted on all pieces of equipment and connected by a communication bus.

9.2.1 Occupancy Period Scheduling

For VAV systems that do not use a DDC control system, the central air source unit determines when the building's occupied period begins and ends in one of two ways: sensing a set of contacts wired to a remote time clock or energy management system; or responding to its own internal programmable time clock. The user programs the clock's occupied and unoccupied periods.

When a DDC control system is used, the occupancy period control typically resides in the zone controllers at the zone level. Any zone's controller can go occupied and overwrite the central unit's occupancy schedule, thus asking the central unit to start. Likewise, the last zone to go unoccupied determines when the central unit stops.

Each zone sensor can be provided with a manual override button. Pushing the button allows the occupant to manually start the system when the zone is unoccupied. The system will start and remain occupied for a pre-configured period of time, and then return to its unoccupied mode.

9.2.2 Occupied Cooling

When an occupied period begins, the central unit fan starts. Inlet guide vanes or a variable-speed drive controls the amount of air delivered to the system by maintaining a supply duct static pressure setpoint. When the VAV terminals throttle to match a falling building load, the supply duct pressure will increase. The air source control module modulates the supply fan to maintain the duct pressure setpoint. Along with the start of the

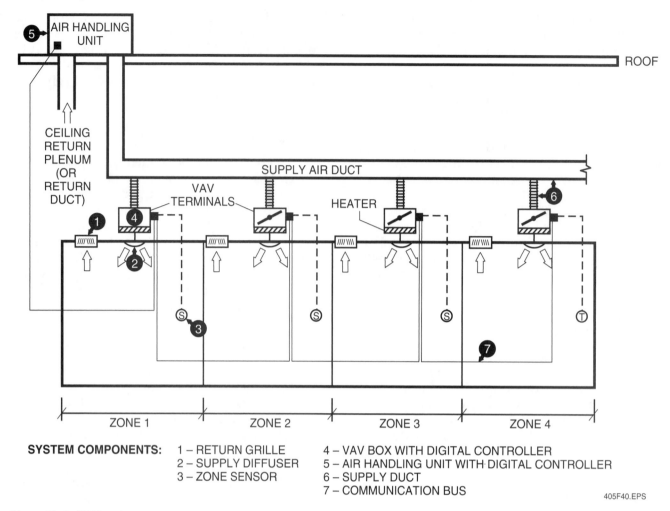

System Components:
1 – RETURN GRILLE	4 – VAV BOX WITH DIGITAL CONTROLLER
2 – SUPPLY DIFFUSER	5 – AIR HANDLING UNIT WITH DIGITAL CONTROLLER
3 – ZONE SENSOR	6 – SUPPLY DUCT
	7 – COMMUNICATION BUS

405F40.EPS

Figure 40 ◆ VAV system.

fan, the central unit's outdoor air dampers open to a pre-configured minimum ventilation position. If the central unit is an air handler, the associated chiller or condensing unit is also started at this time.

The central unit will then add cooling capacity stages as necessary to maintain an adjustable supply air temperature setpoint, typically 55°F. Most VAV systems feature a supply air reset function, which normally resets the supply air to either 65°F, or the temperature needed to maintain building comfort levels. If the unit is equipped with a dry-bulb or enthalpy economizer, and the outdoor air is suitable, the first stage of cooling will be using up to 100 percent outdoor air to maintain this setpoint. If more cooling is desired, stages of compression will be added to maintain the supply air temperature setpoint.

Figure 41 shows how a VAV terminal controls its damper position during cooling and heating. The Y axis shows zone cfm settings (percent airflow), while the X axis shows zone temperature setpoints. In response to a zone sensor, the zone controller modulates the primary air cfm to meet the control zone occupied cooling setpoint, typically 70°F to 74°F. The controller is set to limit the primary air to a maximum cooling cfm to match the zone's design cooling load. The controller is also set to limit the primary air to a minimum cfm to maintain the design ventilation rate for the zone. When the central unit is in the economizer mode, the percentage of ventilation air to the zone may vary from the minimum setting up to 100 percent outdoor air.

9.2.3 Heating

Some DDC control systems have the optional ability to use the central unit as a heat source for the building. If the average zone temperature or night low-limit controller indicates heating is needed, the air source control module can be configured to switch the central unit into the heating mode. Zones requiring heat simply modulate their air terminal dampers accordingly, and all

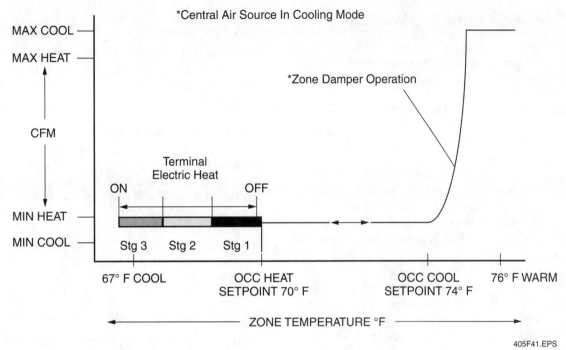

Figure 41 ◆ VAV damper operation.

zone heaters are deactivated. When the zones have been heated sufficiently, the central unit returns to its normal mode.

9.2.4 Unoccupied Period

When the master zone controller determines that all zones have become unoccupied, it stops the central air source. The central unit supply fan is stopped and the outside air dampers close. If the zone controller is equipped with setback temperatures, the central unit operates just like the occupied period except that the outside air dampers remain closed. Central chillers and boilers stop based on their individual time schedules. In some systems, the stopping of the central chillers and boiler is tied to the stopping of the air handlers.

9.2.5 Simultaneous Cooling and Heating

VAV systems are truly designed for buildings that have simultaneous cooling and heating needs. With the central unit in the cooling mode, any zone requiring cooling can be satisfied. Likewise, any zone needing heat at the same time can also be satisfied.

9.2.6 Morning Warm-Up

Before the building is occupied, the VAV central unit enters into a morning warm-up cycle when the building is cold. During morning warm up, the central air handler's digital controller modulates the heating coil to match the building heating needs. This is accomplished by maintaining the return air temperature from the zones at a predetermined setpoint.

When the zone controller is informed that the central unit is in morning warm-up, it cycles the air terminal primary air damper between maximum and minimum heating cfm limits to control the space to the occupied heating setpoint.

9.3.0 Night-Time Free Cooling

Both the newer VVT systems and VAV systems can use an energy-saving routine called night-time free cooling (NTFC). During the unoccupied cycle (typically between 3 AM and 7 AM), the central air handler's digital controller can initiate the NTFC routine if there is sufficient difference between the outside air and the inside building temperatures. When activated, the central unit opens the outdoor and exhaust dampers 100 percent, starts the supply fan, and prevents mechanical cooling from being activated.

When activated by the central unit's NTFC mode, the zone controllers position their primary air dampers to the maximum cooling cfm limit until the NTFC setpoint is reached. When the NTFC setpoint is reached, each zone controller repositions its primary air damper to the minimum cooling cfm limit. Each zone's primary damper is thus cycled between these two cfm settings until the building becomes occupied or all zone NTFC setpoints have been reached.

9.4.0 Demand-Controlled Ventilation

Both the newer VVT systems and VAV systems are capable of controlling indoor air quality (IAQ) at the zone level by employing a control strategy called demand controlled ventilation (DCV). Employing DCV with a VVT system, however, requires that the zone dampers be single-duct or fan-powered VAV-type terminals instead of the normal low-pressure VVT dampers.

With DDC controls, a CO_2 sensor is mounted in each zone. The sensor is typically wall-mounted, equipped with a display, and wired back to a zone controller. The sensor is also available as a combination temperature and CO_2 sensor to minimize installation costs.

When occupied zone CO_2 levels exceed an adjustable setpoint (measured in parts per million), the zone controller modulates the primary air damper between the maximum and minimum cool cfm limits to satisfy the CO_2 setpoint. During this time, control of zone temperature is suspended. If overcooling of the zone occurs, the primary air damper is temporarily positioned to a minimum IAQ reheat cfm limit and terminal reheat is added to return the room temperature to setpoint.

The central air handling unit controller continually monitors the zone CO_2 levels and adjusts the outdoor air dampers accordingly. If the zone CO_2 levels are satisfactory, the central unit maintains the base ventilation cfm. If the discharge temperature falls below the supply air setpoint (typically 55°F) in some central packaged units during DCV operation, the central unit heater is activated to provide reheat and prevent operational problems. If the central unit is equipped with economizer control, the economizer routine overrides DCV and modulates the outdoor air dampers to provide cooling. DCV operation is disabled whenever the central unit is in the unoccupied mode.

9.5.0 Humidity Control

The newer VVT systems, as well as VAV systems, are capable of controlling relative humidity at the zone level instead of IAQ control. Both zone relative humidity and zone IAQ control cannot be provided simultaneously.

When the zone is occupied, the operation of the central unit normally prevents humidity buildup in the zone by dehumidifying the supply air. However, this method may be insufficient at part load. Humidity control may be improved by sensing humidity at the zone level. An optional relative humidity sensor may be wired to the zone controller. With DDC controls, zone controllers equipped with humidity sensors send humidity information back to the central unit controls. In response, the central unit adds cooling capacity to lower the supply air temperature, creating drier supply air.

At the zone level, each zone controller modulates its primary air damper between minimum and maximum cool cfm limits to maintain a zone humidity setpoint. If the maximum zone temperature setpoint is exceeded while controlling zone relative humidity, the zone controller positions the primary damper to a reheat cfm limit and adds terminal reheat to return the zone to normal temperature.

9.6.0 Building Pressurization

Most VVT and VAV systems require some form of building pressurization control. The central unit is typically provided with a powered exhaust system to accomplish this. Responding to a space pressure sensor located in a key building space, the central unit controls activate a powered exhaust fan or VFD-controlled fan to modulate building exhaust air accordingly.

Unlike VVT central units, VAV central units are sometimes equipped with a return fan. When this is the case, some form of fan tracking control is required to keep the cfm leaving the conditioned space in a controlled relationship to the cfm being supplied. In one method, the central unit controls maintain a constant cfm differential between the supply and return airflow. This maintains a positive pressure in the conditioned space. In another method, the central controls vary the return fan cfm to maintain a predetermined space pressure. The designer must choose the method based on the application needs. Air terminal controls do not participate in this function.

9.7.0 Smoke Control

It is common practice to use the central unit supply and exhaust fan to provide zone smoke control in the event of a building fire. Under the command of a fire marshal's panel (part of the building's fire/life safety system), the central unit may be manually placed in one of three modes:

- *Pressurization* – The DDC control system provides excess air into the zones served by the central unit by opening outdoor air dampers

100 percent and running the supply fan. During this mode the return and exhaust dampers are closed to prevent or delay smoke entry from adjacent zones. Each zone damper is positioned to its maximum cool cfm limit. If an air terminal series fan is installed, it is started.

- *Evacuation* – This mode removes smoke from the control zones by running the powered exhaust fan only. The return and outdoor air dampers are held closed. Each zone damper is moved to its closed position. This position will be maintained until the central unit changes operating mode. If an air terminal series fan is installed, it is stopped.

- *Smoke purge* – This mode flushes smoke and/or contaminated air from the controlled zones by running both the supply and powered exhaust fans, utilizing 100 percent outdoor air. Return dampers are closed. Each zone damper is positioned to its maximum cool cfm limit. This cfm will be maintained until the central unit changes operating mode. If an air terminal series fan is installed, it is started.

9.8.0 Interoperability Strategies

In the current market, there are several prevalent control strategies that stretch across building systems. To illustrate this trend, a couple of examples of interoperability are provided in the following paragraphs.

9.8.1 Real-Time HVAC Usage

A building owner only wants to operate the lights and HVAC system when the building is occupied by tenants. Thus, when a tenant's employee security access card is scanned upon entrance to the building, an open communication protocol is used to monitor the card access, and then command the lights and HVAC system on in that portion of the building associated with the tenant. At the end of the day, when all tenant employees associated with a particular portion of the building leave, the associated lighting and HVAC system shuts down.

9.8.2 Demand Response

The critical shortage of electricity in the United States is driving utilities to offer significant incentives to curtail electrical loads at peak times. In addition, utility companies are developing rate structures that vary by time of day, day of week, etc., and through digital web-based communication has the capability of providing building owners with real-time utility rates. Demand response is the ability of a BMS system to read the incoming real-time utility rate structure and to modify the operation of HVAC and other building systems to optimize the building's energy consumption and operational cost. Thus, by integrating Demand Response as an interoperable control strategy, a building can be optimized for both utility cost and comfort while resulting in the lowest possible energy consumption.

1. A digital controller consists of three major components: input module, microprocessor, and _____.
   a. output module
   b. actuator
   c. valve
   d. sensor

2. A general-purpose controller is _____.
   a. also known as a product-integrated controller
   b. typically installed on equipment by the manufacturer
   c. typically installed on components with simple operating schemes
   d. used for a chiller or boiler

3. A discrete input or output is one that _____.
   a. is proportional
   b. has two states (conditions)
   c. varies within a wide range
   d. is used only with pneumatic devices

4. The type of sensor that uses a long resistance-sensing element to sense average duct temperature is a(n) _____.
   a. thermistor
   b. RTD
   c. thermometer
   d. PID

5. A standard analog output signal used in computer-controlled systems is _____.
   a. 24VAC
   b. 4–20mA
   c. 120VAC
   d. 20VDC

6. The most accurate algorithm used for processing information when the controlled variable is changing rapidly over time is _____.
   a. proportional
   b. proportional-integral
   c. proportional-integral-derivative
   d. proportional-derivative

7. The rate at which information travels on a network is known as the _____ rate.
   a. bit
   b. baud
   c. bus
   d. LAN

8. In *Figure 9*, the device that connects all the building digital controllers together is the _____.
   a. local area network
   b. network communication bus
   c. phone line
   d. GPC

9. For a residential HVAC system serving several physical locations, the device that modulates the airflow to each zone is the _____.
   a. zone damper
   b. bypass damper
   c. room sensor
   d. deflector

10. Because a VVT system uses a constant volume packaged unit as the central air source, a bypass damper is used to _____.
    a. modulate airflow to each control zone
    b. bypass indoor air around the packaged unit
    c. maintain a constant airflow through the packaged unit
    d. bypass outdoor air around the packaged unit

11. Among the information a zone controller (secondary controller) is likely to receive from a primary controller (monitor thermostat) is _____.
    a. time of day
    b. indoor temperature
    c. zone temperature
    d. average of indoor and outdoor temperatures

12. The purpose of load-shedding is to _____.
    a. avoid exceeding the utility company's demand limit during peak power usage periods
    b. make sure equipment is turned off when the building is unoccupied
    c. increase power consumption during off-peak periods
    d. make sure energy use is kept at a steady, constant level

13. In order for remote computers to access a local BMS network today, all that is needed is a simple _____.
    a. bridge
    b. web browser
    c. gateway
    d. router

14. A locally connected PC is a valid user interface device to a building automated system.
    a. True
    b. False

15. The term *interoperability* applies to _____.
    a. an energy management system that can be accessed from phone lines
    b. a system that allows the use of both handheld local interface devices and permanently located interface devices
    c. the transfer of information and commands across building systems and different protocols
    d. the controllers that use identical protocols to function within a control system

16. LAN is a valid candidate for use as a standard open protocol.
    a. True
    b. False

17. The _____ protocol allows the user to use a web browser program, like Internet Explorer, to view a BMS system's web pages.
    a. BACnet
    b. LON
    c. http
    d. Modbus

18. Which portion of the front-end software deals with uploading all the files from the network controllers and storing them according to the networks in which they exist?
    a. Report management
    b. Alarm management
    c. Operator management
    d. Database management

19. In *Figure 30*, the operator could click on the _____ to change a time schedule or setpoint schedule for the air handler.
    a. image of the supply fan
    b. image of the room sensor
    c. image of the cooling coil
    d. point value of the room temperature

20. Which of the following is a correct statement about installing sensor, power, and communication wiring?
    a. Use solid copper wire to prevent the wire from breaking.
    b. Run all communication and power wiring in the same conduit to minimize installation labor and costs.
    c. Use daisy chain wiring connections and avoid T-tap connections to prevent unwanted noise.
    d. Ground all sensor and communication wires at both ends to prevent unwanted noise.

21. When comparing general-purpose controllers (GPCs) and product-specific controllers, _____.
    a. product-specific controllers require much more programming than GPCs to activate their algorithms
    b. GPCs require much more programming than product-specific controllers to activate their algorithms
    c. actual code must be written for each application and downloaded into product-specific controllers at job sites
    d. blanks must be filled in with coded algorithms when configuration is needed for either type of controller

22. The linkage coordinator in a VVT system regularly polls the individual zone controllers and calculates the _____.
    a. minimum cooling and heating demand
    b. maximum cooling and heating demand
    c. total average cooling and heating demand
    d. total average cooling demand only

23. In a VVT system, simultaneous cooling and heating at the zone level can be provided _____.
    a. if the zone terminals are equipped with zone heaters
    b. never; the central packaged unit always provides heating
    c. only if the central packaged unit is in heating
    d. only if the zone heaters are VAV air terminals

24. In a VAV system, the central unit normally remains in the _____ mode as long as the building is occupied.
    a. heating
    b. night-time free cooling
    c. cooling
    d. economizer

25. The ability of a BMS system to read the incoming real-time utility rate structure and modify the operation of HVAC and/or other building systems to optimize the building's energy consumption and operational costs is called _____.
    a. load-shed
    b. demand response
    c. night-time free cooling
    d. time guard

Summary

In today's environment, it is absolutely essential for the HVAC technician to become familiar with computer-controlled systems. At some point in the future, all HVAC systems will be controlled by digital controllers.

Not only will heating, cooling, and ventilation be controlled, but all building systems will be tied together with control strategies that cross over between each system's controls. As you have seen, integrated building management systems are commonplace in commercial applications. In the future, it will be that way in homes as well. Lighting, entertainment, cooking, and comfort systems will be controlled from a central computer accessed by the homeowner from anywhere within reach of the world wide web. This capability was already a reality in some high-end homes at the beginning of the 21st century.

Notes

Trade Terms Introduced in This Module

Algorithm: A mathematical equation consisting of a series of logic statements used in a computer or microprocessor to solve a specific kind of problem. In HVAC applications, algorithms are typically used in microprocessor-controlled equipment to control a wide range of control function operations based on the status of various system sensor input signals.

Application-specific controller: A digital controller installed by a manufacturer on a specific product at the factory.

Baud: The rate at which information is transmitted across communication lines.

Bit: Short for binary digit. The smallest element of data that a computer can handle. It represents an off or on state (zero or one) in a binary system.

Building management system (BMS): A centralized, computer-controlled system for managing the various systems in a building. Also known as a building automation systems (BAS).

Bus: A multi-wire communication cable that links all the components in a hard-wired computer network.

Control point: The name for each input and output device wired to a digital controller.

Data collection: The collection of trend, runtime, and consumable data from the digital controllers in a building.

Digital controller: A digital device that uses an input module, a microprocessor, and an output module to perform control functions.

Direct digital control (DDC): The use of a digital controller is usually referred to as direct digital control or DDC.

Ethernet: A family of frame-based computer networking technologies for local area networks (LANs). The name comes from the physical concept of the ether. It defines a number of wiring and signaling standards.

Firmware: Computer programs that are permanently stored on the computer's memory during a manufacturing process.

Gateway: A link between two computer programs allowing them to share information by translating between protocols.

Green building: A sustainable structure that is designed, built, and operated in a manner that efficiently uses resources while being ecologically friendly.

Hypertext transfer protocol (http): The base protocol used by the world wide web.

Indoor air quality (IAQ): Measure of the quality of interior air that could affect health and comfort of a building's occupants.

Integrated building design: A collaborative decision-making process that uses a project design team that extends from a project's inception through its design and construction phases.

Internet: A worldwide communication network used by the public to interface computers.

Internet protocol (IP): A data-oriented protocol used for communicating data across a packet-switched internetwork.

Internet protocol address (IP address): A unique address (computer address) that certain electronic devices use in order to identify and communicate with each other on a computer network utilizing the internet protocol (IP) standard.

Interoperability: The ability of digital controllers with different protocols to function together accurately.

Load-shedding: Systematically switching loads out of a system to reduce energy consumption.

Local area network (LAN): A server/client computer network connecting multiple computers within a building or building complex.

Local interface device: A keypad with an alphanumeric data display and keys for data entry. It is connected to the digital controller with a phone-type cable that provides power and communication.

Modem: Short for modulator-demodulator. A device that links a computer to communication lines. It converts voice to electronic signals and vice versa.

Network: A means of linking devices in a computer-controlled system and controlling the flow of information among these devices.

Personal digital assistant (PDA): An electronic device which can include some of the functionality of a computer, cell phone, music player and camera.

Product-integrated controller (PIC): A digital controller installed by a manufacturer on a product at the factory.

Protocol: A convention that governs the format and timing of communication between devices in a computer network.

Software: Computer programs transferred to the computer from various media and stored in an erasable memory.

Sustainable: Designed to reduce impact on the environment.

Tenant billing: The ability to charge a building tenant for after-hours use of the building's HVAC system.

Transients: Short-duration interference signals that are coupled to and transmitted on power lines, communication lines, and/or computer network bus lines. Transient signals can be caused by natural or man-made electrical or electromagnetic disturbances, signals, or emissions. When present in computer network bus lines, transients can cause computer program glitches and/or interrupt operational sequences.

User interface module: A keypad with an alphanumeric data display and keys for data entry. It is connected to a network communication bus.

Web browser: A software application which enables a user to display and interact with text, images, videos, music and other information typically located on a web page at a web site on the world wide web or a local area network.

Web page: A resource of information that is suitable for the world wide web and can be accessed through a web browser.

Wide area network (WAN): A server/client computer network spread over a large geographical area.

World wide web: A system of interlinked hypertext documents accessed via the internet. Commonly shortened to "the web."

Additional Resources and References

Additional Resources

This module is intended to be a thorough resource for task training. The following reference works are suggested for further study. These are optional materials for continued education rather than for task training.

ASHRAE Educational Manual – Fundamentals of HVAC Control Systems, Latest Edition. Atlanta, GA: American Society of Heating, Refrigerating, and Air Conditioning Engineers.

ASHRAE Handbook – HVAC Applications, Latest Edition. Atlanta, GA: American Society of Heating, Refrigerating, and Air Conditioning Engineers.

HVAC Controls and Systems. Troy, MI: Business News Publishing Co.

Figure Credits

NCCER makes every effort to keep these textbooks up-to-date and free of technical errors. We appreciate your help in this process. If you have an idea for improving this textbook, or if you find an error, a typographical mistake, or an inaccuracy in NCCER's Contren® textbooks, please write us, using this form or a photocopy. Be sure to include the exact module number, page number, a detailed description, and the correction, if applicable. Your input will be brought to the attention of the Technical Review Committee. Thank you for your assistance.

Instructors – If you found that additional materials were necessary in order to teach this module effectively, please let us know so that we may include them in the Equipment/Materials list in the Annotated Instructor's Guide.

Write: Product Development and Revision
National Center for Construction Education and Research
3600 NW 43rd St, Bldg G, Gainesville, FL 32606

Fax: 352-334-0932

E-mail: curriculum@nccer.org

Craft

Module Name

Copyright Date

Module Number

Page Number(s)

Description

(Optional) Correction

(Optional) Your Name and Address

03406-09

System Startup and Shutdown

03406-09
System Startup and Shutdown

Topics to be presented in this module include:

Overview

When boilers, chillers, and other commercial heating or cooling systems are brought back on line after an extended shutdown, a specific routine must be followed to properly prepare the system for operation. This routine includes maintenance activities that will protect the equipment and prevent premature failure. Similar actions must be taken when equipment is to be shut down for an extended period. This module provides an overview of the startup and shutdown requirements for boilers, chillers, cooling towers, air handlers, and packaged units. Keep in mind, however, that the equipment manufacturer's recommended startup and shutdown procedures take precedence.

Objectives

When you have completed this module, you will be able to do the following:

1. Prepare a boiler for dry storage.
2. Prepare a boiler for wet storage.
3. Clean, start up, and shut down a steam boiler.
4. Clean, start up, and shut down a hot-water boiler.
5. Start up and shut down a reciprocating liquid chiller and related water system.
6. Start up and shut down a selected centrifugal or screw liquid chiller and related water system.
7. Start up and shut down an air handler and related forced-air distribution system.
8. Test compressor oil for acid contamination.
9. Add or remove oil from a semi-hermetic or open reciprocating compressor.
10. Inspect and clean shell and tube condensers/evaporators and other water-type heat exchangers.

Trade Terms

Bleed-off
Deadband
Desiccant

Layup
Recycle shutdown mode

Required Trainee Materials

1. Pencil and paper
2. Appropriate personal protective equipment

Prerequisites

Before you begin this module, it is recommended that you successfully complete *Core Curriculum*; *HVAC Level One*; *HVAC Level Two*; *HVAC Level Three*; and *HVAC Level Four*, Modules 03401-09 through 03405-09.

This course map shows all of the modules in the fourth level of the *HVAC* curriculum. The suggested training order begins at the bottom and proceeds up. Skill levels increase as you advance on the course map. The local Training Program Sponsor may adjust the training order.

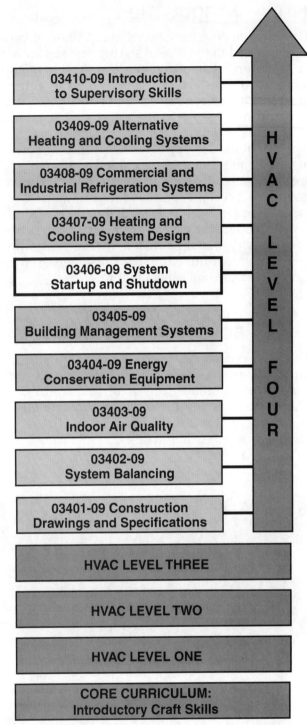

406CMAP.EPS

1.0.0 ◆ INTRODUCTION

This module describes special maintenance procedures related to the typical startup and shutdown of commercial heating/cooling systems. The focus is on the startup of major items of equipment that have been turned off for ex-tended periods of time because of major repairs, seasonal shutdown, or some other reason. These maintenance actions also apply to the startup of newly installed equipment. Also included in this module are special maintenance actions needed to prepare and protect certain commercial heating/cooling system components for extended periods of shutdown. Proper startup and shutdown procedures are necessary to make sure that the system continues to operate efficiently. They also help to prevent premature equipment failure.

2.0.0 ◆ STEAM/HOT-WATER BOILERS AND SYSTEMS

This section describes the tasks needed to prepare larger commercial/industrial steam and hot-water heating boilers for startup and shutdown. The startup and shutdown of small commercial and residential hot-water or steam heating boilers was studied earlier in *HVAC Level Three*.

2.1.0 Shutdown

CAUTION

A boiler should be shut down as directed in the boiler manufacturer's service instructions. No attempt should be made to operate, repair, or dismantle a boiler until the applicable manufacturer's service literature is read and thoroughly understood.

Many large boilers used mainly for heating or as standby units may have extended periods of time when they are not used. When idle, these boilers require special attention to make sure that their water-side and fire-side components are not allowed to deteriorate from corrosion and other problems. Unless proper procedures are followed before taking boilers off line, severe corrosion may occur. Oxygen from the atmosphere can enter an idle boiler and combine with condensed moisture to produce extensive pitting of metal surfaces. The importance of properly preparing a boiler for shutdown cannot be stressed enough. If a boiler is not prepared properly for shutdown, more damage can happen in one month of sitting idle than during an entire heating season of operation.

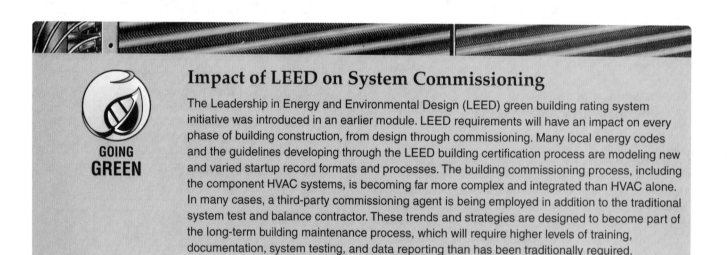

GOING GREEN

Impact of LEED on System Commissioning

The Leadership in Energy and Environmental Design (LEED) green building rating system initiative was introduced in an earlier module. LEED requirements will have an impact on every phase of building construction, from design through commissioning. Many local energy codes and the guidelines developing through the LEED building certification process are modeling new and varied startup record formats and processes. The building commissioning process, including the component HVAC systems, is becoming far more complex and integrated than HVAC alone. In many cases, a third-party commissioning agent is being employed in addition to the traditional system test and balance contractor. These trends and strategies are designed to become part of the long-term building maintenance process, which will require higher levels of training, documentation, system testing, and data reporting than has been traditionally required.

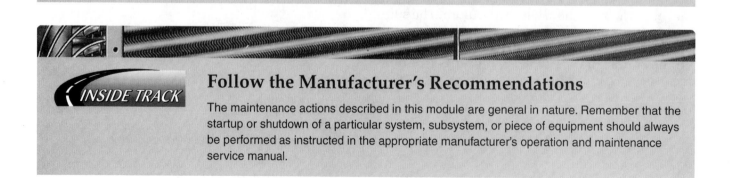

INSIDE TRACK

Follow the Manufacturer's Recommendations

The maintenance actions described in this module are general in nature. Remember that the startup or shutdown of a particular system, subsystem, or piece of equipment should always be performed as instructed in the appropriate manufacturer's operation and maintenance service manual.

The specific shutdown procedures for different gas-fired and oil-fired boilers vary, but the tasks and their sequences are essentially the same. The shutdown procedure typically used with an oil-fired steam boiler (*Figure 1*) is outlined here.

2.1.1 Boiler Shutdown Concerns

One area of concern during the prolonged shutdown (**layup**) of an oil-fired boiler is the damage from corrosion that can be caused by the sulphur content of fuel oils in contact with moisture or other residues. During layup, the fire-side components are exposed to moisture caused by the condensation of air as it cools below its dew point. This moisture and any sulphur residue form an acid solution that attacks the boiler metals. With high-humidity conditions, the corrosive effect of the acid can be serious enough to erode through or damage the boiler tubes and other heating surfaces. This acid/moisture condition usually does not exist during normal boiler operation because the temperatures are high enough to vaporize any condensation.

Corrosion that could occur during the layup period can be greatly reduced by thoroughly cleaning the soot and other products of combustion from the boiler's fire-side components immediately after shutdown of the boiler. Following this cleaning, any remaining moisture should be removed by drying the cleaned areas with a suitable heating device.

2.1.2 Boiler Preparation for Dry Storage

Dry storage of a boiler is typically done when the boiler is subjected to freezing temperatures or a humid environment, or when it will be out of service for more than a month. Dry storage involves completely cleaning the water-side and fire-side components and surfaces to remove all deposits, scale, and soot. After cleaning, the fire-side surfaces are coated with an anticorrosive compound, and heaters are used to dry any remaining moisture from within the boiler. A moisture-absorbing material (**desiccant**) is then placed inside, and the boiler is sealed to maintain a low level of moisture during storage. The specific procedures for the dry storage of the different kinds of gas-fired and oil-fired boilers vary, but the tasks and their sequences are essentially the same. The procedure that follows outlines the preparation of a typical oil-fired steam boiler for dry storage.

> **WARNING!**
>
> Preparing a boiler for storage involves working on electrically energized equipment containing pressurized and/or hot components. Follow all applicable manufacturer's instructions, OSHA regulations, and job-site requirements relating to safety and the use of safety equipment.

406F01.EPS

Figure 1 ◆ Steam boiler.

Avoiding Thermal Shock During Boiler Shutdown

When shutting down a boiler, reduce the load slowly so that the boiler cools at a rate that avoids harmful stresses caused by damaging temperature differentials, also called thermal shock.

Step 1 Shut down the boiler per the manufacturer's service instructions and as briefly outlined below.

   a. Lock out and tag the boiler power and controls.

   b. Shut off and secure the fire.

   c. After shutting off the fire, cut down on the draft to keep the refractory or brickwork from cooling too fast. Rapid cooling can cause flaking or corrosion.

   d. Watch the water level.

   e. After the boiler has stopped steaming, close the main stop valve.

   f. After the pressure has dropped to one or two pounds, open the air cock so that no vacuum forms in the boiler. In extreme cases, a vacuum can cause boiler tubes and shells to collapse.

   g. Do not immediately close the main steam valve or the safety valve may still pop open.

 WARNING!
Failure to safely discharge water or steam can result in serious burns, eye injury, or equipment damage. Piping connected to safety, blowdown, and other drain valves must be routed so that the flow of the discharged wastewater or steam goes into the local sewer system or other safe disposal area.

Do not drain the boiler until all the pressure is relieved.

Step 2 After the boiler is completely cooled, drain the boiler as directed in the manufacturer's service literature. Make sure you follow all federal, state, and local laws, rules, and regulations that govern the discharge of the wastewater into the environment.

Step 3 After the boiler is drained, flush out the water-side with fresh high-pressure water.

Step 4 Thoroughly clean the fire-side surfaces of all soot and deposits from combustion. Brush out and/or vacuum up the loosened materials. Some oil-fired boilers have a water washing device used to clean convection surfaces. Use this device as instructed by the boiler manufacturer.

 WARNING!
Make sure that all steam and other system valves, blowdown valves, and electrical switches are turned off before opening boiler handholes and manhole covers. Always adequately vent the boiler before entering.

Step 5 Inspect all the fire-side metal surfaces for damage or corrosion.

Step 6 Remove any scale or deposits from the water-side surfaces. Check for internal leakage or corrosion.

Force-Drying a Boiler

During the time spent inspecting the boiler components and piping, place a suitable heating device within the boiler to dry it. Be careful not to overheat the boiler; only warm it enough to dry it completely.

Step 7 Check and clean the following:
- Low-water cutoff piping
- Water level controls and cross piping connections
- Blowdown piping, valves, and drain

Step 8 Check all water and steam piping, valves, and other components for leaks, wear, corrosion, or other damage. Replace or repair any components, if needed.

Step 9 Brush the refractories clean and inspect for damage. If cracks over ¼" wide exist, clean and fill them with high-temperature bonding mortar.

Step 10 Wash and coat the refractories using the material recommended by the boiler manufacturer. Usually this is a high-temperature, air-dry mortar diluted with water to the consistency of light cream.

Step 11 Coat the fire-side surfaces with an anti-corrosive material recommended by the boiler manufacturer.

Step 12 After making sure that the boiler is dry and the fire-side is properly coated with anticorrosive, place a desiccant inside as recommended by the boiler manufacturer. Some boiler manufacturers recommend using the following:
- Quick lime at 2 pounds per 3 cubic feet of volume
- Silica gel at 5 pounds per 30 cubic feet of volume
- Calcium chloride at 3 pounds per 100 square feet of surface area

Put the desiccant in half-filled trays to allow room for the water absorbed by the desiccant. Via the boiler access manholes and handholes, place half the amount of desiccant inside the firebox and the other half on top of the tubes.

Step 13 Close and seal all boiler openings, including handholes and manholes. Where needed, use new gaskets. Close all feed-water and steam valves. Close the dampers and vents to prevent air from reaching the fire-side surfaces of the boiler.

Step 14 Maintain lockout/tagout so that no one can start the unit during shutdown.

Step 15 At six-week intervals, open the boiler handholes and manholes and inspect the desiccant in the boiler. Renew any desiccant that is saturated with moisture.

2.1.3 Boiler Preparation for Wet Storage

Wet storage of a boiler is typically done when the boiler is held in standby, where temperatures are above freezing, or where dry storage is impractical. The specific procedures for the wet storage of the different kinds of gas-fired and oil-fired boilers vary, but the tasks and their sequences are essentially the same. The following procedure outlines the preparation of a typical oil-fired steam boiler for wet storage:

Step 1 Drain and clean the boiler according to Steps 1 through 10 of the dry storage procedure.

Step 2 Valve the boiler off from the rest of the system, and then refill it to overflowing with treated water. If deaerated water is not available, fire the boiler for a short time until the water reaches a temperature of about 200°F to drive off most of the dissolved gases. Let the boiler water cool down to room temperature, and then add water until it overflows the top.

Step 3 Add the water treatment chemicals needed to condition the added amount of water that is now above the boiler's normal water line. Close all boiler openings. The water should be circulated in the boiler to prevent stratification and to make sure the added chemicals are thoroughly distributed in order to be in contact with all surfaces.

Step 4 Maintain an internal water pressure above atmospheric pressure. Nitrogen gas is sometimes used for this purpose.

Step 5 Protect the exterior surfaces and components from rust by coating them with mineral oil or boiler paint. Cover all tube and firebox surfaces with a coating of mineral oil or other rust inhibitor.

Step 6 Keep the control circuit energized to prevent condensation from forming in the control cabinet or on the flame safeguard control device.

Step 7 Leave the flue and firebox doors wide open during the period of shutdown. This helps to keep them dry.

Step 8 Keep the boiler room dry and well ventilated.

2.2.0 Steam Boiler System Startup for Normal Operation

A steam boiler should be started up as directed in the boiler manufacturer's service instructions. The specific startup procedures for the different kinds of gas-fired and oil-fired boilers vary, but the tasks and their sequences are essentially the same. The startup procedure typical of that used with an oil-fired steam boiler is outlined here.

 WARNING!

Startup and operation of a boiler system involves working on electrically energized equipment containing pressurized and/or hot components. Follow all applicable manufacturer's instructions, OSHA regulations, and job-site requirements relating to safety and the use of safety equipment.

Step 1 If not previously done, inspect the steam piping system to verify the following:

    a. All strainers have been cleaned.

    b. The settings of all safety valves are correct.

    c. All manual and automatic valves are in the required positions.

Step 2 Check that all tools, equipment, desiccant trays, or other debris are removed from the boiler and firebox. Make sure that all boiler openings, including handholes and manholes, are properly secured.

Step 3 Inspect the boiler and accessories to verify the following:

    a. Fuel and prime electrical power are supplied to the boiler.

    b. The manual reset buttons of all starters and other controls have been pushed to ensure that the controls have been reset.

    c. The linkages of all dampers, metering valves, and cams have full stroke and free motion.

    d. The direction of rotation for all motors is correct when momentarily energizing the control switch.

    e. The pressure control settings are set slightly above the highest steam pressure needed. This setting must be at least 10 percent below the setting of the safety valve(s). On a low-pressure boiler, this setting is typically two or three psig above the operating limit.

    f. The float in the low-water cutoff and pump controls can move freely. Also, make sure the controls are level and the piping is plumb.

    g. Discharge piping connected to safety, blowdown, and other drain valves is routed so that the flow of the wastewater or steam goes into the local sewer system or other safe disposal area.

    h. If not a new startup, replace the oil burner nozzle. Make sure that the electrode gap and positioning of the electrode relative to the nozzle are correct.

    i. All combustion openings and barometric or draft control dampers for the boiler are the proper size for the fuel being used.

    j. Any feedwater treatment equipment, such as filters, chemical feeders, demineralizers, softeners, and deaeraters, are operational and prepared for use.

 CAUTION

If starting up a newly installed boiler or one that has been in dry storage, do not attempt to start up the boiler as described below until the boiler has first been thoroughly cleaned to remove any accumulated dirt and oil.

Step 4 Make sure that the main steam stop valve is closed.

Step 5 If not previously done, fill the boiler with properly treated water up to the normal operating level. Vent the boiler so that no air is trapped in the steam space.

Read Boiler Operation Literature Before Operation

INSIDE TRACK

No attempt should be made to operate a boiler until after the manufacturer's operating instructions have been read and are thoroughly understood.

Step 6 Set the boiler controls for a cold startup, and then fire the boiler as directed by the manufacturer's operating instructions. Note that on initial startup, it may be necessary to attempt igniting the burners several times in order to bleed air from the main and/or pilot fuel lines.

Step 7 The boiler burner and control system should allow the boiler water to warm up slowly via the use of a burner low-fire flame. In some boilers, this may be done automatically. In others, it is done manually by setting the MANUAL/AUTOMATIC control to the MANUAL position and the MANUAL FLAME control to the LOW-FIRE position. During the boiler warmup, monitor the boiler water level frequently to make sure that it stays at the normal operating level.

Step 8 After the boiler water is thoroughly warmed, the boiler burner and control system should function to switch from the low-fire to high-fire mode of burner operation. In nonautomatic boilers, this is done manually by setting the MANUAL FLAME control to the HIGH-FIRE position.

Step 9 When the steam gauge shows pressure in the boiler, blow down the gauge glass, water column, and low-water cutoff as applicable.

Step 10 Using suitable instruments, such as those included in Bacharach or Dwyer test kits, perform combustion efficiency and analysis tests. Make these tests and any related burner and/or draft regulator adjustments as instructed by the test instrument and boiler manufacturer's service literature.

- Oxygen (O_2)
- Carbon dioxide (CO_2)
- Excess air
- Stack temperature
- Carbon monoxide (CO)
- Other gases

Step 11 When the boiler reaches about two or three pounds below the steam header pressure, very slowly open the main boiler stop valve until it is fully open. If opened too quickly, the header can rapidly lose pressure. Also, by opening the valve slowly, carryover can be prevented.

Step 12 With the boiler switched to the automatic mode of operation, allow it to operate at normal pressures and temperatures until it is shut down normally by the operation of the boiler burner control system.

Step 13 Monitor the boiler through enough cycles of operation to make sure that the boiler, including ignition and control program sequences, is functioning correctly.

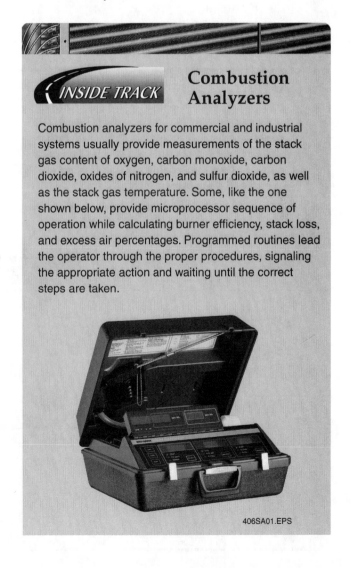

INSIDE TRACK

Combustion Analyzers

Combustion analyzers for commercial and industrial systems usually provide measurements of the stack gas content of oxygen, carbon monoxide, carbon dioxide, oxides of nitrogen, and sulfur dioxide, as well as the stack gas temperature. Some, like the one shown below, provide microprocessor sequence of operation while calculating burner efficiency, stack loss, and excess air percentages. Programmed routines lead the operator through the proper procedures, signaling the appropriate action and waiting until the correct steps are taken.

406SA01.EPS

Flue Gas Relationship to Excess Combustion Air

GOING GREEN

The *Appendix* shows the typical relationships between O_2, CO_2, and excess air for fuel oil, natural gas, and other common fuels. As shown in the chart, when the percentage of excess air increases, the percentage of CO_2 decreases. If it were possible to have perfect combustion, CO_2 would be maximized and O_2 would be at (or close to) zero in the flue gas stream. Because perfect combustion is not practically possible, most combustion equipment is set up to have a small percentage of excess O_2 present. Check the manufacturer's recommendation for excess air. Carbon monoxide (CO) should always be minimized. Note that the maximum allowable CO level in flue gas (on a free air basis) is 400 ppm according to the EPA and AGA.

Step 14 Confirm that the safety controls and protective circuits listed below are functioning properly as directed in the boiler manufacturer's service instructions.

- Pressure control
- Electronic and/or mechanical low-water cutoff
- Flame sensor
- Flame failure
- Ignition failure
- Power failure

NOTE

If above-normal amounts of water treatment chemicals were added to a boiler during a period of wet storage, it may be necessary to use more frequent blowdowns when the boiler is returned to service. This is necessary to reduce the chemical concentrations to normal operating levels.

Step 15 Inspect the steam piping system to verify the following:

- The system is free of leaks.
- All pressure-reducing valves/stations are operating properly.
- All steam traps are operating properly on all equipment, at the ends of mains, and at all drip points.

Step 16 Record boiler and system parameters (*Table 1*).

2.2.1 Gas-Fired Steam Boiler Startup Checks

Except for some of the operating and safety controls and their adjustment, the general procedure for the startup of a gas-fired steam boiler is similar to that described above for an oil-fired steam boiler. The startup and checkout of a gas-fired steam boiler should always be done as directed in the boiler manufacturer's service instructions. As applicable, the operation of the boiler controls

Table 1 Boiler Data and Operating Parameters

Boiler Data	
Location: _____	Manufacturer: _____
Model Number: _____	Serial Number: _____
Type/Size: _____	Fuel Type: _____
Ignition Type: _____	Burner Control: _____
Volts/Phase/Hertz: _____	

Boiler Operating Data	
Pressure/Temperature _____	Voltage _____
No. Safety Valves/Size _____	Amperage T1/T2/T3 _____
Safety Valve Setting _____	Draft Fan Volts/Amps _____
Operating Control Setting _____	Manifold Pressure _____
High-Fire Setpoint _____	Output-MBH (kW) _____
Low-Fire Setpoint _____	Safety Control Check _____

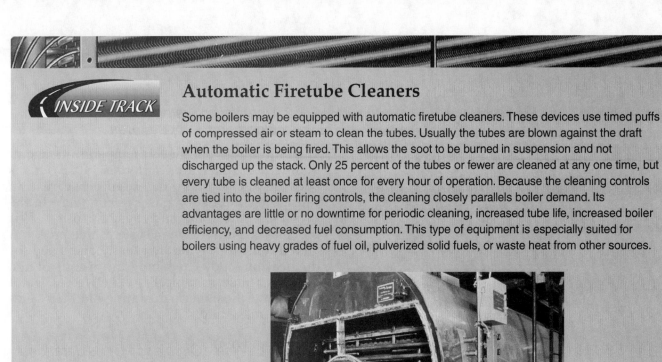

Automatic Firetube Cleaners

Some boilers may be equipped with automatic firetube cleaners. These devices use timed puffs of compressed air or steam to clean the tubes. Usually the tubes are blown against the draft when the boiler is being fired. This allows the soot to be burned in suspension and not discharged up the stack. Only 25 percent of the tubes or fewer are cleaned at any one time, but every tube is cleaned at least once for every hour of operation. Because the cleaning controls are tied into the boiler firing controls, the cleaning closely parallels boiler demand. Its advantages are little or no downtime for periodic cleaning, increased tube life, increased boiler efficiency, and decreased fuel consumption. This type of equipment is especially suited for boilers using heavy grades of fuel oil, pulverized solid fuels, or waste heat from other sources.

406SA02.EPS

and safety devices should be checked in accordance with the boiler manufacturer's service literature:

- Automatic gas valve
- Makeup and/or feedwater controls
- Low-water cutoff
- Flow switches
- Pressure controls
- Safety devices

Once the boiler has been turned on, the following parameters should be measured and recorded. If necessary, adjust the appropriate equipment controls to meet the parameter values recommended by the boiler manufacturer.

- Oxygen (O_2)
- Carbon dioxide (CO_2)
- Carbon monoxide (CO)
- Stack temperature
- Stack draft
- Manifold and supply gas pressures

Refer to the *Appendix* for the typical relationships between O_2, CO_2, and excess air for natural gas and other common fuels. Check the manufacturer's recommendation for excess air. Carbon monoxide (CO) should always be minimized. Note that the maximum allowable CO level in flue gas (on a free air basis) is 400 ppm according to the EPA and AGA.

2.3.0 Hot-Water Boiler Startup

Hot-water boilers are constructed and operated similarly to steam boilers, except some of the operating and safety controls used with hot-water boilers are different from those used with steam boilers. Also, hot-water boilers and related system piping are entirely filled with water, while steam boiler systems are not.

A hot-water boiler (*Figure 2*) should be started up as directed in the boiler manufacturer's service instructions.

Figure 2 ◆ Hot-water boiler.

406F02.EPS

 WARNING!
Do not attempt to operate a boiler until you have read the manufacturer's operating instructions and thoroughly understand them.

With the exceptions covered in the following paragraphs, the tasks and sequence for starting gas-fired and oil-fired hot-water boilers are similar to those previously described for steam boilers.

2.3.1 Filling and Venting a Hot-Water System

When filling a hot-water system with treated water, the entire boiler, piping system, and terminals must be filled with water and the manual system air vents opened (if used) to expel any air that has accumulated in the system. If the system uses automatic air vents, they should be checked for leakage. Because a hot-water boiler is completely filled with water, the hot-water outlet usually includes a dip tube that extends 2" or 3" into the boiler. This dip tube traps any air or oxygen that is released from the water during heating at the top

of the boiler shell and routes the trapped air into the expansion tank where the air is properly released.

2.3.2 Hot-Water System Operating Temperatures

The minimum boiler water temperature recommended for hot-water boilers is 140°F. This is because operation at lower temperatures can cause the combustion gases to condense in the fire-side of the boiler and cause corrosion. This problem is more severe in systems that are operated intermittently or are greatly oversized for the heating load. The temperature control in a hot-water boiler is typically set between 5°F and 10°F above the boiler's operating limit temperature control setting.

2.3.3 Hot-Water Circulation Considerations

The system piping and controls are arranged to prevent the possibility of pumping large volumes of cold water into a hot boiler. If 140°F boiler water is replaced with 80°F water in a short period of time, it causes thermal stress or shock. Thermal shock is the condition in which nonuniform thermal expansion or contraction occurs in the boiler components as a result of a sudden, marked change in the temperature of the water. After a boiler has been drained, or on the initial firing of a new boiler, you must prevent thermal shock by gradually warming the water in the boiler and the water being circulated through the piping system. This means that when starting up a hot-water boiler, the entire water content of the boiler and piping system must first be completely warmed at the low-fire level before the fuel input can be switched to the high-fire level.

During initial startup and subsequent operation, it is important to maintain the pressure/temperature relationships in a hot-water boiler as directed by the boiler manufacturer. *Figure 3* shows the recommended pressure/temperature relationships for one manufacturer's hot-water boilers. The water pressure and temperature gauges on the boiler should reflect boiler internal

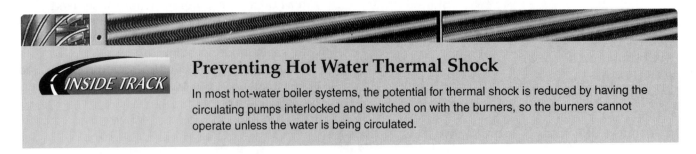

Preventing Hot Water Thermal Shock

INSIDE TRACK

In most hot-water boiler systems, the potential for thermal shock is reduced by having the circulating pumps interlocked and switched on with the burners, so the burners cannot operate unless the water is being circulated.

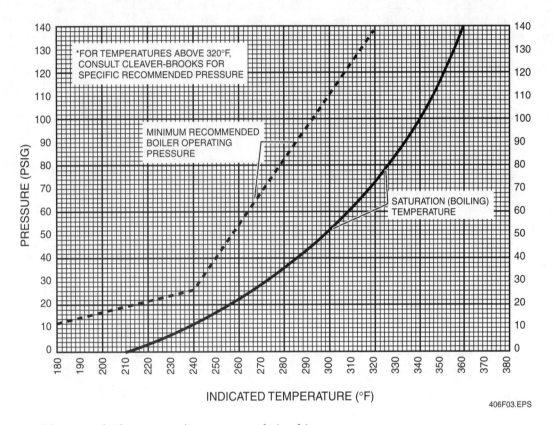

Figure 3 ◆ Typical hot-water boiler pressure/temperature relationships.

pressure and water temperature relationships typical of those shown in the chart.

A constant rate of water circulation through a hot-water system helps eliminate water stratification within the boiler and results in more even water temperatures being delivered to the system. In order to establish an accurate water circulation rate for the system, monitor the temperature drop that occurs as the water flows through the system. This can be done by installing a temperature gauge on the water return line, then determining the difference between the supply and return water temperatures.

Boiler manufacturers usually specify a set of recommended minimum and maximum circulating rates based on the system temperature drop and full boiler output. *Figure 4* shows a chart used by one manufacturer to determine the circulation rate in a system based on temperature drop.

2.4.0 Boiler Cleaning Prior to Startup for Normal Operation

Before a newly installed boiler or one that has been placed in dry storage can be placed in full operation, it must first be cleaned. This is necessary to remove any dirt and oil accumulated during the installation or dry storage. The chemicals used in the boiler cleaning process should be those recommended by a reputable water treatment specialist or the boiler manufacturer. These chemicals should meet all local, state, and federal laws pertaining to the discharge and disposal of the spent cleaning solution. The cleaning (boil-out) of a boiler water-side should always be done as directed in the boiler manufacturer's service literature.

Boiler Size (BHP)	Boiler Output (1000) Btu/Hr	System Temperature Drop – °F									
		10	20	30	40	50	60	70	80	90	100
		Maximum Circulating Rate – GPM									
15	500	100	50	33	25	20	17	14	12	11	10
20	670	134	67	45	33	27	22	19	17	15	13
30	1,005	200	100	67	50	40	33	29	25	22	20
40	1,340	268	134	89	67	54	45	38	33	30	27
50	1,675	335	168	112	84	67	56	48	42	37	33
60	2,010	402	201	134	101	80	67	58	50	45	40
70	2,345	470	235	157	118	94	78	67	59	52	47
80	2,680	536	268	179	134	107	90	77	67	60	54
100	3,350	670	335	223	168	134	12	96	84	75	67
125	4,185	836	418	279	209	168	140	120	105	93	84
150	5,025	1,005	503	335	251	1201	168	144	126	112	100
200	6,695	1,340	670	447	335	268	224	192	168	149	134
250	8,370	1,675	838	558	419	335	280	240	210	186	167
300	10,045	2,010	1,005	670	503	402	335	287	251	223	201
350	11,720	2,350	1,175	784	587	470	392	336	294	261	235
400	13,400	2,680	1,340	895	670	535	447	383	335	298	268
500	16,740	3,350	1,675	1,120	838	670	558	479	419	372	335
600	20,080	4,020	2,010	1,340	1,005	805	670	575	502	448	402
700	23,430	4,690	2,345	1,565	1,175	940	785	670	585	520	470
800	26,780	5,360	2,680	1,785	1,340	1,075	895	765	670	595	535

406F04.EPS

Figure 4 ◆ Typical hot-water boiler circulation chart.

WARNING!
Do not attempt to clean a boiler until you have read the manufacturer's operating and maintenance manuals and thoroughly understand them.

WARNING!
Starting up and cleaning a boiler system involves working on electrically energized equipment containing pressurized and/or hot components. Follow all applicable manufacturer's instructions, OSHA regulations, and job-site requirements relating to safety and the use of safety equipment.

2.4.1 Boiler Water Treatment

A water treatment specialist normally determines the specific water treatment programs and equipment used with a particular steam or hot-water boiler. You should review the material on water treatment covered in the *HVAC Level Three* module, *Water Treatment*. Some important points about water treatment of boilers are reviewed here.

WARNING!
Use the proper face mask, goggles, rubber gloves, and protective clothing when handling and mixing chemicals.

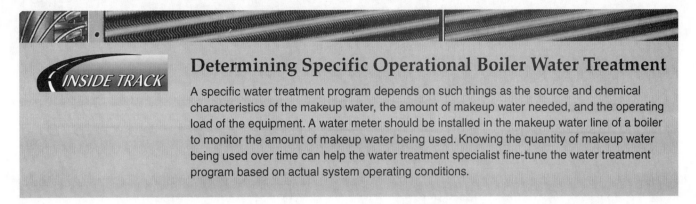
The objectives of boiler water treatment follow:

• The prevention of scale deposits
• The prevention of sludge deposits that impair the rate of heat transfer
• The elimination of corrosive gases and dissolved and suspended solids in the feedwater
• The prevention of caustic embrittlement or intercrystalline cracking of the boiler
• The prevention of carryover and foaming

Water used in a boiler must be kept clean for efficient operation. Never add dirty or rusty water to a boiler. Hard water may eventually interfere with the efficient operation of a boiler. Scale, corrosion, fouling, and foaming can all cause boiler problems.

Pretreatment of makeup water before the water is fed to the boiler can include filtering, demineralizing or softening, deaerating, and preheating. After-treatment usually involves the chemical treatment of the water in a steam boiler or water circulating in a hot-water boiler and system.

3.0.0 ◆ RECIPROCATING CHILLERS AND WATER SYSTEMS

Packaged reciprocating chillers are typically used on jobs requiring chiller capacities ranging from 2 to 200 tons. These chillers can use one or more reciprocating compressors. The chiller's condenser can be water cooled or air cooled. The evaporator (cooler) is usually a direct expansion type in which refrigerant evaporates while it is flowing inside the tubes, and the chilled water is cooled as it flows over the outside of the tubes. *Figure 5* shows a typical installation of a roof-mounted, packaged reciprocating chiller with an air-cooled condenser. As shown, the chiller is supplying chilled water to the coil of an air handling unit on an all-air system.

Depending on the capacity of the chiller, several air handling units are often supplied by a single chiller. The paragraphs below describe the tasks needed to start up and shut down a reciprocating chiller and related chilled-water system like the one shown in *Figure 5*.

3.1.0 Startup

Complete the startup of a reciprocating liquid chiller as directed in the chiller manufacturer's service instructions. The specific startup procedures for the different kinds of reciprocating liquid chillers vary, but the tasks and their sequences are essentially the same. The tasks in the procedure are those that are done when starting up a new chiller after installation or when it has been shut down for an extended period of time. The procedure presumes that the chiller is correctly installed and is charged with the correct type and amount of refrigerant.

WARNING!

Do not attempt to operate a chiller until you have read the manufacturer's operating instructions and thoroughly understand them.

WARNING!

Preparing a chiller for startup involves working on electrically energized equipment containing pressurized and/or hot components. Follow all the applicable manufacturer's instructions, OSHA regulations, and job-site requirements relating to safety and the use of safety equipment.

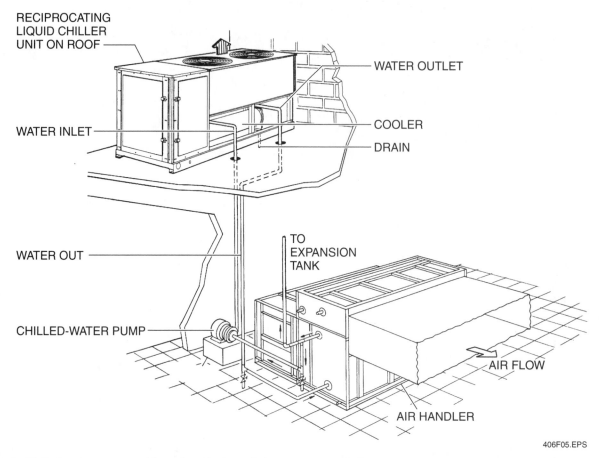

RECIPROCATING
LIQUID CHILLER
UNIT ON ROOF

WATER OUTLET

COOLER

DRAIN

WATER INLET

WATER OUT

TO
EXPANSION
TANK

CHILLED-WATER PUMP

AIR FLOW

AIR HANDLER

406F05.EPS

Figure 5 ◆ Chilled-water system with packaged, air-cooled reciprocating chiller.

Step 1   Check that all tools, equipment, and debris are removed from the chiller and the chilled-water system components.

CAUTION

Apply control power to the chiller unit at least 24 hours before attempting chiller startup. This allows the crankcase heaters sufficient time to warm the oil in the compressor crankcase.

Step 2   If not previously done, perform an inspection of the chilled-water circuit piping to verify the following:

   a. All system strainers are installed and clean.
   b. All piping connections are connected and tightened.
   c. All coil fins and tubing are undamaged, and the coils are free of debris or other obstructions.

   d. All safety and pressure-relief valves are set to the correct setpoint.
   e. All balance or isolating valves are set or opened to the required positions.
   f. Stop valves in the chiller unit evaporator (cooler) circuit are open.

Step 3   Fill the components and piping of the chilled-water system with clean water that has been treated with inhibitors, as needed. If starting up a system after it has been shut down during the previous heating season, drain the ethylene glycol or other antifreeze solution from the chilled-water components and piping, then flush and refill the system with clean water.

Step 4   Turn on the chilled-water circuit circulating pump and other equipment. Verify the following:

   a. No leaks are present in the system coils or piping.

b. All air is purged from the piping and system terminal units using the vents usually located at the highest point in the circuit.

c. The water flow, water level, and pressures are correct for the height of the highest terminal unit.

Step 5   Check the chilled-water system pumps and fans for the following:

a. Rotation is in the proper direction.

b. Operation is quiet, and lubrication is adequate.

c. Driveshaft alignments are correct.

d. Drive belt type and tightness are correct.

e. Setscrews on the driveshaft coupling and/or fan blades are tight.

f. All belt and/or fan blade guards are in place and secured.

Step 6   Check the pH of the chilled water to determine its acidity or alkalinity. If necessary, use the chemicals recommended by the water treatment specialist to maintain the water pH between 7 and 8.

Note that pH levels below 7 are acidic and those above 7 are alkaline. A pH of 7 is considered neutral.

Step 7   Perform a preliminary inspection of the air-cooled chiller to verify the following:

a. The electrical power source voltage meets the unit nameplate requirements.

b. The compressor crankcase is warm, indicating that the crankcase heaters are energized.

c. Oil is visible in the compressor sight-glass.

d. The compressor floats freely on its mounting springs.

e. The setpoint of the low-water temperature cutout is correct.

f. The setpoint of the chilled-water temperature controller is correct.

g. The condenser coils are free of residue from trees and plants.

h. The condenser fan to venturi (orifice) ring adjustment is correct per the manufacturer's specifications.

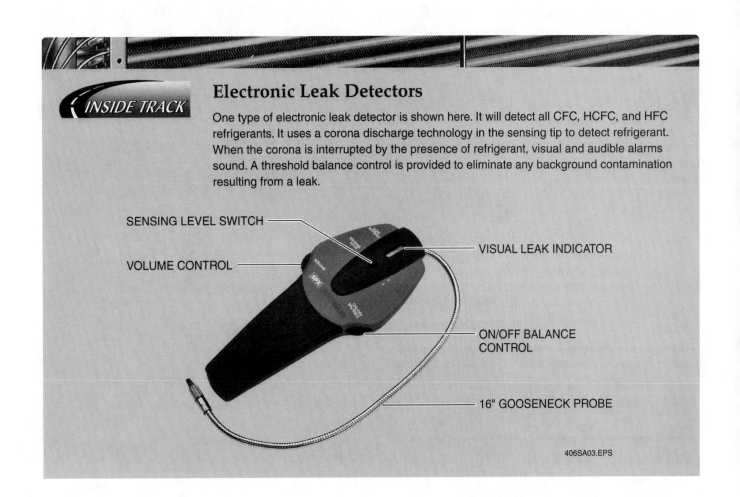

Electronic Leak Detectors

INSIDE TRACK

One type of electronic leak detector is shown here. It will detect all CFC, HCFC, and HFC refrigerants. It uses a corona discharge technology in the sensing tip to detect refrigerant. When the corona is interrupted by the presence of refrigerant, visual and audible alarms sound. A threshold balance control is provided to eliminate any background contamination resulting from a leak.

SENSING LEVEL SWITCH

VOLUME CONTROL

VISUAL LEAK INDICATOR

ON/OFF BALANCE CONTROL

16" GOOSENECK PROBE

406SA03.EPS

Step 8 Open the chiller compressor suction and discharge valves. Open the liquid line valve.

Step 9 Using an electronic leak detector, completely test the chiller components and piping for refrigerant leaks. An oily film in the area of the leak, such as on valves or flanged gasket connections, may indicate a possible leak. Make any necessary repairs, being sure to use the proper refrigerant containment procedures and equipment.

NOTE

Most chillers have a short-cycle protection circuit that provides a delay of about five or six minutes from the time a compressor is stopped before it can be restarted. This delay is activated when the compressor stops at the end of a normal cooling cycle, if a safety device opens, or if there is a power interruption. If applicable, be sure to allow enough time for the short-cycle control circuit to time out before attempting to start the compressor.

Step 10 Turn on the chiller. The chiller compressor should start after about a 10- to 15-second delay, during which the condenser fan operates to purge any residual heat from the area of the condenser coil.

Step 11 If the unit is equipped with a sightglass moisture indicator in the liquid line, check that it is clear with no sign of bubbles—this normally indicates a sufficient charge of refrigerant.

Step 12 Check to see if the moisture indicator shows the presence of moisture in the refrigerant circuit. Should it indicate only a slight amount of moisture, shut down the chiller and change the filter-drier(s). If the moisture indicator shows wetness, the source of moisture must be determined and eliminated before it results in a chiller failure.

NOTE

A moisture indicator sightglass may not be accurate. Always use an approved moisture test kit to verify the presence of moisture.

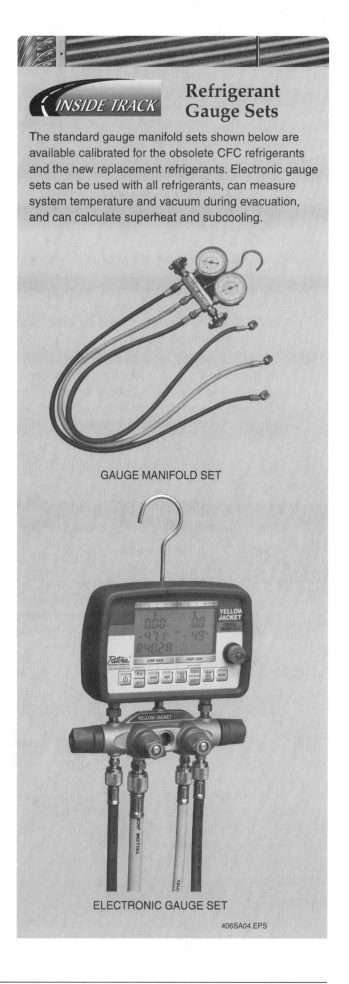

INSIDE TRACK

Refrigerant Gauge Sets

The standard gauge manifold sets shown below are available calibrated for the obsolete CFC refrigerants and the new replacement refrigerants. Electronic gauge sets can be used with all refrigerants, can measure system temperature and vacuum during evacuation, and can calculate superheat and subcooling.

GAUGE MANIFOLD SET

ELECTRONIC GAUGE SET

406SA04.EPS

Step 13 After the compressor is started and running, check that the temperature of the chilled water leaving the chiller's cooler continues to drop until the chiller automatically shuts off. The temperature at which this occurs should agree with the setting of the temperature controller. If it is not the same as the controller setting, adjust the control point as necessary per the manufacturer's service instructions.

Step 14 With the chiller operating, measure the high-side and low-side operating pressures with a gauge manifold set. Compare these pressures with the unit's superheat charging table to determine if the amount of superheat leaving the cooler is correct. Typically, the superheat level should be between 8°F and 10°F.

If necessary, follow the manufacturer's instructions to add or remove refrigerant from the unit in order to obtain the correct level of superheat.

Step 15 Confirm that the chiller's low-water and temperature cutout safety controls are functioning properly as directed in the manufacturer's service instructions.

Step 16 Check the cooler heater cable. Make sure that the outdoor thermostat causes the heater cable to be energized when the ambient temperature drops below 35°F.

Step 17 Record the chiller unit and chilled-water circulating pump data and operating parameters on suitable forms. Records should include the information summarized in *Tables 2* and *3*.

Table 2 Packaged Chiller Data and Operating Parameters

Chiller Data	
Location: _____	Manufacturer: _____
Model Number: _____	Serial Number: _____
Compressor Operating Data	
Suction Pressure/Temperature _____	Crankcase Heater Amperes _____
Discharge Pressure/Temperature _____	Chilled-Water Control Setting _____
Oil Pressure/Temperature _____	Condenser Water Control Setting _____
Voltage _____	Low-Pressure Cutout Setting _____
Amperes T1/T2/T3 _____	High-Pressure Cutout Setting _____
KW Input _____	Safety Control Check _____
Evaporator (Cooler) Operating Data	
Entering/Leaving Water Pressure _____	Flow in Gallons Per Minute (gpm) _____
Water Pressure Difference _____	Refrigerant Temperature _____
Entering/Leaving Water Temperature _____	Refrigerant Pressure _____
Water Temperature Difference _____	
Condenser Operating Data	
Water-Cooled Condenser	**Air-Cooled Condenser**
Entering/Leaving Water Pressure _____	Entering/Leaving Dry-Bulb Temperature _____
Water Pressure Difference _____	Air Temperature Difference _____
Entering/Leaving Water Temperature _____	Refrigerant Temperature _____
Water Temperature Difference _____	Refrigerant Pressure _____
Flow in Gallons Per Minute (gpm) _____	
Refrigerant Pressure _____	
Refrigerant Temperature _____	

Table 3 Circulating Pump Data and Operating Parameters

Circulating Pump Data	
Location: _____	Manufacturer: _____
Model Number: _____	Serial Number: _____
Circulating Pump Operating Data	
Pump Off Pressure _____	Final Suction Pressure _____
Valve Shutoff Differential _____	Final Pressure Difference _____
Valve Open Differential _____	Final gpm _____
Valve Open gpm _____	Voltage _____
Final Discharge Pressure _____	Amperage T1/T2/T3 _____

3.2.0 Shutdown

Except for turning off the chiller and related chilled-water circuit equipment, only a few tasks are required to prepare a reciprocating chiller with an air-cooled condenser for an extended period of shutdown. Be sure that the chiller unit compressor power is turned off, but do not shut off the control power during the shutdown period. If the chiller is being shut down for the winter season, the chiller's evaporator (cooler) should be isolated from the chilled-water circuit by closing the chilled-water input and output valves. After the cooler is isolated, remove the drain plug from the cooler and drain the cooler of all water. Put ethylene glycol or another approved permanent antifreeze solution in the cooler to prevent any residual water from freezing. Use the amount of antifreeze recommended by the manufacturer for the specific unit being winterized.

The chiller refrigerant should be tested for the presence of moisture and/or acid contamination. A sealed-tube acid/moisture test kit or an oil acid test kit (*Figure 6*) can be used for this purpose. A sealed-tube acid/moisture test kit is connected to a system service port to obtain a sample of the refrigerant. Oil acid test kits require that the test be performed on a sample of the compressor oil. In either case, follow the test kit manufacturer's instructions for using the kit to determine the amount of acid/moisture contamination in the system, if any. Should moisture and/or acid be detected, make repairs as necessary during the shutdown period to eliminate the cause of the problem.

TEST TUBES

ACID TEST KIT

406F06.EPS

Figure 6 ◆ Acid and moisture test kits.

Some manufacturers recommend that the bulk of the chiller system refrigerant charge be isolated in the condenser or receiver during shutdown. This minimizes the amount of refrigerant that might be lost due to any small leak on the low-pressure side of the system. If the system uses an open compressor, it reduces the amount of any refrigerant that might leak through the crankshaft oil seal. Isolation of the refrigerant in the condenser or receiver usually involves bypassing the low-pressure protective switch and readjusting the chilled-water temperature controller settings. This is necessary in order to operate the system during the procedure. For these reasons, it should always be done only as directed in the chiller manufacturer's service instructions.

4.0.0 ◆ CENTRIFUGAL CHILLERS AND WATER SYSTEMS

Packaged centrifugal chillers are typically used on jobs requiring chiller capacities over 100 tons. They use a centrifugal compressor with one or more stages. The condenser is normally water cooled via a cooling tower with the refrigerant condensing on the outside of the tubes. The evaporator (cooler) can be a direct expansion type in smaller units, but is usually a flooded type in larger units. *Figure 7* shows a packaged centrifugal chiller with a water-cooled condenser. Cooling water for the condenser is supplied by a cooling tower. The section that follows describes the tasks needed to start up and shut down a centrifugal chiller and related chilled-water system, such as the one shown in *Figure 7*.

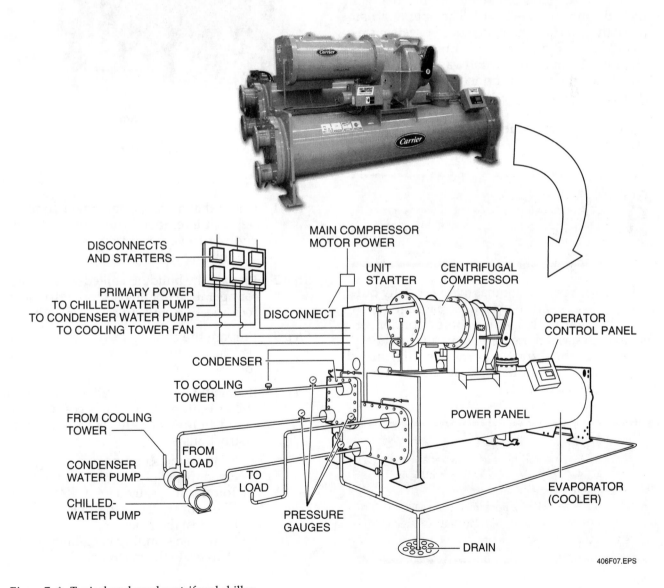

406F07.EPS

Figure 7 ◆ Typical packaged centrifugal chiller.

4.1.0 Startup

The startup and operation procedures of a centrifugal chiller are normally much more automated than with reciprocating chillers. Most have a microprocessor-based control center that monitors and controls equipment operation. The microprocessor control system matches the cooling capacity of the chiller to the cooling load. It also monitors system operating conditions to execute capacity overrides or safety shutdowns.

Operator interface with the chiller's control system is usually done at an interface panel (*Figure 8*). There, the operator can input machine setpoints, schedules, setup functions, and options. Typically, these inputs are entered in response to software-driven prompts shown on the panel's displays (screens). The responses are entered using softkeys located on the panel. On some chillers, the responses may be entered using a standard computer keyboard. The operator also can use the interface panel to monitor system status while the chiller is running or not running.

Because the operational software and the formats of the operator interface display screens vary widely among manufacturers, it is essential to complete the startup of a centrifugal chiller as directed in the manufacturer's service instructions.

WARNING!
Do not attempt to operate a chiller until you have read the manufacturer's operating instructions and thoroughly understand them.

The startup procedure that follows briefly outlines the tasks used to start up a packaged centrifugal chiller with a water-cooled condenser. The tasks described are those that are commonly performed when starting a new chiller after installation or when it has been shut down for an extended period of time.

WARNING!
Preparing a chiller for startup involves working on electrically energized equipment containing pressurized and/or hot components. Follow all applicable manufacturer's instructions, OSHA regulations, and job-site requirements relating to safety and the use of safety equipment.

TYPICAL STATUS SCREEN

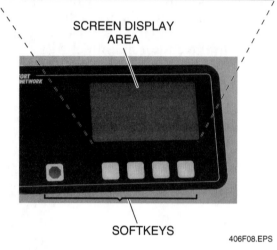

Figure 8 ◆ Typical operator interface panel and display screen.

Step 1 Check that all tools, equipment, or other debris are removed from the chiller, the cooling tower, and the components of the chilled-water system.

Step 2 Prepare the building chilled-water circuit for operation per Steps 2 through 6 of the reciprocating chiller startup procedure.

Step 3 Prepare the cooling tower for operation per the cooling tower startup procedure (described in detail later in this module).

Step 4 If starting a newly installed centrifugal chiller unit, perform the necessary tasks listed below per the manufacturer's instructions:

a. Remove the shipping packaging.
b. Open the oil circuit valves.
c. Torque all gasketed joints.
d. Check the machine tightness.
e. Leak test the machine.
f. Perform a standing vacuum test.
g. Dehydrate the machine.

h. Inspect the water piping including the pumpout compressor water piping (if so equipped).

i. Check the relief devices.

j. Inspect the wiring.

k. Check the starter.

l. Check the oil charge.

m. Energize and check the oil heater operation.

n. Set up the machine control configuration.

o. Check the optional pumpout system controls and compressor.

p. Charge refrigerant into the machine.

Step 5   If starting a chiller after an extended shutdown, make sure that the chilled-water, condenser water, and any other water drains are closed. If required, flush the water circuits to remove any soft rust that may have formed.

Step 6   Before attempting machine startup, check the following:

a. Power is turned on to the main starter, tower fan starter, oil pump and heater relays, and the machine control center.

b. Cooling tower water is at the proper level, and its temperature is at or below the machine's design entering temperature.

c. Machine is charged with refrigerant, and all refrigerant and oil valves are in the proper operating position.

d. Oil is at the proper level in the reservoir sightglass.

e. Oil reservoir temperature is within the range recommended by the manufacturer (typically 150°F to 160°F).

f. Valves in the evaporator and condenser water circuits are open. If the circulating pumps are not automatic, make sure they are turned on and water is circulating properly.

g. Solid-state starter checks (if so equipped) are performed.

Step 7   As applicable, energize the unit and perform the following tasks per the manufacturer's instructions:

a. Dry run to test the startup sequence.

b. Perform a compressor motor rotation check.

c. Conduct oil pressure and compressor stop checks.

d. Calibrate the motor current demand setting.

Step 8   Start the chiller per the manufacturer's instructions. If the chiller fails to start, check for conditions that prohibit starting, such as activated machine safety devices, no building occupied mode schedule selected for the current time period, or short-cycle time delays not timed out from the last shutdown or startup.

Assuming none of these conditions exist, the automatic sequence for the startup of a typical centrifugal chiller is similar to the one shown in *Figure 9* and is described as follows.

> **NOTE**
>
> Note that the times and other values given in the sequence are not specific, but are typical times and values used for descriptive purposes only. Failure to achieve any of the events described in sequence A through E will result in the control system automatically aborting the startup sequence and displaying the applicable failure message(s) on the control panel display.

Event A – The chilled-water pump starts when startup is initiated if the equipment protective safety limits and control settings are okay. These safety limits and control settings typically include the following:

- Temperature sensors out of range
- Pressure transducers out of range
- Compressor discharge temperature
- Motor winding temperature
- Evaporator refrigerant temperature
- Transducer voltage
- Condenser pressure
- Oil pressure
- Line voltage
- Compressor motor load
- Starter acceleration time

Event B – The condenser water pump starts after a short delay of about five seconds.

Event C – The system controller begins monitoring the status of the chilled-water and condenser water flow switches. It waits up to five minutes to verify water flow. Note that the waiting period allocated to monitor the water flow usually can be changed at the control panel, if desired. After water flow is verified, the chiller is enabled.

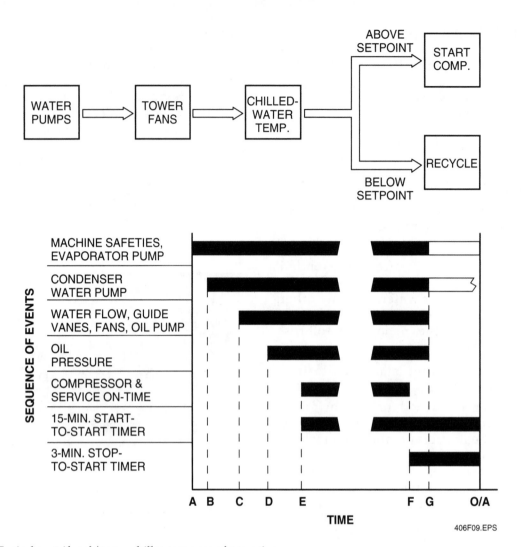

Figure 9 ◆ Typical centrifugal/screw chiller sequence of operation.

As the sequence continues, the chilled-water temperature is compared to the chiller's leaving (or entering) water temperature control point and **deadband**. The deadband is the tolerance on the chilled-water temperature control point. For example, a 1°F deadband controls the water temperature to within ±0.5°F. Should the chilled-water temperature be 5°F or more below the control point and deadband, the startup sequence stops and the machine control system initiates a normal recycle shutdown mode. This mode is described later in this section.

If the chilled-water temperature is equal to or higher than the control point and deadband, the control system continues the startup sequence and checks the position of the capacity control guide

vane and the oil pump pressure. If the vane is closed and the oil pump pressure differential is less than 3 psi, the oil pump relay is energized and the oil pump is turned on.

Event D – The system waits until the oil pressure differential reaches a minimum pressure of 18 psi.

Event E – 15 seconds after the minimum oil pressure is reached, the compressor relay energizes and the compressor starts. The 15-minute start-to-start timer is also activated.

Step 9 After the compressor starts and system conditions have stabilized, monitor the control panel displays and record the chiller data and operating parameters.

4.2.0 Normal Shutdown

During normal chiller operation, the chiller can shut down for a variety of reasons, such as when the stop control is pressed, the time schedule has moved to a building unoccupied mode, or the machine is recycling to maintain chilled-water temperature. When a stop command occurs, the automatic sequence for the shutdown of a typical centrifugal chiller is similar to the one shown for events F and G in *Figure 9* and described below:

Event F – The compressor start relay is de-energized, causing the compressor to stop. Also, the guide vanes are brought to the closed position and the 3-minute stop-to-start timer is activated.

Event G – 60 seconds after the compressor turns off, the oil pump, chilled-water pump, and condenser pumps are turned off. However, if the condenser pressure or temperature is high, the condenser water pump and the tower fan may continue to run.

During normal operation, the chiller may automatically cycle off and wait until the load increases before it restarts. This mode of operation is called the **recycle shutdown mode**. It occurs when the compressor is running under a light load and the chilled-water temperature drops 5°F below the chilled-water control setpoint (*Figure 10*). The sequence of shutdown in this mode is the same as described above, except the chilled-water pump remains running so that the temperature of the chilled water continues to be monitored. The chiller control system automatically restarts the chiller when it senses that the chilled-water temperature has increased to 5°F above the chilled-water control setpoint.

4.3.0 Extended Shutdown

When the centrifugal chiller is going to be shut down for an extended period, transfer the refrigerant into a storage tank. This helps reduce machine pressure and the possibility of leaks. A holding charge recommended by the manufacturer should be maintained in the machine to prevent air from leaking in. Many chillers are equipped with isolation valves, a pumpout system, and/or an optional storage tank. Following the manufacturer's instructions, use this equipment to transfer the refrigerant charge into the storage tank. For chillers equipped with isolatable condenser and cooler vessels, transfer the refrigerant into the condenser vessel.

If the machine may become exposed to freezing temperatures, drain the chilled-water, condenser

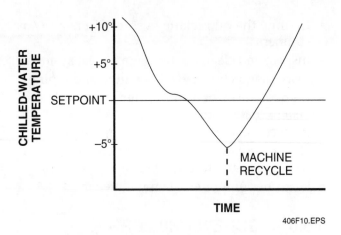

Figure 10 ◆ Recycle shutdown mode setpoints.

water, and pumpout condenser unit water circuits to avoid freeze-up. Also, leave the water box drains open. Leave the oil charge in the machine with the oil heater and controls energized to maintain the minimum oil reservoir temperature.

During extended periods of shutdown, perform the scheduled maintenance procedures recommended by the chiller manufacturer as directed in the service literature. These procedures can include the following:

- Test the refrigerant for acid/moisture contamination.
- Inspect and clean the operator control panel.
- Check the safety and operating controls.
- Change the oil and oil filters.
- Change the refrigerant filter-driers.

INSIDE TRACK

Preventing Freeze-Ups

In closed systems, manufacturers usually allow an appropriate amount of proper antifreeze to be added to the chilled water and/or condenser water to prevent freeze-up during operation or shutdown. A benefit of most recommended antifreeze is that it usually contains inhibitors that prevent corrosion, biological growth, and sludge formation.

Most manufacturers recommend using auxiliary electric heating of piping and/or open-system water basins to prevent freeze-up in systems that must operate in the winter. For short term below-freezing temperatures, some manufacturers allow the fans to be temporarily reversed to prevent ice formation on the tower. If the system is shut down for the winter but not drained, auxiliary heating with provisions for water circulation must be installed.

- Change the oil reclaim system filters and/or strainers.
- Inspect and clean the refrigerant float system.
- Inspect and clean the relief valves and piping.
- Inspect and clean the cooler and condenser heat exchanger tubes.
- Inspect and clean the starting equipment.
- Check the calibration of the oil, condenser, and cooler pressure transducers.
- Perform pumpout system maintenance.

5.0.0 ◆ SCREW CHILLERS AND WATER SYSTEMS

Packaged screw chillers (*Figure 11*) are typically used on jobs requiring chiller capacities over 100 tons. They use an oil-injected screw compressor.

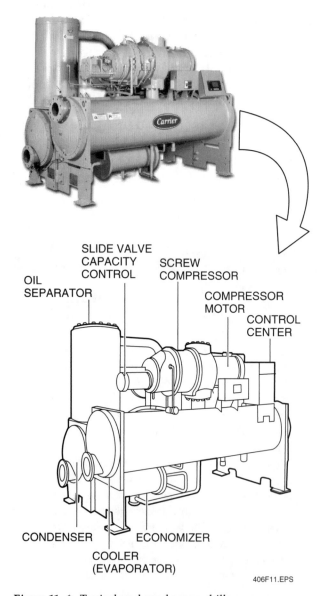

Figure 11 ◆ Typical packaged screw chiller.

The condenser is normally water cooled via a cooling tower with the refrigerant condensing on the outside of the tubes. When the chiller is used with a remotely located air-cooled or evaporative condenser, a liquid receiver may replace the condenser on the package assembly. The evaporator (cooler) can be either a direct expansion or flooded type. When the direct expansion type is used, the refrigerant usually flows through the tubes. With a flooded type, the refrigerant is normally outside the tubes. Because the screw compressor is oil injected, an oil separator is also part of the package.

Like centrifugal chillers, the startup, operation, and shutdown of most screw chillers is highly automated. Operation is under the control of a microprocessor-based control system that monitors and controls all operations of the machine. The screw chiller operator interface control panel and menu-driven displays (screens) are also similar to those used with centrifugal chillers. The sequence and tasks involved in the startup and shutdown of a screw chiller are basically the same as those described earlier for a centrifugal chiller.

6.0.0 ◆ COOLING TOWER WATER SYSTEMS

Correct startup and operation of a cooling tower (*Figure 12*) water system is important to efficient and troublefree water-cooled chiller and tower operation. It is even more important because of the health and safety issues related to tower water systems. Poorly prepared and maintained cooling towers are known to support the growth of Legionella bacteria, which cause Legionnaires' disease. The aerosol produced has the potential of infecting not only the people at the equipment site, but also the surrounding community. Note that the sequence and tasks performed to start up, operate, and maintain an evaporative condenser are the same as described for a cooling tower.

WARNING!

Preparing a cooling tower for startup involves working on electrically energized equipment containing pressurized and/or hot components. Follow all applicable manufacturer's instructions, OSHA regulations, and job-site requirements relating to safety and the use of safety equipment.

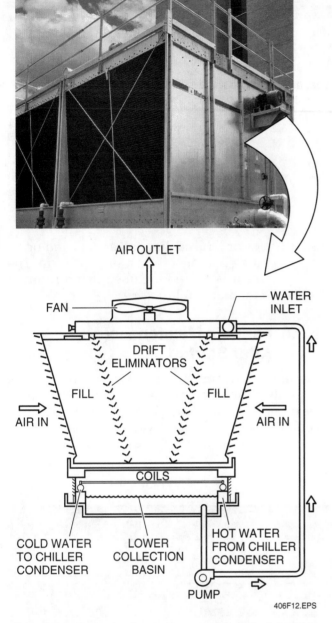

Figure 12 ◆ Typical indirect-contact induced-draft cooling tower.

6.1.0 Startup

Step 1   Check that all tools, equipment, or other debris are removed from the cooling tower and other components of the tower water system.

Step 2   If not previously done, perform an inspection of the cooling tower and circuit piping to verify the following:

    a. All system strainers are installed and clean.

    b. All piping connections are connected and tightened.

c. All balance or isolating valves are set or opened to the required positions.

d. Motor and pump driveshafts are correctly aligned.

e. Motor and/or fan drive belt tension is correct.

f. Driveshaft couplings and/or fan blade setscrews are tight.

g. All belt and/or fan blade guards are in place and secured.

h. Stop valves in the chiller unit water-cooled condenser circuit are open.

i. There is no mechanical damage, and the tower components are free of visible algae growths and deposits.

j. There is no slimy feel to the tower fill pack, tower sump, or sidewalls.

k. There is no visible corrosion on the outside of the sump heater.

l. The motors and bearings are lubricated.

CAUTION

After installation or major repair of a tower system, you must thoroughly clean the internal water system of protective oils, films, grease, welding flux, dirt, and other debris. Perform this cleaning only as recommended by a qualified water treatment specialist.

Step 3   Fill the components and piping of the tower water system with clean water that has been treated with inhibitors as needed. If starting up a system after it has been shut down during the previous heating season, first drain any antifreeze solution from the system components and piping, then flush and refill the system with clean, treated water.

If the system remained filled with treated water during shutdown, clean and replace or recharge the water treatment chemical feeding devices. Test the water to determine its condition, and treat as necessary to achieve an acceptable quality.

Step 4   Turn on the tower water system and adjust the float valve for the correct water level. Make sure that the water level is high enough to prevent the sump pump from drawing air into the water.

Step 5   Check that no leaks are present in the tower components and system piping.

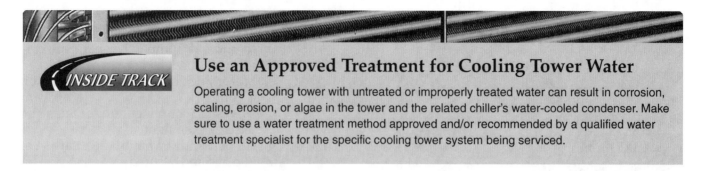

Step 6 Check the pumps and fans for proper direction of rotation and quiet operation.

Step 7 Check that the sump immersion heaters turn on when the setpoint on the thermostat is reached.

Step 8 Check the operation of spray nozzles and drift eliminators. The bypass of aerosol droplets should be minimal when the fans are operating.

Step 9 Adjust the **bleed-off** rate to match the rate set by the system water treatment program. A bleed-off of about two gallons per hour per ton of cooling is typical. Bleed-off is a method used to control corrosion and scaling in the system by periodically draining and disposing of a small amount of the water circulating in a system. This helps limit the buildup of impurities caused by the continuous adding of makeup water.

Step 10 Record the cooling tower data and operating parameters on a suitable form. The records should include the information summarized in *Table 4*.

6.2.0 Periodic Maintenance After Startup

The efficient operation of a cooling tower depends not only on mechanical maintenance, but also on cleanliness. Recirculating and spraying water in a cooling tower allows the water to come into contact with air. As a result, the composition of the

Table 4 Cooling Tower Data and Operating Parameters

Cooling Tower Data	
Location: _____	Manufacturer: _____
Model Number: _____	Serial Number: _____
Pump Data	
Make/Model _____	Motor HP/rpm _____
Serial Number _____	Volts/Phase/Hertz _____
Motor Make/Frame _____	Gallons per Minute (gpm) _____
Fan Data	
Number of Fan Motors _____	Motor Sheave Diameter/Bore _____
Motor Make/Frame _____	Fan Sheave Diameter/Bore _____
Motor HP/rpm _____	Sheave C/L Distance _____
Volts/Phase/Hertz _____	Number Belts/Make/Size _____
Operating Water Flow Parameters	
Entering/Leaving Water Pressure _____	Water Temperature Difference _____
Water Pressure Difference _____	Flow in gpm _____
Entering/Leaving Water _____	Temperature Bleed-Off gpm _____

water is drastically changed by evaporation, aeration, and other chemical and/or physical processes. Also, acidic gases and other contaminants in the air are absorbed into the water. Cooling towers are highly susceptible to the growth of algae, bacteria, and other living organisms, especially if they are located where the water surface is exposed to sunlight. Because of these harmful conditions, periodic maintenance on cooling towers must be done faithfully to prevent the buildup of corrosion and scale and to eliminate the growth of algae, bacteria, and other living organisms in the system. The tower manufacturer often furnishes service manuals that recommend scheduled maintenance tasks that should be performed on the tower. If manufacturer information is not available, *Table 5* can be used as a guide.

6.3.0 Winter Operation

When a cooling tower is used in subfreezing temperatures, the open recirculating water, closed recirculating water, and sump water systems need to be winterized. Equipment used for this winterization should be already installed in the system. It should be operated as directed by the tower manufacturer's service literature. During cold weather operation, more frequent visual

inspections should be made and routine maintenance performed. This helps to make sure all the controls are operating properly and aids in the detection of icing conditions before they become serious.

Table 5 Cooling Tower/Open Recirculating Water System Maintenance

Scheduled Maintenance Task	Frequency
Test and record bacteriological quality of the system water.	Biweekly
Test and record biocide and inhibitor reserves, pH, and conductivity of the system water.	Biweekly
Check that dosing equipment containers are full, pumps are operating properly, and supply lines are not blocked.	Weekly
Check the bleed-off control equipment to make sure it is operating properly, and be sure the controller is in calibration. The solenoid valve should be manually operated to confirm that the flow of water to the drain is clear.	Weekly
Check the system for growths and deposits. There should be no algae growth in the towers or slimy feel to the fill pack, tower sump, or sidewalls.	Weekly
Check the operation of sump immersion heaters. There should be no visible corrosion on the outside of the heater, and the unit should activate when the setpoint on the thermostat is reached.	Monthly
Check the operation of sprays, fans, and drift eliminators. There should be no mechanical damage and the components should be free of visible deposits. The bypass of aerosol droplets should be minimal when the fans are operating. The distribution system should have no deposits, with an even flow of water to all areas of the tower.	Monthly
Lubricate motors and/or fan drives.	Semiannually
Drain, clean, and disinfect cooling towers and associated pipe work in accordance with the method approved for the site. The chlorination period should be a minimum of five hours. Free chlorine residuals should be checked regularly. If possible, the tower pack should be removed for cleaning. Post-cleaning chlorination should be monitored to make sure that free chlorine residuals are maintained.	Semiannually
Review maintenance and water treatment program performance, including the quality of results obtained and the cost of system operation.	Annually

6.4.0 Shutdown

After the system chiller and cooling tower equipment have been shut down, the cooling tower and related water piping system can be prepared for an extended period of shutdown, as follows:

 WARNING!
Use the proper face mask, goggles, rubber gloves, and protective clothing as applicable when draining tower wastewater and working on tower equipment and surfaces that have been in contact with water treatment chemicals.

Step 1 Turn off power to all cooling tower equipment, including the water treatment equipment.

Step 2 If exposed to freezing temperatures during shutdown, turn off the supply water and drain the supply water line. Drain all recirculating water lines. Be sure to follow all local, state, and federal laws pertaining to the discharge and disposal of the drained wastewater.

Step 3 Remove all chemicals from the various water treatment equipment, and store or dispose of them as directed by the applicable equipment or chemical manufacturer.

Prepare the water treatment equipment for extended shutdown as directed by the applicable equipment manufacturer.

Step 4 Remove and clean the screens, strainers, and strainer baskets.

Step 5 Clean the makeup water system float valve assembly.

Step 6 Drain the water circulation pumps. Inspect and perform any scheduled maintenance on the pumps and motors specified in the pump manufacturer's service literature. When completed, add a corrosion inhibitor and lubricant to the pumps so that they will restart easily.

Step 7 Inspect and perform any scheduled maintenance on the fans and fan motors specified in the fan manufacturer's service literature.

Step 8 Inspect and clean the distribution basins or spray nozzles, cold water basin, and pipelines.

Step 9 Inspect for clogging and clean the drift eliminators and fill and/or decking.

Step 10 As needed, paint metal parts and components exposed to weather or wetting during operation. Do not paint spray nozzles or heat exchanger surfaces. Use a paint that is suitable for contact with water and the water treatment chemicals being used.

Step 11 To reduce the airborne dirt and debris that can be carried into the tower during shutdown, cover or screen the fan and louver openings.

7.0.0 ◆ AIR HANDLING UNIT/AIR DISTRIBUTION SYSTEM

Efficient operation of HVAC systems requires more than properly operating heating and cooling equipment. Of equal importance is the delivery of the correct quantity of conditioned air to the occupied space. This requires that any related water-to-air, steam-to-air, or air-to-air distribution system be properly installed and balanced to achieve efficient operation. Poor operation of the air distribution system can result in some building areas being overcooled or overheated while others are undercooled or underheated. Some areas may be drafty and others stuffy. Also, air pressures within the building can be out of balance, making exterior doors hard to open or close.

7.1.0 Startup

The startup procedure outlined in this section describes the startup of a typical constant volume air handling unit (*Figure 13*) connected to a single-duct air distribution system. The sequence and tasks described are basic to the startup and checkout of almost all types of air distribution systems. Procedures used to balance airflow in the different kinds of air distribution systems were studied in *HVAC Level Three* module, *Air Properties and Air System Balancing*. The startup of an air handling unit should be performed as directed in the manufacturer's service instructions.

 WARNING!
Do not attempt to operate an air handling unit until you have read the manufacturer's operating instructions and thoroughly understand them.

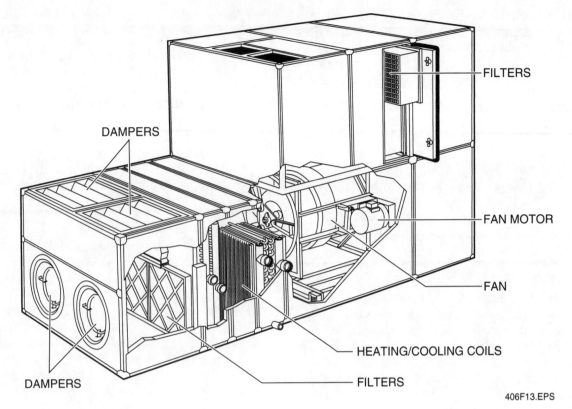

DAMPERS

FILTERS

DAMPERS

FAN MOTOR

FAN

HEATING/COOLING COILS

FILTERS

406F13.EPS

Figure 13 ◆ Typical air handling unit.

 WARNING!

Startup of an air distribution system involves working on electrically energized equipment with rotating and moving parts. It also involves working on the related cooling, heating, or air handling equipment with pressurized and/or hot components. Follow all applicable manufacturer's instructions, OSHA regulations, and job-site requirements relating to safety and the use of safety equipment.

7.1.1 Preliminary Checks

Step 1   Obtain the building plans and specifications, then locate all the system components.

Step 2   Make sure that all electrical power to the equipment is turned off. Open, lock, and tag disconnects.

Step 3   Check all fans and blowers. Verify that:

   a. All shipping restraints and protective covers are removed.
   b. All bearings are lubricated.
   c. Fan wheels clear the housings and are correctly positioned. Improper clearance can greatly affect the fan/ blower performance, especially with backward-inclined fans.

   d. Motors and bases are securely fastened.
   e. Motor fan driveshafts are correctly aligned.
   f. All setscrews are tight.
   g. Belt tensions are correct.
   h. Belt guards are in place.

Step 4   Inspect the complete supply air system from the supply air fan through the main supply ducts and through all branch duct runs and outlet terminals. Also inspect the complete return air duct system. Verify that:

   a. All tools, equipment, and debris are removed.
   b. The ductwork is complete with all openings in the duct sealed, all end caps in place, and all access doors closed and secured.
   c. All balancing, volume, and terminal dampers are in their marked positions.

Step 5   Check the air handling unit filters and all other air filters. Verify that:

   a. The size and types of filters are correct.
   b. The filters are clean.
   c. The filters and filter frames are properly installed and are airtight.

Step 6 Check that the air handling unit coils are undamaged and clean. Make sure that all piping connections are connected and tightened. Verify that the cooling coil condensate drain pan opening and drain line are clear.

Step 7 Check for conditions external to the system that can affect the airflow distribution to the occupied space. Make sure:

   a. Windows or outside doors are closed.
   b. Doors within the building are open.
   c. All ceiling tiles are in place.
   d. All air distribution grilles, registers, and diffusers are unobstructed.

CAUTION

Proper operation of the system dampers during startup is critical. If dampers are closed, restricting airflow, the casings, housings, and ductwork can be damaged. On the initial startup of a new system, a temperature control specialist or system designer should be consulted to make sure all automatic temperature control dampers are properly positioned.

Step 8 Check that all dampers are open or set so that air travels through the correct components of the system. Normally, dampers should not cause a blocked or restricted condition.

7.1.2 Startup and Air Distribution System Balance

Step 1 Turn on the air handling unit per the manufacturer's service literature. Select the mode of operation that calls for maximum airflow. Normally, this is the cooling mode with wetted coils.

Step 2 Check that the supply air fan is operating and its direction of rotation is correct.

Step 3 Check that all automatic dampers are being controlled and are in the proper position.

Step 4 Measure and record the voltage and current of the supply air fan motor to check for load conditions. If the amperage exceeds the motor nameplate full load amperage, stop the fan to determine the cause or to make necessary adjustments.

Step 5 Measure and record the speed of the supply air fan. If the fan is not running at the specified rpm, or as close as standard

pulleys will allow, the problem must be corrected before proceeding further.

Step 6 With the supply air fan running and the outdoor air dampers closed, measure and record the supply and return duct static pressures. Be sure to make the measurements in the ducts close to the air handling unit, but not in the unit itself. Calculate and record the total external static pressure drop across the supply air fan.

NOTE

Supply duct static pressure readings are normally positive, while return duct static pressure readings are normally negative. To calculate the total external static pressure across the supply fan, add the absolute values of the supply and return duct static pressure readings. For example, a supply duct static pressure of 0.75 in. w.g. and a return duct static pressure of –0.25 in. w.g. yields a total external static pressure of 1.00 in. w.g. (Absolute values ignore positive or negative signs.)

Step 7 Obtain the manufacturer's fan charts or performance curves for the specific supply fan. Use the values of supply fan rpm and total static pressure measured in Steps 5 and 6 and a manufacturer's fan curve chart like the one shown in *Figure 14* to determine the total volume of air being moved by the supply fan. *Figure 14* shows an example of a typical manufacturer's fan curve chart. For the purpose of an example, assume that the static pressure is 1.40 in. w.g. and the fan is running at 900 rpm. The cfm is found by locating the intersection point of the 1.40 in. w.g. static pressure line and the 900 rpm curve on the chart. From this point, drop down vertically to the cfm scale, and then read the value of 7,500 cfm.

Step 8 If the volume of airflow measured in Step 7 is not within ±10 percent of the fan's design capacity at the related design rpm, adjust the fan speed to obtain the required volume of airflow. Before making any rpm adjustments, make sure that the cause of insufficient airflow is not the result of an excessive air pressure drop occurring across filters, coils, or elsewhere in the system. Record the fan data and operating parameters on a suitable form. The records should include the information summarized in *Table 6*.

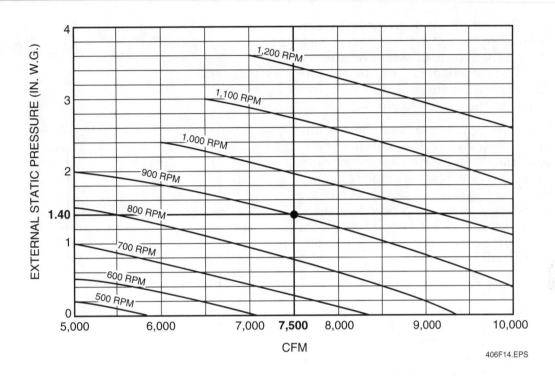

Figure 14 ◆ Example of a typical fan curve chart.

Table 6 Supply Fan Data and Operating Parameters

Supply Fan Data	
Location: _____	Manufacturer: _____
Model Number: _____	Serial Number: _____
Type/Class: _____	Motor Make/Style: _____
Motor HP/rpm: _____	Volts/Phase/Hertz: _____
F.L. Amps/S.F.: _____	Fan Sheave Make: _____
Motor Sheave Diameter/Bore: _____	Motor Sheave Diameter/Bore: _____
Motor Sheave Make: _____	Sheave C/L Distance: _____
No. Belts/Make/Size: _____	

Fan Operating Data	Required	Actual
cfm	_____	_____
Fan rpm	_____	_____
Total Static Pressure	_____	_____
Voltage	_____	_____
Amperage T1/T2/T3	_____	_____

Step 9 Use the building plans and specifications to determine the airflow in cfm specified for the air distribution system main supply and return ducts, each branch duct, and all outlet terminals. If not previously done, prepare suitable report forms listing all of the above items and their required airflow.

Step 10 Make a pitot traverse as outlined below to determine the actual volume of airflow being delivered to each branch duct. As necessary, adjust the branch dampers to deliver the required cfm to each of the branches. Repeat the procedure one or more times, making finer adjustments each time because the adjustment of one branch will affect the others.

    a. Make a pitot tube traverse of velocity pressure readings on the first branch duct, and record the readings on a suitable form. The information that should be recorded is shown in *Table 7*. For the purpose of explanation, the table also shows an example of data for a traverse taken on an 18" × 12" rectangular duct (*Figure 15*).

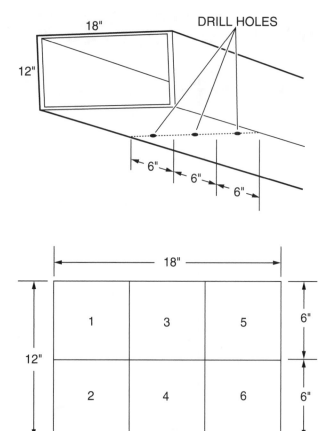

406F15.EPS

Figure 15 ◆ Traverse readings.

Table 7 Rectangular Duct Traverse Data

Rectangular Duct Data		
Location/Zone: _____	Size (in.): ___18 × 12___	Area (ft.): ___1.5___
Static Pressure: _____	Required fpm: ___850___	Actual fpm: ___900___

Traverse Position No.	Measured Velocity Pressure (in. w.g.)	Velocity (fpm) from Chart (*Figure 16*)
1	0.04	750
2	0.05	850
3	0.06	950
4	0.07	1,050
5	0.05	850
6	0.06	950
Total Velocity		5,400
average velocity (fpm) = 5,400 ÷ 6 = 900 fpm		
cfm = area × average velocity = 1.5 × 900 = 1,350 cfm		

Traverse Readings

To determine the average velocity in air ducts, you must take a series of velocity pressure readings, called a traverse, at points of equal area. To do this, make a pattern of pitot tube sensing points across the duct cross-section. For the example shown in *Figure 15* and *Table 7*, the longest duct dimension (18") is divided into 6" increments, and a hole is drilled into the center of each increment. The aim is to take a velocity pressure measurement at the center of each 6" square marked on the figure. For the duct shown, six measurements are needed. Regardless of the duct size, 6" should be the maximum distance between measurement points. The pitot tube is marked with insertion depth graduations to aid in positioning the pitot tube in the duct to the desired depth.

b. Use a velocity pressure/velocities chart (*Figure 16*) to convert each of the velocity pressure readings into velocity in feet per minute (fpm). By finding the velocity for each reading, and then averaging the results, determine the average velocity of airflow in the duct in fpm.

c. Calculate and record the airflow in cfm for the branch duct using the following formula:

$$\text{cfm} = \text{duct area (ft}^2) \times \text{avg. velocity (fpm)}$$

d. Repeat a. through c. at each of the remaining branch ducts.

Velocity Pressure in. wg.	Velocity fpm	Velocity Pressure in. wg.	Velocity fpm	Velocity Pressure in. wg.	Velocity fpm	Velocity Pressure in. wg.	Velocity fpm	Velocity Pressure in. wg.	Velocity fpm
0.01	300	0.26	2050	0.90	3800	1.92	5550	3.32	7300
0.01	350	0.27	2100	0.92	3850	1.95	5600	3.37	7350
0.01	400	0.29	2150	0.95	3900	1.99	5650	3.41	7400
0.01	450	0.30	2200	0.97	3950	2.02	5700	3.46	7450
0.02	500	0.32	2250	1.00	4000	2.06	5750	3.51	7500
0.02	550	0.33	2300	1.02	4050	2.10	5800	3.55	7550
0.02	600	0.34	2350	1.05	4100	2.13	5850	3.60	7600
0.03	650	0.36	2400	1.07	4150	2.17	5900	3.65	7650
0.03	700	0.37	2450	1.10	4200	2.21	5950	3.70	7700
0.04	750	0.39	2500	1.13	4250	2.24	6000	3.74	7750
0.04	800	0.41	2550	1.15	4300	2.28	6050	3.79	7800
0.05	850	0.42	2600	1.18	4350	2.32	6100	3.84	7850
0.05	900	0.44	2650	1.21	4400	2.36	6150	3.89	7900
0.06	950	0.45	2700	1.23	4450	2.40	6200	3.94	7950
0.06	1000	0.47	2750	1.26	4500	2.43	6250	3.99	8000
0.07	1050	0.49	2800	1.29	4550	2.47	6300	4.04	8050
0.08	1100	0.51	2850	1.32	4600	2.51	6350	4.09	8100
0.08	1150	0.52	2900	1.35	4650	2.55	6400	4.14	8150
0.09	1200	0.54	2950	1.38	4700	2.59	6450	4.19	8200
0.10	1250	0.56	3000	1.41	4750	2.63	6500	4.24	8250
0.11	1300	0.58	3050	1.44	4800	2.67	6550	4.29	8300
0.11	1350	0.60	3100	1.47	4850	2.71	6600	4.35	8350
0.12	1400	0.62	3150	1.50	4900	2.76	6650	4.40	8400
0.13	1450	0.64	3200	1.53	4950	2.80	6700	4.45	8450
0.14	1500	0.66	3250	1.56	5000	2.84	6750	4.50	8500
0.15	1550	0.68	3300	1.59	5050	2.88	6800	4.56	8550
0.16	1600	0.70	3350	1.62	5100	2.92	6850	4.61	8600
0.17	1650	0.72	3400	1.65	5150	2.97	6900	4.66	8650
0.18	1700	0.74	3450	1.69	5200	3.01	6950	4.72	8700
0.19	1750	0.76	3500	1.72	5250	3.05	7000	4.77	8750
0.20	1800	0.79	3550	1.75	5300	3.10	7050	4.83	8800
0.21	1850	0.81	3600	1.78	5350	3.14	7100	4.88	8850
0.22	1900	0.83	3650	1.82	5400	3.19	7150	4.94	8900
0.24	1950	0.85	3700	1.85	5450	3.23	7200	4.99	8950
0.25	2000	0.88	3750	1.89	5500	3.28	7250	5.05	9000

406F16.EPS

Figure 16 ◆ Velocity pressure/velocities chart.

Step 11 Measure the airflow at each branch duct diffuser or outlet terminal and record on a suitable form. The record should include the information summarized in *Table 8*. Make sure to follow the diffuser or terminal manufacturer's recommendations for taking measurements on the different devices. If necessary, adjust the outlets in each branch as outlined below:

    a. In the branch duct, measure the farthest outlet first. If the actual airflow at the outlet is below the required airflow, leave the damper fully open and go on to the next outlet upstream. If it is above the required airflow, throttle it down and go on to the next outlet upstream.

    b. Continue adjusting the outlets as necessary until all outlets in the branch have been adjusted.

    c. Repeat the procedure one or more times, making finer adjustments each time because the adjustment of one outlet affects the others.

    d. Readjust the branch duct damper if the outlet damper throttling process requires closing the outlet dampers to the extent that they generate noise.

Step 12 Adjust the amount of fresh outdoor air to provide for minimum building ventilation as follows:

    a. Check that the outdoor air damper and damper motor are both in the fully closed position. If necessary, adjust and tighten the drive rod to achieve this position.

    b. Find the percent minimum fresh air requirement from the building specifications. If the specifications are given in cfm instead of percent, divide the fresh air cfm requirement by the total cfm of the system to obtain the percentage.

    c. Set the fresh air control potentiometer to the minimum fresh air percent value (for example, 20 percent).

    d. With the system operating and stabilized, perform the following:

      • Measure the return air temperature in the return air duct at a location upstream from the fresh air inlet.

      • Measure the fresh outdoor air temperature at the outdoor intake to the ductwork.

    e. Calculate the mixed air temperature needed using the following formula:

$$MAT = [(OAT - RAT) \times OA\% \div 100] + RAT$$

Where:

    MAT = Mixed outdoor air temperature

    OAT = Outdoor air temperature

    RAT = Return air temperature

    f. Compare the calculated mixed air value to the measured mixed air temperature. The mixed air temperature should be measured downstream from the fresh air inlet. Several measurements of the mixed air should be made and an average calculated. Adjust the outdoor air damper minimum position as necessary to achieve the calculated mixed air temperature.

Table 8 Air Outlet Data

System:					Test Instrument:		
Area Served		**Outlet**				**Required cfm**	**Actual cfm**
	No.	**Type**	**Size**	**Area Factor**			

7.2.0 Shutdown

The shutdown of an air handler does not usually require any special service tasks. Do not leave outdoor air dampers open when the unit is not in use, especially during cold weather when freeze-ups can occur. During extended periods of shutdown, perform the scheduled maintenance procedures recommended by the air handler unit manufacturer as directed in the service literature.

Shutdown provides an opportunity to examine and clean the air distribution system ductwork and components. Increased emphasis has been placed on duct cleaning as a means of controlling indoor air quality. When duct cleaning is done in conjunction with the recommended HVAC equipment scheduled maintenance, it can help reduce the threat of indoor air pollution. Clean duct using portable and/or truck-mounted HEPA-filtered vacuuming and power brushing equipment to dislodge dirt and debris in the ductwork. The method used should follow the guidelines given in the National Air Duct Cleaners Association (NADCA) *Standard 1992-01*. Before and after cleaning the ductwork, make a visual inspection of the internal ductwork and components using borescopes, video cameras, and VCRs to provide a video record of the duct condition and cleanliness. It also may detect mechanical damage to the components that is not visible from the outside of the duct. Any wet fiberglass ductboard and/or insulation internal to the ductwork should be replaced so it does not become a breeding ground for biological contaminants. Detailed information on duct cleaning was provided in the module *Indoor Air Quality*.

8.0.0 ◆ PACKAGED YEAR-ROUND AIR CONDITIONING UNITS

Packaged year-round air conditioning (YAC) units provide gas or fuel oil heating, electric cooling, and air handling in a single unit. Most are sold as rooftop units similar to the one shown in *Figures 17* and *18*. This unit has a 20-ton cooling/235,000-Btuh heating capacity. In most applications, more than one YAC is installed on a building. Like air

ELECTRIC DISCONNECT SWITCH   FLUE GAS VENT FROM COMBUSTION AIR INDUCER

OPTIONAL ECONOMIZER DAMPERS AND HOOD

CONDENSATE TRAP AND VENT   GAS SUPPLY SHUTOFF VALVE

OPTIONAL GRAVITY EXHAUST DAMPERS AND HOOD   CONDENSER COIL

406F17.EPS

Figure 17 ◆ Typical packaged YAC unit.

INSIDE TRACK

Cleaning Grease from Heating/Cooling or Evaporator Coils

Alkaline-based cleaners are always better than acid cleaners for degreasing and cleaning. To help coils stay cleaner longer, coat them with a dry Teflon® coil protectant after cleaning. This will help the coil repel moisture, dirt, grease, and grime.

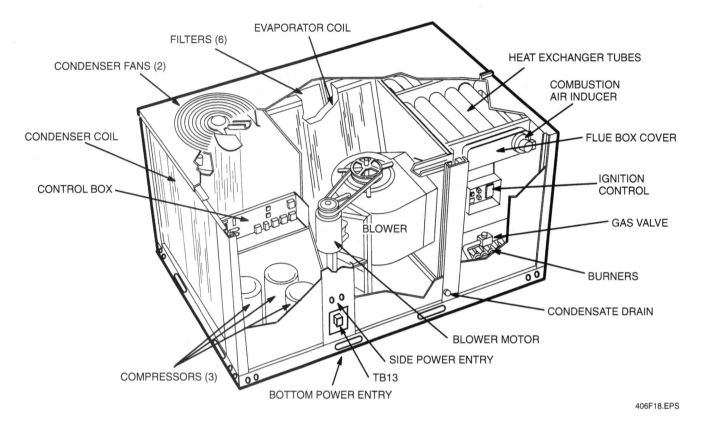

FILTERS (6)

EVAPORATOR COIL

CONDENSER FANS (2)

CONDENSER COIL

CONTROL BOX

HEAT EXCHANGER TUBES

COMBUSTION AIR INDUCER

FLUE BOX COVER

IGNITION CONTROL

GAS VALVE

BLOWER

BURNERS

CONDENSATE DRAIN

BLOWER MOTOR

SIDE POWER ENTRY

COMPRESSORS (3)

TB13

BOTTOM POWER ENTRY

406F18.EPS

Figure 18 ◆ Typical YAC unit parts arrangement.

handling units, these units must be properly installed and balanced to achieve efficient operation. Poor operation of the air distribution system can result in some building areas being overcooled or overheated, while others are undercooled or underheated. Some areas may be drafty and others stuffy. In addition, air pressures within the building can be out of balance, making exterior doors hard to open or close.

8.1.0 Startup

The startup procedure outlined here describes the startup of a typical YAC unit connected to a supply air and return air duct distribution system. The sequence and tasks described are basic to the startup and checkout of almost all types of heating, cooling, and air distribution systems. Procedures used to balance airflow in the different kinds of air distribution systems were studied in the module *System Balancing*. The startup of a YAC unit must be performed as directed in the manufacturer's service instructions.

WARNING!

Startup of a packaged year-round air conditioning unit involves working on electrically energized equipment with rotating and moving parts. It also involves working on the related cooling, heating, or air handling equipment with pressurized and/or hot components. Follow all applicable manufacturer's instructions, OSHA regulations, and job-site requirements relating to safety and the use of safety equipment.

8.1.1 Preliminary Checks

Use the following procedure to perform preliminary checks:

Step 1   Obtain the building plans and specifications, and then locate all the system components.

Step 2   Make sure that all electrical power to the equipment is turned off. Open, lock, and tag disconnects.

Step 3   Check all fans and blowers and compressors. Verify the following:

a.   All shipping restraints, lock-down bolts or nuts, and protective covers are removed.

b.   All bearings are lubricated.

c.   All fan wheels clear housings and are correctly positioned. Improper clearance can greatly affect the fan/ blower performance, especially with backward-inclined fans.

d.   Motors and bases are securely fastened.

e.   Motor and fan driveshafts are correctly aligned.

f.   All setscrews are tight.

g.   Belt tensions are correct.

h.   Belt guards are in place.

Step 4   Inspect the complete supply air system from the supply air fan through the main supply ducts and through all branch duct runs and outlet terminals. Also, inspect the complete return air duct system. Verify the following:

a.   All tools, equipment, and debris are removed.

b.   The ductwork is complete with all openings in the ducts sealed, all end caps in place, and all access doors closed and secured.

c.   All balancing, volume, and terminal dampers are in their marked positions.

Step 5   Check the air handling unit filters and all other air filters. Verify the following:

a.   The size and type of filters are correct.

b.   The filters are clean.

c.   The filters and filter frames are properly installed and are airtight.

Step 6   Check that the air handling unit coils are undamaged and clean. Check that refrigerant lines do not rub against the cabinet or against other refrigerant lines. Verify that the cooling coil condensate drain pan opening and drain line are clear.

Step 7   Check for conditions external to the system that can affect the airflow distribution to the occupied space. Check the following:

a.   Windows or outside doors are closed.

b.   Doors within the building are open.

c.   All ceiling tiles are in place.

d.   All air distribution grilles, registers, and diffusers are unobstructed.

Step 8   Check that all dampers are open or set so that air travels through the correct components of the system. Normally, dampers should not cause a blocked or restricted condition.

Step 9   Turn on internal and external heating fuel valves. If heating is gas-fired, check for and correct any leaks. With the internal gas valve on, check for proper manifold gas pressure at the manifold pressure port on the gas valve.

Step 10   If equipped with electronic ignition, check for a proper spark gap at the igniter and proper spacing and location of the flame sensor per the manufacturer's instructions.

8.1.2 Startup and Air Distribution System Balance

Step 1 Set external control(s) to activate cooling mode and turn on YAC unit power.

Step 2 If more than one compressor is contained in the unit, check that all three-phase compressors are rotating in the correct direction by monitoring system suction and discharge pressures. Normal suction and discharge pressures usually indicate correct phasing. A phasing meter can also be used to check for correct rotation.

Step 3 Check for proper refrigerant charge per the manufacturer's instructions with all external outdoor air dampers closed and the system stabilized. This is normally accomplished by checking for normal pressures at each refrigerant circuit.

Step 4 Set external controls to activate the YAC unit in the heating mode.

Step 5 Observe correct burner firing sequence and proper burner flame per the manufacturer's instructions.

Step 6 Set external controls to activate a mode that calls for maximum airflow. Normally, this is the cooling mode with wetted coils.

Step 7 Check that the supply air blower is operating and its direction of rotation is correct.

Step 8 Check that all automatic dampers are being controlled and are in the proper position.

Step 9 Check the blower motor current, rpm, and supply/return duct static pressures as described in the previous section for air distribution systems. If necessary, set initial blower speed with an adjustable pulley or, if so equipped, with a variable frequency drive (VFD) as described in the manufacturer's instructions. Then perform the air balancing procedures as described for air distribution systems.

8.2.0 Shutdown

If required, periodic shutdown of a YAC unit usually does not require any special service tasks. Do not leave outdoor air dampers open when the unit is not in use, especially during cold weather when freeze-ups can occur. If freezing temperatures will occur during shutdown or during the heating season, the condensate drain traps and condensate pans must be drained or siphoned.

During extended periods of shutdown, perform scheduled maintenance procedures recommended by the YAC unit manufacturer as directed in the service literature. These tasks can include the following:

- Motor and/or blower bearing lubrication
- Blower belt replacement
- Filter replacement
- Condenser, evaporator, and blower wheel cleaning

INSIDE TRACK

YAC Operating Sequence in the Heating Mode

Information describing the proper sequence of operation for a YAC in the heating mode is normally contained in the service literature for the equipment. A normal sequence for a YAC with pilot reignition and a single-stage burner system is given here.

1. Thermostat (or controller) calls for heat.
2. Induced draft motor starts.
3. Pilot gas and spark turn on.
4. Pilot ignites, flame is proved, and pilot spark turns off.
5. Gas is turned on to burners, and burners ignite.
6. Evaporator fan turns on.
7. Thermostat is satisfied.
8. Burners and induced draft motor turn off.
9. Evaporator fan turns off.
10. Unit is ready for next heating cycle.

- Inspection of the burner flame and cleaning of burners
- Inspection and cleaning of the combustion air inducer and vent cap
- Inspection and cleaning of the flue box and any flue tube baffles in the heat exchanger tubes

As mentioned earlier, a shutdown period provides an opportunity to examine and clean the air distribution system ductwork and components as a means of controlling indoor air quality. Clean duct using portable and/or truck-mounted HEPA-filtered vacuuming and power brushing equipment to dislodge dirt and debris in the ductwork. The method used should follow the guidelines given in the National Air Duct Cleaners Association (NADCA) *Standard 1992-01*. Before and after cleaning the ductwork, visually inspect the internal ductwork and components using borescopes, video cameras, and VCRs to provide a video record of the duct condition and cleanliness. This also may detect mechanical damage to the components that is not visible from the outside of the duct. Any wet fiberglass duct board and/or insulation internal to the ductwork should be replaced so it does not become a breeding ground for biological contaminants. Detailed information on duct cleaning was provided in the module, *Indoor Air Quality*.

9.0.0 ◆ POST-SHUTDOWN MAINTENANCE

During extended periods of shutdown, the scheduled maintenance procedures recommended by the various equipment manufacturers should be performed as directed in their service literature. Some of the more common tasks performed during shutdown include the following:

- Changing and/or removing the oil in a semi-hermetic or open reciprocating compressor
- Inspecting and cleaning shell and tube condensers and evaporators (coolers)
- Inspecting and cleaning heat exchangers

9.1.0 Oil Charging/Removal in a Semi-Hermetic or Open Reciprocating Compressor

When removing oil from a compressor, test a sample of the oil for the presence of contamination. Use an acid test kit and follow the test kit manufacturer's instructions to determine the amount of acid contamination in the system, if any.

Should acid be detected, perform the maintenance actions necessary in the system and/or compressor to find and repair the cause of the problem.

9.1.1 Charging Oil

Oil is added (charged) into a compressor when it has been drained or when oil must be added to make up for a residual oil loss. One recommended method uses a refrigeration oil pump (*Figure 19*) to charge oil into an operating compressor crankcase in a closed system. The refrigeration pump helps prevent contaminants from entering the compressor. Before charging oil into the compressor, always refer to the manufacturer's instructions for the compressor being serviced. Make sure to use the correct type of oil and the amount specified by the manufacturer. It is very important to use the exact oil specified by the manufacturer. Do not use substitutes unless authorized to do so. This is especially true with the synthetic oils that are used with the newer refrigerants. Their characteristics can vary widely from one manufacturer to another. Charging is outlined in the following procedure:

Step 1 Run the compressor fully loaded, then close the suction service valve and reduce the crankcase pressure to 0 psig. Note that the low-pressure switch may have to be bypassed.

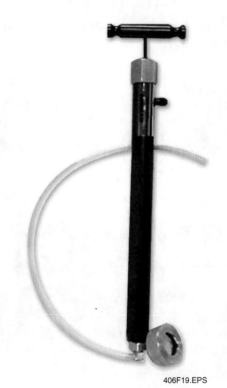

406F19.EPS

Figure 19 ◆ Refrigeration oil pump.

Step 2 Stop the compressor and isolate it from the system by closing the discharge service valve.

Step 3 If the compressor does not have an oil charging valve, remove the oil fill plug (*Figure 20*) and install an angle service valve in the oil fill plug hole.

CAUTION

POE oil in a closed container is free of moisture, but will readily absorb moisture if exposed to air. POE oil should not be used from a container that has been open for any significant length of time or from one that contains used oil. POE oil should never be stored in an open container.

Step 4 Mount and secure the refrigeration oil pump on the container of new oil. Loosely connect the pump to the angle service valve using a refrigerant charging hose or copper tubing. Purge the oil pump line by operating the pump until oil appears at the loose connection, and then tighten the connection.

Step 5 Open the service valve. While watching the oil level in the sightglass, pump oil into the compressor crankcase from the container of oil as needed.

Step 6 Close the service valve and disconnect the pump.

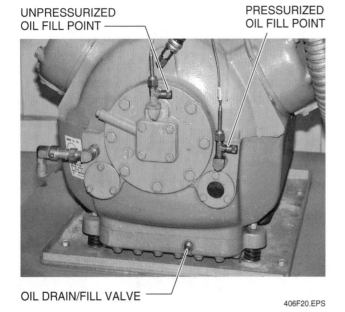

UNPRESSURIZED OIL FILL POINT

PRESSURIZED OIL FILL POINT

OIL DRAIN/FILL VALVE

406F20.EPS

Figure 20 ◆ Typical semi-hermetic compressor.

Step 7 To remove any moisture that may have entered the compressor, evacuate the compressor to 500 microns (29.89 inches of mercury).

Step 8 Open the suction and discharge service valves, and then restart the compressor.

Step 9 Run the system for about 20 minutes fully loaded, and then recheck the oil level at the sightglass.

NOTE

The correct sightglass level varies from one manufacturer to another. Check the manufacturer's instructions.

9.1.2 Removing Oil Through the Compressor Drain Plug

Remove oil from a compressor with a drain plug using the following general guidelines:

WARNING!

When removing oil, be sure to wear rubber gloves and eye protection to prevent possible injury. Contaminated refrigerant oil may contain heavy concentrations of acid. Do not allow contact with your skin or eyes as severe burns may result.

Step 1 If possible, run the compressor fully loaded. Then close the suction service valve and reduce the crankcase pressure to 0 psig. Note that the low-pressure switch may have to be bypassed.

Step 2 Stop the compressor and isolate it from the system by closing the discharge service valve.

NOTE

If removing only some of the oil through the compressor drain plug, do not completely remove the plug because the full oil charge may be lost.

Step 3 Loosen the plug until the oil just seeps around the plug threads. Drain the oil in this manner until the oil is lowered to the correct level in the compressor. Tighten the plug. If removing all the oil from the compressor, remove the drain plug and drain all the oil. After the oil has been completely drained, replace the plug.

CAUTION

If all the oil was removed from the compressor, make sure to refill the compressor with new oil. Use the closed system method for charging oil previously described in this section.

Step 4 Open the suction and discharge service valves. Restart the compressor.

Step 5 Run the system for about 20 minutes with the compressor fully loaded (working to capacity), and then recheck the oil level at the sightglass.

9.1.3 Removing Oil Through the Oil Fill Plug Hole

Remove oil from a compressor with an oil fill plug hole using the following general guidelines:

Step 1 If possible, run the compressor fully loaded. Then close the suction service valve and reduce the crankcase pressure to 0 psig. Note that the low-pressure switch may have to be bypassed.

Step 2 Stop the compressor and isolate it from the system by closing the discharge service valve.

Step 3 Remove the oil fill plug. Use the refrigeration oil pump with the siphon kit to remove the oil from the compressor. Continue to remove the oil in this manner until the oil is lowered to the proper level in the compressor, or if applicable, all the oil is removed. Replace the plug and add clean oil.

Step 4 To remove any moisture, evacuate the compressor to 500 microns (29.89 inches of mercury).

Step 5 Open the suction and discharge service valves. Restart the compressor.

Step 6 Run the system for about 20 minutes fully loaded, and then recheck the oil level at the sightglass.

9.2.0 Inspecting and Cleaning Shell and Tube Condensers and Evaporators (Coolers)

Water-cooled shell and tube condensers that use water supplied from open cooling towers are subject to contamination and scaling. These condensers should be periodically disassembled,

inspected, and cleaned per the manufacturer's instructions. Typically, they should be cleaned at least once a year and more often if the water is contaminated.

Normally, the tubes and heat exchanger surfaces will require cleaning using brushes designed to prevent scraping and scratching of the tube walls. Wire brushes should not be used. If cleaning the condenser tubes and heat exchanger surfaces with brushes fails to adequately remove the corrosion or scaling, chemical cleaning may be required. In this situation, the condenser manufacturer and/or a water specialist should be contacted in order to determine what chemicals should be used and how they are used to clean the condenser.

Inspect the entering and leaving condenser water sensors also for signs of corrosion and scaling. If corroded, the sensors should be replaced. If scale is found, they should be cleaned.

The water-side inspection and cleaning process used with shell and tube evaporators (coolers) in chilled-water systems is similar to the inspection and cleaning process used for water-cooled condensers.

9.3.0 Inspecting and Cleaning Heat Exchangers

In general, heat exchangers require maintenance only in the form of periodic inspection of their heating surfaces to determine if excessive fouling has occurred. The amount of fouling usually depends on water temperatures and local feedwater conditions. If either side of the heat exchanger is part of a closed system, it is unlikely that any cleaning of that portion of the heat exchanger will be needed, as long as the water has been properly treated.

If the heat exchanger is used to heat fresh water, fouling can occur. The amount of fouling depends on the water temperature and the amount of dissolved and entrained minerals contained in the water. Some of the entrained minerals break down into insoluble compounds when heated. They precipitate out of the water and adhere to the heater tubing. The higher the temperature to which the water is heated, the more rapidly scaling occurs.

Heat exchanger tubing does not scale up in a uniform manner. Most of the scaling takes place on the tubing immediately adjacent to the cold-water inlet. As the scale builds up at the entry of the tubes, it becomes an insulator, and heating will take place further back in the tubing, causing progressive scaling. Inspection of the tubing at the cold-water inlet will indicate when scale has formed.

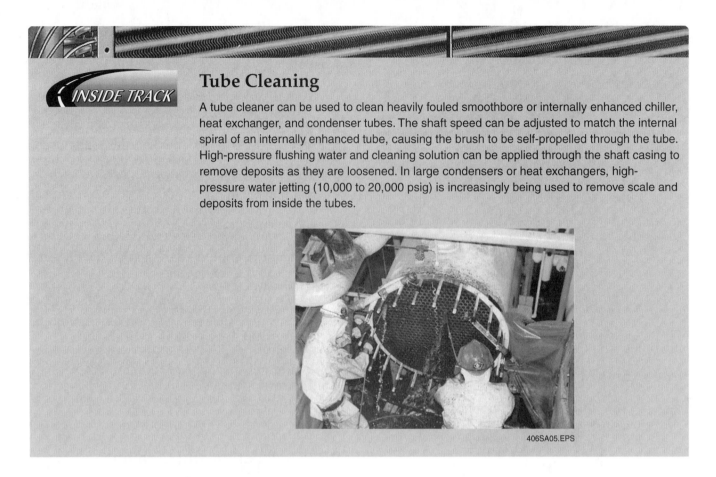

Tube Cleaning

A tube cleaner can be used to clean heavily fouled smoothbore or internally enhanced chiller, heat exchanger, and condenser tubes. The shaft speed can be adjusted to match the internal spiral of an internally enhanced tube, causing the brush to be self-propelled through the tube. High-pressure flushing water and cleaning solution can be applied through the shaft casing to remove deposits as they are loosened. In large condensers or heat exchangers, high-pressure water jetting (10,000 to 20,000 psig) is increasingly being used to remove scale and deposits from inside the tubes.

406SA05.EPS

Any one of the many available chemical coil cleaners may be used for cleaning if the manufacturer's recommendations on concentration and application are followed. When scaling is not too severe, standpipes can be installed in the head openings and the tubes filled with cleaning solution (*Figure 21*). It may take several applications to remove all the scale.

After cleaning, these tubes should be flushed with clean water to remove any chemical residue. Larger, multiple-pass heads will have a drain opening that can be used to eliminate spent cleaning solution.

Visual inspection determines when cleaning is complete. On severely fouled tubes, it may be necessary to pump the cleaning solution through the tubes continuously. If the tubes become completely plugged or are severely corroded, a new tube bundle may have to be installed.

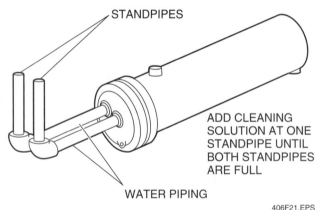

STANDPIPES

ADD CLEANING SOLUTION AT ONE STANDPIPE UNTIL BOTH STANDPIPES ARE FULL

WATER PIPING

406F21.EPS

Figure 21 ◆ Typical heat exchanger cleaning setup.

1. If not properly prepared for an extended period of shutdown, _____ inside an idle boiler can combine with condensed moisture to produce pitting of metal surfaces.
   a. oxygen
   b. bacteria
   c. algae
   d. scaling

2. When an oil-fired boiler is shut down, corrosion can be greatly reduced by _____.
   a. cleaning away soot and drying cleaned areas with a heating device
   b. applying a heavy-grade lubricant to all surfaces
   c. drying the boiler with a soft cloth and covering it with a protective tarp
   d. applying protective paint to the beginning areas of corrosion

3. A boiler is usually placed into dry storage if it is subjected to freezing temperatures or _____ or when it will be out of service for more than a month.
   a. excessive scaling
   b. acidic conditions
   c. corrosion
   d. a humid environment

4. When preparing a steam boiler for wet storage, the boiler is valved off from the rest of the system and filled with treated water _____.
   a. and cleaning chemicals
   b. then drained
   c. to the normal water line
   d. to overflowing

5. When preparing a steam boiler for startup, the main steam stop valve should be _____.
   a. fully open
   b. fully closed
   c. opened to the mid-position
   d. just cracked open

6. When initially firing a boiler at startup, the burners should be operated in the _____ mode.
   a. low-fire
   b. mid-fire
   c. high-fire
   d. standby

7. When the steam gauge just begins to register pressure during the startup of a steam boiler, _____.
   a. fully open the main stop valve
   b. vent the boiler so that no air is trapped in the steam space
   c. blow down the gauge glass and water column
   d. just crack open the main stop valve

8. When a hot-water boiler includes a dip tube, its purpose is to _____.
   a. indicate the level of hot water in the boiler
   b. allow personnel to check on the level of fuel used to heat the boiler
   c. allow personnel to directly pour in water treatment chemicals
   d. trap any air or oxygen released from the water during heating

9. The minimum boiler temperature that should be used with a hot-water boiler is _____.
   a. 140°F
   b. 180°F
   c. 190°F
   d. 200°F

10. When shutting down a hot-water boiler in preparation for storage, the load should be reduced slowly to avoid _____.
    a. foaming
    b. stresses caused by thermal shock
    c. corrosion
    d. rapid system depressurization

11. Crankcase heaters in a reciprocating chiller compressor should be energized for at least _____ hours before attempting chiller startup.
    a. 8
    b. 12
    c. 24
    d. 36

12. The pH level of the chilled water flowing in a chilled-water system should be between _____.
    a. 1 and 2
    b. 7 and 8
    c. 9 and 10
    d. 13 and 14

13. The deadband on a centrifugal chiller's water temperature control is 2°F. This means the chilled-water temperature is controlled within _____ of the temperature control setpoint.
    a. –2°
    b. ±1°
    c. ±2°
    d. ±4°

14. All of the following events occur in the automatic sequence for shutdown of a typical centrifugal chiller *except* the _____.
    a. compressor start relay de-energizes
    b. guide vanes are brought to the open position
    c. three minute stop-to-start timer is activated
    d. fans and pumps stop after 60 seconds

15. When preparing a screw chiller for extended shutdown at temperatures below 32°F, _____.
    a. the oil charge should be drained from the machine
    b. a holding charge of nitrogen should be maintained in the machine
    c. the refrigerant should be transferred to a storage tank or vessel
    d. the water circuits should be drained

16. If a cooling tower water system remained filled with treated water during shutdown, the system _____.
    a. can be started up without the need for water testing or treatment
    b. must be drained and refilled with clean water before startup
    c. water treatment chemical feeders should be cleaned and the chemicals changed or recharged
    d. should be thoroughly cleaned by a qualified water treatment specialist

17. The typical bleed-off rate that should be used for a 100-ton capacity cooling tower is _____ gallons per hour.
    a. 25
    b. 50
    c. 100
    d. 200

18. Cooling towers and piping systems should be drained, cleaned, and disinfected at least _____.
    a. once a year
    b. twice a year
    c. monthly
    d. every two years

19. When operating a cooling tower in subfreezing temperature conditions, it is recommended that _____.
    a. salt be added to the water
    b. the frequency of routine maintenance be increased
    c. the treated water be heated
    d. the cooling tower not be operated after sundown

20. During startup of an air handling unit, you measure a supply duct static pressure of 1.0 in. w.g. and a return duct static pressure of –0.25 in. w.g. The total external static pressure is _____ in. w.g.
    a. –0.75
    b. –1.25
    c. 0.75
    d. 1.25

21. Using the fan curve chart shown in *Figure 14*, find the cfm delivered by a fan operating at 1,000 rpm with an external static pressure of 1.6 in. w.g. The cfm delivered is _____.
    a. 7,500
    b. 8,500
    c. 9,000
    d. 9,500

22. Velocity pressure readings of 0.07, 0.09, 0.10, and 0.12 in. w.g. are measured while making a duct traverse. Using the chart in *Figure 16*, calculate the velocity in fpm.
    a. $(0.07 + 0.09 + 0.10 + 0.12) \div 4 = 0.095$
    b. $0.07 + 0.09 + 0.10 + 0.12 = 0.38$
    c. $(1,050 + 1,200 + 1,250 + 1,400) \div 4 = 1,225$
    d. $1,050 + 1,200 + 1,250 + 1,400 = 4900$

23. What is the volume of air in cfm being delivered by a duct that has an area of 6 ft^2 and an average air velocity of 1,200 fpm?
    a. 36 cfm
    b. 200 cfm
    c. 216 cfm
    d. 7,200 cfm

24. While checking the airflow in an air distribution system, it is necessary to make a major adjustment to one of the supply dampers in order to obtain the required system airflow. The next thing you should check is the _____.
    a. supply fan rpm
    b. supply fan motor current draw to make sure the motor is not overloaded
    c. total external static pressure
    d. supply duct static pressure

25. Duct cleaning should be done using _____.
    a. steel brushes
    b. HEPA-filtered vacuums and power brushing equipment
    c. pressure washers and squeegees
    d. high-pressure blowers

Summary

Proper startup and shutdown of HVAC systems is necessary in order to make sure the systems operate efficiently without premature failure. Proper startup and shutdown of HVAC systems involves the following procedures:

- All dirt and debris are removed, and the system is cleaned.
- All components in the system are put into operation and are functioning normally.
- Where required, testing, balancing, and adjustment are completed.
- The actual operating performance of the system meets the required (design) performance for the system.
- The system interacts properly with related systems and subsystems.
- All pertinent test and operating data for the system at startup is recorded for possible future use.

Notes

Bleed-off: A method used to help control corrosion and scaling in a water system. It involves the periodic draining and disposal of a small amount of the water circulating in a system. Bleed-off aids in limiting the buildup of impurities caused by the continuous addition of makeup water to a system.

Deadband: In a chiller, the tolerance on the chilled-water temperature control point. For example, a 1°F deadband controls the water temperature to within ±0.5°F of the control point temperature (0.5°F + 0.5°F = 1°F deadband).

Desiccant: A moisture-absorbing material.

Layup: An industry term referring to the period of time a boiler is shut down.

Recycle shutdown mode: A chiller mode of operation in which automatic shutdown occurs when the compressor is operating at minimum capacity and the chilled-water temperature has dropped below the chilled-water temperature setpoint. In this mode, the chilled-water pump remains running so that the chilled-water temperature can be monitored.

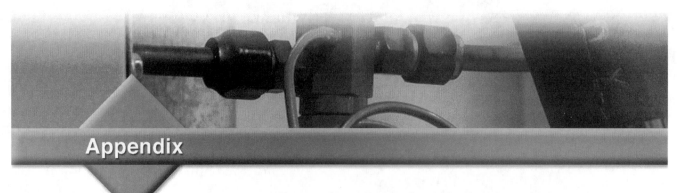

Typical Relationships Between O_2, CO_2, and Excess Air

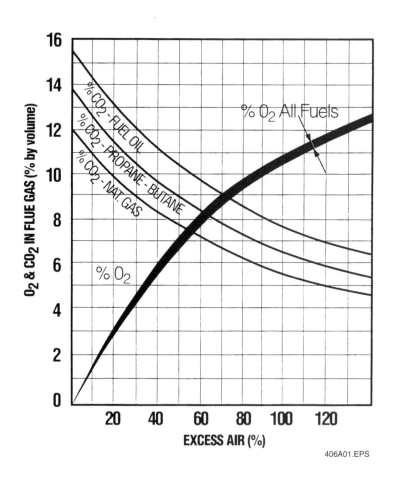

406A01.EPS

Additional Resources and References

Additional Resources

This module is intended to be a thorough resource for task training. The following reference works are suggested for further study. These are optional materials for continued education rather than for task training.

ASHRAE Handbook—HVAC Applications. Atlanta, GA: American Society of Heating, Refrigerating, and Air-Conditioning Engineers, Inc.

ASHRAE Handbook—HVAC Systems and Equipment. Atlanta, GA: American Society of Heating, Refrigerating, and Air-Conditioning Engineers, Inc.

Boilers Simplified. Troy, MI: Business News Publishing Company.

HVAC Systems Testing, Adjusting & Balancing. Chantilly, VA: Sheet Metal and Air Conditioning Contractors National Association (SMACNA).

Water Treatment Specification Manual. Troy, MI: Business News Publishing Company.

Figure Credits

Cleaver-Brooks, Inc., 406F01–406F04

Bacharach, Inc., 406SA01

Image courtesy of Fuel Efficiency, LLC, 406SA02

Robinair, a business unit of SPX Corporation, 406SA03, 406F19

Topaz Publications, Inc., 406SA04 (top), 406F06 (top), 406F07 (photo), 406F08 (photo), 406F17, 406F20

Ritchie Engineering company, Inc., 406SA04 (bottom)

Highside Chemicals, Inc., 406F06 (bottom)

Carrier Corporation, 406F11

Courtesy of SPX Cooling Technologies, Inc., 406F12

Lennox International, 406F18

Hennigan Engineering Co., Inc., Hingham, MA, 406SA05

NCCER makes every effort to keep these textbooks up-to-date and free of technical errors. We appreciate your help in this process. If you have an idea for improving this textbook, or if you find an error, a typographical mistake, or an inaccuracy in NCCER's Contren® textbooks, please write us, using this form or a photocopy. Be sure to include the exact module number, page number, a detailed description, and the correction, if applicable. Your input will be brought to the attention of the Technical Review Committee. Thank you for your assistance.

Instructors – If you found that additional materials were necessary in order to teach this module effectively, please let us know so that we may include them in the Equipment/Materials list in the Annotated Instructor's Guide.

Write: Product Development and Revision
National Center for Construction Education and Research
3600 NW 43rd St, Bldg G, Gainesville, FL 32606

Fax: 352-334-0932

E-mail: curriculum@nccer.org

Craft _____   Module Name _____

Copyright Date _____   Module Number _____   Page Number(s) _____

Description _____

_____

_____

_____

(Optional) Correction _____

_____

_____

(Optional) Your Name and Address _____

_____

_____

03407-09

Heating and Cooling System Design

03407-09
Heating and Cooling System Design

Topics to be presented in this module include:

Overview

Selection of the proper heating and cooling equipment along with proper design of the air distribution and refrigerant piping systems are critically important to the success of an installation. Once you delve into the system design process, you will realize how many variables can influence system design and how little must be left to chance in doing the design. For example, the color of the roofing material, the position of a building on its lot, and even the types of window treatments, are all factors that can affect equipment selection. Even if you never intend to work on the sales and design side of the industry, an understanding of the design process will help you in trouble-shooting a system that is not performing to the satisfaction of the building occupants. As an example, if a space is reconfigured during a remodeling project, it can have an affect on HVAC performance. Something as simple as replacing lined draperies with sheer curtains on a south-facing wall can affect both cooling and heating performance. This module will provide you with a better understanding of the factors that influence HVAC system design.

Objectives

When you have completed this module, you will be able to do the following:

1. Identify and describe the steps in the system design process.
2. From construction drawings or an actual job site, obtain information needed to complete heating and cooling load estimates.
3. Identify the factors that affect heat gains and losses to a building and describe how these factors influence the design process.
4. With instructor supervision, complete a load estimate to determine the heating and/or cooling load of a building.
5. State the principles that affect the selection of equipment to satisfy the calculated heating and/or cooling load.
6. With instructor supervision, select heating and/or cooling equipment using manufacturers' product data.
7. Identify the various types of duct systems and explain why and where each type is used.
8. Demonstrate the effect of fittings and transitions on duct system design.
9. Use a friction loss chart and duct sizing table to size duct.
10. Install insulation and vapor barriers used in duct systems.
11. Following proper design principles, select and install refrigerant and condensate piping.
12. Estimate the electrical load for a building and calculate the effect of the comfort system on the electrical load.

Trade Terms

Cubic feet per minute (cfm)	Register
Conductance	Static pressure
Conductivity	Temperature differential
Diffuser	Thermal conductivity
External static pressure	Throw
Fan brake power	Total pressure
Grille	U-factor
Infiltration	Velocity
R-value	Velocity pressure
	Volume

Required Trainee Materials

1. Pencil and paper
2. Appropriate personal protective equipment

Prerequisites

Before you begin this module, it is recommended that you successfully complete *Core Curriculum*; *HVAC Level One*; *HVAC Level Two*; *HVAC Level Three*; and *HVAC Level Four*, Modules 03401-09 through 03406-09.

This course map shows all of the modules in the fourth level of the *HVAC* curriculum. The suggested training order begins at the bottom and proceeds up. Skill levels increase as you advance on the course map. The local Training Program Sponsor may adjust the training order.

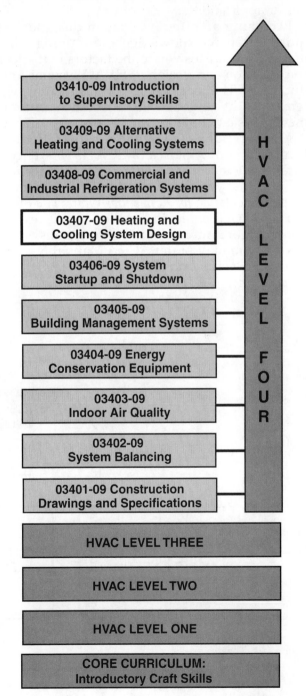

407CMAP.EPS

1.0.0 ◆ INTRODUCTION

This module discusses heating and cooling system design. The source of discomfort, noise, or inefficient operation of an air conditioning system can sometimes be traced to design flaws, poor installation practices, or changes that occurred in the building after the equipment was placed into service. In order to effectively troubleshoot a comfort air conditioning system, you must understand the principles of system design. In addition, if you move into management or sales in the HVAC industry, you must know how to determine the size and type of system needed for a particular application.

Becoming an effective system designer requires extensive knowledge and training. This module introduces you to the factors that affect the selection and design of HVAC systems and ductwork. The focus will be on residential systems, but the coverage will also touch on commercial system design.

Material published by the Air Conditioning Contractors of America (ACCA) is referenced extensively in this module. It is strongly recommended that you obtain and read ACCA Manual J, *Load Calculation for Residential Winter and Summer Air Conditioning*, in addition to this text. If you are more interested in commercial system design, you may prefer ACCA Manual N, *Load Calculation for Commercial Winter and Summer Air Conditioning*.

2.0.0 ◆ OVERVIEW OF THE DESIGN PROCESS

The design process has two important goals:

1. Selecting the size and type of equipment needed to deliver the correct amount of conditioned air to the building
2. Determining the type and size of ductwork needed to support the selected equipment.

The first step in this process, as shown in *Figure 1*, is to collect all the information you can about the building. If you are dealing with new construction, you may have to use the building and site blueprints and specifications. If it is a building expansion or a replacement for existing equipment, you can survey the actual building. A floor plan showing building dimensions, overhangs, number and types of windows and doors, and other factors, is absolutely essential.

The load estimate (Step 2) can be very time consuming and is usually done at the office rather than at the site. Therefore, you must collect all the information you need during the site survey. The purpose of Step 2 is to calculate the heating and cooling loads (also referred to as heat loss and heat gain, respectively) and the airflow requirements for heating and cooling. If both heating and cooling are being installed, the load estimating process will yield five values:

- Sensible cooling load (in Btuh)
- Latent cooling load (in Btuh)
- Cooling **cubic feet per minute (cfm)**

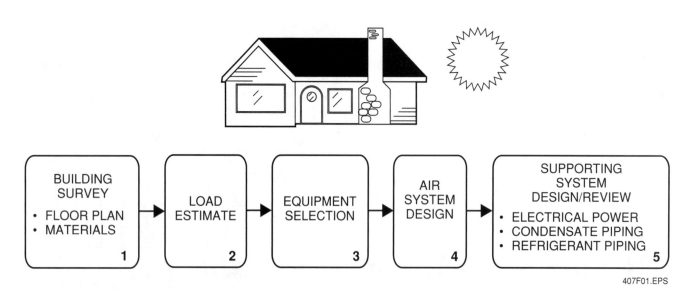

Figure 1 ◆ The design process.

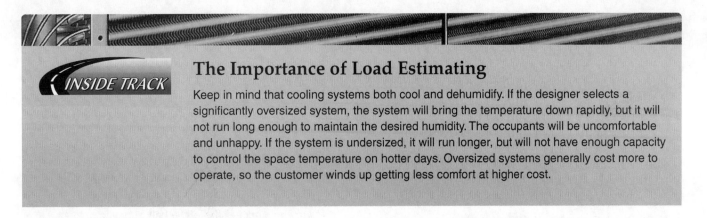

The Importance of Load Estimating

Keep in mind that cooling systems both cool and dehumidify. If the designer selects a significantly oversized system, the system will bring the temperature down rapidly, but it will not run long enough to maintain the desired humidity. The occupants will be uncomfortable and unhappy. If the system is undersized, it will run longer, but will not have enough capacity to control the space temperature on hotter days. Oversized systems generally cost more to operate, so the customer winds up getting less comfort at higher cost.

- Sensible heating load (in Btuh)
- Heating cfm

In Step 3, the heating and cooling loads and cfm values calculated in Step 2 are translated into cooling and heating equipment capacity using manufacturers' product data. For example, if the cooling load is 52,000 Btuh (about 4½ tons), select equipment of that capacity, or as close to it as possible, from the manufacturer's catalog, as long as it meets the cfm requirements.

A similar process is used to select heating equipment; that is, a furnace or other heating appliance is selected from the manufacturer's product literature based on the heating Btuh and cfm. The type of appliance selected is a function of structural, geographical, and economic factors, which will be covered later.

Once the equipment is selected, design the ductwork system or verify that an existing ductwork system will support the new equipment (Step 4). This is a very important step; if the duct system is not correctly designed, the system will not perform well and could create noise or comfort problems for the occupants.

Finally, in Step 5, the effect of the new system on other building systems is considered. This step is partly intended to make sure that existing

systems will support the proposed equipment and also to make sure that all the necessary parts and materials are available when the installation is started. Here are some of the things to consider in this step:

- Will the electrical service support the load?
- How much wire is needed for power and control wiring?
- How much and what size refrigerant and condensate piping is needed?

3.0.0 ◆ BUILDING EVALUATION/SURVEY

As you will see when the load estimating phase is discused, many variables affect the design of an HVAC system. Some of them have a very dramatic effect. If any factor is ignored or miscalculated, the system may not perform properly because it may be undersized or oversized. For that reason, it is very important to obtain as much information as possible about the building. Some of this information is obtained from the blueprints or by observing the building itself. If possible, it is also important to talk to the building owner or occupants, especially if a replacement or expansion is being considered.

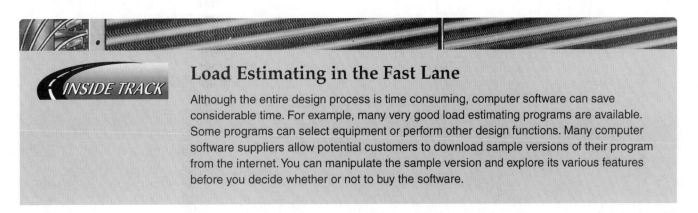

Load Estimating in the Fast Lane

Although the entire design process is time consuming, computer software can save considerable time. For example, many very good load estimating programs are available. Some programs can select equipment or perform other design functions. Many computer software suppliers allow potential customers to download sample versions of their program from the internet. You can manipulate the sample version and explore its various features before you decide whether or not to buy the software.

Occupants can tell you about problems they have experienced with the current system or about special considerations that must be factored into the design. Unless, for example, there are special health considerations, residential designs are pretty straightforward. In a commercial job it may be more difficult, especially if the building will be occupied by more than one business.

The types of information you will need to prepare the load estimate and select the equipment and ductwork are as follows:

- Type of roof (material and color)
- Amount and type of insulation in the attic, walls, and basement
- General tightness of the building
- Type of construction material used (frame, masonry, brick)
- Existence, size, and insulation of basements or crawl spaces
- Number of floors
- Direction the building faces (orientation to the sun)
- Number, sizes, and types of windows and doors
- When known, the type(s) of window covering, such as draperies, sheer curtains, or blinds
- Exterior dimensions of the building
- Color of the exposed exterior walls

- Length, width, and height of each room
- Shading from trees, adjacent buildings, or large hills
- Size(s) of roof overhang(s)
- Type and size of electrical service
- Number of walls, floors, and ceilings exposed to the outdoors or to uninsulated areas such as garages and crawl spaces
- Special design considerations such as zoning

These factors, as they affect heat loss and heat gain, are summarized in *Figures 2* and *3*, respectively.

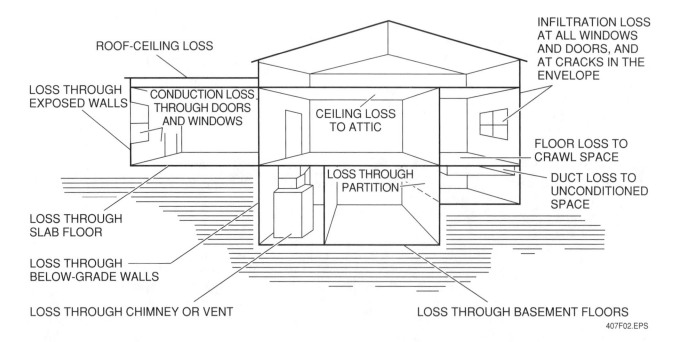

Figure 2 ◆ Winter heating loads (heat losses).

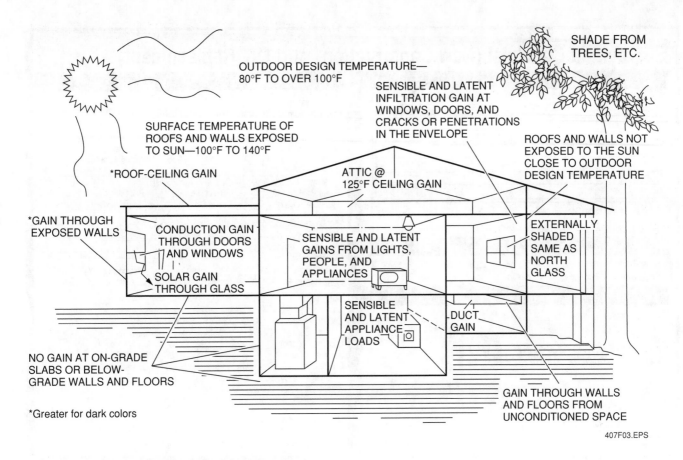

SHADE FROM TREES, ETC.

OUTDOOR DESIGN TEMPERATURE— 80°F TO OVER 100°F

SENSIBLE AND LATENT INFILTRATION GAIN AT WINDOWS, DOORS, AND CRACKS OR PENETRATIONS IN THE ENVELOPE

SURFACE TEMPERATURE OF ROOFS AND WALLS EXPOSED TO SUN—100°F TO 140°F

ROOFS AND WALLS NOT EXPOSED TO THE SUN CLOSE TO OUTDOOR DESIGN TEMPERATURE

*ROOF-CEILING GAIN

ATTIC @ 125°F CEILING GAIN

*GAIN THROUGH EXPOSED WALLS

CONDUCTION GAIN THROUGH DOORS AND WINDOWS

SENSIBLE AND LATENT GAINS FROM LIGHTS, PEOPLE, AND APPLIANCES

EXTERNALLY SHADED SAME AS NORTH GLASS

SOLAR GAIN THROUGH GLASS

SENSIBLE AND LATENT APPLIANCE LOADS

DUCT GAIN

NO GAIN AT ON-GRADE SLABS OR BELOW-GRADE WALLS AND FLOORS

GAIN THROUGH WALLS AND FLOORS FROM UNCONDITIONED SPACE

*Greater for dark colors

407F03.EPS

Figure 3 ◆ Summer cooling loads (heat gains).

If the job is an equipment replacement, you need to know the type and capacity of existing air conditioning and/or heating equipment and whether it is providing adequate comfort. You also need to know what remodeling has been done since the original system was installed. If rooms have been added or the windows replaced, the heating and cooling loads will be affected significantly. It's also very important to know the type and size of existing ductwork. However, do not assume that the original equipment and ductwork were properly sized, because oversizing is very common. If the building has a gas-fired or oil-fired furnace that you plan to replace, you need to see how it is vented. Keep in mind that old vents may not meet requirements for today's high-efficiency furnaces.

During this step, the designer will make a preliminary decision about the equipment configuration and zoning arrangements, if any. Even a residence can be too large to be efficiently handled by a single furnace or cooling system. Depending on the orientation of the building, amount of glass, and other factors, a single-zone approach for a smaller building may not make sense. As discussed in the *Building Management*

Systems module, a building with one large, unzoned system depends on a single room thermostat. The room containing the thermostat will be the most comfortable. Other rooms may be too hot or too cold. If a cooling thermostat is located in a room with a large picture window and a southern exposure, for example, the rooms on the north side of the building will probably be very cool during hot summer afternoons because the cooling system will be trying to keep up with the demand created by the solar heat pouring in through the picture window.

Using a specially designed checklist to collect information about the building is a good way to make sure you leave the site with all the information you need. *Figure 4* shows one manufacturer's form, which is used in conjunction with their residential load estimating process. The system is designed so that the information collected can be directly transferred to the load estimating forms.

Figure 5 shows an annotated floor plan. Between the survey and checklist (*Figure 4*) and the floor plan, enough information should be available to do the load estimate and equipment selection.

Survey and Checklist (New Construction, Add-On, Replacement)

1. Design Conditions

Location: City ANYTOWN   State U.S.   Latitude 34°N

Temperature °F	Summer	Winter
Outside Design DB/WB	96/77°F	20°F
Daily Range	22°F	xxxxxx
Inside Design DB/%RH	75/50%	72/30%
Difference	21°F	52°F
Swing	3°F	xxxxxx

Special Internal Loads (Computers, Etc.) _____NONE_____

Frequent Entertaining ☐   Doors Opened Often ☐

2. Orientation and Type

House Faces: N   NE   E   SE   (S)   SW   W   NW
Single Story ☐   Two Story ☑   Split Level ☐

3. Construction

Ceiling Height: Basement 7½'   1st Floor 8'   2nd Floor 7'

Insulation (Inches)

Walls: 0 1 2 3½ 6 8 12   R Value
Frame ☑ Masonry Above Grade☑ Bsmt   ☐ ☐ ☐ ☐ (Frame) R- 11
Masonry Below Grade: 0-5'☑ >5'☐   ☑ ☐ ☐ ☐   R- 0

Roof:
Ceiling under Ventilated Attic ☑   ☐ ☐ ☐ ☐ ☑ ☐   R- 25
Roof-Ceiling Combination ☐   ☐ ☐ ☐ ☐ ☐   R-___
Roof on Exposed Beams or Rafters ☐   ☐ ☐ ☐ ☐ ☐   R-___
Roof Color: Dark ☑   Light ☐

Floor:
Slab on Grade ☐   Edge Insulation: ☐ ☐ ☐   ___
Floor over Garage or Vented Crawl space ☐
    Floor Finish: Hardwood ☐   Carpeted ☐
    Ceiling Below: Yes ☐ No ☐   ☐ ☐ ☐ ☐ ☐   ___ R-___
Floor over Unheated Basement or Enclosed Crawl space ☐
    Ceiling Below: Yes ☐ No ☐   ☐ ☐ ☐ ☐ ☐   ___ R-___
Basement Floor >2' Below Grade ☑

4. Windows & Doors

Windows:
Type: Movable ☑   Fixed ☐   Jalousie ☐
Glass: Single ☐   Double (Single + Storm) ☑   Triple ☐
         Clear ☐   Tinted ☐   Reflective ☐   Low "e" ☐
Frame: Wood ☑   Metal ☐   Metal w/TB ☐

Number	Width	Size Height	Sq. Ft.	Type
2	3'	3'	9.0	24" O.H. - SOUTH
9	3'	4.5'	13.5	SHADED
3	3'	4.5'	13.5	24" O.H. - SOUTH
6	3'	4.5'	13.5	0" O.H.

Skylight: Wood ☐   Metal ☐   Metal w/TB ☐

Doors:
Type PANEL   Storms METAL

Shading:
Internal: None ☑   Full ☐   Half ☐
Overhangs: 0" ☑   12" ☐   24" ☑   36" ☐   48" ☐   (24" - SOUTH)
Permanent External Shading   NONE

5. Infiltration

Weatherstripping: Windows ☐   Doors ☑
Building Tightness: Loose ☐   Medium ☑   Tight ☐
Ventilation Fan ☐   Location _____   CFM_____
Fireplaces: No. 1   Loose ☐   Medium ☑   Tight ☐

6. Present Equipment Survey

Indoor System:
Forced Air Furnace: Heat Only ☐   Heat/Cool ☑
    Furnace w/Cooling ☐   Heat Pump w/Fan Coil ☐   Hydronic ☐
Fuel: Natural Gas ☑   LP Gas ☐   Oil ☐   Electricity ☐
Unit: Upflow ☑   Downflow ☐   Horizontal ☐   Package ☐
Location/Condition: W. END, BSMT.   Good ☐   Avg ☑   Poor ☐   Age 15
Make: BRAND "X"   Model No: XYZ
Capacity: Input 100,000 Output 76,000 BTU ☑   KW ☐
Blower: Motor HP 1/3   Direct Drive ☐   Belt Drive ☐
    Multiple Speeds ✓   Blower Dia. 10½"   Blower Width 10½"

Vent System: Condition: Good ☐   Avg ☑   Poor ☐   Age_____
Metal: Single Wall ☐   Double Wall Type "B" ☐   Dia._____
Masonry: Unlined ☐   Lined ☑   Liner Size _____ "x_____"
PVC Plastic: ☐
    Vent Connector: Dia._____ 6" Length_____ ft. Corroded: Yes ☐ No ☑
Water Heater: 40,000 BTUH Common Vent: Yes ☑ No ☐ Replace: Yes ☐ No ☐

Outdoor:
Unit: Split System ☐   Package Unit ☐   Room Air ☐
Condensing Method: Air Cooled ☐   Water Cooled ☐
    Water Supply: City ☐   Well ☐   Max Summer Temp_____°F
Location/Condition:_____   Good ☐   Avg ☐   Poor ☐   Age_____
Ratings: Capacity_____   BTUH; EER or SEER_____
Condensate: Gravity Drain ☐   Pump ☐   Sump ☐   Drywell ☐   Floor ☐
    Emergency Overflow Pan in Attic: Yes ☐   No ☐
Refrigerant Lines: Length_____ ft.
    Diameter: Suction_____"   Liquid_____"

Other Equipment:
Central Humidifier: Yes ☐   No ☑   Electronic Air Cleaner: Yes ☐   No ☑
Zoning: Yes ☐   No ☑   Heat Recovery Vent: Yes ☐   No ☑   Other: _____

7. Utilities

Natural Gas Meter Location N.E. CORNER, BASEMENT
LP Gas Tank Location_____
Oil Tank Location_____; Pump Above ☐   Below ☐
    Tank Size_____; Distance from Pump_____
Electrical Service: Volts 240   Phase 1   Hz 60   Amps 100
    Location of Entrance Panel N.E. CORNER, BASEMENT
Major Elec. Loads: Range/Self Clean ☐ Range/Oven ☑ Range Top ☐ Single Oven ☐
    Double Oven ☐ Dryer ☐ Dish _____ KW   Other_____KW
                                    └→ FUTURE

8. Controls

Zones: Single ☑   Multi ☐   Number 1
Thermostat Type: Heating ☑   Heat & Cool ☐   Continuous Fan ☑
    Auto Changeover ☐   Clock-type w/Night Setback ☐   Programmable ☐
Location of Master Thermostat: LIVING ROOM - INSIDE WALL

9. Air Distribution System

Supply:
Location: Basement ☑   In Slab ☐   Crawl space ☐   Ceiling ☐   Attic ☐   Soffit ☐
Exposure: In Unconditioned Space ☐   To Outdoor Temp. ☐
Insulation: 0" ☐   1/2" ☐   1" ☐   1 1/2" ☐   2" ☐   (1" @ Wall Stacks)
Plenum: Width 16"   Depth 19"   Height 42   Clearance to Rafters 2"
Main Trunk Duct: Width 20"   Height 8"
Runout Diameter: 6"
Outlets: Floor Perimeter ☑   Baseboard ☐   Ceiling ☐   High Sidewall ☐   Low Sidewall ☐

Return:
Location: Basement ☑   In Slab ☐   Crawl space ☐   Ceiling ☐   Attic ☐   Soffit ☐
Exposure: In Unconditioned Space ☐   To Outdoor Temp. ☐
Insulation: 0" ☑   1/2" ☐   1" ☐   1 1/2" ☐   2" ☐
Main Trunk Duct: Width 16"   Height 8"

407F04.EPS

Figure 4 ◆ Example of a residential data takeoff form.

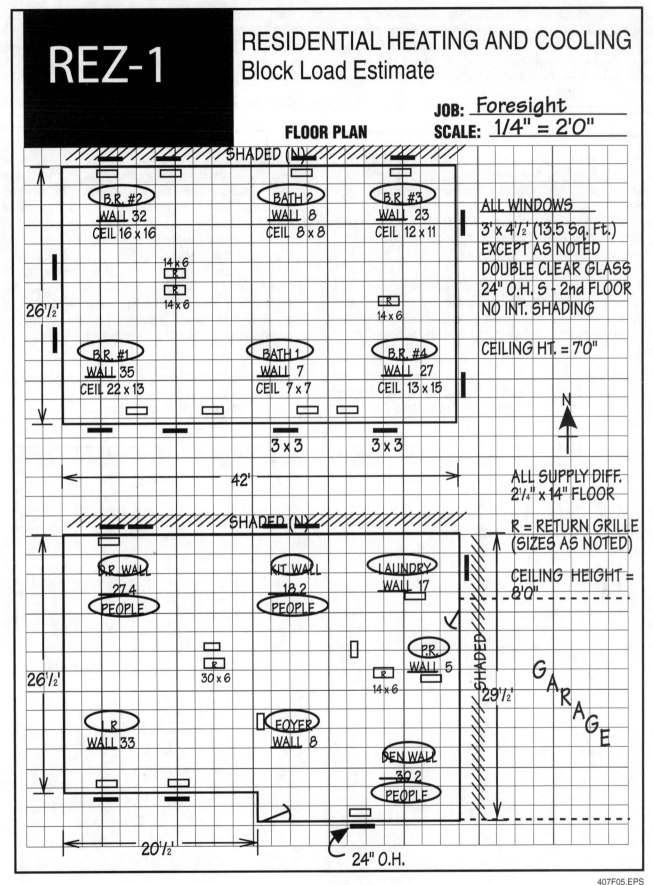

Figure 5 ◆ Example of an annotated floor plan.

The most common estimating system for residences is Manual J, which is published by the Air Conditioning Contractors of America (ACCA). ACCA also publishes Manual N, which is a load estimating system for commercial structures. Computer-based load estimating programs are also available, including a computerized version of Manual J. Most commercial load estimating, and a significant amount of residential load estimating, is now done with computer programs. Load estimating is a number-crunching process with many variables and therefore is ideal for computer applications. One major advantage of computer-based programs is that it is easy to compare the way different options might affect the load.

For example, if a building design plan shows standard window glass, it is relatively easy to see how much the load would be reduced if energy-efficient glass were used. This process is sometimes known as playing what-if games.

Computer-based load analysis tools are often combined with cost-estimating tools that can show how a particular change will affect operating cost and payback time.

4.0.0 ◆ LOAD ESTIMATING

It is generally impossible to measure either the actual peak load or the partial load in a given space; therefore, these loads must be estimated. The load estimate is based on design conditions inside and outside the building. The outside design conditions, which are the usual extremes of temperature based on National Weather Service data, are readily available. *Figure 6* shows an example of this data taken from ACCA Manual J. The significant factors in the table are the design temperature and the daily range. You can see that the location has a lot to do with the type and intensity of the load. The design temperature is not the same as the maximum temperature that

OUTDOOR DESIGN CONDITIONS FOR UNITED STATES AND CANADA
DESIGN GRAINS BASED ON AN INSIDE DESIGN TEMPERATURE OF 75°F

Location	Latitude Degrees	WINTER		SUMMER				
		97½% Design db	Heating D.D. Below 65°F	2½% Design db	Coincident Design wb	Grains Difference 55% RH	Grains Difference 50% RH	Daily Range
ALABAMA								
Alexander City	33	22	. . . .	93	76	37	44	21 M
Anniston AP	33	22	2810	94	76	35	42	21 M
Auburn	32	22	. . . .	93	76	37	44	21 M
Birmingham AP	33	21	2710	94	75	30	37	21 M
Decatur	34	16	3050	93	74	25	32	22 M
Dothan AP	31	27	1400	92	76	39	46	20 M
Florence AP	34	21	3199	94	74	23	30	22 M
Gadsden	34	20	3000	94	75	30	37	22 M
Huntsville AP	34	16	3190	93	74	25	33	23 M
Mobile AP	30	29	1620	93	77	44	51	18 M
Mobile CO	30	29	1620	93	77	44	51	16 M
Montgomery AP	32	25	2250	95	76	33	40	21 M
Selma-Craig AFB	32	26	2160	95	77	38	47	21 M
Talladega	33	22	. . . .	94	76	33	42	21 M
Tuscaloosa AP	33	23	2590	96	76	32	39	22 M
ALASKA								
Anchorage AP	61	-18	10860	68	58	0	0	15 L
Barrow (S)	71	-41	20265	53	50	0	0	12 L
Fairbanks AP (S)	64	-47	14290	78	60	0	0	24 M
Juneau AP	58	1	9080	70	58	0	0	15 L
Kodiak	57	13	8860	65	56	0	0	10 L
Nome AP	64	-27	14170	62	55	0	0	10 L
ARIZONA								
Douglas AP	31	31	2630	95	63	0	0	31 H
Flagstaff AP	35	4	7290	82	55	0	0	31 H
Fort Huachuca AP (S)	31	28	2551	92	62	0	0	27 H
Kingman AP	35	25	. . . .	100	64	0	0	30 H
Nogales	31	32	2150	96	64	0	0	31 H
Phoenix AP (S)	33	34	1680	107	71	0	0	27 H
......tt AP	34	9	. . . .	94	60	0	0	30 H
	32	32	1700	102	66	0	0	26 H
	35	10	4780	95	60	0	0	32 H
	32	39	970	109	72		0	27 H
		15	3760	94				
		23	. . . .	96				
		23	2300	9.				
			3840					

407F06.EPS

Figure 6 ◆ Example of design conditions data.

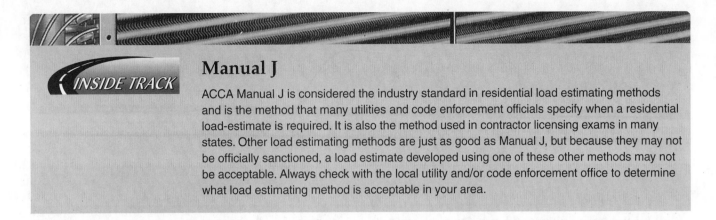
may be encountered in an area. If a maximum temperature were to be used instead of the design temperature, the heat loss or heat gain would be exaggerated, resulting in oversizing the equipment. Temperature extremes tend to be isolated and of short duration, so they are not used as a design temperature. The design temperature is closer to the actual temperature that can be expected on a fairly regular basis.

The inside design conditions are the temperature and humidity to be maintained for comfort conditioning or for material processing. The purpose of a load estimate is to determine the size and balance of the conditioning equipment necessary to maintain the inside design conditions during anticipated extremes in outside temperature and humidity.

4.1.0 Heat Transfer

Heat is a form of energy; therefore, it cannot be created or destroyed. It can, however, be moved from one place to another through a variety of mediums. Heat always flows in one direction—from a position of higher temperature to one of lower temperature. The greater the temperature difference, the greater the quantity of heat that will flow in a given unit of time. For example, if it is 70°F inside and 10°F outside, heat will move more rapidly out of the structure. If it is 70°F inside and 65°F outside, the heat loss will be insignificant. As you will recall, there are three main ways that heat transfer takes place: conduction, radiation, and convection.

4.1.1 Conduction

Conduction is the transfer of heat energy in a substance from particle to particle from the warmer region to the colder region. If, for example, a rod is heated over a flame, heat travels by

conduction from the hot end to the cooler end. Heat transfer by conduction occurs not only within an object or substance but also between different substances that may be in contact with one another.

An example can be found in a building constructed of a combination of brick or concrete, insulation, wood, and plaster (*Figure 7*). These materials are often in contact with each other. If it is warmer inside the building than outdoors, heat will pass through these materials by conduction (heat loss). If it is warmer outdoors than inside, heat will be conducted into the building (heat gain). Certain building materials, such as metal, will conduct heat faster than other materials, such as wood.

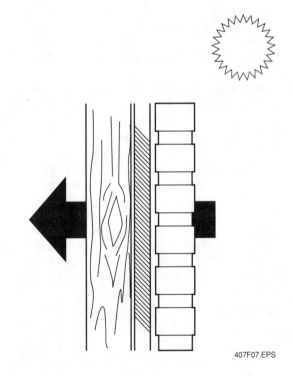

407F07.EPS

Figure 7 ◆ Conduction.

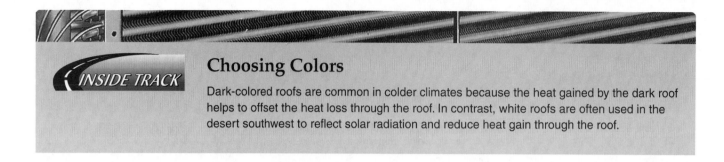

Choosing Colors

Dark-colored roofs are common in colder climates because the heat gained by the dark roof helps to offset the heat loss through the roof. In contrast, white roofs are often used in the desert southwest to reflect solar radiation and reduce heat gain through the roof.

4.1.2 Radiation

Radiant heat does not need a substance to carry it from one object to another. It can travel through a vacuum. Radiant heat exhibits many properties of light. It cannot pass through an opaque object, but it can pass through transparent materials. It can also be reflected from a bright surface, just as light is reflected by a mirror. Radiant heat passing through air does not warm the air through which it passes. For example, a roof with the sun shining on it might be heated to 180°F by the rays of the sun, but the air through which the radiant heat travels may maintain a temperature of only 80°F.

All objects radiate heat. The higher the temperature of the object or substance, the greater the quantity of heat it radiates. The amount of radiant heat energy given off in a unit of time depends on the temperature of the radiating body and also on the type and extent of its surface. A rough, dark surface, for example, radiates much more heat than a smooth, light surface of the same dimensions. The darker a surface, the more solar radiation it will absorb. Thus, dark surfaces always have higher heat gains than light surfaces exposed to the same amount of sunlight.

4.1.3 Convection

Convection is the transfer of heat energy due to the movement of fluid. A fluid has been defined as anything that flows; therefore, gases and liquids are both fluids. Air, being a mixture of several gases, can be considered a fluid. As the air moves or circulates, it carries heat from one place to another. In a gas furnace, for example, the energy is released by the combustion process inside the heat exchanger. The walls of the heat exchanger are considerably cooler than the burner flames, so the heat is absorbed by the walls of the heat exchanger. Air from the conditioned space is forced over the heated walls of the heat exchanger by the blower. Because the air is cooler than the walls of the heat exchanger, heat will flow from the warm heat exchanger to the

cooler air. The heat exchanger is designed in such a manner that the air passes over its entire surface in a wiping motion (*Figure 8*), causing a large amount of the heat energy to transfer to the moving air.

4.2.0 Heat Gain and Loss

In the summer, exterior walls transfer heat to the air in a room because they are warmer than the room air. Theoretically, a wall continually losing heat to a room would eventually cool down to the room air temperature. In reality, this does not happen; the heat that the wall loses to the room air is continuously replaced from the exterior heat. There is a steady flow of heat to the outer face of the wall. This heat is equal to the heat the room air gets from the inner face of the wall.

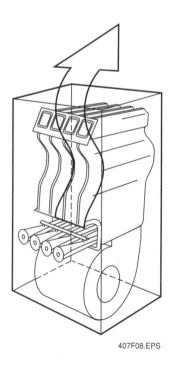

407F08.EPS

Figure 8 ◆ Convection.

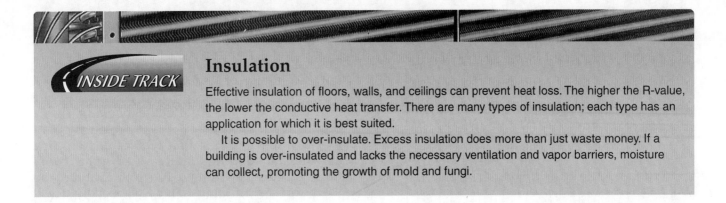
Heat flow through a wall separating two spaces at different temperatures depends on three factors:

- The area of the wall
- The temperatures of the two spaces (**temperature differential**)
- The heat-conducting properties of the wall

The larger the area of a wall, the more heat it can conduct. For example, a wall with an area of 200 square feet can potentially conduct twice as much heat as a wall with an area of 100 square feet. If the difference in temperature between the two spaces is 50°F, only one-half as much sensible heat will flow through the wall compared to a temperature difference of 100°F. These heat flow principles also apply to windows, roofs, and other building surfaces. In summary, the flow of heat through any surface is directly proportional to its area and the difference in temperature of the spaces separated by the particular surface. A third factor affecting heat flow through walls involves the wall material and its thickness.

The terms **conductivity** and **conductance** are used to describe heat flow through building materials. Conductivity (*Figure 9*) is the ability of a material to conduct heat. This ability varies from one material to another. The best conductors of heat are metals; the least effective heat conductors are wood, inert gases, and cork. The ability of a substance to transmit heat by conduction is a physical property of the particular material. This physical property is called **thermal conductivity**, which is defined as the heat flow per hour (Btuh) through one square foot of one-inch thick homogeneous material when the temperature difference between the two faces is 1°F.

Conductance is the term used to denote heat flow through materials such as glass blocks, hollow clay tile, concrete blocks, and other composite materials. In such material, each succeeding

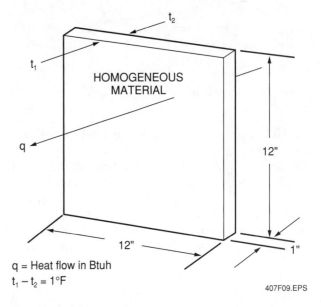

q = Heat flow in Btuh
$t_1 - t_2 = 1°F$

407F09.EPS

Figure 9 ◆ Conductivity.

inch of thickness is not identical with the preceding inch. Therefore, conductivity cannot be used to define the heat flow process. The conductance of a material is defined as the heat flow rate in Btuh through one square foot of a nonhomogeneous material of a certain thickness when there is a 1°F temperature difference between the two surfaces of the material (*Figure 10*).

Conductivity and conductance are not interchangeable terms. Conductivity is the heat flow through one inch of a homogeneous material; conductance is the heat flow through the entire thickness of a nonhomogeneous material.

Air space conductance is another factor that must be considered when calculating the heat gain of a structure via conductance. Air space conductance is defined as the heat flow in Btuh through one square foot of air space for a temperature difference of 1°F between the bounding surfaces.

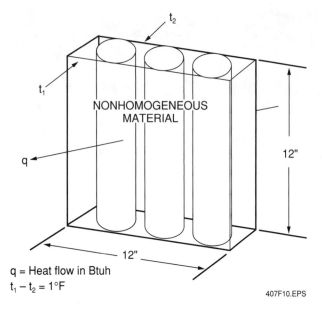

q = Heat flow in Btuh

$t_1 - t_2 = 1°F$

407F10.EPS

Figure 10 ◆ Conductance.

All the preceding factors need to be considered when calculating overall heat transfer. It is inconvenient to find all the surface temperatures for a wall made up of four or five materials, yet it is easy to find the temperature on both sides of a wall with an ordinary thermometer. Therefore, for nonhomogeneous materials, it is much more convenient to use a heat flow equation written with air temperatures.

The term **U-factor** is used to simplify this calculation. The U-factor works for homogeneous and nonhomogeneous materials and for a wall or roof made up of several materials. It is defined as the heat flow per hour through one square foot of the material(s) when the temperature difference is 1°F between the air on the two sides of the wall or roof. U-factor tables are available from the American Society of Heating, Refrigeration, and Air Conditioning Engineers (ASHRAE) that list the U-factors for various types of ceiling, floor, wall, and roofing materials and finishes.

The overall heat transfer formula is:

$$q = A \times U \times (t_2 - t_1)$$

or

$$q = A \times U \times td$$

Where:

q = heat flow in Btuh

A = area in square feet

U = overall heat transfer coefficient in Btuh/sq ft/°F (U-factor)

td = $(t_2 - t_1)$ = difference in temperature between the air on each side of the wall or roof

4.3.0 Cooling and Heating Load Factors

The materials used in constructing a building are not the only factors that affect heat gain and heat loss. The location, design, and positioning of a building also have an effect, especially on the cooling load.

A residence that is shaded by tall trees or a commercial building that lives in the shadow of taller buildings will have less of a cooling load than exposed buildings. This is especially true if the exposed building is oriented so that it has a lot of glass exposed to the sun.

The pie charts in *Figure 11* show the impact of various construction factors on cooling and heating loads. *Table 1* shows how various construction factors affect cooling and heating loads. These factors will be covered in sections that fol-

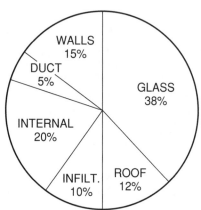

COOLING LOAD COMPONENTS

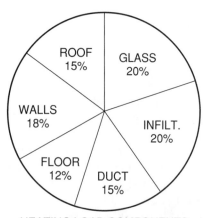

HEATING LOAD COMPONENTS (UNCONDITIONED BASEMENT OR ENCLOSED CRAWL SPACE)

TYPICAL LOAD PERCENTAGES

407F11.EPS

Figure 11 ◆ Load factors.

Table 1 Effects of Construction Factors on Cooling and Heating Loads

Factor	Cooling Load	Heating Load	Comments
Windows and glass doors	X	X	Major impact on cooling load; significant impact on heating load
Shaded glass	X		
Wood and metal doors	X	X	Greater effect on heating load
Exterior walls	X	X	Degree of impact depends on insulation thickness
Below-grade masonry walls		X	
Floors		X	
Roof	X		Darker colors create more load
Infiltration	X	X	Greater effect on heating load
Unconditioned spaces	X	X	
Duct gain/loss	X	X	Greater effect on heating load
Temperature swing	X		
Internal factors	X		People, cooking, and bathing
Insulation (wall & attic/ceiling)	X	X	Greater effect on heating load

low. The designer and builder can help provide a more economical installation and better system operation by specifying the use of insulation, close-fitting windows and doors, and double glazing, storm sashes, or reflective glass. Window glass is a significant contributor to both the cooling and heating loads in many buildings. For this reason, load estimating tools place great emphasis on calculating the glass-related load.

4.3.1 Window Glass

Before examples of load calculations are presented, some of the terms used in the load estimating tables should be clarified. Most of the examples used in this module were taken from ACCA Manual J. Other load calculation methods may use different methods and terms to achieve the same (or similar) results. For example, note the differences between the tables in *Figures 12* and *13*. *Figure 12* is the Manual J table and *Figure 13* comes from an ASHRAE manual.

Although the data in the tables track pretty closely, you can see that the ASHRAE table uses the outdoor design temperature, while the Manual J table uses the difference between the outdoor and indoor design temperatures. In the Manual J table, the result is called the heat transfer multiplier (HTM); in the ASHRAE table, it is called the window glass load factor (GLF).

In cooling, window glass is the single largest load factor (see *Figure 11*). To determine the glass-related load, the area of each type of window glass is multiplied by load factors to determine the amount of load in Btuh. Therefore, the more windows you have and the less energy efficient the windows are, the greater the cooling and heating load.

For example, assume that a house has ten single-pane, unshaded 3' × 5' windows (15 ft² each) with east or west exposure. Refer to *Figure 12* and assume a 25°F temperature difference. The multiplier is 93 and the related load is 13,950 Btuh (150 × 93). That means these ten windows alone create a cooling load of more than 1 ton. Note that if the same windows used reflective coated glass (lower part of the table), the load would be reduced by about one-third.

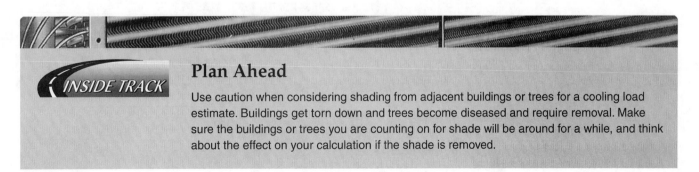

Plan Ahead

Use caution when considering shading from adjacent buildings or trees for a cooling load estimate. Buildings get torn down and trees become diseased and require removal. Make sure the buildings or trees you are counting on for shade will be around for a while, and think about the effect on your calculation if the shade is removed.

Glass Heat Transfers Multipliers (Cooling)
No External Shade Screen
Clear Glass

Design Temperature Difference	Single Pane						Double Pane Single Pane & Low-e Coating						Triple Pane Double Pane & Low-e Coating					
	10	15	20	25	30	35	10	15	20	25	30	35	10	15	20	25	30	35
DIRECTION WINDOW FACES	NO INTERNAL SHADING																	
N	23	27	31	35	39	43	19	21	23	25	27	29	17	18	19	20	21	22
NE and NW	56	60	64	68	72	76	47	49	51	53	55	57	43	44	45	46	47	48
E and W	81	85	89	93	97	101	68	70	72	74	76	78	62	63	64	65	66	67
SE and SW	70	74	78	82	86	90	59	61	63	65	67	69	53	54	55	56	57	58
S	40	44	48	52	56	60	34	36	38	40	42	44	30	31	32	33	34	35
	DRAPERIES OR VENETIAN BLINDS																	
N	14	18	22	26	30	34	12	14	16	18	20	22	10	11	12	13	14	15
NE and NW	33	37	41	45	49	53	29	31	33	35	37	39	25	26	27	28	29	30
E and W	48	52	56	60	64	68	42	44	46	48	50	52	37	38	39	40	41	42
SE and SW	41	45	49	53	57	61	37	39	41	43	45	47	32	33	34	35	36	37
S	24	28	32	36	40	44	21	23	25	27	29	31	18	19	20	21	22	23
	ROLLER SHADES — HALF DRAWN																	
N	17	21	25	29	33	37	16	18	20	22	24	26	14	15	16	17	18	19
NE and NW	41	45	49	53	57	61	38	40	42	44	46	48	34	35	36	37	38	39
E and W	60	64	68	72	76	80	55	57	59	61	63	65	49	50	51	52	53	54
SE and SW	52	56	60	64	68	72	47	49	51	53	55	57	42	43	44	45	46	47
S	30	34	38	42	46	50	27	29	31	33	35	37	24	25	26	27	28	29
	AWNING, PORCHES, OR OTHER EXTERNAL SHADING																	
ALL DIRECTIONS	23	27	31	35	39	43	19	21	23	25	27	29	17	18	19	20	21	22

Tinted (Heat Absorbing) Glass

Design Temperature Difference	Single Pane						Double Pane Single Pane & Low-e Coating						Triple Pane Double Pane & Low-e Coating					
	10	15	20	25	30	35	10	15	20	25	30	35	10	15	20	25	30	35
DIRECTION WINDOW FACES	NO INTERNAL SHADING																	
N	16	20	24	28	32	36	12	14	16	18	20	22	9	10	11	12	13	14
NE and NW	39	43	47	51	55	59	29	31	33	35	37	39	22	23	24	25	26	27
E and W	57	61	65	69	73	77	42	44	46	48	50	52	32	33	34	35	36	37
SE and SW	49	53	57	61	65	69	36	38	40	42	44	46	28	29	30	31	32	33
S	28	32	36	40	44	48	21	23	25	27	29	31	16	17	18	19	20	21
	DRAPERIES OR VENETIAN BLINDS																	
N	12	16	20	24	28	32	9	11	13	15	17	19	6	7	8	9	10	11
NE and NW	30	34	38	42	46	50	22	24	26	28	30	32	15	16	17	18	19	20
E and W	43	47	51	55	59	63	31	33	35	37	39	41	21	22	23	24	25	26
SE and SW	37	41	45	49	53	57	27	29	31	33	35	37	18	19	20	21	22	23
S	21	25	29	33	37	41	15	17	19	21	23	25	10	11	12	13	14	15
	ROLLER SHADES — HALF DRAWN																	
N	14	18	22	26	30	34	10	12	14	16	18	20	7	8	9	10	11	12
NE and NW	34	38	42	46	50	54	25	27	29	31	33	35	18	19	20	21	22	23
E and W	49	53	57	61	65	69	36	38	40	42	44	46	26	27	28	29	30	31
SE and SW	42	46	50	54	58	62	31	33	35	37	39	41	22	23	24	25	26	27
S	24	28	32	36	40	44	18	20	22	24	26	28	13	14	15	16	17	18
	AWNING, PORCHES, OR OTHER EXTERNAL SHADING																	
ALL DIRECTIONS	16	20	24	28	32	36	12	14	16	18	20	22	9	10	11	12	13	14

Reflective Coated Glass

Design Temperature Difference	Single Pane						Double Pane Single Pane & Low-e Coating						Triple Pane Double Pane & Low-e Coating					
	10	15	20	25	30	35	10	15	20	25	30	35	10	15	20	25	30	35
DIRECTION WINDOW FACES	NO INTERNAL SHADING																	
N	14	18	22	26	30	34	10	12	14	16	18	20	6	7	8	9	10	11
NE and NW	34	38	42	46	50	54	24	26	28	30	32	34	15	16	17	18	19	20
E and W	49	53	57	61	65	69	34	36	38	40	42	44	21	22	23	24	25	26
SE and SW	43	47	51	55	.59	63	29	31	33	35	37	39	18	19	20	21	22	23
S	24	28	32	36	40	44	17	19	21	23	25	27	10	11	12	13	14	15
	DRAPERIES OR VENETIAN BLINDS																	
N	11	15	19	23	27	31	8	10	12	14	16	18	5	6	7	8	9	10
NE and NW	28	32	36	40	44	48	20	22	24	26	28	30	12	13	14	15	16	17
E and W	40	44	48	52	56	60	30	32	34	36	38	40	17	18	19	20	21	22
SE and SW	35	39	43	47	51	55	26	28	30	32	34	36	15	16	17	18	19	20
S	20	24	28	32	36	40	15	17	19	21	23	25	9	10	11	12	13	14
	ROLLER SHADES — HALF DRAWN																	
N	12	16	20	24	28	32	9	11	13	15	17	19	5	6	7	8	9	10
NE and NW	30	34	38	42	46	50	21	23	25	27	29	31	13	14	15	16	17	18
E and W	44	48	52	56	60	64	31	33	35	37	39	41	18	19	20	21	22	23
SE and SW	38	42	46	50	54	58	27	29	31	33	35	37	16	17	18	19	20	21
S	22	26	30	34	38	42	15	17	19	21	23	25	9	10	11	12	13	14
	AWNING, PORCHES, OR OTHER EXTERNAL SHADING																	
ALL DIRECTIONS	14	18	22	26	30	34	10	12	14	16	18	20	6	7	8	9	10	11

407F12.EPS

Figure 12 ◆ Manual J load factors.

Window Glass Load Factors (GLF) for Single-Family Detached Residences[a]

Design Temperature, °F	Regular Single Glass						Regular Double Glass						Heat-Absorbing Double Glass						Clear Triple Glass		
	85	90	95	100	105	110	85	90	95	100	105	110	85	90	95	100	105	110	85	90	95
No inside shading																					
North	34	36	41	47	48	50	30	30	34	37	38	41	20	20	23	25	26	28	27	27	30
NE and NW	63	65	70	75	77	83	55	56	59	62	63	66	36	37	39	42	44	44	50	50	53
E and W	88	90	95	100	102	107	77	78	81	84	85	88	51	51	54	56	59	59	70	70	73
SE and SW[b]	79	81	86	91	92	98	69	70	73	76	77	80	45	46	49	51	54	54	62	63	65
South[b]	53	55	60	65	67	72	46	47	50	53	54	57	31	31	34	36	39	39	42	42	45
Horizontal skylight	156	156	161	166	167	171	137	138	140	143	144	147	90	91	93	95	96	98	124	125	127
Draperies, venetian blinds, translucent roller shades fully drawn																					
North	18	19	23	27	29	33	16	16	19	22	23	26	13	14	16	18	19	21	15	16	18
NE and NW	32	33	38	42	43	47	29	30	32	35	36	39	24	24	27	29	29	32	28	28	30
E and W	45	46	50	54	55	59	40	41	44	46	47	50	33	33	36	38	38	41	39	39	41
SE and SW[b]	40	41	46	49	51	55	36	37	39	42	43	46	29	30	32	34	35	37	35	36	38
South[b]	27	28	33	37	38	42	24	25	28	31	31	34	20	21	23	25	26	28	23	24	26
Horizontal skylight	78	79	83	86	87	90	71	71	74	76	77	79	58	59	61	63	63	65	69	69	71
Opaque roller shades, fully drawn																					
North	14	15	20	23	25	29	13	14	17	19	20	23	12	12	15	17	17	20	13	13	15
NE and NW	25	26	31	34	36	40	23	24	27	30	30	33	21	22	24	26	27	29	23	23	26
E and W	34	36	40	44	45	49	32	33	36	38	39	42	29	30	32	34	35	37	32	32	35
SE and SW[b]	31	32	36	40	42	46	29	30	33	35	36	39	26	27	29	31	32	34	29	29	31
South[b]	21	22	27	30	32	36	20	20	23	26	27	30	18	19	21	23	24	26	19	20	22
Horizontal skylight	60	61	64	68	69	72	57	57	60	62	63	65	52	52	55	57	57	59	56	57	59

[a]Glass load factors (GLFs) for single-family detached houses, duplexes, or multi-family, with both east and west exposed walls or only north and south exposed walls, Btuh·ft².

[b]Correct by +30% for latitude of 48° and by −30% for latitude of 32°. Use linear interpolation for latitude from 40 to 48° and from 40 to 32°.

To obtain GLF for other combinations of glass and/or inside shading: $GLF_a = (SC_a/SC_t)(GLF_t - U_t D_t) + U_a D_t$, where the subscripts a and t refer to the alternate and table values, respectively. SC_t and U_t are given in Table 5. $D_t = (t_a - 75)$, where $t_a = t_o - (DR/2)$; t_o is the outdoor design temperature and DR is the daily range.

407F13.EPS

Figure 13 ◆ ASHRAE load factors.

These are some of the tradeoffs that go into system design. If this were a new construction project, the owner could be given the option of installing energy-efficient windows. The initial cost would be much higher, but the savings in the size of the cooling system, combined with lower operating costs, would pay back the difference in a few years. These tradeoffs are an important aspect of the design and selling process. You can also see the effect that window coverings, such as draperies or blinds, can have. If a project is sized for windows with full, insulated draperies and the homeowner later switches to light, open-style curtains, it can affect the ability of the system to meet the cooling or heating demand on design temperature days, especially if there is a large amount of east- and west-exposure glass.

If windows are fully or partially shaded by roof overhangs, awnings, or shade screens, that fact must be taken into account when calculating the cooling load. A fully shaded east-, west-, or south-exposure window is treated like a north-exposure window. If a window is partially shaded, the percent of shading must be factored in. Manual J provides a special calculation in which the shaded and unshaded portions are treated as separate entities, and then the results are added to determine the Btuh load for the entire window. In other methods, the amount of overhang or interior shading may be factored into the load multiplier table (see *Figure 14*). French doors, sliding glass doors, and other construction features that are primarily glass are treated as windows.

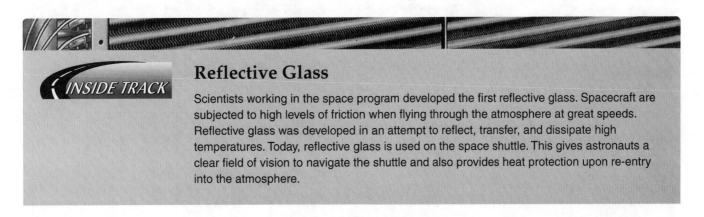

INSIDE TRACK

Reflective Glass

Scientists working in the space program developed the first reflective glass. Spacecraft are subjected to high levels of friction when flying through the atmosphere at great speeds. Reflective glass was developed in an attempt to reflect, transfer, and dissipate high temperatures. Today, reflective glass is used on the space shuttle. This gives astronauts a clear field of vision to navigate the shuttle and also provides heat protection upon re-entry into the atmosphere.

CLEAR - DOUBLE GLASS

EXPOSURE	OVERHANG (INCHES)														
	0			12			24			36			48		
	INTERNAL SHADING			INTERNAL SHADING			INTERNAL SHADING			INTERNAL SHADING			INTERNAL SHADING		
N	23	16	20	NO SIGNIFICANT EFFECT BY OVERHANG ON THESE EXPOSURES											
NE/NW	51	33	42												37
E/W SE/SW -	72	46	59	55	36	44	40	26	33	25	18	22	23	16	20
30°N. LAT	63	41	51	58	38	47	47	31	38	36	24	30	25	17	21
SE/SW - 40°N. LAT	63	41	51	60	39	49	51	34	42	42	28	35	33	22	28
SE/SW - 50°N. LAT	63	41	51	23	16	20	23	16	20	23	16	20	23	16	20
S - 30°N. LAT	38	25	31	32	21	26	23	16	20	23	16	20	23	16	20
S - 40°N. LAT	38	25	31	35	23	28	29	20	24	23	16	20	23	16	20
S - 50°N. LAT	38	25	31												

407F14.EPS

Figure 14 ◆ Glass load table with overhangs and shading factored in.

4.3.2 Walls, Roofs, Ceilings, and Floors

Any exposed or partially exposed surface must be figured into the load calculation because these surfaces gain and lose heat as described earlier. The amount of load a surface creates depends on the construction material used and is calculated in a similar way to that of window glass. In Manual J, tables are provided showing the heat transfer multiplier for each type of construction material. *Figure 15* shows one page of a typical load table for cooling.

A separate table containing similar data is also provided for determining heating load factors. When determining the load represented by exposed walls, the square footage for applicable windows and doors must be subtracted from the wall area. Otherwise, those areas will be counted twice.

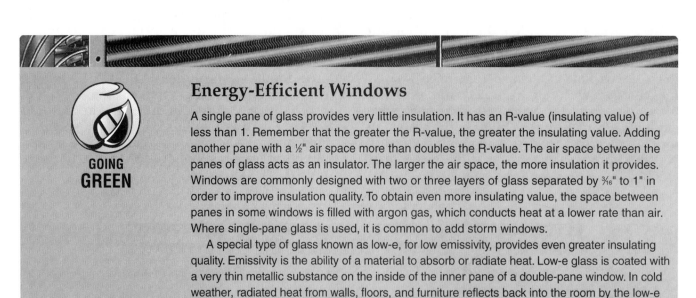

GOING GREEN

Energy-Efficient Windows

A single pane of glass provides very little insulation. It has an R-value (insulating value) of less than 1. Remember that the greater the R-value, the greater the insulating value. Adding another pane with a ½" air space more than doubles the R-value. The air space between the panes of glass acts as an insulator. The larger the air space, the more insulation it provides. Windows are commonly designed with two or three layers of glass separated by ³⁄₁₆" to 1" in order to improve insulation quality. To obtain even more insulating value, the space between panes in some windows is filled with argon gas, which conducts heat at a lower rate than air. Where single-pane glass is used, it is common to add storm windows.

A special type of glass known as low-e, for low emissivity, provides even greater insulating quality. Emissivity is the ability of a material to absorb or radiate heat. Low-e glass is coated with a very thin metallic substance on the inside of the inner pane of a double-pane window. In cold weather, radiated heat from walls, floors, and furniture reflects back into the room by the low-e coating instead of escaping through the windows. This reduces the heat loss, which in turn saves heating costs. In summer months, radiated heat from outdoor sources, such as the sun, roads, and parking lots, is reflected away from the building by the low-e coating. Although windows with low-e glass are considerably more expensive than standard windows, they usually pay for themselves in reduced heating and cooling costs within three or four years.

No. 13 - Partitions Between Conditioned and Unconditioned Space - Wood Frame Partitions	Summer Temperature Difference and Daily Temperature Range												
	10		15			20			25		30	35	U
	L	M	L	M	H	L	M	H	M	H	H	H	
	HTM (Btuh per sq. ft.)												
A. None ½" Gypsum Board (R-0.5)	2.4	1.4	3.8	2.7	1.4	5.1	4.1	2.7	5.4	4.1	5.4	6.8	.271
B. None ½" Asphalt Board (R-1.3)	2.0	1.1	3.0	2.2	1.1	4.1	3.3	2.2	4.3	.3.3	4.3	5.4	.217
C. R-11 ½" Gypsum Board (R-0.5)	.8	.4	1.3	.9	.4	1.7	1.3	.9	1.8	1.3	1.8	2.2	.090
D. R-11 ½" Asphalt Board (R-1.3) R-11 ½" Bead Brd. (R-1.8) R-13 ½" Gypsum Brd. (R-0.5)	.7	.4	1.1	.8	.4	1.5	1.2	.8	1.6	1.2	1.6	2.0	.080
E. R-11 ½" Extr Poly Brd. (R-2.5) R-11 ¾" Bead Brd. (R-2.7) R-13 ½" Asphalt Brd. (R-1.3) R-13 ½" Bead Brd. (R-1.8)	.7	.4	1.0	.8	.4	1.4	1.1	.8	1.5	1.1	1.5	1.9	.075
F. R-11 1" Bead Brd. (R-3.6) R-11 ¾" Extr Poly Brd. (R-3.8) R-13 ½" Extr Poly Brd. (R-2.5) R-13 ¾" Bead Brd. (R-2.7)	.6	.4	1.0	.7	.4	1.3	1.0	.7	1.4	1.0	1.4	1.8	.070
G. R-13 ¾" Extr Poly Brd. (R-3.8) R-13 1" Bead Brd (R-3.6)	.6	.3	.9	.6	.3	1.2	1.0	.6	1.3	1.0	1.3	1.6	.065
H. R-11 1" Extr Brd. (R-5.0) R-13 1" Extr Poly Brd. (R-5.0) R-19 ½" Gypsum Brd. (R-0.5)	.5	.3	.8	.6	.3	1.1	.9	.6	1.2	.9	1.2	1.5	.060
I. R-19 ½" Asphalt Brd. (R-1.3) R-19 ½" Bead Brd. (R-1.8)	.5	.3	.8	.5	.3	1.0	.8	.5	1.1	.8	1.1	1.4	.055
J. R-11 R-8 Sheathing R-13 R-8 Sheathing R-19 ½" or ¾" Extr Poly R-19 ¾" or 1" Bead Brd.	.4	.2	.7	.5	.2	.9	.7	.5	1.0	.7	1.0	1.2	.050
K. R-19 1" Extr Poly Brd. (R-5.0)	.4	.2	.6	.4	.2	.9	.7	.4	.9	.7	.9	1.1	.045
L. R-19 R-8 Sheathing	.4	.2	.6	.4	.2	.8	.6	.4	.8	.6	.8	1.0	.040

No. 13 - Partitions Between Conditioned & Unconditioned Space. Brick or Brick Partitions	10		15			20			25		30	35	U
	L	M	L	M	H	L	M	H	M	H	H	H	
	HTM (Btuh per sq. ft.)												
M. 8" Brick, No Insul., Unfinished	1.3	0	3.8	1.8	0	6.4	4.3	1.8	6.9	4.3	6.9	9.4	.510
N. 8" Brick R-5	.4	0	1.1	.5	0	1.8	1.2	.5	1.9	1.2	1.9	2.7	.144
O. 8" Brick R-11	.2	0	.6	.3	0	1.0	.7	.3	1.0	.7	1.0	1.4	.077
P. 8" Brick R-19	.1	0	.4	.2	0	.6	.4	.2	.6	.4	.6	.9	.048
Q. 4" Brick 8" Block, No Insul.	1.0	0	3.0	1.4	0	5.0	3.4	1.4	5.4	3.4	5.4	7.4	.400
R. 4" Brick 8" Block R-5	.3	0	1.0	.5	0	1.7	1.1	.5	1.8	1.1	1.8	2.5	.133
S. 4" Brick 8" Block R-11	.2	0	.6	.3	0	.9	.6	.3	1.0	.6	1.0	1.4	.074
T. 4" Brick 8" Block R-19	.1	0	.4	.2	0	.6	.4	.2	.6	.4	.6	.9	.047

No. 14 - Masonry Walls, Block or Brick Finished or Unfinished - Above Grade	10		15			20			25		30	35	U
	L	M	L	M	H	L	M	H	M	H	H	H	
	HTM (Btuh per sq. ft.)												
A. 8" or 12" Block, No Insul., Unfinished	5.3	3.2	7.8	5.8	3.2	10.4	8.3	5.8	10.9	8.3	10.9	13.4	.510
B. 8" or 12" Block + R-5	1.5	.9	2.2	1.6	.9	2.9	2.3	1.6	3.1	2.3	3.1	3.8	.144
C. 8" or 12" Block + R-11	.8	.5	1.2	.9	.5	1.6	1.3	.9	1.6	1.3	1.6	2.0	.077
D. 8" or 12" Block + R-19	.5	.3	.7	.5	.3	1.0	.8	.5	1.0	.8	1.0	1.3	.048
E. 4" Brick + 8" Block, No Insul.	4.1	2.5	6.1	4.5	2.5	8.1	6.5	4.5	8.5	6.5	8.5	10.5	.400
F. 4" Brick + 8" Block + R-5	1.4	.8	2.0	1.5	.8	2.7	2.2	1.5	2.8	2.2	2.8	3.5	.133
G. 4" Brick + 8" Block + R-11	.8	.5	1.1	.8	.5	1.5	1.2	.8	1.6	1.2	1.6	1.9	.074
H. 4" Brick + 8" Block + R-19	.5	.3	.7	.5	.3	1.0	.8	.5	1.0	.8	1.0	1.2	.047

407F15.EPS

Figure 15 ◆ Example of Manual J load estimating cooling table.

Energy Performance Ratings

The National Fenestration Rating Council (NFRC) is a nonprofit organization created by the window, door, and skylight industry that includes manufacturers, architects, code officials, government agencies, and others as members. The primary goal of the organization is to provide accurate information to measure and compare the energy performance of windows, doors, and skylights. NFRC has established a voluntary energy performance rating and labeling system for fenestration (window and door) products.

The NFRC label contains vital information such as U-factor, solar heat gain coefficient, visible transmittance, and air leakage. Individual window and door manufacturers also provide performance data for their products. This information is available from the manufacturer in printed form or is available over the Internet.

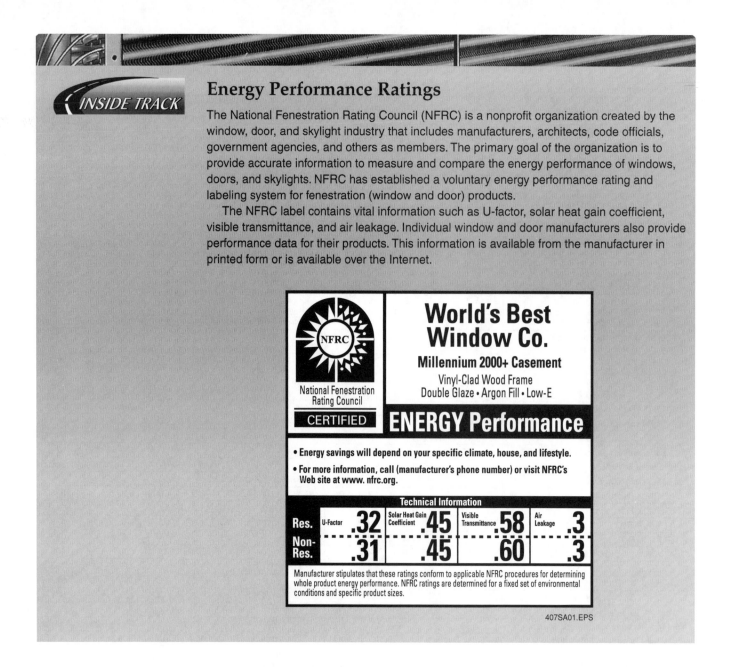

407SA01.EPS

As you can see in *Figure 15*, insulation (**R-value**) usually plays a significant role in the amount of load presented by exposed surfaces. Money invested in insulation will generally pay for itself over time. Other tables are provided so that the estimator can determine the insulation values of various construction materials.

When calculating heating loads, special consideration must be given to walls, floors, and ceilings that adjoin unconditioned spaces such as garages, crawl spaces, and attics. Even with insulation, these surfaces can create as much load as insulated, exposed walls.

Basements also require special treatment when calculating a heating load. The load created by an uninsulated block or brick wall above grade is about equal to that of double-pane glass. Even with insulation, the load is substantial. The portion of the wall that is below grade has significantly less heat loss. Therefore, the portions above and below grade must be treated separately, like partially shaded windows. Basement floors must also be calculated separately. If the first floor is above an unheated basement or crawl space, it can represent a significant heating load, especially if there is no insulation.

Roof area can have a large affect on cooling load. In Manual J and with other load estimating methods, the reflective qualities of the roof are important. As we discussed earlier, light-colored roofing material will reflect radiant heat better than dark material; light-colored material therefore creates less of a load. Again, insulation is an important factor in the amount of load presented by the roof.

4.3.3 Infiltration

Infiltration affects both heating and cooling loads. Air can enter a building through cracks around windows and doors and other construction joints. Fireplaces and undampered vents are another source of infiltration. As previously shown, infiltration contributes more to the heating load than the cooling load, especially if a fossil-fuel furnace is used. Caulking and weatherstripping can make a huge difference in cold climates.

Infiltration must be factored into the load estimate. The amount of infiltration is a function of the building's tightness. Tightness is a function of several factors, as stated in the ACCA manual:

Best – Continuous infiltration barrier, all cracks and penetrations sealed; tested leakage of windows and doors less than 0.25 cfm per running foot of crack; vents and exhaust fans dampered, recessed ceiling lights gasketed or taped; no combustion air required or combustion air from outdoors; no duct leakage.

Average – Plastic vapor barrier, major cracks and penetrations sealed; tested leakage of windows and doors between 0.25 and 0.50 cfm per running foot of crack; electrical fixtures that penetrate the envelope not taped or gasketed; vents and exhaust fans dampered; combustion air from indoors; intermittent ignition and flue damper; some duct leakage to unconditioned space.

Poor – No infiltration barrier or plastic vapor barrier; no attempt to seal cracks and penetrations; tested leakage of windows and doors greater than 0.50 cfm per running foot of crack; vents and exhaust fans not dampered; combustion air from indoors; standing pilot; no flue damper; considerable duct leakage to unconditioned space.

Fireplace evaluation:

Best – Combustion air from outdoors, tight glass doors and damper.
Average – Combustion air from indoors, tight glass doors or damper.
Poor – Combustion air from indoors, no glass doors or damper.

Figure 16 shows how the effect of infiltration is determined. Infiltration is apportioned to the load estimate based on the square footage of windows and doors in each space. *Appendix A* contains a sample of the Manual J load estimating form for heating loads. Item 12 on the form is the infiltration factor (HTM), which is determined by the calculation procedure shown in *Figure 16*. Infiltration is a function of the cfm of air entering the building. This factor must be converted to a multiplier that can be used with loads that are stated in Btuh.

4.3.4 Duct Losses

Duct losses affect both heating and cooling loads. However, it has a greater effect on heating loads. The United States Department of Energy states that typical residential duct systems lose 25 to 40 percent of the heating or cooling energy fed into them by a furnace or air conditioner. Adding insulation and sealing of duct leaks can significantly reduce those losses. *Figure 17* shows how the use of insulation reduces duct losses.

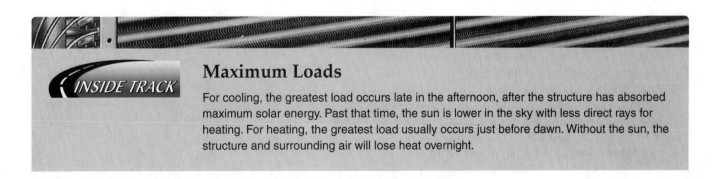

INSIDE TRACK

Maximum Loads

For cooling, the greatest load occurs late in the afternoon, after the structure has absorbed maximum solar energy. Past that time, the sun is lower in the sky with less direct rays for heating. For heating, the greatest load usually occurs just before dawn. Without the sun, the structure and surrounding air will lose heat overnight.

Infiltration Evaluation

Winter Air Changes Per Hour

Floor Area	900 or less	900-1500	1500-2100	over 2100
Best	0.4	0.4	0.3	0.3
Average	1.2	1.0	0.8	0.7
Poor	2.2	1.6	1.2	1.0
For each fireplace add:		Best 0.1	Average 0.2	Poor 0.6

Summer Air Changes Per Hour

Floor Area	900 or less	900-1500	1500-2100	over 2100
Best	0.2	0.2	0.2	0.2
Average	0.5	0.5	0.4	0.4
Poor	0.8	0.7	0.6	0.5

Procedure A · Winter Infiltration HTM Calculation

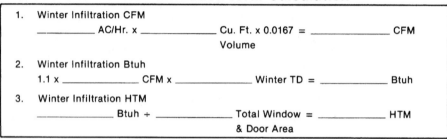

1. Winter Infiltration CFM
   _____ AC/Hr. x _____ Cu. Ft. x 0.0167 = _____ CFM
   Volume

2. Winter Infiltration Btuh
   1.1 x _____ CFM x _____ Winter TD = _____ Btuh

3. Winter Infiltration HTM
   _____ Btuh ÷ _____ Total Window = _____ HTM
   & Door Area

Procedure B · Summer Infiltration HTM Calculation

1. Summer Infiltration CFM
   _____ AC/Hr. x _____ Cu. Ft. x 0.0167 = _____ CFM
   Volume

2. Summer Infiltration Btuh
   1.1 x _____ CFM x _____ Summer TD = _____ Btuh

3. Summer Infiltration HTM
   _____ Btuh ÷ _____ Total Window = _____ HTM
   & Door Area

407F16.EPS

Figure 16 ◆ Determining infiltration HTM.

4.3.5 Cooling Load Factors

Refer to *Table 1*. You can see there are a few load factors that affect only cooling. You have already learned about shaded glass and roofs. Temperature swing and internal factors also contribute to the cooling load.

The temperature swing multiplier (*Figure 18*), referred to as RSM in Manual J, represents the amount of deviation that is allowed from the thermostat setpoint over the course of a summer design day. The swing multiplier relates the load to the actual equipment capacity. The larger the multiplier, the smaller the capacity of the cooling equipment, and the more discomfort occupants might experience on the occasional design temperature day. It's a tradeoff between comfort and cost.

People and appliances (internal factors) also affect the cooling load. People give off body heat, so the more occupants in the building, the greater the cooling load. If the occupants entertain large groups, that must also be factored into the load estimate. In the absence of the specific number of occupants, a rule of thumb is two people per bedroom.

Duct Loss Multipliers

Case I - Supply Air Temperatures Below 120°F	Duct Loss Multipliers	
Duct Location and Insulation Value	Winter Design Below 15°F	Winter Design Above 15°F
Exposed to Outdoor Ambient		
Attic, Garage, Exterior Wall, Open Crawl Space - None	0.30	0.25
Attic, Garage, Exterior Wall, Open Crawl Space - R2	0.20	0.15
Attic, Garage, Exterior Wall, Open Crawl Space - R4	0.15	0.10
Attic, Garage, Exterior Wall, Open Crawl Space - R6	0.10	0.05
Enclosed In Unheated Space		
Vented or Unvented Crawl Space or Basement - None	0.20	0.15
Vented or Unvented Crawl Space or Basement - R2	0.15	0.10
Vented or Unvented Crawl Space or Basement - R4	0.10	0.05
Vented or Unvented Crawl Space or Basement - R6	0.05	0.00
Duct Buried In or Under Concrete Slab		
No Edge Insulation	0.25	0.20
Edge Insulation R Value = 3 to 4	0.15	0.10
Edge Insulation R Value = 5 to 7	0.10	0.05
Edge Insulation R Value = 7 to 9	0.05	0.00
Case II - Supply Air Temperatures Above 120°F		
Duct Location and Insulation Value	Winter Design Below 15°F	Winter Design Above 15°F
Exposed to Outdoor Ambient		
Attic, Garage, Exterior Wall, Open Crawl Space - None	0.35	0.30
Attic, Garage, Exterior Wall, Open Crawl Space - R2	0.25	0.20
Attic, Garage, Exterior Wall, Open Crawl Space - R4	0.20	0.15
Attic, Garage, Exterior Wall, Open Crawl Space - R6	0.15	0.10
Enclosed In Unheated Space		
Vented or Unvented Crawl Space or Basement - None	0.25	0.20
Vented or Unvented Crawl Space or Basement - R2	0.20	0.15
Vented or Unvented Crawl Space or Basement - R4	0.15	0.10
Vented or Unvented Crawl Space or Basement - R6	0.10	0.05
Duct Buried In or Under Concrete Slab		
No Edge Insulation	0.30	0.25
Edge Insulation R Value = 3 to 4	0.20	0.15
Edge Insulation R Value = 5 to 7	0.15	0.10
Edge Insulation R Value = 7 to 9	0.10	0.05

407F17.EPS

Figure 17 ◆ Manual J duct loss factors.

RATING AND TEMPERATURE SWING MULTIPLIER (RSM)

METHOD USED TO SELECT EQUIPMENT	SUMMER DESIGN	TEMP. 4.5	SWING 3.0
Selection made at the actual Summer Design condition using manufacturer's performance data		0.90	1.00
Selection made at the ARI Standard Rating Design Condition	85–90	0.85	0.95
	95	0.90	1.00
	100	0.95	1.05
	105	1.00	1.10
	110	1.05	1.15

407F18.EPS

Figure 18 ◆ Temperature swing multipliers.

Manual J uses 300 Btuh per person to represent the sensible load. In a room-by-room load estimate, the people are apportioned to the rooms they would normally occupy during the peak load hours, generally the living room, family room, or dining room/kitchen.

Stoves, water heaters, baths, showers, washers, dryers, and lights also give off heat. A rule of thumb for this factor is 1,200 Btuh for the total appliance load. However, if the washer and dryer are located in the conditioned space, an additional 500 Btuh should be included.

Part of the load created by people, appliances, and infiltration is latent heat (moisture). The load factors discussed so far represent sensible heat, and the load estimate itself deals with the sensible heat load only. The latent heat load adds to the relative humidity and must be accounted for in selecting the equipment. This will be discussed in the next section.

4.4.0 Preparing the Load Estimate

The initial phase of the load estimating process yields two kinds of information. One is the size of the building and its parts organized into convenient groups, such as windows, walls, and floors. This information is further divided into subgroups that make it easier to perform the load calculations. The second type of information is the multipliers (HTMs) that represent heat gains and losses that this particular building is likely to experience based on its construction and location. To obtain the total sensible load for the building, the window, wall, door, floor, and ceiling areas for each room are added together and then multiplied by their respective HTMs. Examples can be seen in the load estimating forms provided in *Appendix A* (heating load) and *Appendix B* (cooling load).

NOTE

Although the heating and cooling load calculations are shown on separate forms in the *Appendixes*, they can be done on the same form if the building is being sized for both heating and cooling.

4.5.0 Load Estimating Software

Calculating a heat loss or heat gain using a manual, paper-based method can be time consuming and tedious. Various charts and tables have to be referenced constantly and manual calculations have to be made, which increases the possibility of error. In today's business environment, many sales professionals do not want to, or do not have the time to, perform a manual load calculation. This often leads to the reliance on old, ineffective rules of thumb that can lead to grossly oversized equipment.

The availability of personal computers revolutionized all aspects of system design, including load estimating. There are dozens of excellent computer-based load estimating programs from which to choose. All of them offer significant time savings, as well as accurate results. A system designer can take a laptop computer to the job site and enter the information as it is collected. All the background information needed to process the input is inherent in the software. Results are available instantly. There is no longer a valid reason for not performing a load calculation.

Figure 19 shows a typical input screen from a major equipment manufacturer's residential load estimating software. In this example, once the city is selected, the outside design, daily range, and latitude information are automatically available. The system designer only has to choose inside design and inside swing factors.

Across the top of the screen are other areas that require input. Once all the information is entered, the summary printout (*Figure 20*) provides the completed load estimate for heating and cooling. When printed, the completed load estimate carries great legitimacy when presented to a customer.

The various computerized load estimating programs may contain unique features such as equipment selection, operating cost comparisons, and airflow requirements, which enhance the programs and make them more user-friendly. Many residential load estimating programs are based on the information in ACCA Manual J.

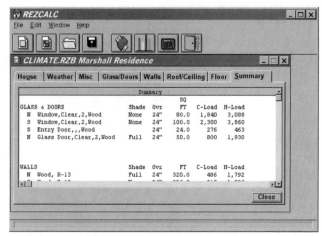

407F19.EPS

Figure 19 ◆ Example of a data screen from a load estimating software program.

REZCALC
RESIDENTIAL BLOCK LOAD ESTIMATE

Name   Marshall Residence
Address   4476 Oak Place
Desc.   3 BR Ranch

DESIGN CONDITIONS	Outdoor:	Summer DB/WB	Winter	Inside:	Summer	RH	Winter	Range	Swing
		95 / 67	29		76	55%	70	34	3.0F

GLASS & DOORS

Dir	Type	Glass	Panes	Frame	Shade	Ovr.	C-F	H-F	SQ FT	C-Load	H-Load
N	Window	Clear	2	Wood	None	24"	23.00	38.60	80.0	1,840	3,088
S	Window	Clear	2	Wood	None	24"	23.00	38.60	100.0	2,300	3,860
S	Entry Door			Wood		24"	11.50	19.30	24.0	276	463
N	Glass Door	Clear	2	Wood	Full	24"	16.00	38.60	50.0	800	1,930

WALLS

Dir	Frame			R	Shade	Ovr.					
N	Wood			13	Full	24"	1.52	5.60	320.0	486	1,791
S	Wood			13	None	24"	1.58	5.60	326.0	515	1,825
E	Wood			13	Full	24"	1.52	5.60	270.0	410	1,511
W	Wood			13	None	24"	1.80	5.60	270.0	486	1,511

ROOF/CEILING

Description			R						
Ceiling under vented attic			30	1.50	2.31	1,500.0	2,250	3,465

FLOOR

Description	Cover	Ceiling Below	R					
Slab without duct			0"	0.00	60.00	160.0	0	9,600

COOLING LOAD SUMMARY

		COOLING BTUH
SUBTOTAL (All C-LOADS)		9,363
x SWING MULT.	1.00	
x DUCT MULT.	1.05	
x SUMMER CLIMATE MULT	0.91	
= SUBTOTAL		8,946
+ # PEOPLE x 530	3	1,590
+ APPLIANCES		1,200
= SUBTOTAL (Less Outside Air Load)		11,736
+ OUTSIDE AIR LOAD infil.&vent. cfm x load (cfm) 142 x 22		3,124
= TOTAL COOLING LOAD		**14,860**
- # PEOPLE x 230	3	690
- OUTSIDE AIR LATENT LOAD infil.&vent. cfm x latent load (cfm) 142 x 0		0
= LATENT COOLING LOAD		**690**
= SENSIBLE COOLING LOAD		**14,170**

HEATING LOAD SUMMARY

		HEATING BTUH
SUBTOTAL (All H-LOADS)		29,044
+ INFIL. & VENT. CFM x 77	195	15,015
= SUBTOTAL		44,059
x DUCT MULT.	1.05	
x WINTER CLIMATE MULT.	0.58	
= TOTAL HEATING LOAD		**26,832**

AIR QUANTITY (CFM)

COOLING OR HP CFM sensible cool. load / cooling CFM factor	24.30	583
FURNACE CFM total heating load / heating CFM factor	38.00	706
COOLING SENSIBLE HEAT FACTOR sensible cool. load / total cooling load		0.9536

Figure 20 ◆ Example of a printout from a load estimating software program.

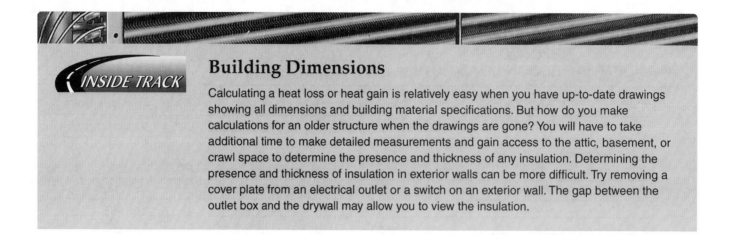

Building Dimensions

Calculating a heat loss or heat gain is relatively easy when you have up-to-date drawings showing all dimensions and building material specifications. But how do you make calculations for an older structure when the drawings are gone? You will have to take additional time to make detailed measurements and gain access to the attic, basement, or crawl space to determine the presence and thickness of any insulation. Determining the presence and thickness of insulation in exterior walls can be more difficult. Try removing a cover plate from an electrical outlet or a switch on an exterior wall. The gap between the outlet box and the drywall may allow you to view the insulation.

As with any load estimating method, whether manual or computer-based, always make sure the method selected complies with local codes and/or utility requirements.

5.0.0 ◆ EQUIPMENT SELECTION

Four major items of information are necessary to select the equipment for a particular application:

- The type of system
- The sensible heating and/or cooling load
- The latent heat load (for cooling)
- The cooling and/or heating airflow needed (cfm)

Detailed load estimates are performed to identify the equipment that comes as close as possible to matching the true load. In general, the selected equipment should range from no more than 5 percent undersized to no more than 15 percent oversized. (Some designers think 20 percent oversized is acceptable.) Undersized equipment will not be able to handle peak loads and will take longer to return the space to the comfort level. Oversized equipment will have higher first cost and operating costs and will cycle more often. Oversized cooling equipment will pull down the load more quickly. In doing so, it may not run long enough to reduce the humidity, especially on design or nearly design days. Thus, the comfort level will not be satisfactory. Oversized heating equipment will not be efficient and will not provide good comfort. Oversizing a standard efficiency induced-draft furnace may allow condensation to form in the vent or heat exchanger, which can cause corrosion and failure.

Residential cooling and heat pump equipment is generally sized in ½-ton increments in smaller sizes and 1-ton increments in larger sizes. For example, most manufacturers provide sizes of 1½, 2, 2½, 3, 3½, 4, and 5 tons. With this selection,

first determine the heat gain, and then select the size that best matches it. If a 32,000 Btuh heat gain were calculated, a 3-ton (36,000 Btuh) unit would be an appropriate match. Furnace size increments may vary by as much as 20,000 Btuh in output. For example, a manufacturer might offer 40,000, 60,000, 80,000, and 100,000 Btuh sizes. After determining the heat loss, select an output size that matches or is slightly higher than the heat loss.

The airflow requirement must be calculated separately for cooling and heating. In order to make a preliminary selection of equipment, it is necessary to approximate the heating and/or cooling cfm. However, the final determination on fan size and speed is based on the duct design process, which is discussed later in this module. Cooling cfm is generally higher than heating cfm. One method of estimating cfm uses the following formula:

Cooling:

$$cfm = \frac{\text{sensible load (Btuh)}}{1.08 \times (t_1 - t_2)}$$

Heating:

$$cfm = \frac{\text{sensible load (Btuh)}}{1.08 \times (t_2 - t_1)}$$

Where:

t_1 = outdoor temperature
t_2 = indoor temperature

5.1.0 Cooling Equipment Selection

In most parts of the country, split systems (*Figure 21*) are used to satisfy residential cooling requirements. Although packaged units would generally serve this purpose, their major disadvantage is that the ductwork must penetrate the building.

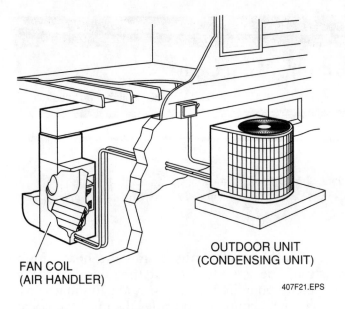

FAN COIL
(AIR HANDLER)

OUTDOOR UNIT
(CONDENSING UNIT)

407F21.EPS

Figure 21 ◆ Split system.

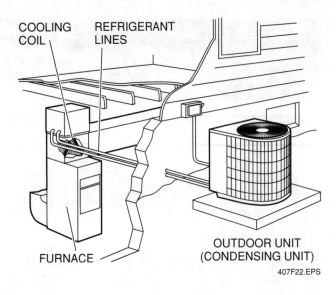

COOLING   REFRIGERANT
COIL   LINES

FURNACE

OUTDOOR UNIT
(CONDENSING UNIT)

407F22.EPS

Figure 22 ◆ Furnace with cooling coil.

In the western and southwestern U.S., they are sometimes rooftop-mounted and used to service high-wall or ceiling outlets in residential and light commercial applications. They may also be used in homes with basements and crawl spaces. With a split system, only the small refrigerant lines penetrate the wall. Openings for the refrigerant lines can be easily drilled in an existing building, while duct openings would be very difficult to make.

The condensing unit is placed outside the building, as close to the evaporator unit as possible, but generally not more than 50' away. It contains the controls, compressor, condensing coil, and fan. The cost of the condensing unit is primarily a function of its operating efficiency. A high-SEER unit may contain a high-efficiency or two-speed compressor and a larger condensing coil. A deluxe model may also contain high-pressure and low-pressure safety switches, sound-deadening shields, quick-start gear, a crankcase heater, a filter-drier, and other special items. The high-efficiency components add to the initial cost, but will pay back the extra expense over time in reduced energy consumption. Other devices provide comfort, convenience, and equipment safety in return for the extra investment.

The indoor unit of a split system contains a cooling coil, blower, and metering device. In cold climates where a furnace is available, a cooling coil can be added to the furnace (*Figure 22*). Circulating air is supplied by the furnace blower. Modern furnaces are specifically designed for combination heating-cooling applications.

The components of a split system must be closely matched. They should be made by the same company and designed to work together. In fact, when piston-type metering devices are used, the correct piston may be shipped with the condensing unit. The installer then installs the correct piston in the indoor section. Selection of the outdoor unit is made from manufacturer's product data based on capacity in Btuh. The indoor unit is selected from a compatible group based on its capacity, ability to meet the cfm requirement, and efficiency rating. Cooling equipment must be selected to match both the sensible and latent heat loads.

In areas with low humidity and high dry-bulb temperatures, such as in the Southwest, it is not uncommon to see an indoor unit with a higher capacity than the outdoor unit. This approach, which is sometimes called mix-matching, trades off higher efficiency against less humidity control. It can be a problem during periods when the moisture level of the air is high. Typically, the indoor unit capacity will be about a half-ton (6,000 Btuh) greater than that of the condensing unit. In desert conditions, the evaporator may have a ton more capacity than the condensing unit.

5.2.0 Heating Equipment Selection

As you have seen, there are many ways to deliver heat. There are gas, electric, and oil furnaces, along with heat pumps. In addition, there are packaged cooling units that have built-in gas furnaces or electric heating elements.

The selection of heating equipment depends on the local climate, as well as prevailing energy conditions. If a furnace is needed, and natural gas is available, it is generally a good choice. In some locations, especially in rural areas, fuel oil is used

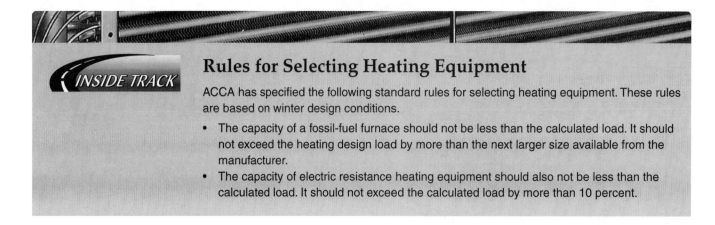

because natural gas is not available. Liquid propane (LP) gas may be an option in such areas, but it also means having a large, exposed tank on the property.

5.3.0 Heat Pump Selection

A heat pump is an excellent choice when electric heat is the only option available. For design heating days, it can be supplemented with auxiliary electric heaters. Because a heat pump provides both heating and cooling with the same equipment, the selection of a heat pump is often a tradeoff process. As you can see from the charts in *Appendix A* and *Appendix B*, the heating and cooling loads for a building are likely to be very different. In the example shown, the sensible heating load is nearly three times the cooling load. In colder climates, the heating load drives the equipment selection process. One of the possibilities in these situations is that there will be too much cooling capacity for the application. This will result in inefficiency and poor dehumidification. To prevent this problem, size the heat pump to the cooling load and make up the difference with auxiliary electric heat. When the heat pump is selected for heating, the cooling capacity should not exceed the cooling load by more than 15 or 20 percent.

Another basic rule is that the heat pump should be sized to provide the lowest possible balance point in the heating mode. In some areas, the power company specifies the minimum heat pump balance point. This approach limits the amount of electric resistance heat that can be used. In some cases, it could force the use of a larger heat pump. The utility may also specify the staging of resistance heaters to improve energy efficiency and reduce instantaneous power drains.

When electric resistance heaters are used, the combined capacity of the heat pump and the electric heaters should not exceed 115 percent of the calculated heating load. Electric resistance heaters are also used to provide emergency heat in the event of a heat pump compressor failure. In such cases, the heaters should be sized to provide 80 percent of the heating load. The auxiliary electric heaters can provide part of the emergency heat in these situations. Emergency heat requirements may be specified by the local utility.

A heat pump can also be combined with a fossil-fueled furnace. In this add-on configuration, the heat pump provides heat above the balance point. Below the balance point, the system switches to oil or gas heat.

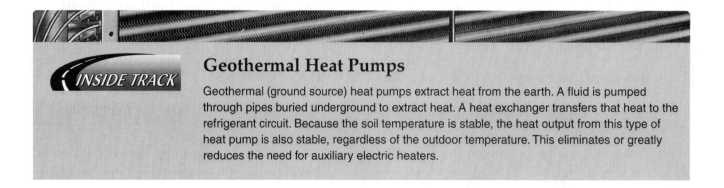

6.0.0 ◆ AIR DISTRIBUTION SYSTEM DUCT DESIGN

Knowing the basics of air distribution system duct design will help you when you are installing an HVAC system. Knowledge of duct design will also help you recognize and solve duct system problems, such as noise, vibration, or incorrect air distribution. The first part of this section reviews some basic air system operating principles. The second part covers factors relating to air system design. The desired design goals for all air distribution systems include the following:

- Supply the right quantity of air to each conditioned space.
- Supply the air in each space so that stratification is minimized and air motion is adequate but not drafty.
- Condition the air to maintain the proper comfort zones or the necessary conditions for a commercial or manufacturing process.
- Provide for the return of air from all conditioned areas to the air handler.
- Operate efficiently without excessive power consumption or noise.
- Operate with minimum maintenance.

6.1.0 Duct System Basics

The resistance to airflow caused by the components of a duct system is overcome by the system fan. The fan supplies the energy needed to overcome the duct resistance and maintain the necessary airflow. Air can be moved by positive pressure (above atmospheric) or negative pressure (below atmospheric). All fans (blowers) produce both conditions. The air inlet to a fan is below atmospheric pressure, while the exhaust of the fan is above atmospheric pressure.

With the exception of the fan, air flows through a duct system naturally from a higher pressure area to a lower pressure area. As shown in *Figure 23*, normal atmospheric pressure exists in the conditioned space at the return **grille** and supply **diffuser**. At the face of the return air grille, the pressure is slightly lower than atmospheric pressure; therefore, air moves into the duct. The pressure decreases to its lowest point at the input to the blower. Through the action of the fan, the air pressure is increased to its highest level at the blower discharge. From there, the air resumes its normal natural flow from the higher pressure area at the fan discharge to the lower pressure area of the diffuser in the conditioned space.

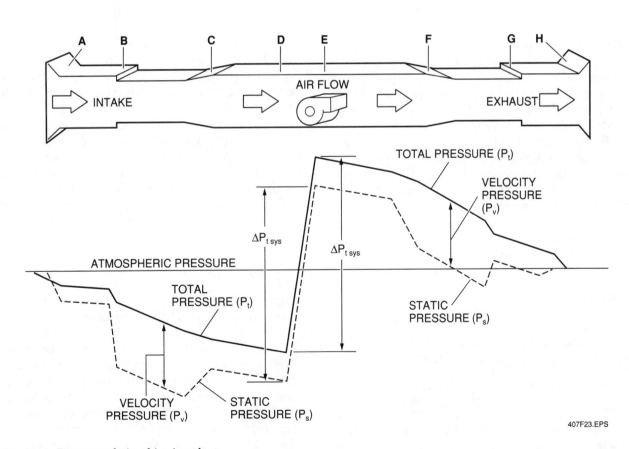

Figure 23 ◆ Pressure relationships in a duct.

The amount of pressure difference needed to move air through a duct system depends on the **velocity** and **volume** of air, the cross-sectional area of the duct, and the length of the duct. Velocity is how fast the air is moving and is usually measured in feet per minute (fpm). Volume is a measure of the amount of air in cubic feet that flows past a point in one minute. Volume in cfm can be calculated by multiplying the air velocity in fpm, by the area it is moving through (in square feet) as follows:

cfm = area × velocity

This formula can be rearranged as follows:

Velocity = cfm ÷ area

Area = cfm ÷ velocity

Cubic feet per minute can also be measured directly with various test instruments.

6.1.1 Pressure Relationships Within a Duct

Three pressures exist in an operating duct system: **total pressure**, **static pressure**, and **velocity pressure**. Total pressure determines how much energy must be supplied to the system by the fan to maintain airflow. Total pressure always decreases in the direction of airflow. For any cross-section of the duct, the total pressure is the sum of the static pressure and the velocity pressure. Static pressure and velocity pressure are present when the duct system is in operation. Static pressure is a stationary air pressure that is exerted uniformly in all directions within the duct. It is the same kind of pressure that is applied equally on the internal walls of a balloon or inflated tire. Velocity pressure is the pressure caused by the velocity and weight of the moving air. It acts in the direction of airflow only. It is the difference between the total pressure and the static pressure. Static and velocity pressures can either increase or decrease in the direction of airflow. The magnitudes of the total, static, and velocity pressures can be calculated as follows:

Total pressure (P_t) = static pressure (P_s) + velocity pressure (P_v)

Static pressure (P_s) = total pressure (P_t) − velocity pressure (P_v)

Velocity pressure (P_v) = total pressure (P_t) − static pressure (P_s)

The levels of the static, velocity, and total pressures in a duct system are very small. Because of this, the scale used to measure them must be numerically large to be accurate. Inches of water column (in. w.c.) or inches of water gauge (in. w.g.) are the units of measure commonly used to express air pressures in a duct system and other systems with very low pressures. The terms in. w.c. and in. w.g. are interchangeable. Inches of water column is the height, in inches, to which the pressure will lift a column of water. The instrument used to measure air pressures in a duct system in inches of water column is the manometer.

6.1.2 Friction Losses

The inside surface of the duct offers resistance to the flow of air. The velocity of airflow within a duct is not uniform. It varies from zero at the duct walls to a maximum at the center of the duct. This variation in velocity is caused by the resistance encountered by the air molecules as they are dragged over the duct surfaces. The resistance to airflow or velocity in straight duct sections is called friction loss. Pressure drop in a straight duct is caused by surface friction. It varies with the air velocity, the duct size and length, and the interior surface roughness. The amount of friction loss in straight duct is normally found by using air friction charts. These charts are reviewed later in this section.

As the cross-sectional area of a straight duct section becomes smaller (*Figure 23*, Sections BC and FG), an increase in the airflow velocity and velocity pressure occurs. As a result, the static and total pressure lines drop more rapidly than in the larger cross-sectional area ducts. This drop occurs because the pressure losses increase as the square of the velocity increases. As the air flows within a constant-area straight duct section, the static pressure and total pressure losses increase at the same rate. This is because the velocity and velocity pressure (P_v) of the air flowing within the duct are constant. *Figure 24* is an example of a chart used to determine friction loss in a duct.

6.1.3 Dynamic Losses

Dynamic losses occur when there are changes in the direction or velocity of the air, such as at transitions, elbows, and other fittings. Dynamic losses also occur at duct obstructions such as dampers. When duct cross-sectional areas are reduced, either abruptly or gradually (*Figure 23*, Points B and F), turbulent airflow occurs. This is because of the sudden change in airflow velocity. Both the velocity and velocity pressure increase in the direction of airflow, while the absolute values for both the total and static pressures decrease.

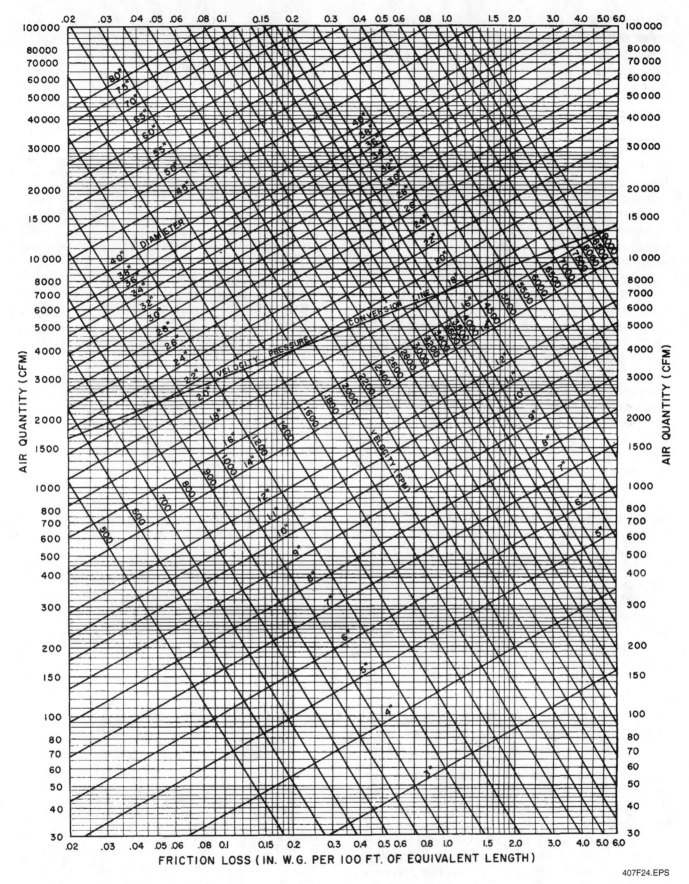

Figure 24 ◆ Friction loss chart for round duct.

407F24.EPS

The result is that a greater loss in total pressure takes place than would occur in a steady flow through an equal length of straight duct with a uniform cross-section. The amount of loss in excess of the straight duct friction at Points B and F is called the dynamic loss.

Dynamic pressure losses can be expressed as a loss coefficient value or C-value. A loss coefficient value is a known, dimensionless value assigned to each type of duct elbow, transition, or other fitting. To determine the pressure loss through a specific kind of duct fitting, the C-value is multiplied by the velocity pressure in the fitting. The dynamic pressure loss of a specific type of fitting can also be expressed by its equivalent length of straight duct value. For example, a radius elbow has the same resistance as 25' of the same size straight duct. C-values are more accurate values of pressure loss and are typically used when designing large commercial duct systems. Equivalent length of straight duct values are commonly used when designing residential or light commercial duct systems. C-value and/or equivalent length of straight duct values for the various kinds of fittings are found using published tables. These tables are available from the Sheet Metal and Air Conditioning Contractors' National Association (SMACNA), ASHRAE, and most duct component manufacturers. *Figure 25* shows examples of the loss coefficient data and equivalent length of straight duct data given in a table for a rectangular elbow.

6.1.4 Static Regain

Abrupt or gradual increases in duct cross-sectional area (*Figure 23*, Points C and G) cause a decrease in airflow velocity and velocity pressure. There is also a decrease in total pressure accompanied by an increase in static pressure. This increase in static pressure is called static regain. It is caused by the conversion of velocity pressure to static pressure.

6.1.5 Duct System External Static Pressure and Supply Fan Relationship

The total pressure that the system fan must supply is the sum of the friction losses in the supply duct system, return duct system, and all the components not included in the fan rating (*Figure 26*). These include all straight duct sections; the dynamic losses of each duct fitting or obstruction, such as **registers** and grilles; and the pressure loss of each duct component, such as coils, filters, and dampers, in the system. The pressure loss or

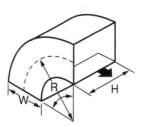

COEFFICIENTS FOR 90°:

R/W	COEFFICIENTS C								
	H/W								
	0.25	0.5	0.75	1.0	1.5	2.0	4.0	6.0	8.0
0.5	1.5	1.4	1.3	1.2	1.0	1.0	1.1	1.2	1.2
0.75	0.57	0.52	0.48	0.44	0.40	0.39	0.40	0.43	0.44
1.0	0.27	0.25	0.23	0.21	0.19	0.18	0.19	0.27	0.21
1.5	0.22	0.20	0.19	0.17	0.15	0.14	0.15	0.17	0.17
2.0	0.20	0.18	0.16	0.15	0.14	0.13	0.14	0.15	0.15

ELBOW, RECTANGLE, SMOOTH RADIUS WITHOUT VANES FITTING LOSS (P_t) = C × V USE THE VELOCITY PRESSURE V OF THE UPSTREAM SECTION

LOSS COEFFICIENT (C-FACTOR)

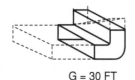

G = 30 FT

EQUIVALENT LENGTH OF DUCT

407F25.EPS

Figure 25 ◆ Examples of loss coefficient and equivalent length of straight duct data.

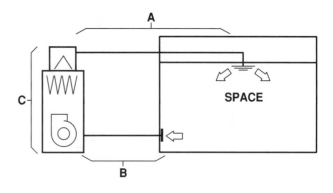

TOTAL FRICTION LOSS = A + B + C
A = SUPPLY SYSTEM LOSS
B = RETURN SYSTEM LOSS
C = COMPONENTS NOT INCLUDED IN FAN RATING

407F26.EPS

Figure 26 ◆ System external static pressure.

drop for a given system component is available from the equipment manufacturer. The chart in *Figure 27* lists pressure drop values for a square ceiling diffuser. The total pressure loss of the duct system components external to the fan assembly is called the **external static pressure**. The size of the system fan is based on the external static pressure losses resulting from the system ductwork and its components. Resistances internal to the fan assembly resulting from any components, such as a filter; losses in the fan itself; and losses in any other components of the assembly, are accounted for by the manufacturer in the design of the fan assembly.

The fan, often called the blower (*Figure 28*), provides the pressure difference that forces the air into the supply duct system, through the grilles and registers, and into the conditioned space. It also must overcome the pressure loss involved in the return of the air as it flows into the return air grilles and through the return ductwork system back to the fan. **Fan brake power** is the actual power required to drive the fan when delivering the required volume of air through a duct system. It is greater than the power to deliver the air because it includes losses due to turbulence and other inefficiencies in the fan, plus bearing losses. The performance of all fans and blowers is governed by three rules commonly known as the Fan Laws. Because cubic feet per minute (cfm), revolutions per minute (rpm), static pressure (P_s), and brake horsepower (bhp) are all related, when one changes, all the others change.

407F28.EPS

Figure 28 ◆ Typical residential blower assembly.

Fan Law 1 states that the amount of air delivered by a fan will vary in direct proportion to the speed of the fan. Stated mathematically:

New cfm = (new rpm × existing cfm) ÷ existing rpm

or

New rpm = (new cfm × existing rpm) ÷ existing cfm

Fan Law 2 states that the static pressure (resistance) of a system varies directly with the square of the fan speed. Stated mathematically:

New P_s = existing P_s × (new rpm ÷ existing rpm)^2

Fan Law 3 states that the horsepower varies directly with the cube of the fan speed. Stated mathematically:

New hp = existing hp × (new rpm ÷ existing rpm)^3

24" SQUARE CEILING DIFFUSER									
FACE VELOCITY		300	400	500	600	700	800	900	1000
PRESSURE LOSS		0.006	0.010	0.016	0.022	0.031	0.040	0.050	0.062
6 Ak 0.165	cfm	50	65	85	100	115	130	150	165
	Throw	3.5	4.5	5.5	6.5	8	9	10	11
8 Ak 0.280	cfm	85	110	140	170	195	225	250	280
	Throw	4.5	5.5	7	8.5	10	11	12	14
10 Ak 0.420	cfm	125	170	210	250	295	335	380	420
	Throw	5	6.5	8	9.5	11.5	13	15	16
12 Ak 0.595	cfm	180	240	300	355	415	475	535	595
	Throw	6	8	10	11.5	13.5	15.5	17.5	19
14 Ak 0.820	cfm	245	330	410	490	575	655	740	820
	Throw	7	9	11.5	13.5	16	18	20	22.5
16 Ak 1.03	cfm	310	410	515	620	720	825	925	1030
	Throw	7.5	10	12.5	15	18	20	22	25
18 Ak 1.33	cfm	400	530	665	800	930	1065	1200	1330
	Throw	8.5	11	14	17	20	23	26	28
20 Ak 1.60	cfm	480	640	800	960	1120	12801	1440	1600
	Throw	9.5	12	16	18	22	25	28	31
22 Ak 1.90	cfm	570	760	950	1140	1330	1520	1710	1900
	Throw	10.5	13.5	17	19	24	27	30	33
24 Ak 2.30	cfm	690	920	1150	1380	1610	1840	2070	2300
	Throw	11	14.5	18.5	22	26	30	33	36

Terminal Velocity of 50 fpm

407F27.EPS

Figure 27 ◆ Example of a ceiling diffuser pressure drop chart.

Fan curve charts or fan performance charts are normally used to find the relationships that exist for a set of system conditions involving P_s, fan rpm, fan bhp, and cfm. These charts are produced by equipment manufacturers for the specific model of equipment being used.

6.1.6 Airflow in a Typical System

Figure 29 shows a simplified building air distribution system. We will discuss the airflow through this system to review some of the concepts and pressure relationships you have learned, as well as some new ones. The volume of air that an air distribution system must deliver is normally based on the mode of operation (cooling or heating) that needs the most airflow. Generally, this is the cooling mode.

In *Figure 29*, the airflow shown is for the cooling mode related to a cooling unit with a 3-ton capacity. Therefore, the system fan (blower) must be able to supply the volume of air needed for 3 tons of cooling. Using a rule of thumb that cooling requires about 400 cfm of air per ton, the

blower in the example system must supply 1,200 cfm (3 × 400 cfm) or more of air. Further, the external static pressure of the duct system that the fan must work against is 0.4 in. w.c. This is derived by adding the absolute values of the supply and return external static pressures of 0.2 in. w.c. and –0.2 in. w.c., respectively. As shown, the system has 11 air supply outlets, each delivering 100 cfm and two smaller outlets, each delivering 50 cfm. The return air is taken into the system through two centrally located grilles.

While studying this diagram, consider the entire building as part of the system. The supply air leaves all the supply registers and sweeps the walls of the building. Then, it travels through the conditioned spaces within the building as it flows toward the return air grilles. The air is at room temperature at this time. The duct system begins at the two return air grilles. Relative to the atmospheric pressure of the rooms, there is a slight negative pressure at the grilles. As shown, the pressure on the blower side of the return air grille filters is about –0.03 in. w.c., which is lower than the pressure in the rooms. This results in the

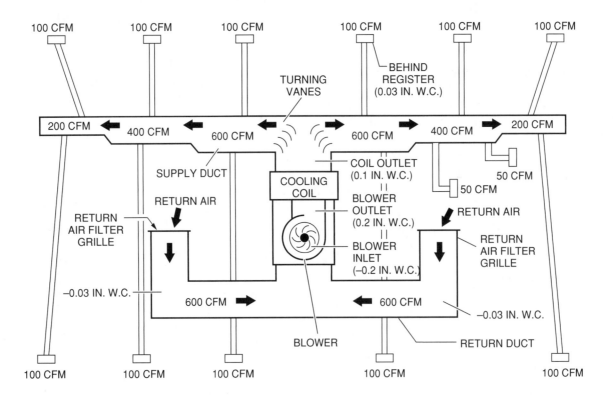

SYSTEM CAPACITY 3 TONS
CFM REQUIREMENT 400 CFM PER TON = 400 × 3 = 1,200 CFM
BLOWER STATIC PRESSURE (0.4 IN. W.C.)
SUPPLY DUCT STATIC PRESSURE (0.2 IN. W.C.)
RETURN DUCT STATIC PRESSURE (–0.2 IN. W.C.)

407F29.EPS

Figure 29 ◆ Simplified air distribution system.

higher room pressures pushing the air through the return air filters and into the return duct. As the air flows down the return duct towards the blower, the pressure continues to decrease as a result of friction losses in the duct. At the inlet to the blower, the air pressure is at its lowest pressure in the system. For our example, it is at –0.20 in. w.c. below the room pressure. The return air is forced through the blower, and at the blower output, is increased to the highest level in the duct system. For our example, this pressure is 0.20 in. w.c. above room pressure. The difference in static pressure between the input and output of the blower is 0.40 in. w.c.

The air at the blower output is pushed through the furnace heat exchanger and the cooling coil, where it encounters a pressure drop of 0.10 in. w.c. At the input to the supply duct, the air enters at a pressure of 0.10 in. w.c. After the air enters the supply duct, it undergoes a slight pressure drop at the tee where the duct is split into two reducing trunks, one to feed each end of the building.

Each first section of the reducing trunk must handle 600 cfm of air. Two branch duct outlets, each with an air capacity of 100 cfm, are supplied from the first trunk section on each side. This reduces the quantity of air supplied to the next sections of the trunk to 400 cfm for each side. These sections each supply 200 cfm of air to the conditioned space. This reduces the quantity of air supplied to the last sections of the trunk to 200 cfm for each side. This allows another reduction in the trunk size for each of these sections. The last section of trunk on each side of the system supplies the remaining 200 cfm of air to the last outlets on each side. In this example, smaller reducing trunks were used to save materials. Also, reducing the duct size as air was distributed off the trunk keeps the pressure in the duct system at the desired level all along the duct.

Normally, dampers are installed in each branch to balance the quantity of air supplied to each room. The system in the example will furnish 100 cfm to each of 11 outlets and 50 cfm to 2 outlets, but if a room did not need that much air, the dampers could be adjusted to reduce the quantity.

6.2.0 Air Distribution Duct Systems

There are many air distribution system designs. However, most of them consist of two duct systems: the supply duct system and the return duct system. The supply duct system receives air from the output of the system air fan, then distributes

INSIDE TRACK — Duct Leakage

All duct system designs assume a tight system with no losses due to leakage of the conditioned air. In fact, however, most duct systems, especially sheet metal systems, are inherently leaky. According to the U.S. Department of Energy, 25 to 40 percent of the energy in the conditioned air in ducts is lost through leakage or lack of insulation. Much of the leakage can be attributed to poor workmanship when assembling the system.

The joints in sheet metal systems will generally have some leakage, even if properly assembled. To correct this problem, the system designer should specify that all duct joints and connections be sealed at the time of installation. The sealing method will vary according to the duct material used. Fiberglass ductboard and flexible duct joints can be sealed with aluminum foil tape. Sheet metal ducts can be sealed with aluminum foil tape or with mastic. Local codes may dictate the type of material to be used for duct sealing.

the air to the terminal units and through the registers or diffusers into the conditioned space. The return duct system collects and routes air contained in the conditioned space for return to the input of the system air fan. Some systems use a return air fan to aid in this process. The design of air distribution systems for commercial and industrial structures varies widely, depending on the structure and its intended use. Because the designs of air distribution systems for residential applications are more uniform, they will be used as the basis for discussion in the remainder of this section. Although the size of the heating and cooling loads, the system layout, and the physical size of the system components will vary, the principles of operation and types of components are basically the same in all duct systems.

6.2.1 Duct Systems Used in Cold Climates

The type of duct system used in a building is mainly determined by the climate. In cold climates, most buildings use perimeter duct systems, which have floor or baseboard supply diffusers along the exterior walls of the building. Use of floor or baseboard supply diffusers provides a good tradeoff for heating and cooling performance.

In winter, the warm air supplied by the furnace blankets the outside walls and windows. This compensates for the cold downdrafts that tend to develop at the outside walls, windows, and doors. The return air grilles are located on the interior partition walls, at or near the floor. Central returns may be used, or for better performance, individual returns can be installed in each room. Locating return grilles on the interior walls near floor level helps remove any cool air from the floor where it tends to collect or stratify.

Figure 30 shows the pattern of room airflow during the heating and cooling modes of system operation. During the heating mode, the heated air blankets the outside walls and windows. Because it is warmer and lighter than the room air, it spreads across the ceiling and down the inside wall. Room air is drawn (induced) into the warm airstream and mixes with it. A resulting stratified zone of cool air tends to collect near the floor then leaves the room through a low sidewall return.

During the cooling mode, cold supply air travels up the outside wall and windows and strikes the ceiling. Because it is cooler and heavier than the room air, it travels a short distance along the ceiling and then drops back down into the room as shown. The cold air mixes with the room air, leaving only a small stratified layer of warm air near the ceiling. High sidewall returns would minimize this problem, but would result in a loss of heating performance. In this situation, the use of a ceiling fan during the heating and cooling seasons would help break up the stratified air, resulting in better indoor comfort.

Perimeter systems can have various layouts. The common ones include the following:

- Loop perimeter
- Extended plenum
- Reducing plenum

Loop perimeter duct systems are seen in structures built on concrete slabs in colder climates (*Figure 31*). They are easily used with centrally located, downflow air handlers. The perimeter loop is a continuous round duct of constant size imbedded in the slab. It runs close to the outer walls, with the outlets next to the wall. The perimeter loop is fed by several branches from the plenum. When the furnace fan is running, warm air is in the whole loop, which helps to keep the slab at an even temperature. Heat loss to the outside is minimized by the use of insulation around the slab. The loop has a constant pressure and provides the same pressure to all outlets.

The extended plenum duct system (*Figure 32*) uses rectangular trunk ducts as the main supply and return ducts. The supply and return trunk ducts are a constant size over the whole length. This is the reason it is called an extended plenum system. These systems are commonly used in below-floor (basement or crawl space) or ceiling (attic) installations.

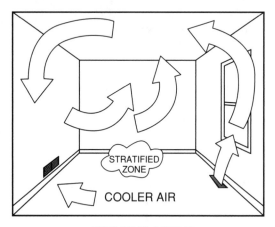

HEATING MODE

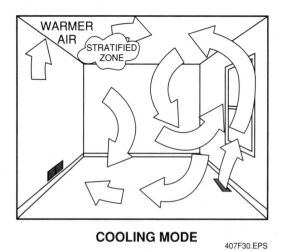

COOLING MODE

407F30.EPS

Figure 30 ◆ Room air distribution patterns for a perimeter duct system.

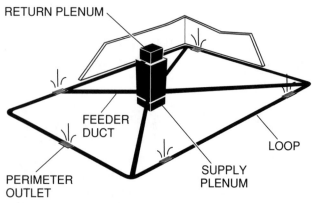

407F31.EPS

Figure 31 ◆ Loop perimeter duct system.

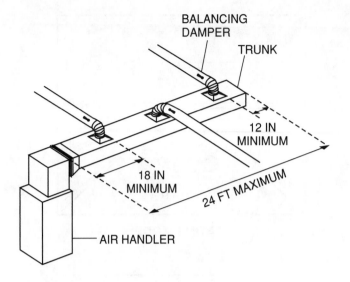

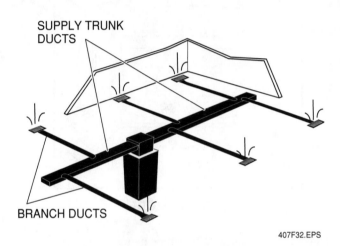

Figure 32 ◆ Extended plenum duct system.

Separate branch ducts run from the trunk duct to each supply outlet. The extended plenum system works best when the air handler is located at the center of the main duct. However, it can be run in one direction. The trunk ducts are normally installed near the center line of the building, and their dimensions are constant over their entire length. The branch ducts are normally round but can be rectangular. An air volume damper is usually installed in each branch duct near the trunk. This allows the airflow to be balanced with all supply air outlets fully open. Recommendations for the design of an extended plenum duct system are as follows:

• The main trunk duct should extend no more than 24' from the air handler.
• The first branch duct should be at least 18" from the beginning of the main duct. This helps to achieve the best balancing of the branch ducts.
• The main trunk should extend at least 12" from the last branch duct.

A reducing extended plenum system is similar to an extended plenum system. *Figure 33* shows an example of a reducing extended plenum duct system. It works well in larger buildings that require longer duct runs. It is also a better choice for systems where the supply fan assembly is installed on one end of the main trunk duct rather than in the middle. When the system is properly designed, the same pressure drop is maintained from one end of the duct system to the other. This allows each branch duct to have

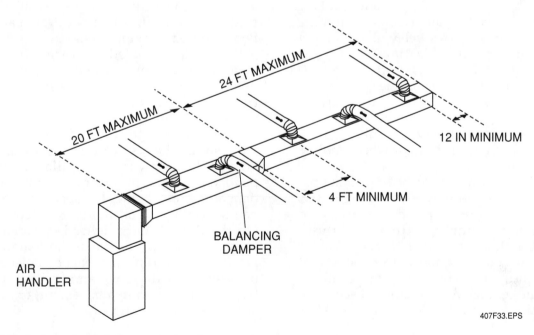

Figure 33 ◆ Reducing extended plenum duct system.

about the same pressure pushing the air into its takeoff from the trunk duct. Recommendations for designing a reducing extended plenum duct system are as follows:

- The first main duct section should be no longer than 20'.
- The length of each uniformly sized reducing section should not exceed 24'.
- The first branch duct connection down from a single-taper transition should be at least four feet from the beginning of the transition fitting. This distance allows the air turbulence caused by the fitting to die down before the air is sent into the next branch duct. If the distance is less than four feet, the branch ducts near the transition can be hard to balance and may cause the system to be noisy.

6.2.2 Duct Systems Used in Warm Climates

In warm climates, buildings should have duct systems that favor cooling over heating. Perimeter systems like those used in cold climates can work reasonably well in some warm areas. However, their use is normally limited to buildings constructed over a basement or crawl space. Because cold floors and downdrafts from the outside walls are not normally a problem in warm climates, the air supply outlets do not need to be located at the building perimeter. In warm climates, supply and return air openings can be mounted high on the interior walls or in the ceiling to intensify cooling.

Figure 34 shows the room airflow with high sidewall outlets. In the cooling mode, cool air moves across the ceiling and wraps around the far wall. The room air mixes well with the supply air, and almost no stratification occurs. Air motion throughout the room is good. In the heating mode, the supply air remains near the ceiling and moves partway down the outside wall. Because of its buoyancy, the warm air does not descend down the wall very far. This causes a large stratified area near the floor where cool air tends to build up.

Ceiling diffusers are one of the best air supply methods used for cooling, but they are not as effective for heating in perimeter areas. In the cooling mode, supply air from the diffuser mixes well with the room air (*Figure 35*). Air motion in the room is good with no stagnant areas. In the heating mode, warm air tends to rise toward to the ceiling. Very little of it reaches the lower portions of the occupied space. Also,

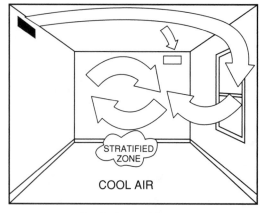

HEATING MODE

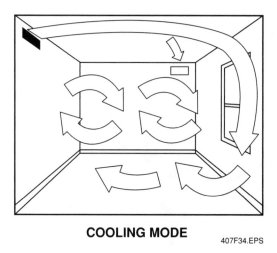

COOLING MODE   407F34.EPS

Figure 34 ◆ Room air distribution patterns for high sidewall outlets.

because of convection currents moving down the cold outside wall, cold air is drawn across the floor and up the inside wall.

Duct systems used in warm climates can have various layouts. The three typical layouts used in buildings on concrete slabs are overhead trunk, overhead or attic radial, and overhead extended plenum.

Regardless of the type of system, the ductwork should be insulated if it is installed in an attic. If the ductwork is installed above a suspended ceiling, it may or may not be insulated, depending on whether the space above the suspended ceiling is considered to be conditioned. Overhead trunk and overhead extended plenum systems should be laid out as those shown in *Figures 32* and *33*. Trunk and extended plenum types of systems are designed and laid out the same, whether installed in a basement, crawl space, attic, or above a ceiling.

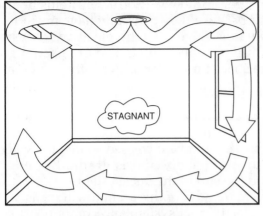

HEATING MODE

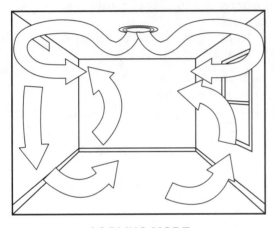

COOLING MODE

407F35.EPS

Figure 35 ◆ Room air distribution patterns for ceiling outlets.

In the overhead or attic radial duct system (*Figure 36*), separate branch ducts are run from a common supply air plenum to each outlet. The branch ducts are metal, ductboard, or flexible duct and may be insulated, depending on whether or not they are in the conditioned space. Overhead radial systems often use a central return.

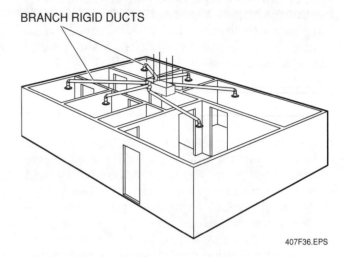

BRANCH RIGID DUCTS

407F36.EPS

Figure 36 ◆ Overhead (attic) radial duct system.

6.3.0 Duct System Components

Building code requirements pertaining to the design of air distribution systems vary widely across the country. Each new building or renovation job is governed by the applicable federal, state, or local standards and codes. These standards and codes require compliance with health, safety, and energy conservation regulations. Almost all localities have minimum standards or codes that determine the type of materials and methods that must be used. Also, the materials and methods used in residential, commercial, and industrial air distribution systems vary. You should become familiar with the codes that apply to each job and always follow them.

Areas that experience earthquakes have stringent local codes that cover all aspects of building construction, including the HVAC equipment installation. System designers must be aware of those code requirements and comply with them. On a national level, SMACNA has a 200-page publication titled *Seismic Restraint Manual: Guidelines for Mechanical Systems*. This publication is

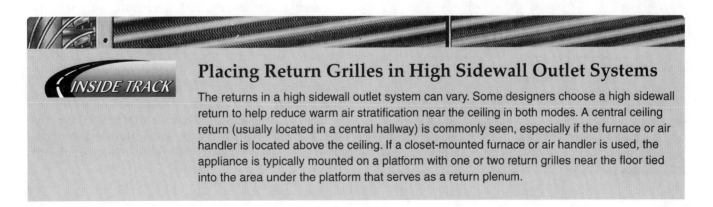

Placing Return Grilles in High Sidewall Outlet Systems

The returns in a high sidewall outlet system can vary. Some designers choose a high sidewall return to help reduce warm air stratification near the ceiling in both modes. A central ceiling return (usually located in a central hallway) is commonly seen, especially if the furnace or air handler is located above the ceiling. If a closet-mounted furnace or air handler is used, the appliance is typically mounted on a platform with one or two return grilles near the floor tied into the area under the platform that serves as a return plenum.

approved by the American National Standards Institute (ANSI) as a national standard. The manual shows designers and contractors how to determine quakeproof restraints for sheet metal ducts, piping, and conduit. ASHRAE devotes a chapter of their handbook to seismic and wind restraint design.

Duct systems are classified by their use and static pressure. *Table 2* lists the pressure levels normally used in residential, commercial, and industrial systems. Duct systems in public assembly, business, educational, general factory, and mercantile buildings are normally classified as commercial systems. Industrial systems are those systems that are outside the pressure range of commercial systems. Industrial duct systems also include air pollution control systems and industrial exhaust systems.

Duct systems are also classified as low-velocity or high-velocity systems. *Table 3* lists the duct velocity levels and supply outlet pressure levels used with low-velocity and high-velocity systems. When space is available, it is recommended that larger, low-velocity duct systems be used. This is because they generally have much lower system noise levels. High-velocity duct systems have higher noise levels and are used mainly where space limitations prevent the use of larger ducts. Some installations may have areas with space restrictions that require the use of smaller duct operating at higher velocities. As soon as these areas are passed and space is available, the duct size should be increased so that the velocity rate drops sharply and is gradually reduced toward the end of the duct system.

The remainder of this section will focus on the low-velocity duct system and components commonly used in residential and light commercial applications. *Table 4* lists the recommended maximum duct velocities for common low-velocity applications.

The main components of an air distribution duct system are as follows:

- Main trunk and branch ducts
- Fittings and transitions
- Supply air outlets and return air inlets
- Dampers
- Insulation and vapor barriers

6.3.1 *Main Trunk and Branch Ducts*

Duct systems can be installed in basements, crawl spaces, attics, and concrete floors (slabs). In commercial jobs, they are often installed in open areas, such as warehouses and garages. Ductwork can be made from metal, fiberglass ductboard, ceramic, or plastic materials. Galvanized sheet metal or fiberboard ducts are typically used for heating/cooling air distribution. When installed in a concrete slab, ducts are usually made of metal, plastic, or ceramic. Spiral metal and flexible ducts are also commonly used. Where weight is a factor, aluminum duct is sometimes used.

Galvanized steel duct can be round, square, or rectangular. Popular sizes of round and rectangular steel ducts, along with an assortment of standard fittings, can be obtained from HVAC supply houses. *Figures 37* and *38* show typical steel rectangular duct and round duct systems, respectively. For large commercial jobs involving customized ductwork, the ducts and fittings are often made separately in a metal shop or are fabricated at the job site. Because sheet metal duct is rigid, the layout must be well planned, and all the pieces exactly cut, or the duct system will not fit together.

Table 2 Duct System Classification by Pressure

System	Pressure Level
Residential	±0.5 in. w.c.
	±1 in. w.c.
Commercial	±0.5 in. w.c.
	±1 in. w.c.
	±2 in. w.c.
	±3 in. w.c.
	+4 in. w.c.
	+6 in. w.c.
	+10 in. w.c.
Industrial	Pressure varies with application

Table 3 Duct System/Supply Outlet Classifications

Air Outlet Pressures	Pressure
Low-velocity duct	Main duct 1,000 to 2,400 fpm
	Branches 600 to 1,600 fpm
High-velocity duct	Main duct 2,500 to 4,500 fpm
	Branches 2,000 to 4,000 fpm

Duct Velocities	Velocity
Low-pressure outlets	Static pressure 0.1 to 0.5 in. w.c.
High-pressure outlets	Static pressure 1 to 3 in. w.c.

Table 4 Recommended Maximum Duct Velocities for Common Low-Velocity Applications

	Main Duct		Branch Ducts	
Application	**Supply (fpm)**	**Return (fpm)**	**Supply (fpm)**	**Return (fpm)**
Apartments, residences	1,000	800	600	600
Auditoriums, theaters	1,300	1,100	1,000	800
Banks, meeting rooms, libraries, offices, restaurants, retail stores	2,000	1,500	1,600	1,200
Hospital rooms, hotel rooms	1,500	1,300	1,200	1,000

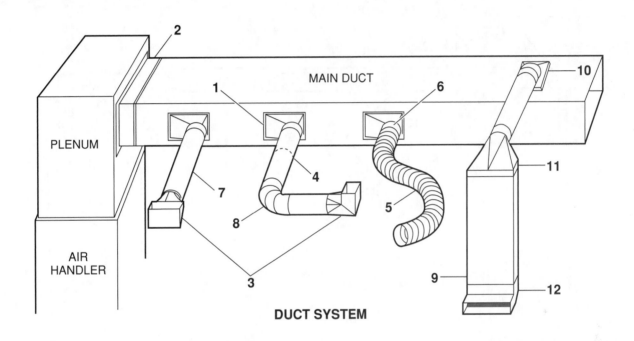

DUCT SYSTEM

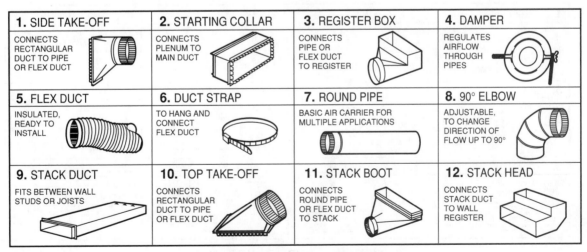

FITTINGS AND TRANSITIONS

407F37.EPS

Figure 37 ◆ Typical rectangular duct system and components.

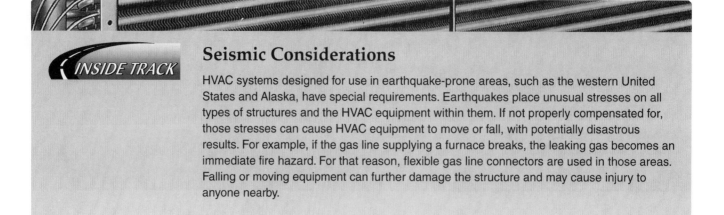

Seismic Considerations

HVAC systems designed for use in earthquake-prone areas, such as the western United States and Alaska, have special requirements. Earthquakes place unusual stresses on all types of structures and the HVAC equipment within them. If not properly compensated for, those stresses can cause HVAC equipment to move or fall, with potentially disastrous results. For example, if the gas line supplying a furnace breaks, the leaking gas becomes an immediate fire hazard. For that reason, flexible gas line connectors are used in those areas. Falling or moving equipment can further damage the structure and may cause injury to anyone nearby.

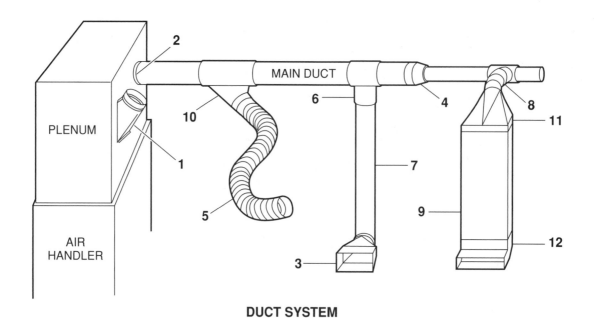

DUCT SYSTEM

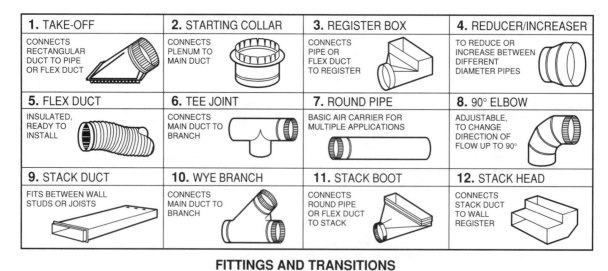

FITTINGS AND TRANSITIONS

407F38.EPS

Figure 38 ◆ Typical round duct system and components.

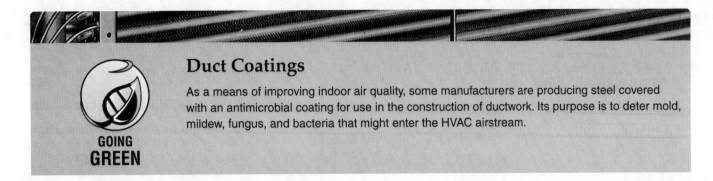

Duct Coatings

As a means of improving indoor air quality, some manufacturers are producing steel covered with an antimicrobial coating for use in the construction of ductwork. Its purpose is to deter mold, mildew, fungus, and bacteria that might enter the HVAC airstream.

The thickness of galvanized steel and other metal duct is expressed as a gauge thickness (*Figure 39*). When a duct is made of 28-gauge sheet metal, this means that the thickness of the duct walls is $\frac{1}{28}$ of an inch. Likewise, a sheet metal duct made out of 24-gauge metal has a wall thickness of $\frac{1}{24}$ of an inch, and so on. Larger ducts are made from thicker metal and are more rigid than smaller ducts. This prevents them from swelling and making popping noises when the system blower starts and stops. Also, lines or ridges, normally called cross-breaks, are used on large sheet metal panels or ducts to make them more rigid. *Figure 40* shows some typical gauge thicknesses used for rectangular and round metal ducts.

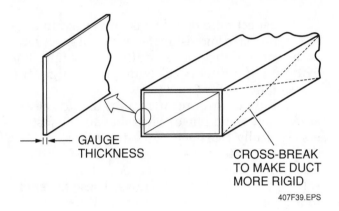

Figure 39 ◆ Metal ducts.

RECTANGULAR DUCT WIDTH IN INCHES	COMMERCIAL		RESIDENTIAL
	SHEET METAL GALVANIZED	ALUMINUM	SHEET METAL GALVANIZED
UP TO 12	26	0.020	28
13–23	24	0.025	26
24–30	24	0.025	24
31–42	22	0.032	–
43–54	22	0.032	–
55–60	20	0.040	–
61–84	20	0.040	–
85–96	18	0.050	–
OVER 96	18	0.050	–

RECTANGULAR DUCT

ROUND DUCT DIAMETER IN INCHES	COMMERCIAL SHEET STEEL GALVANIZED GAUGE	RESIDENTIAL SHEET STEEL GALVANIZED GAUGE
UP TO 12	26	30
13–18	24	28
19–28	22	–
27–36	20	–
35–52	18	–

ROUND DUCT

407F40.EPS

Figure 40 ◆ Typical metal duct gauge thicknesses.

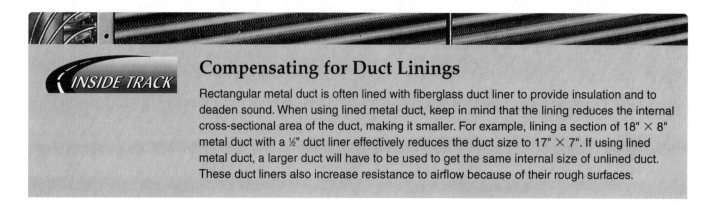
The aspect ratio of a duct is often used to classify a duct size and estimate its cost. Aspect ratio is the ratio of the duct's width to its height. For example, if a duct is 18" wide and 6" high, the aspect ratio is 18 ÷ 6 or 3 to 1.

Sections of square or rectangular duct are assembled using one of several available fasteners. Typically, S-fasteners and drive clips, and/or snap-lock fasteners are used (*Figure 41*). Round duct sections are normally fastened together with self-tapping sheet metal screws. These fasteners make a nearly airtight connection. When further sealing is needed, the joint can be taped using aluminum foil duct tape. Duct sealing mastic that can be brushed on or applied with a caulking gun is also available. Leaking joints cut down on the amount of air available for delivery to the outlets.

A ductwork system must be well supported so that it does not move. If it is not properly supported, movement can occur when the fan starts or as a result of expansion and contraction as the duct heats and cools. This type of movement can be contained by using flexible or fabric joints at different points in the system. Sheet metal ductwork systems also transmit vibrations from the air handling equipment. Transmission of these vibrations to the duct system can be prevented by using flexible connectors or fabric joints at the main supply and return ductwork connections to the air handler. *Figure 42* shows some typical ductwork noise and vibration control devices.

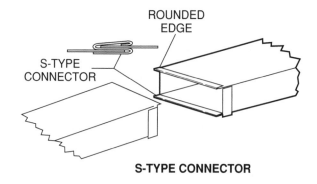

S-TYPE CONNECTOR

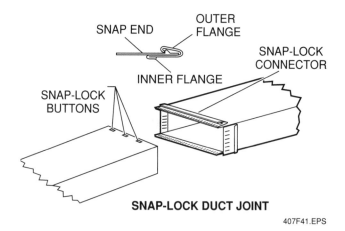

SNAP-LOCK DUCT JOINT

407F41.EPS

Figure 41 ◆ Typical square and rectangular sheet metal duct fasteners.

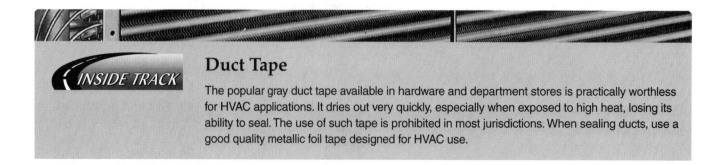

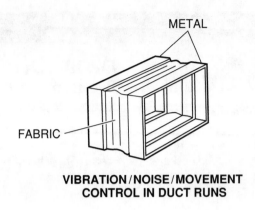

METAL

FABRIC

**VIBRATION/NOISE/MOVEMENT
CONTROL IN DUCT RUNS**

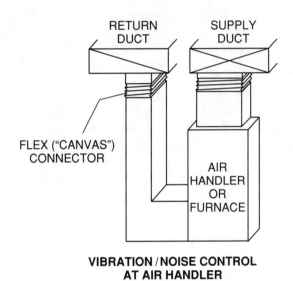

RETURN DUCT

SUPPLY DUCT

FLEX ("CANVAS") CONNECTOR

AIR HANDLER OR FURNACE

**VIBRATION/NOISE CONTROL
AT AIR HANDLER**

407F42.EPS

Figure 42 ◆ Ductwork vibration and noise control devices.

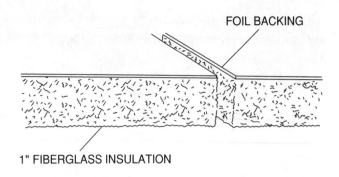

FOIL BACKING

1" FIBERGLASS INSULATION

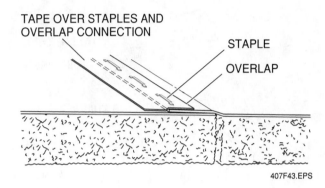

TAPE OVER STAPLES AND
OVERLAP CONNECTION

STAPLE

OVERLAP

407F43.EPS

Figure 43 ◆ Fiberglass ductboard.

Fiberglass ductboard can be used almost anywhere that metal duct is used in applications involving velocities up to 2,400 fpm and pressures of ±2 in. w.c. Molded round fiberglass ducts are available to handle higher pressures. Fiberglass duct has more friction loss than metal duct, but it is quieter because the ductboard absorbs blower and air noises better. Fiberglass duct is available in flat sheets for fabrication or as prefabricated round duct. Fiberglass duct is normally 1" thick with an aluminum foil backing. This backing is reinforced with fiber to make it strong. The inside surface of the ductboard is coated with plastic or a similar substance to prevent the erosion of duct fibers into the supply air and to provide smoother airflow. Fiberglass particles released into the air can be harmful to health.

Ductwork is made from sheets of fiberglass ductboard using special knives or cutting machines. When two pieces are fastened together, an overlap of foil is left so that one piece can be stapled to the other using special staples (*Figure 43*). The joint is then taped to make it airtight. Round fiberglass duct is also easy to install

because it can be cut to size with a knife. Fiberglass ductwork systems must be properly supported or they will sag over long runs. Special hangers must be used that do not damage the cover of the ductboard.

Flexible round duct (*Figure 44*) is available in sizes of up to about 24" in diameter. Flexible duct is available with a reinforced aluminum foil backing for use in conditioned areas. It also comes wrapped with insulation protected by a vapor barrier made of fiber-reinforced vinyl or foil backing for use in unconditioned areas.

Flexible duct is typically used in spaces where obstructions make the use of rigid duct difficult or impossible. Flexible duct is easy to route around corners and other bends. Duct runs should be kept as short and straight as possible. Gradual bends should be used because tight turns can greatly reduce the airflow and can cause the duct to collapse. If a connection to a ceiling diffuser needs an elbow, it is better to use an insulated metal elbow at the input to the diffuser than to bend flexible duct tightly to form the connection. This is because diffuser performance is disrupted far less by use of the metal elbow.

Long runs of flexible duct are not recommended unless the friction loss is taken into account. Even when properly installed, most flex ducts cause at least two to four times as much resistance to airflow as the same diameter sheet

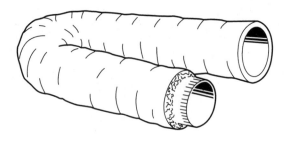

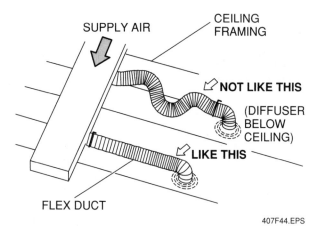

Figure 44 ◆ Flexible duct.

metal duct. To avoid sags in the run, flexible duct should be amply supported with one-inch wide or wider bands to keep the duct from collapsing and reducing the inside dimension (*Figure 45*). Some flexible duct comes with built-in eyelet holes for hanging the duct.

NOTE

In commercial applications, flexible duct runs are limited to 6 feet.

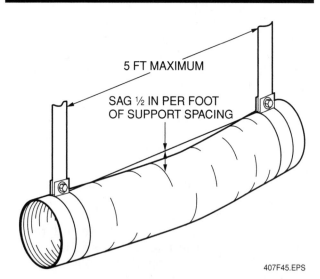

Figure 45 ◆ Supporting flexible duct.

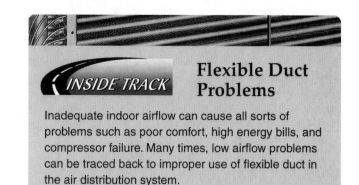

Flexible Duct Problems

Inadequate indoor airflow can cause all sorts of problems such as poor comfort, high energy bills, and compressor failure. Many times, low airflow problems can be traced back to improper use of flexible duct in the air distribution system.

Fittings in duct systems, such as elbows, angles, takeoffs, and boots, change the direction of airflow or change its velocity. Transitions are typically used to change from one size or shape of duct to another.

Air moving in a duct has inertia that makes it tend to continue flowing in a straight line. Each fitting in a duct run adds friction and decreases the quantity of air the duct can carry. It takes energy to overcome the resistance (friction) of a fitting. Adding fittings to a duct run has the same effect on the pressure loss of a duct as increasing its overall length. This is why duct runs should be made as straight as possible. It is also why using unnecessary fittings or ones not best suited for the job must be avoided.

The pressure loss that results from using a specific type of fitting can be found by multiplying its loss coefficient value (C-value) by the velocity pressure in the fitting. C-values give more accurate values of pressure loss and are typically used when designing large commercial duct systems. C-values for the various kinds of fittings are found in published tables readily available in SMACNA, ASHRAE, and duct component manufacturers' published literature.

Equivalent length of straight duct values assigned to a fitting are commonly used when designing residential or light commercial duct systems. This means that a specific fitting produces a pressure drop equal to a certain number of feet of straight duct length of the same size. For each standard type of fitting, the pressure drop is known and has been converted to the equivalent feet of duct length. This information is available in a set of charts published in air duct system related SMACNA and ASHRAE documents and in some duct component manufacturers' product literature. These charts show the standard types of fittings and/or transitions and the value for the equivalent feet of length used for each one. The total equivalent feet of length for a duct run is calculated by adding all the equivalent lengths for fittings in the run to the actual length of straight duct used. Each type of fitting or transition presented in the charts is identified by a letter and a number. The letter identifies the type of fitting and the number indicates the equivalent length of straight duct, in feet, for that fitting. In the example shown in *Figure 46*, an elbow (G) with an equivalent length of 30' is added to a duct with 100' of straight length. The resulting pressure drop is the same as that of 130' of straight duct. If elbow (F) had been chosen, the equivalent length of the duct would be 110' instead of 130'. The required turn in the duct would be made, but with a lower pressure drop.

Supply air outlets control the air pattern within a conditioned area to obtain proper air motion and to blend the supply air with the room air so that the room is comfortable without excess noise or drafts. Supply air outlets are selected for each area in a building based on the volume of air required in cfm, the distance available for **throw** or radius of diffusion, the building structure, and the decor. These components were discussed in detail in the *HVAC Level Two* module, *Air Distribution Systems*.

Smudging, a possible problem, is a pattern of discoloration that exists around the ceiling diffuser. It is caused by the dirt particles held in suspension in the room air being subjected to turbulence at the face of the outlet. If smudging occurs, anti-smudge rings can be used.

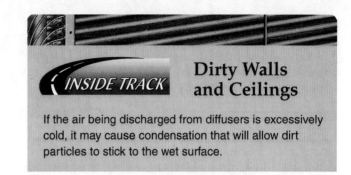

Return air inlets are located to be compatible with supply outlets and ductwork. Generally, return inlets are positioned to return room air of the greatest temperature differential that collects in stagnant areas. This results in the warmest air being returned during cooling and the coolest air being returned during heating. In general, the same types of grilles and diffusers used for supplying air are used for returning air. Normally, they do not have any deflection or volume control devices.

Volume dampers are used to control the volume of airflow through the various sections of an air distribution system. They do this by introducing a resistance to airflow in the system. Volume dampers are normally installed in each branch duct serving a conditioned area. Without volume dampers, air distribution systems cannot be properly balanced, causing some rooms to

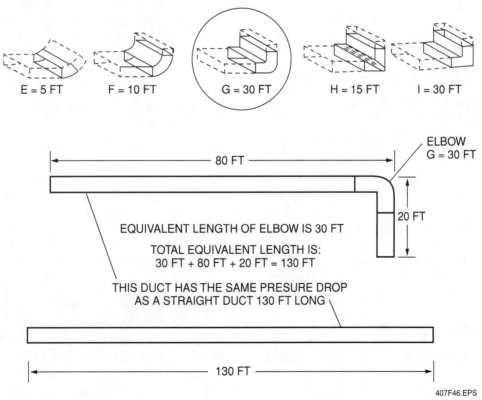

Figure 46 ◆ Example of equivalent length.

407F46.EPS

receive too much air and others not enough. In high-pressure systems, volume dampers are commonly called pressure-reducing valves. They are usually installed between a high-pressure trunk duct and a low-pressure branch duct.

Dampers are usually manually adjustable. When used in electronically controlled zoned systems, however, dampers are automatically controlled. Sometimes dampers are used to mix two airflows, such as fresh and recirculated air. By code requirements, commercial and industrial buildings normally have automatic fire dampers installed in all the vertical duct runs because all ducts, especially vertical ducts, will carry smoke and flames from fires.

A damper used to balance a system should be installed in an accessible place in each branch supply duct. The closer the dampers are to the main duct or supply air plenum, the better. They should be tight-fitting with minimum leakage. The built-in, single-leaf damper that is part of a manufactured air supply grille or register should not be used to balance an air system. It should only be used to apply or shut off airflow to a room. When partially closed, it disrupts the performance of the grille or register and also makes it noisy. Generally, single-leaf volume dampers are used in branch ducts where the total static pressure is less than 0.5 in. w.c. Multi-vane or multi-blade volume dampers are normally used in ductwork where the total static pressure exceeds 0.5 in. w.c.

6.4.0 Duct System Design

The design of an air distribution duct system is based on the heat loss and/or heat gain for the entire building and for each of the rooms or areas within the building. In order to make an initial equipment selection, the required airflow (cfm) is determined using a simple formula or a rule of thumb such as 400 cfm per ton. This value is further refined in designing the duct system or evaluating an existing duct system. It is entirely possible that the initial equipment selection will have to be changed because its air delivery does not support the volume of air actually required by the particular building or because it cannot overcome the static pressure losses in the ductwork system.

6.4.1 General Design Procedure

Armed with the total building cfm requirement and the heat gain and losses for each area in the building, you can begin the duct design process:

Step 1 Study the building plans, and locate the supply outlets and return inlets to provide proper distribution and circulation of the air within each space.

Step 2 Calculate the cfm delivered by each supply outlet and returned by each return inlet. Size the supply outlets and return inlets using the cfm and the grille manufacturer's product data sheets.

Step 3 Sketch the duct system, connecting the supply outlets and return inlets with the air handling units/air conditioners. Note that the physical layout of the duct system is driven by the building construction and the placement of other building system components. To avoid problems in duct design, the final duct system should be laid out after the building has a roof, the partitions are in place, and the plumbing and electrical piping is installed.

Step 4 Divide the duct system layout into sections and identify each of the sections. The duct system should be divided at all points where flow direction, size, or shape changes.

Step 5 Size the main and branch ducts by the selected method. Methods commonly used are the equal friction method, static regain method, and velocity reduction method. There are also several duct design calculators produced by SMACNA and the major HVAC equipment manufacturers that can be used for this purpose.

Step 6 Calculate total system pressure loss, and then select the fan.

Step 7 Lay out the system in detail. It often takes several attempts to find a duct and air distribution layout that fits the building's construction and also works well. If duct routing and fittings vary greatly from the original plan, it may be necessary to recalculate the pressure losses and reselect the fan.

6.4.2 Selecting the Number and Location of Supply Outlets and Return Inlets

Because the heating and cooling loads are not distributed evenly throughout a building, the number and location of supply outlets and return inlets must be accurately determined for each conditioned area; otherwise, the air system might perform poorly. Once installed, it can be difficult and costly to make corrections. One manufacturer recommends that at least one outlet be used for each 4,000 Btuh heat gain, or 8,000 Btuh heat loss, whichever is greater, in any room or area.

A good duct design provides for as much capacity for return air as it does for supply air. Individual return grilles can be placed in rooms for good air circulation and to avoid stratification. Normally, this approach provides better performance than a central return. Return air inlets should not be located directly in the primary airstream from supply outlets. If this occurs, the supply air can be short circuited back into the return air system without mixing with the room air. Also, return inlets should not be installed in areas, such as bathrooms, that can have undesirable odors. Otherwise, the odors will spread throughout the building by way of the ductwork system.

NOTE

There must be a return air path in every space. If there are no return air grilles in a space, the door can be undercut, or transfer grilles can be used.

6.4.3 Calculating Supply Outlet CFM and Selecting Outlet Size

In order to determine the size of the supply outlets for each room (or area), it is first necessary to calculate and record the airflow for each room in cfm. Make sure to calculate the cfm for both cooling and heating. The cfm for cooling and heating are calculated using the following formulas:

$$\text{cfm (cooling)} = \frac{\text{sensible load (Btuh)}}{1.08 \times (t_1 - t_2)}$$

$$\text{cfm (heating)} = \frac{\text{sensible load (Btuh)}}{1.08 \times (t_2 - t_1)}$$

Where:

cfm = volume of air in cubic feet per minute

Sensible load = value in Btuh for heat gain or heat loss for a room as determined by a room-by-room load estimate

t_1 = design outdoor temperature

t_2 = design indoor temperature

Once the cooling and heating cfm for each room have been calculated, the larger of the two values (cooling cfm or heating cfm) is used to size the supply outlets for that room. If more than one outlet is being used to supply a room, the total room cfm must be divided by the number of outlets to determine the size of the individual outlets. For example, if a room requiring 184 cfm is supplied by two outlets, the air that must be delivered by each outlet is 92 cfm.

Once the cfm for each supply outlet is known, the types of diffusers and their sizes can be selected using the manufacturer's product data. *Figure 47* shows an example of the cfms for various rooms in a house and the selected sizes of supply outlets. *Appendix D* lists the cfm ratings and sizes for the most common types of ductwork and supply/return grilles.

Compromises are almost always made when sizing supply outlets. In the example shown, each of the room outlets could have been custom-sized to match its cfm. However, to give a uniform appearance in the rooms, a standard size is commonly used for all outlets. In our example, both the 2½" × 14" and 4" × 10" floor diffusers can supply the needed cfm for all outlets. However, the 4" × 10" outlet size was selected because the 2½" × 14" floor diffusers are more difficult to fit between the floor joists. Also, the 4" × 10" outlet is a standard size stocked by most suppliers.

6.4.4 Calculating Return Inlet CFM and Selecting Inlet Size

A good design provides as much capacity for return air as it does for supply air. Sizing return inlet grilles is done the same way as sizing supply outlets. *Figure 47* shows the return air cfm requirements for the various rooms used in the previous supply outlet example and the selected sizes of return inlets and grilles. As with the supply outlets, the return air grille sizes were also standardized. The sizes of various return air grilles are listed in *Appendix D*.

NOTE

When the air handler is installed in a crawlspace or attic, or other limited-access location, it is common to include filters in the return air grilles, rather than having a single filter at the return side of the air handler. If very high efficiency filters, such as HEPA filters, are used, they can impede airflow. This fact must be taken into account during the design.

AIR SYSTEM DESIGN WORKSHEET

Job Name __EXAMPLE_____ Job No. _____ Date _____

Address_____ Estimator _____ Salesman _____

SUPPLY SIZING

ROOM NAME	CFM FROM LOAD ESTIMATE		GREATER CFM**	CFM PER OUTLET	SUPPLY OUTLET TYPE	OUTLET NO.	ACTUAL SUPPLY OUTLET SIZE (IN)	SLECTED SUPPLY OUTLET SIZE (IN)	BRANCH DUCT SIZE	REMARKS
	HEAT	COOL								
L.R.	102	154	154	77	FLOOR	2, 3	2 1/4 x 12	4 x 10	6"	
KIT.	155	184	184	92	FLOOR	11, 12	2 1/4 x 14	4 x 10	6"	
D.R.	73	52	73	73	FLOOR	1	2 1/4 x 12	4 x 10	6"	
LAV.	13	14	14	14	FLOOR	9	2 1/4 x 10	4 x 10	6"	
BATH	32	23	32	32	FLOOR	5	2 1/4 x 10	4 x 10	6"	
BR #1	90	71	90	90	FLOOR	6	2 1/4 x 14	4 x 10	6"	
BR #2	80	65	80	80	FLOOR	7	2 1/4 x 12	4 x 10	6"	
BR #3	52	58	58	58	FLOOR	8	2 1/4 x 10	4 x 10	6"	
FOYER	30	36	36	36	FLOOR	4	2 1/4 x 10	4 x 10	6"	
B.FOYER	53	63	63	63	FLOOR	10	2 1/4 x 10	4 x 10	6"	
			784							

RETURN SIZING

RETURN LOCATION	RETURN NO.	CFM**	GRILLE SIZE (IN)	NO. STUD SPACES REQ'D	NO. PANNED JOIST REQ'D	DUCT SIZE	REMARKS
BR #1		90	12 x 6	1	1	PANNED	LOW WALL
BR #2		80	12 x 6	1	1		RETURNS
BR #3		58	12 x 6	1	1		
HALL S		111	12 x 6	1	1	↓	
HALL N		111	12 x 6	1	1		↓
D.R.		334	30 x 8	2	2		
		784					

* The larger of the heating or cooling cfm is used in this column.
** Return sizing should be based on supply cfm.

407F47.EPS

Figure 47 ◆ Example of room cfm requirements used to determine supply outlet sizes.

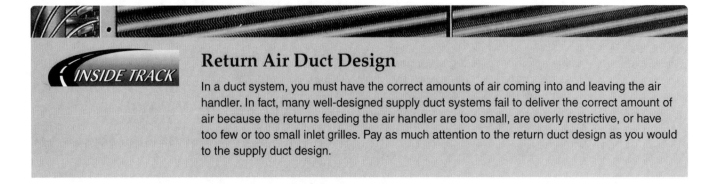

Return Air Duct Design

INSIDE TRACK

In a duct system, you must have the correct amounts of air coming into and leaving the air handler. In fact, many well-designed supply duct systems fail to deliver the correct amount of air because the returns feeding the air handler are too small, are overly restrictive, or have too few or too small inlet grilles. Pay as much attention to the return duct design as you would to the supply duct design.

6.4.5 Sizing Ductwork

There are several acceptable methods you can use to size ductwork. This module will use a method called the equal friction method. This method was selected because it is the most common means of sizing low-pressure supply air, return air, and exhaust air duct systems. Also, most of the manufacturers' duct design calculators, including the SMACNA calculator, are based on this method. Equal friction does not mean that the total friction remains constant throughout the system. It means that a specific friction loss or static pressure loss per 100' of duct is selected before the ductwork is laid out. Then, this loss value per 100' is used constantly throughout the design. The equal friction method automatically reduces the air velocities in the direction of airflow so that by using a reasonable initial airflow velocity, the chances of introducing noise are reduced or eliminated. The duct static regain (or loss) caused by airflow velocity changes is included in the loss coefficient values (C-values) or equivalent length of straight duct values assigned to the different duct fittings.

The equivalent length of fittings in a duct run equals the sum of the equivalent lengths for all fittings used in the run. *Figure 48* shows an example of a simplified ductwork system. For the purpose of explanation, the method used to find the equivalent length of fittings in the duct run to outlet S5 is explained in this section. To aid in understanding this example, the types of fittings in the

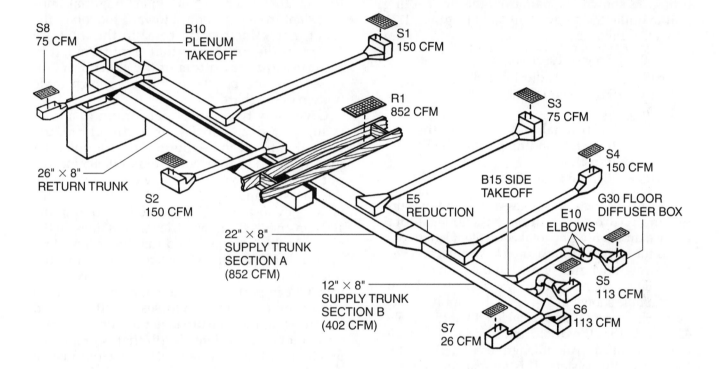

OUTLET	EQUIVALENT LENGTH OF FITTINGS (FT)	MEASURED LENGTH OF DUCT (FT)	TOTAL EQUIVALENT LENGTH OF RUN (FT)	BRANCH DUCT SIZE (ROUND)
S1	105	29	134	7"
S2	105	19	124	7"
S3	80	43	123	6"
S4	130	49	179	7"
S5	115	42	157	6"
S6	105	42	147	6"
S7	105	42	147	6"
S8	125	12	137	6"
RETURN R1	140	24	164	

407F48.EPS

Figure 48 ◆ Example of a duct system.

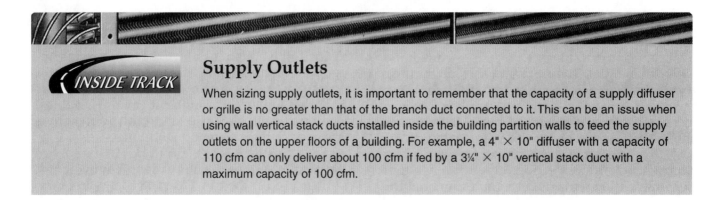

Supply Outlets

When sizing supply outlets, it is important to remember that the capacity of a supply diffuser or grille is no greater than that of the branch duct connected to it. This can be an issue when using wall vertical stack ducts installed inside the building partition walls to feed the supply outlets on the upper floors of a building. For example, a 4" × 10" diffuser with a capacity of 110 cfm can only deliver about 100 cfm if fed by a 3¼" × 10" vertical stack duct with a maximum capacity of 100 cfm.

duct run to outlet S5 have been marked on the figure. For each duct fitting, the capital letter identifies the type of fitting and the number indicates the equivalent length of straight duct in feet for that fitting. As shown, the path from the air handling unit to outlet S5 has an equivalent length of 115', derived as follows:

1 G30	Floor diffuser box = 30'
3 E10	90-degree adjustable elbows = 30'
1 B15	Side takeoff = 15'
1 E5	Reduction = 5'
	Additional loss for one of the first three takeoff fittings after the E5 transition = 25'
1 B10	Plenum takeoff = 10'

To size the ducts, it is necessary to find the supply and return runs that have the longest total equivalent length. The total equivalent length of a run is the sum of the measured horizontal and vertical straight lengths of duct in the path plus the equivalent length of all fittings in the path. Using the example shown in *Figure 48*, the supply run for outlet S5 has a total equivalent length of 157' (115' + 42'). Note that the duct run to outlet S4 has the longest total equivalent length of 179' (130' + 49') for the example system. This comparison between the S5 and S4 duct runs is stressed to emphasize that if the duct run with the greatest resistance is not obvious, it may be necessary to calculate the resistance of several or all of the runs in order to make absolutely sure that the longest equivalent length is used to establish the system pressure loss.

As shown in the example, the return duct run to inlet R1 has a measured length of 24' and an equivalent length of fittings of 140'; therefore, its total equivalent length is 164'.

The sizing of trunk and branch ducts can be done using different methods. One method uses a friction chart (*Appendix E*) for round duct as the basis for determining duct size. Using a conversion table, a round duct size taken from the chart can be converted into an equivalent square or rectangular duct size. The friction chart shows values of air quantity in cfm at the left. Round duct sizes are shown as diagonal lines that slant from left to right toward the top of the chart. Air velocities (in fpm) in the duct are shown as diagonal lines that slant from left to right toward the bottom of the chart. Friction losses in in. w.c. per 100' of duct are shown at the bottom of the chart.

Given any two values pertaining to a duct (cfm, duct size, velocity, or friction loss), the related values can be found. For example, if a duct with a design friction loss of 0.085 in. w.c. per 100' must carry 100 cfm, we can determine the duct size and related velocity in the duct. This is done by locating the intersection point of the 100 cfm and 0.085 in. w.c. friction loss lines on the chart. At this point, we find that the intersecting round duct size line shows a 6" duct, and the intersecting velocity line shows a velocity of 500 fpm. One important point to remember is that the chart is based on friction losses per 100' of duct. When using the chart to size the duct for a run with a total equivalent length that is greater or less than 100', a corrected friction factor must be used for the value of friction loss. This factor can be calculated as follows:

Corrected friction factor =

$$\frac{\text{design friction factor (per 100')}}{100} \times \text{total equivalent length of duct}$$

For example, to find the corrected friction factor when using a design friction loss of 0.085 in. w.c. per 100' to size a duct run with a total equivalent length of 50':

Corrected friction factor =

$$\frac{0.085 \text{ in. w.c.}}{100} \times 50 = 0.0425 \text{ in. w.c.}$$

To find the corrected friction factor when using a design friction loss of .085 in. w.c. per 100' to size a duct run with a total equivalent length of 200':

Corrected friction factor =

$$\frac{0.085 \text{ in. w.c.}}{100} \times 200 = 0.17 \text{ in. w.c.}$$

Failure to use a corrected friction factor when sizing a duct run that is under 100' long will oversize the duct and result in too much airflow. This is because the duct does not provide much resistance to airflow. Similarly, failure to use a corrected friction factor when sizing a duct run that is longer than 100' will undersize the duct and result in too little airflow because the duct provides too much resistance.

Ductwork can be sized using tables (*Appendix D*) that show the specific round and/or rectangular duct sizes used for various cfm rates. Typically, for low-velocity systems, these charts are based on a predetermined design supply duct friction loss of 0.08 in. w.c. per 100'. This means that the duct system will use up 0.08 in. w.c. of static pressure for each 100' of equivalent length that the air travels through. A predetermined design friction loss of 0.05 in. w.c. per 100' is used for the return duct.

Ductwork is most often sized starting at the air handler or furnace and working toward the end of the system. The rectangular duct sizes shown in *Figure 48* for the example duct system were selected using the duct sizing table in *Appendix D*. *Section A* of the supply trunk must carry the total system air quantity of 852 cfm to feed outlets S1 through S8. Therefore, the duct size needed is 22" × 8". *Section B* of the supply trunk is a 12" × 8" duct. This is based on the need to deliver the 402 cfm of air used to feed outlets S4 through S7. The return duct is sized at 22" × 8" because it must return 852 cfm to the system. As shown, the branch ducts are custom-sized using round duct based on their cfm output. Depending on the application, compromises are sometimes made when sizing the branch ducts. For example, outlets S5 and S6 are shown supplied with 6" round duct, even though they must carry 113 cfm. This was done for the sake of duct size standardization. Some designers may select a 7" diameter duct that would also handle 113 cfm. The advantage of the larger diameter would be slightly quieter operation.

In addition to using friction charts and duct sizing tables, duct calculators (*Figure 49*) can also

be used to size ducts. Duct calculators are available from all major HVAC equipment manufacturers, as well as from trade organizations such as SMACNA and ACCA. All duct calculators come with a set of instructions describing how to use them to size duct. Most can be used to design an average duct system without additional references and can save time.

Flexible insulated duct is widely used, especially in new residential construction. Flex duct has a much higher resistance to airflow than sheet metal duct, so a larger diameter of flex duct is required to get the same airflow. *Table 5* shows that a 6" sheet metal duct can carry 100 cfm, while a 6" flex duct can only carry 55 cfm. To carry 100 cfm with flex duct, an 8" diameter would be required.

Calculate the total loss of the air system to be sure that the selected system fan can produce the energy needed to move the required air quantity. As previously shown, the total friction loss for an air system is the sum of the supply system loss, return system loss, and components not included in the system fan rating. The friction loss for the supply system ductwork is found by taking the equivalent length of the longest run and multiplying it by the system design friction rate. In the example system in *Figure 48*, the duct run feeding outlet S4 is the longest run, with a total equivalent length of 179'. Also, the design friction rate used for the supply duct system per the duct sizing table in *Appendix D* was 0.08 in. w.c. per 100' of duct. Therefore, the supply friction loss for the example system is:

Friction loss = 179' × 0.08 in. w.c./100'

= 179' × 0.0008

= 0.143 in. w.g.

The pressure loss of the supply diffuser at the end of the run must be added to the supply duct loss. This value can be found using the manufacturer's product data sheets for the grille. A standard figure to use for a residential or light commercial diffuser is 0.05 in. w.c. For our example, the total supply loss is 0.193 in. w.c. (0.143 duct + 0.05 diffuser).

The friction loss for the return system duct is found the same way as for the supply system duct. This means you take the equivalent length of the longest return run and multiply it by the return system design friction rate. In the example system in *Figure 48*, the duct run for return R1 is the only return run. It has a total equivalent length of 164'.

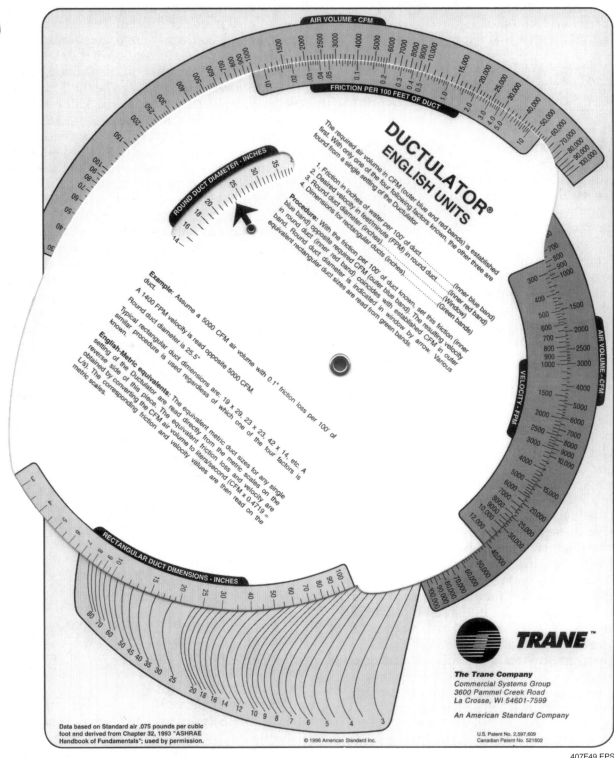

Figure 49 ◆ Duct design calculator.

Table 5 Maximum CFM Through Runout Ductwork

Runout Size	Supply CFM	Return CFM
Sheet Metal or Ductboard		
5" diameter	60	45
6" diameter	100	75
7" diameter	140	110
8" diameter	210	160
3¼" × 8" stack	70	55
3¼" × 10" stack	100	75
3¼" × 14" stack	140	110
2¼" × 12" stack	70	55
2¼" × 14" stack	90	70
Flex Duct*		
6" diameter	55	40
8" diameter	120	90
10" diameter	200	160
12" diameter	320	250
14" diameter	480	375
16" diameter	660	530
18" diameter	880	680
20" diameter	1,200	900

* The maximum duct capacity varies depending upon length, bends, and sags. The numbers shown assume straight runs cut to the proper length.

Also, the design friction rate per the duct sizing table of *Appendix D* used for the return duct system was 0.05 in. w.c. per 100' of duct. Therefore, the return friction loss for the example system is:

Friction loss = 164' × 0.05 in. w.c./100'

= 164' × 0.0005

= 0.082 in. w.c.

The pressure loss of the return grille at the end of the run must be added to the return duct loss. This can be found using the manufacturer's product data sheets for the specific grille; however, for the example, use the same value of 0.05 in. w.c. that was used with the supply system. For the example, the total return loss is 0.132 in. w.c. (0.082 duct + 0.05 grille).

The last component of the air system friction loss accounts for the losses for the components not included in the system fan rating, such as a wet evaporator coil or an electronic air cleaner. The friction loss of each of these items is normally included in the equipment product literature. We will arbitrarily use an equipment friction loss of 0.25 in. w.c. in order to complete the calculation for the total system loss of the example system:

Total air system loss = 0.193 supply + 0.132 return

+ 0.25 equipment

= 0.575 in. w.c.

For any system, there should be an operating balance between the energy produced by the fan and the energy consumed by the air system. For the example system, at the design airflow of 852 cfm, the total air system friction loss against which the fan must operate is 0.575 in. w.c. In order to operate at this airflow, the fan must at least be able to generate the same external static pressure at the design cfm as that consumed by the static pressure in the air system. Once the system external static pressure is known, the next step is to refer to the manufacturer's product data sheets for the selected equipment to verify that its fan can meet the requirement.

Figure 50 shows a manufacturer's fan performance chart for a furnace. The 40-B size operating on the medium-low fan speed can deliver a maximum of 945 cfm against an external static pressure of 0.6 in. w.c. The requirement for 852 cfm at a 0.575 in. w.c. external static pressure is within this particular furnace's performance envelope.

6.5.0 Other Duct System Design Considerations

This section provides some additional information to take into consideration when designing a duct system.

6.5.1 System Duct and Supply Outlet Capacity

When sizing an air system for an add-on or replacement job, it is necessary to evaluate the existing duct system to see if it has the capacity to handle the increase in required air volume.

FAN PERFORMANCE – HIGH-EFFICIENCY FURNACE AIR DELIVERY - CFM (WITH FILTER)								
FURNACE		EXTERNAL STATIC PRESSURE (IN. W.G.)						
SIZE	SPEED	0.1	0.2	0.3	0.4	0.5	0.6	0.7
40-A	High	1425	1370	1320	1265	1195	1125	1060
	Med-High	1315	1275	1230	1180	1125	1070	1000
	Med-Low	1145	1115	1090	1045	1010	955	880
	Low	1035	1000	970	930	900	850	800
40-B	High	1515	1465	1405	1345	1275	1210	1115
	Med-High	1350	1310	1275	1220	1160	1100	995
	Med-Low	1155	1140	1110	1070	1020	945	825
	Low	975	960	945	915	870	780	710
60	High	1510	1450	1380	1320	1230	1150	1045
	Med-High	1350	1300	1245	1190	1115	1030	945
	Med-Low	1165	1125	1090	1045	980	910	775
	Low	965	945	915	880	830	735	640
80	High	1590	1530	1470	1390	1315	1225	1140
	Med-High	1425	1390	1345	1285	1200	1135	1040
	Med-Low	1250	1220	1185	1135	1085	1010	930
	Low	1060	1055	1030	980	930	870	805

407F50.EPS

Figure 50 ◆ Fan performance chart.

One method of determining this is to calculate the capacity ratio of new system cfm requirements to total capacity of the existing duct system using the formula given below. In a well-designed ductwork system, the duct capacity ratio of actual airflow in cfm through the trunk ductwork should not exceed the total ductwork maximum capacity in cfm by more than 10 percent. In order to be an acceptable system, it must not exceed it by more than 15 percent.

Duct capacity ratio =

$$\frac{\text{required supply cfm}}{\text{total existing duct capacity (cfm)}}$$

The supply outlets and return grilles should also be checked for the following:

• Check the type of existing supply outlets to make sure they are suitable for the application and climate.

• Compare the capacities of the existing supply outlets for each room with the cfm requirement for the upgraded system. The capacity ratio for each supply outlet or return grille should be calculated using the formula below. In a properly designed system, the required cfm should not exceed the existing supply diffuser capacity by more than 15 percent. If it does, additional outlets should be installed in the room to meet the new requirements.

Outlet capacity ratio =

$$\frac{\text{required supply cfm}}{\text{total existing outlet capacity (cfm)}}$$

• Check the types and capacities of existing return grilles to make sure they are suitable for the application and volume of airflow for the upgraded system.

6.5.2 Insulation and Vapor Barriers

When ductwork passes through an unconditioned space, such as an attic or crawl space, heat transfer takes place between the air in the duct and the air in the unconditioned space. Insulation should be applied to the ductwork if it passes through an unconditioned space or if there is a difference of 15°F from the inside to the outside of the duct. Many installations use preinsulated ductboard for the main supply and return ducts.

Metal duct can be insulated in two ways: on the outside or on the inside. Insulation inside the duct is installed by the duct manufacturer. Insulation with a vapor barrier can be wrapped around the outside of the ductwork after it has been installed. Always use a vapor barrier on duct insulation. Local code requirements may dictate the insulation thickness. Once installed, all joints must be properly sealed with duct tape. To avoid condensation damage, you must also seal punctures, seams, and slits in the vapor barrier.

Sheet metal duct with outside insulation has a lower pressure loss, and is therefore more efficient than sheet metal lined on the inside. Another advantage is that the cost of the unlined metal duct is lower because the physical size of the duct can be smaller. A duct with 1" of internal insulation must have width and height dimensions that are 2" larger in order to deliver the same amount of air.

Disadvantages of using duct with outside insulation are that it takes longer to install, there is a greater chance of damage during installation, and it tends to be noisier than a lined system. *Figure 51* shows a typical installation of insulated ductwork in an unconditioned space.

In attic installations, the ductwork must be insulated to maintain proper cooling and heating in the conditioned rooms of the building. *ASHRAE Standard 90-80* specifies the minimum acceptable value (R-value) of insulation. *Figure 52* shows an example using *ASHRAE Standard 90-80* to calculate the R-value for insulation related to the cooling mode.

The R-value for the system must also be calculated for the heating mode the same way. The amount of insulation eventually used is determined by the mode with the greatest need.

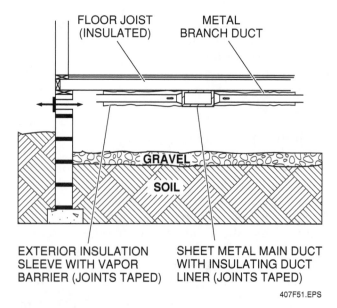

Figure 51 ◆ Duct installation in an unconditioned space.

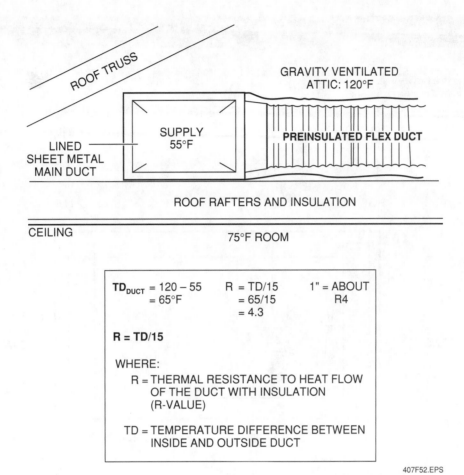

$$TD_{DUCT} = 120 - 55 \qquad R = TD/15 \qquad 1" = ABOUT$$
$$= 65°F \qquad\qquad = 65/15 \qquad\qquad R4$$
$$= 4.3$$

R = TD/15

WHERE:
   R = THERMAL RESISTANCE TO HEAT FLOW
   OF THE DUCT WITH INSULATION
   (R-VALUE)

   TD = TEMPERATURE DIFFERENCE BETWEEN
   INSIDE AND OUTSIDE DUCT

407F52.EPS

Figure 52 ◆ Example of R-value calculation using *ASHRAE Standard 90-80*.

Because attic systems are more common in warm climates, it is the cooling mode that usually determines the amount of insulation required. Local code requirements may call for even thicker insulation.

7.0.0 ◆ SUPPORT SYSTEMS

This section will discuss various support systems associated with heating and cooling systems.

7.1.0 Refrigerant Piping

A comfort air conditioning system has three refrigerant piping runs (*Figure 53*):

• The suction line conveys low-temperature, low-pressure, superheated vapor refrigerant from the evaporator outlet to the compressor inlet.
• The hot gas (discharge) line conveys hot, high-pressure vapor refrigerant from the compressor to the condenser.

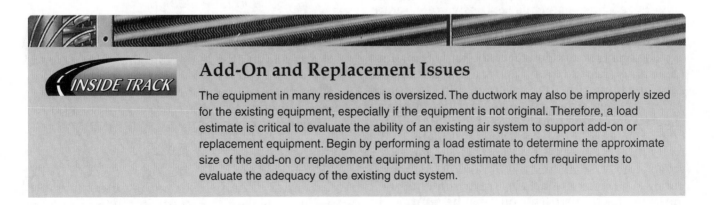

Add-On and Replacement Issues

The equipment in many residences is oversized. The ductwork may also be improperly sized for the existing equipment, especially if the equipment is not original. Therefore, a load estimate is critical to evaluate the ability of an existing air system to support add-on or replacement equipment. Begin by performing a load estimate to determine the approximate size of the add-on or replacement equipment. Then estimate the cfm requirements to evaluate the adequacy of the existing duct system.

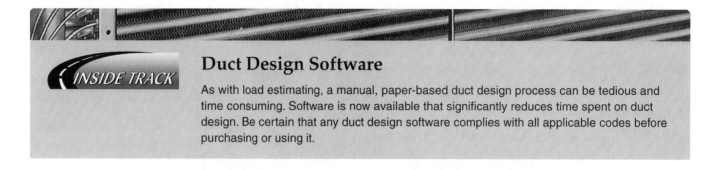

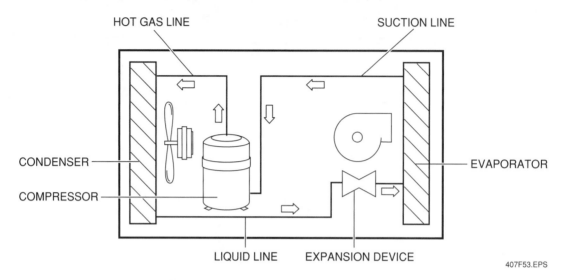

Figure 53 ◆ Refrigeration system piping runs.

- The liquid line conveys high-pressure liquid refrigerant from the condenser to the expansion device.

In residential split system applications, only the suction and liquid lines, which connect the indoor and outdoor units, are installed at the site. The hot gas line is installed in the condensing unit at the factory. The suction and liquid line sizes are specified in the manufacturer's product data. Do not deviate from the recommended sizes. Undersized lines can cause noisy operation, evaporator starvation, and liquid slugging. Oversized lines can cause excessive refrigerant charge, liquid slugging, excess power consumption, and compressor damage.

The diameter and length of the refrigerant lines affect the pressure drop. The smaller the diameter and the longer the line, the greater the pressure drop. In their product data, manufacturers will specify larger suction line diameters if the standard diameter line exceeds a certain length.

Oil return is also important. The compressor discharges a small amount of oil with the refrigerant. During cold startups, larger amounts of oil may be discharged. This oil must be returned to the compressor in order for the compressor to have proper lubrication. This is not a problem in the liquid line because oil mixes easily with liquid refrigerant. In the suction and discharge lines, however, it can be a major problem. If these lines are oversized, there will not be sufficient refrigerant velocity to move the oil up vertical piping sections (risers), and the oil will settle in low spots. In general, lines should be pitched in the direction of flow. This will help maintain oil flow and avoid backward flow during shutdown.

7.1.1 Suction Line Design

The location of the indoor coil relative to the compressor in the outdoor unit must be considered when running the refrigerant lines. See *Figure 54*. If the indoor coil is located above the outdoor unit, the suction line should loop up to the height of the indoor coil (A). This helps prevent liquid refrigerant from migrating to the compressor in the outdoor unit during the compressor off cycle. When the indoor coil and the outdoor unit are at the same level, the suction line should pitch toward the compressor in the outdoor unit (B) with no sags or low spots in any straight run.

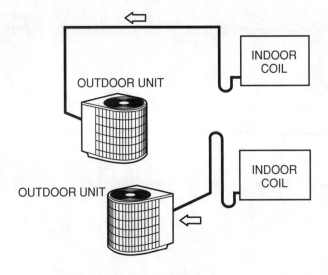

(A) VAPOR LINE WITH INDOOR COIL ABOVE THE OUTDOOR UNIT

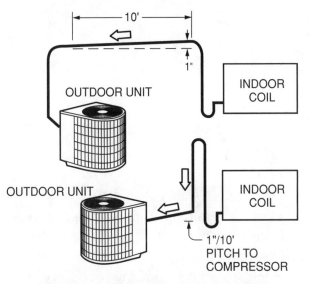

(B) VAPOR LINE WITH INDOOR COIL AT THE SAME LEVEL AS THE OUTDOOR UNIT

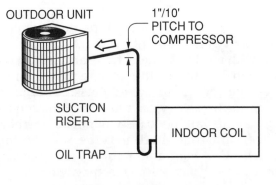

(C) VAPOR LINE WITH INDOOR COIL BELOW OUTDOOR UNIT

407F54.EPS

Figure 54 ◆ Suction line design.

When the indoor unit is below the outdoor unit, oil must be returned to the compressor in the outdoor unit via a vertical riser in the suction line (C). An oil trap should be installed at the start of the riser to collect oil for feedback to the compressor. The horizontal run to the outdoor unit should be pitched toward the outdoor unit. The suction line should always be adequately insulated.

7.1.2 Liquid Line Design

Undersizing the liquid line will cause flash gas, which will affect the performance of the metering device. An oversized liquid line will create the need for extra refrigerant charge. The liquid line should be insulated wherever it is exposed to direct sunlight or excessive heat, such as in an attic. Heat causes flashing and loss of capacity. For vertical runs of more than 25', refer to the manufacturer's instructions.

7.2.0 Condensate Piping

Condensate forms on the coil of an indoor unit and will accumulate in the area beneath the coil. The condensate must be removed to a proper drain, or it can cause damage to the unit or the building. The indoor unit will have a condensate pan under the evaporator coil. A pipe connected to the pan drains the condensate to an indoor drain or to the outside of the building. Condensate drainage is especially important where units are installed in attics or above drop ceilings. In such cases, many local codes require the placement of a separate condensate pan (*Figure 55*) under the unit. This pan catches and removes the condensate in the event the built-in system is plugged or damaged.

Some systems are equipped with two condensate drains. The secondary drain is located above the primary drain. If the primary drain is plugged, the condensate can escape through the secondary drain.

The condensate drain requires a trap to prevent water from being held in the drain pan by the blower. The condensate pipe must be pitched toward the drain at a minimum of ¼" per foot.

Condensate drainage is also required for condensing furnaces. A condensate pipe is usually run to a floor drain from the vent pipe connected to the furnace condensing coil (secondary heat exchanger).

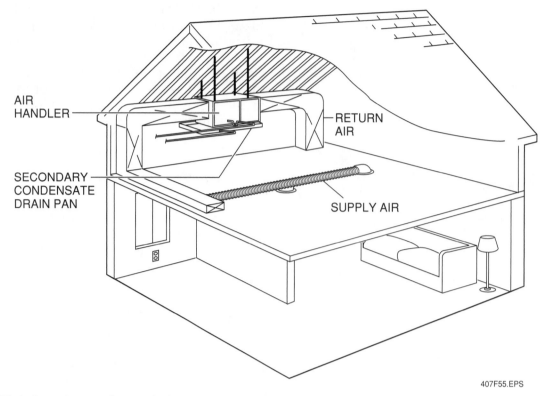

Figure 55 ◆ Secondary condensate drain pan.

7.3.0 Electrical Service

Electrical service is not usually an issue in new construction situations because the electrical service should be designed to handle the load created by the planned heating and cooling equipment. It is likely to be more of an issue in cases where air conditioning is being added and sometimes when existing equipment is being replaced.

Some older homes may be equipped with only a 100-amp electrical service. Depending on the number of appliances, power tools, and other electrical equipment, that capacity can be quickly met. This leaves no room for additional loads, especially high-current loads such as air conditioners. This problem is even more of a concern when electric resistance heaters are being used as the primary heat source or as a supplemental heat source for heat pumps. In these cases, a new, higher-amperage electrical service might have to be installed.

You can estimate the existing electrical load by totaling the current draw in amps (A) for all the appliances and lighting in the building. Here are some examples:

Electric dryer	25A
Washer	5A–10A
Electric oven/range	45A
Electric oven/range (self-cleaning)	60A
Microwave oven	10A
Dishwasher	10A
Small appliances (combined)	10A–20A
Electric hot water heater	15A–40A
Garbage disposal	5A–10A
Lighting (combined)	10A–15A

A new comfort air conditioning system can add anywhere from 15 to 60 amps, depending on the capacity and type of system. A 5-ton heat pump with electric heaters, for example, can add more than 50 amps to the load. In many homes, that can be enough to exceed the capacity of the existing electrical service. It may seem reasonable to assume that all the loads will not be operating at once, and that there may be a difference between rated load and actual load. Consult a qualified electrician before reaching a decision.

407F55.EPS

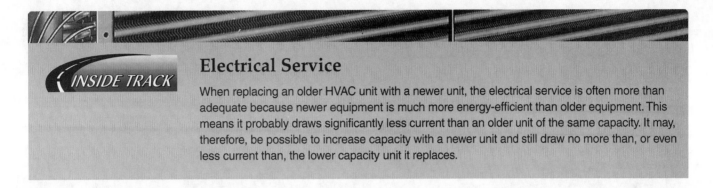

8.0.0 ◆ LOAD ESTIMATING FOR COMMERCIAL BUILDINGS

ACCA Manual J provides load estimating data and instructions for residential buildings. ACCA Manual N provides load estimating data and instructions for small to medium commercial buildings using packaged or split systems, including heat pumps. Neither method is intended for large buildings that use built-up systems or central station equipment. Such designs must be done by qualified engineers and architects using information and procedures from the latest ASHRAE handbook. Although some of the load factors and calculations for commercial systems are similar to those for residential systems, there are several major differences. Some examples are as follows:

- Commercial buildings use different construction materials. While homes lean toward wood frame or masonry construction, commercial buildings often use facades of metal, masonry, and glass.
- It is often necessary to deal individually with core and exterior zones. Because of the effect of radiation and infiltration on the perimeter of a building, the loads at the perimeter and core can vary widely. It is not unusual to have situations during occupied periods when perimeter zones need cooling and the core zone needs heating or just ventilating.
- A commercial building is more likely to have a large, open area, such as a lobby or atrium with vaulted ceilings, and a lot of traffic in and out of the building with its attendant infiltration.
- The load varies significantly at different times of day. The people load is a critical factor in commercial system design. Shopping centers, theaters, and office buildings have some periods when they are heavily occupied and other periods when they may be unoccupied. The load varies widely between these periods. Solar radiation is another major factor. *Appendix C* contains examples of load estimating forms from ACCA Manual N. You can see that, unlike Manual J, the calculations are made for different times of day. You can also see that the load varies greatly between those times. Note especially how different the solar radiation and transmission loads are from noon to 6 PM. Also note the change in sensible cooling load.

- Loads can vary radically from one area to another within the building. (While this is true to some extent in residential work, it is less of a factor because residential buildings are generally much smaller and use much different construction materials.) In a southern exposure room with a lot of glass at mid-afternoon, the load might be three to four times that of a northern exposure room and two to three times that of a southern exposure room with no glass.
- The interiors of commercial buildings are often rearranged to accommodate tenants. The system design must therefore be flexible enough to deal with changing load patterns.

Commercial buildings have many of the same heating and cooling loads as residential buildings. They also have other considerations such as:

- Large latent loads
- Large lighting loads
- Computers
- Printers and copiers

Commercial buildings are also subject to changing loads and interior partitions are added or moved to accommodate changing tenant demand.

1. All of the following are factors used in calculating the heating load *except* the _____.
   a. type of construction material used
   b. exterior dimensions of the building
   c. color of the roof
   d. number, sizes, and types of windows

2. Below-grade (underground) walls are a factor in calculating _____.
   a. heating and cooling loads
   b. heating loads only
   c. cooling loads only
   d. heat gains

3. The term convection refers to heat that is transferred by the _____.
   a. flow of heat through a substance
   b. light shining off reflective objects
   c. movement of a fluid
   d. contact of materials or objects

4. All of these factors affect the amount of heat that will flow through a wall *except* _____.
   a. relative humidity
   b. the area of the wall
   c. the indoor and outdoor temperatures
   d. the heat-conducting properties of the wall

5. The term *U-factor* can represent the _____.
   a. amount of below-grade wall
   b. amount of area on the upper floors of a building
   c. rate at which heat will flow through the walls or roof of a building
   d. amount of ultraviolet light to which an area is exposed

6. Which of these factors affects both the heating and cooling load calculations?
   a. Floors
   b. Infiltration
   c. Swing factor
   d. Appliance load

7. When the area of a window is multiplied by the HTM, the result represents the _____.
   a. heat gain for the window
   b. heat loss for the window
   c. heating or cooling load in Btuh
   d. heating or cooling load in Btus per sq ft

8. All of the following factors must be considered in the selection of a furnace for a heating-only application *except* _____.
   a. latent heat load
   b. sensible heat load
   c. cfm
   d. indoor temperature

9. Which of the following is an effect that might be caused by undersized cooling equipment?
   a. Frequent on-off cycling
   b. Poor humidity control
   c. Inability to handle peak loads
   d. Higher initial cost

10. The indoor unit of a split system contains the _____.
    a. cooling coil, metering device, and blower
    b. cooling coil, condenser, and blower
    c. compressor, condenser, and blower
    d. condenser, metering device, and blower

11. The condensing unit should generally be placed no more than _____ from the evaporator unit.
    a. 30'
    b. 50'
    c. 70'
    d. 90'

12. In colder climates, heat pumps are selected to match the cooling load.
    a. True
    b. False

13. The combined capacity of a heat pump equipped with electric resistance heaters should *not* exceed _____ percent of the calculated heating load.
    a. 100
    b. 105
    c. 110
    d. 115

14. Airflow velocity is usually measured in _____.
    a. feet per minute (fpm)
    b. cubic feet per minute (cfm)
    c. feet per second (fps)
    d. miles per hour (mph)

15. Within an air distribution system, the highest pressure level is found at the _____.
    a. conditioned space
    b. input to the return duct
    c. input to the blower
    d. output of the blower

16. The loss coefficient value (C-value) is used to calculate the _____.
    a. pressure loss through a straight duct section
    b. pressure loss through a specific type of fitting
    c. size of rectangular duct
    d. equivalent length of a specific type of fitting

17. The term *external static pressure* refers to losses caused by _____.
    a. registers and return grilles
    b. conditions external to the duct system
    c. the fan assembly
    d. all components external to the fan assembly

18. Failure to use a corrected friction factor when sizing a duct run under 100' long will _____.
    a. undersize the duct and result in too little airflow
    b. undersize the duct and result in too much airflow
    c. oversize the duct and result in too much airflow
    d. oversize the duct and result in too little airflow

19. Refer to the duct friction chart in *Appendix E*. What size round duct should be used to carry an air quantity of 1,000 cfm at a duct design friction loss of 0.2 in. w.c. per 100'?
    a. 8"
    b. 11"
    c. 12"
    d. 20"

20. Refer to *Appendix D*. What size rectangular duct should be used to carry a return system air quantity of 780 cfm?
    a. 16" × 8"
    b. 18" × 8"
    c. 20" × 8"
    d. 24" × 8"

21. Insulation should be applied to ductwork _____.
    a. if it passes through an air-conditioned space
    b. instead of a vapor barrier
    c. if there is a temperature difference of 15° from the inside to the outside of the duct
    d. and is more efficient when applied inside sheet metal ducts rather than outside

22. The suction line conveys _____.
    a. hot, high-pressure vapor from the compressor to the condenser
    b. low-temperature, low-pressure vapor from the evaporator to the compressor
    c. high-pressure liquid refrigerant from the condenser to the evaporator
    d. high-pressure liquid from the condenser to the metering device

23. When the indoor coil is at a higher level than the outdoor unit, the suction line should _____.
    a. be pitched toward the indoor coil
    b. have a trap at the compressor inlet
    c. have a loop that rises to the height of the indoor coil
    d. be no more than 50' long

24. When liquid lines are being installed, _____.
    a. they need to be insulated if they are exposed to cold weather
    b. they should be insulated wherever they are exposed to direct sunlight or excessive heat
    c. they are best installed using long vertical runs
    d. oversized liquid lines will cause flash gas, affecting metering device performance

25. In regard to condensate piping, _____.
    a. water is collected in a pan beneath the coil and removed through evaporation
    b. the drain requires a trap to prevent water from being held in the drain pan by the blower
    c. local codes require outdoor units to have secondary condensate pans and secondary drains
    d. drainage is not required for condensing furnaces because condensed water is soon transformed to vapor

Summary

Improper sizing of heating and cooling equipment can lead to unsatisfactory operation. If the equipment is undersized or oversized, the building occupants will be uncomfortable on days when design conditions occur. If ductwork is improperly designed, similar problems as well as noisy operation can occur. Even if the original equipment was properly designed, later additions or modifications to the building can cause problems.

It is often up to the service technician to spot design or installation problems that could lead to unsatisfactory operation. By understanding the basic principles of system design, you will be better equipped to recognize these issues.

Notes

Trade Terms Introduced in This Module

Cubic feet per minute (cfm): A measure of the amount or volume of air in cubic feet flowing past a point in one minute. Cubic feet per minute can be calculated by multiplying the velocity of air, in feet per minute (fpm), by the area it is moving through, in square feet. It can also be measured with various test instruments.

Conductance: A measurement of the heat flow through nonhomogeneous materials such as glass blocks, hollow tiles, and concrete blocks. Specifically, it is the heat flow rate through one square foot of a nonhomogeneous material of a given thickness when there is a 1°F temperature difference between the two surfaces of the material.

Conductivity: The ability of a material to conduct heat.

Diffuser: An outlet that discharges supply air into a room in various directions and planes.

External static pressure: The total pressure loss of the duct system ductwork and components external to the supply fan assembly.

Fan brake power: The actual total power needed to drive a fan to deliver the required volume of air through a duct system. It is greater than the expected power needed to deliver the air because it includes losses due to turbulence, inefficiencies in the fan, and bearing losses.

Grille: A louvered covering for any opening through which air passes.

Infiltration: Air that enters or escapes the building though openings such as windows, doors, vents, fireplace chimneys, or structural cracks.

R-value: The thermal resistance of a given thickness of insulating material.

Register: A grille equipped with a damper or control valve.

Static pressure: The pressure exerted uniformly in all directions within a duct.

Temperature differential: The difference in air temperature on two sides of an object.

Thermal conductivity: The ability of a given substance to conduct heat; specifically, it is the heat flow per hour (Btuh) through one square foot of one-inch thick homogeneous material when the temperature difference between the two faces is 1°F.

Throw: The horizontal or vertical axial distance an airstream travels after leaving a supply outlet before the maximum stream velocity is reduced to a specific terminal velocity.

Total pressure: The sum of the static pressure and the velocity pressure for any cross section of an air duct. It determines how much energy must be supplied to the system by the fan to maintain airflow.

U-factor: The heat flow per hour through one square foot of material when the temperature difference between the two surfaces of the material is 1°F.

Velocity: A measurement of how fast the air is moving. The rate of airflow is usually measured in fpm.

Velocity pressure: The pressure in a duct due to the movement of the air. It is the difference between the total pressure and the static pressure.

Volume: The amount of air in cubic feet flowing past a given point in one minute (cfm).

Example of Heating Load Estimate

FIGURE 3-3 EXAMPLE HEAT LOSS CALCULATION
DO NOT WRITE IN SHADED BLOCKS

		Const No.	HTM Htg.	HTM Clg.	Entire House Area or Length	Entire House Btuh Htg.	Entire House Btuh Clg.	Living 1 Area or Length	Living Btuh Htg.	Living Btuh Clg.	Dining 2 Area or Length	Dining Btuh Htg.	Dining Btuh Clg.	Laundry 3 Area or Length	Laundry Btuh Htg.	Laundry Btuh Clg.	Kitchen 4 Area or Length	Kitchen Btuh Htg.	Kitchen Btuh Clg.	Bath-1 5 Area or Length	Bath-1 Btuh Htg.	Bath-1 Btuh Clg.
1 Name of Room					Entire House			Living			Dining			Laundry			Kitchen			Bath-1		
2 Running Ft. Exposed Wall					160			21			25			18			11			9		
3 Room Dimensions Ft.					51 x 29			21 x 14			7 x 18			7 x 11			11 x 11			9 x 11		
4 Ceiling Ht. Ft Directions Room Faces					8			8 West			8 North			8			8 East			8 East		
5 Gross Exposed Walls & Partitions	a	12-d			1280			168			200			144			88			72		
	b	14-b			480																	
	c	15-b			800																	
	d																					
6 Windows & Glass Doors Htg.	a	3-A	41.3		60	2478		40	1652		20	826										
	b	2-C	48.8		20	976																
	c	2-A	35.6		105	3738											11	392		8	285	
	d																					
7 Windows & Glass Doors Clg.	North																					
	E&W																					
	South																					
8 Other Doors		11-E	14.3		37	529								17	243							
9 Net Exposed Walls & Partitions	a	12-d	6.0		1078	6468		128	768		180	1080		127	762		77	462		64	384	
	b	14-b	10.8		460	4968																
	c	15-b	5.5		800	4400																
	d																					
10 Ceilings	a	16-d	4.0		1479	5916		294	1176		126	504		77	308		121	484		99	396	
	b																					
11 Floors	a	21-a	1.8		1479	2662																
	b																					
12 Infiltration HTM			70.6		222	15673		40	2824		20	1412		17	1200		11	777		8	565	
13 Sub Total Btuh Loss = 6+8+9+10+11+12						47808			6420			3822			2513			2115			1630	
14 Duct Btuh Loss			0%																			
15 Total Btuh Loss = 13 + 14						47808			6420			3822			2513			2115			1630	
16 People @ 300 & Appliances 1200																						
17 Sensible Btuh Gain = 7+8+9+10+11+12+16																						
18 Duct Btuh Gain				%																		
19 Total Sensible Gain = 17 + 18																						

From Table 2

ASSUMED DESIGN CONDITIONS AND CONSTRUCTION (Heating):

		Const. No.	HTM
A.	Determing Outside Design Temperature -5° db-Table 1		
B.	Select Inside Design Temperature 70° db		
C.	Design Temperature Difference: 75 Degrees		
D.	Windows: Living Room & Dining Room - Clear Fixed Glass, Double Glazed - Wood Frame - Table 2 .	3A	41.3
	Basement - Clear Glass Metal Casement Windows, with Storm - Table 2	2C	48.8
	Others - Double Hung, Clear, Single Glass and Storm, Wood Frame - Table 2	2A	35.6
E.	Doors: Metal, Urethane Core, no Storm - Table 2	11E	14.3
F.	First Floor Walls: Basic Frame Construction with Insulation (R-11) ½" Board - Table 2	12d	6.0
	Basement wall: 8" Concrete Block - Table 2		
	Above Grade Height: 3 ft (R = 5)	14b	10.8
	Below Grade Height: 5 ft (R = 5)	15b	5.5
G.	Ceiling: Basic Construction Under Vented Attic with Insulation (R-19) - Table 2............	16d	4.0
H.	Floor: Basement Floor, 4" Concrete - Table 2	21a	1.8
I.	All moveable windows and doors have certified leakage of 0.5 CFM per running foot of crack (without storm), envelope has plastic vapor barrier and major cracks and penetrations have been sealed with caulking material, no fireplace, all exhausts and vents are dampered, all ducts taped.		

407A01.EPS

DO NOT WRITE IN SHADED BLOCKS

6 Bedroom 3			7 Bedroom 2			8 Bath 2			9 Bedroom 1			10 Hall			11 Rec. Room			12 Shop & Utility			#
10			24			5			29			8			83			88			2
10 x 11			14 x 10			5 x 5			15 x 14			8 x 14			27 x 29			24 x 29			3
8	East		8	E & S		8	South		8	S & W		8	West		8	E & S		8	East		4
Area or Length	Htg	Clg	Area or Length	Htg	Clg	Area or Length	Htg	Clg	Area or Length	Htg	Clg	Area or Length	Htg	Clg	Area or Length	Htg	Clg	Area or Length	Htg	Clg	
80			192			40			232			64			249			231			5
															415			385			
															16	781		4	195		6
22	783		28	997		8	285		28	997												
																					7
												20	286								8
58	348		164	984		32	192		204	1224		44	264		233	2516		227	2452		9
															415	2283		385	2118		
110	440		140	560		25	100		210	840		112	448								10
															783	1409		696	1253		11
22	1553		28	1977		8	565		28	1977		20	1412		16	1130		4	282		12
	3124			4518			1142			5038			2410			8119			6300		13
	—			—			—			—			—			—			—		14
	3124			4518			1142			5038			2410			8119			6300		15
																					16
																					17
																					18
																					19

407A02.EPS

Example of Cooling Load Estimate

	Name of Room				Entire House			1 Living			2 Dining			3 Laundry			4 Kitchen			5 Bath-1		
2	Running Ft. Exposed Wall				160			21			25			18			11			9		
3	Room Dimensions Ft.				51 x 29			21 x 14			7 x 18			7 x 11			11 x 11			9 x 11		
4	Ceiling Ht. Ft Directions Room Faces				8			8 West			8 North			8			8 East			8 East		
	TYPE OF EXPOSURE	Const. No.	HTM Htg.	HTM Clg.	Area or Length	Htg.	Clg.	Area or Length	Htg.	Clg.	Area or Length	Htg.	Clg.	Area or Length	Htg.	Clg.	Area or Length	Htg.	Clg.	Area or Length	Htg.	Clg.
5	Gross a	12-d			1280			168			200			144			88			72		
	Exposed b	14-b			480																	
	Walls & c	15-b			800																	
	Partitions d	13N			232																	
6	Windows a																					
	& Glass b																					
	Doors Htg. c																					
	d																					
7	Windows North			14	20		280				20		280									
	& Glass E & W			44	115		5060	40		1760							11		484	8		352
	Doors Clg. South			23	30		690															
	Basement			70/36	8/8		848															
8	Other Doors	10-e		3.5	37		130							17		60						
9	Net a	12-d		1.5	1078		1617	128		192	180		270	127		191	77		116	64		96
	Exposed b	14-b		1.6	233		373															
	Walls & c	15-b		0																		
	Partitions d	13-n		0																		
10	Ceilings a	16-d		2.1	1479		3106	294		617	126		265	77		162	121		254	99		208
	b																					
11	Floors a	21-a		0																		
	b	19-f		0																		
12	Infiltration HTM			7.18	218		1565	40		287	20		144	17		122	11		79	8		57
13	Sub Total Btuh Loss = 6+8+9+10+11+12																					
14	Duct Btuh Loss %																					
15	Total Btuh Loss = 13 + 14																					
16	People @ 300 & Appliances 1200						3000	3		900	3		900			—			1200			—
17	Sensible Btuh Gain = 7+8+9+10+11+12+16						16669			3756			1859			535			2133			713
18	Duct Btuh Gain %									—			—			—			—			—
19	Total Sensible Gain = 17 + 18						16669			3756			1859			535						

NOTE: USE CALCULATION PROCEDURE D TO CALCULATE THE EQUIPMENT COOLING LOADS

*Answer for "Entire House" may not equal the sum of the room loads if hall or closet areas are ignored or if heat flows from one room to another room.

	ASSUMED DESIGN CONDITIONS AND CONSTRUCTION (Cooling)	Const. No.	HTM
A.	Outside Design Temperature: Dry Bulb 88 Rounded to 90 db 38 grains - Table 1		
B.	Daily Temperature Range: Medium - Table 1		
C.	Inside Design Conditions: 75F, 55% RH Design Temperature Difference = ... (90-75 = 15)		
D.	Types of Shading: Venetian Blinds on All First Floor Windows - No Shading, Basement		
E.	Windows: All Clear Double Glass on First Floor - Table 3A		
	North		14
	East or West		44
	South		23
	All Clear Single Glass (plus storm) in Basement - Table 3A Use Double Glass		
	East		70
	South		36
F.	Doors: Metal, Urethane Core, No Storm, 0.50 CFM/ft.	11e	3.5
G.	First Floor Walls: Basic Frame Construction with Insulation (R-11) x ½ " board - Table 4	12d	1.5
	Basement Wall: 8" Concrete Block, Above Grade: 3 ft (R-5) - Table 4	14b	1.6
	8" Concrete Block Below Grade: 5 ft (R-5) - Table 4	15b	0
H.	Partition: 8" Concrete Block Furred, with Insulation (R-5), Δ T approx. 0°F - Table 4	13n	0
I.	Ceiling: Basic Construction Under Vented Attic with Insulation (R-19), Dark Roof - Table 4	16d	2.1
J.	Occupants: 6 (Figured 2 per Bedroom, But Distributed 3 in Living, 3 in Dining)		
K.	Appliances: Add 1200 Btuh to Kitchen		
L.	Ducts: Located in Conditioned Space - Table 7B		
M.	Wood & Carpet Floor Over Unconditioned Basement, Δ T approx. 0°F	19	0
N.	The Envelope was Evaluated as Having Average tightness - (Refer to the Construction details at the Bottom of Figure 3-3)		
O.	Equipment to be Selected From Manufacturers Performance Data.		

407A03.EPS

6 Bedroom 3			7 Bedroom 2			8 Bath 2			9 Bedroom 1			10 Hall			11 Rec. Room			12 Shop & Utility			#
10			24			5			29			8			83			88				2
10 x 11			14 x 10			5 x 5			15 x 14			8 x 14			27 x 29			24 x 29				3
8	East		8	E & S		8	South		8	S & W		8	West		8	E & S		8	East			4
Area or Length	Btuh Htg	Clg	Area or Length	Btuh Htg	Clg	Area or Length	Btuh Htg	Clg	Area or Length	Btuh Htg	Clg	Area or Length	Btuh Htg	Clg	Area or Length	Btuh Htg	Clg	Area or Length	Btuh Htg	Clg	
80			192			40			232			64									5
															249			231			
															415			385			
															232						
																					6
22		968	17		748				17		748				8/8		560				7
			11		253	8		184	11		253				8/8		288				
												20		70							8
58		87	164		246	32		48	204		306	44		66							9
															233		373				
110		231	140		294	25		53	210		441	112		236							10
																					11
22		158	28		201	8		57	28		201	20		144	16		115				12
																					13
																					14
																					15
	—			—			—			—			—			—			—		16
		1444			1742			342			1949						516			1336	17
	—			—			—			—			—			—			—		18
		1444			1742			342			1949						516			1336	19

Line 1. Identify each area.

Lines 2 and 3. Enter the pertinent dimensions.

Line 4. For reference, enter the ceiling height and the direction the glass faces.

Lines 5A through 5D. Enter the gross wall area for the various walls. For rooms with more than one exposure, use one line for each exposure. For rooms with more than one type of wall construction, use one line for each type of construction. Find the construction number in the tables in back of this manual. Enter the construction number on the appropriate line.

Example: The gross area of the west living room wall is 168 sq. ft. This wall is listed in Table 4, number 12, line D. The construction number is 12-d.

Line 6. Not required for cooling calculations.

Line 7. Enter the areas of windows and glass doors for the various rooms and exposures. Use the drawings and construction details, or determine by inspection, the types of windows used in each room. Also note the shading and the exposure. Refer to the tables in the back of this manual and select the HTM for each combination of window, shading and exposure. Enter the HTM values in the column designated cooling. Multiply each window area by its corresponding HTM to determine the heat gain through the window. Enter this value in the column Btuh - Cooling.

Example: The living room has 40 sq. ft. of west-facing glass. The window is double pane with drapes or blinds. The design temperature difference is rounded to 15°F. The HTM listed in Table 3A, (double glass, drapes, or venetian blinds, design temperature difference of 13°F), is Btu/(hr. sq. ft.) The heat gain is:

44 Btu/(hr. sq. ft.) x 40 sq. ft. = 1,760 Btuh.

407A04.EPS

ACCA Manual N Commercial Load Estimating Forms

COMMERCIAL LOAD CALCULATIONS
(ROOM, ZONE or BLOCK LOAD)

Air Conditioning Contractors of America
1513 16th Street, N.W., Washington, D.C. 20036

FORM N - 1

For:	Name	Rex Drugstore	City	Jefferson City	Phone	State	MO	Zip
	Address				Phone	State		Zip
By:	Contractor				Phone	State		Zip
	Address		City			State		Zip

1. DESIGN CONDITIONS (COOLING)

(Time of Day __noon__) (Daily Range __23__) (Latitude __38__)

a) Inside db __75__   RH __50%__   b) Outside db __95__   wb __74__   Grains __28__

Outside db @ 3 p.m. __95__ (-) minus time of day corrections __5__ (-) minus inside drybulb __75__ = __15__ T.D.

2. SOLAR RADIATION HEAT GAIN THROUGH GLASS

| | | | | | **COOLING LOAD** | |
| | | | | | Sensible | Notes |

Exposure	Sq. Ft.		Solar Factor		Shading and/or Glass Factor		Sensible	Notes
EAST	24	x	58	x	0.64	=	851	
EAST	24	x	91	x	0.94	=	2,053	
WEST	40	x	30	x	0.94	=	1,128	
WEST	126	x	30	x	0.94	=	3,553	
		x		x		=		
		x		x		=		
		x		x		=		

3. TRANSMISSION GAINS

	Exposure	Sq. Ft.		U Factor		Equivalent or db Temp Diff			Notes
Glass	ALL	214	x	1.04	x	15	=	3,338	noon TD 150°F
			x		x		=		
			x		x		=		
Walls	EAST	252	x	0.077	x	28	=	593	-2 OR correction
	SOUTH	750	x	0.077	x	10	=	578	
	WEST	134	x	0.077	x	10	=	103	
Doors			x		x		=		
			x		x		=		
Partitions			x		x		=		
Floors	✓	2,250	x	0.28	x	5	=	3,150	-2 OR correction
Roof			x		x		=		
Roof/Ceiling	✓	2,250	x	0.10	x	26	=	5,850	
RA Ceiling			x		x		=		

Use Table 9a to determine the Temperature Difference Across a Return Air Ceiling

4. INTERNAL HEAT GAIN

a. Occupants

	Number	Sensible	Latent		Sensible	Latent
	14	x 255		=	3,570	
	9	x 315		=	2,835	
	14		x 325	=		4,550
	9		x 325	=		2,925

b. Lights & Others

NOTE: Use 60% of installed watts for lights in return air ceiling.

	Watts			Sensible	Latent
Incandescent Lights		X 3.4	=		
Flourescent Lights	6,000	X 4.1	=	24,600	

	N.P.H.P.	Btuh	Load Factor	Usage Factor		
Motors	1/4	1,180	x 1.0	x 0.75	885	Exb. Hood
	1/8	900	x 1.0	x 0.75	675	
			x	x		

			Lat	Usage		
Appliances	Coffee Maker	500 Sens	na x —	500	WA
	Hot Plate	2,400	na x —	2,400	NA
	Steam Table	1,000	na x —	1,000	NA
Other	Refrigerator	625	na x —	625	NA

*This form designed to be used with ACCA Manual N   Page 1 Subtotal   **58,677**   **7,475**

407A05.EPS

	Sensible	Latent
Page 1 Subtotal	58,677	7,475

5. INFILTRATION

ft³/min .... 460 .... X db Temp Diff .. 15 .... X 1.1 = **7,590** | **8,758**
ft³/min .... 460 .... X Grains Diff. .. 28 .... X0.68 =

6. SUBTOTAL COOLING LOAD FOR SPACE
66,267 | **16,233**

7. SUPPLY DUCT HEAT GAIN
Gain factor .... 0.03 .... X Line 6 Sensible Gain .. 66,267 .. = **1,988**

8. ROOM, ZONE OR BLOCK DESIGN LOAD
Add duct gain (7) to Subtotal (6)
Use this load to estimate the cooling CFM
68,255 | **16,233**

Cooling CFM = .. 68,255 .. (Line 8 Sensible) ÷ (1.1 X 19) (Supply TD) = 3,266

9. VENTILATION
NOTE: For return air ceilings db difference = (outdoor db - plenum db)
ft³/min .... 600 .... X db Temp Diff .. 15 .... X 1.1 = **9,900** | **11,424**
ft³/min .... 600 .... X Grains Diff. .. 28 .... X0.68 =

10. RETURN AIR LOAD FROM LIGHTING AND ROOF

NOTE: add 40% of the installed watts for lights recessed in a return air ceiling.

Incandescent Lights .... X3.4 = (+) N A
Fluorescent Lights. .... X4.1 = (+) N A

NOTE: Use U value & ETD for roof with no ceiling

(Roof Load)
Sq. Ft.   U-Factor   ETD*
.... X .... X .... = (+) N A
*(ETD correction based on plenum temperature.)

NOTE: Subtract the ceiling load, refer to No. 3.

Ceiling Load Credit .... (-) N A

11. RETURN DUCT HEAT GAIN
Gain factor .... X Line 6 Sensible Gain. .... = **N A**

12. TOTAL LOADS ON EQUIPMENT (Btuh)
78,155 | **27,657**

13. DESIGN CONDITIONS (HEATING)   72 - 7 = 65
Inside db . (-) minus Outside db . = Difference .

14. TRANSMISSION LOSSES

	Exposure	Sq. Ft.	Factor	db Temp Diff	HEATING LOAD Load	Notes
Windows	ALL	214	X 1.10	X 65	15,301	
Walls	EAST	252	X 0.077	X 65	1,261	
	SOUTH	750	X 0.077	X 65	3,754	
	WEST	134	X 0.077	X 65	671	
	BASE	1,080	X 5.65	X NA	6,102	
Doors			X	X		
Partitions Floors	BASE	2,250	X 1.55	X NA	3,488	
Roof Roof/Ceiling		2,250	X 0.10	X 65	14,625	

15. INFILTRATION
ft³/min .... 765 .... X db Temp Diff .. 65 .... X 1.1 = **54,697**

16. SUBTOTAL HEATING LOAD FOR SPACE
99,899

17. SUPPLY DUCT HEAT LOSS
Loss factor .... 5% .... X Line 16 Sensible Gain .. 99,899 .. = **4,995**

18. VENTILATION
ft³/min .... 600 .... X db Temp Diff .. 65 .... x 1.1 = **42,900**

19. HUMIDIFICATION LOAD
Inside RH (Desired .... X(Max 16% .)
ft³/min .. 1,365 .... ÷ 100 X Btu/hr 775 = **10,575**
(water) gal/day 2.1 .... X (air) ft³/min .. 1,365 .. ÷ 100 = 28.7

20. RETURN DUCT HEAT LOSS
Loss factor .... X Line 16 Loss .... = **N A**

21. TOTAL HEATING LOAD ON EQUIPMENT (Btuh)
158,373

407A06.EPS

COMMERCIAL LOAD CALCULATIONS
(ROOM, ZONE or BLOCK LOAD)

Air Conditioning Contractors of America
1513 16th Street, N.W., Washington, D.C. 20036

FORM N - 1

For:	Name _Rex Drugstore_	Phone ___	
	Address _____ City _Jefferson City_	State _MO_	Zip ___
By:	Contractor _____	Phone ___	
	Address _____ City ___	State ___	Zip ___

1. DESIGN CONDITIONS (COOLING)

(Time of Day _6 PM_)   (Daily Range _23_)   (Latitude _38_)

a) Inside db _75_   RH _50%_   b) Outside db _95_   wb _74_   Grains _28_

Outside db @ 3 p.m. _95_ (-) minus time of day corrections _5_ (-) minus inside drybulb _75_ = _15_ T.D.

2. SOLAR RADIATION HEAT GAIN THROUGH GLASS

COOLING LOAD

Exposure	Sq. Ft.		Solar Factor		Shading and/or Glass Factor		Sensible	Notes
EAST	24	x	24	x	0.64	=	369	
EAST	24	x	41	x	0.94	=	925	
WEST	40	x	132	x	0.94	=		
WEST	126	x	132	x	0.94	=	15,633	
		x		x		=		
		x		x		=		
		x		x		=		

3. TRANSMISSION GAINS

	Exposure	Sq. Ft.		U Factor		Equivalent or db Temp Diff		Sensible	Notes
Glass	ALL	214	x	1.04	x	15	=	3,338	6 PM
			x		x		=		TD=15
Walls	EAST	252	x	0.077	x	33	=	640	-2 OR
	SOUTH	750	x	0.077	x	30	=	1,733	Correction
	WEST	134	x	0.077	x	31	=	320	
Doors			x		x		=		
			x		x		=		
Partitions			x		x		=		
Floors	✓	2,250	x	0.28	x	5	=	3,150	
Roof			x		x		=		
Roof/Ceiling	✓	2,250	x	0.10	x	58	=	13,050	
RA Ceiling			x		x		=		

Use Table 9a to determine the Temperature Difference Across a Return Air Ceiling

4. INTERNAL HEAT GAIN

a. Occupants

Latent

Number	Sensible		Latent			Sensible	Latent
10	x	255			=	2,550	
9	x	315			=	2,835	3,250
10			x	325	=		2,925
9			x	325	=		

b. Lights & Others

Watts

NOTE: Use 60% of installed watts for lights in return air ceiling.

	Incandescent Lights		X 3.4	=	
	Flourescent Lights	6,000	X 4.1	=	24,600

	N.P.H.P.	Btuh	Load Factor	Usage Factor	
Motors	√4	1180	x 1.0	x 0.75	885
	√8	900	x 1.0	x 0.75	675
			x	x	

			Sens	Lat	Usage		
	coffee maker			NA	x 1.0	500	NA
Appliances	Hot Plate	2,800		NA	x 1.0	2,800	NA
	Steam Table	1,000		NA	x 1.0	1,000	NA
Other	Refrigerator	625		NA	x 1.0	625	NA

*This form designed to be used with ACCA Manual N	Page 1 Subtotal	60,551 6,175

407A07.EPS

	Sensible	Latent
Page 1 Subtotal	80,591	6,175

5. INFILTRATION

ft³/min .... 460 .... X db Temp Diff ... 15 ....X 1.1 .... = **7,590**

ft³/min .... 460 .... X Grains Diff .... 28 ....X0.68 .... = **8,758**

6. SUBTOTAL COOLING LOAD FOR SPACE — **88,181** | **14,933**

7. SUPPLY DUCT HEAT GAIN

Gain factor ... 0.03 .... X Line 6 Sensible Gain. 88,181 ... = **2,645**

8. ROOM, ZONE OR BLOCK DESIGN LOAD

Add duct gain (7) to Subtotal (6) — **90,826** | **14,933**

Use this load to estimate the cooling CFM

Cooling CFM = 90,826 .... ÷ (1.1×19)(Supply TD) = 4346

9. VENTILATION   NOTE: For return air ceilings db difference = (outdoor db -plenum db)

ft³/min ... 600 .... X db Temp Diff ... 15 ....X 1.1 .... = **9,900** | **11,424**

ft³/min ... 600 .... X Grains Diff .... 28 ....X0.68 ....

10. RETURN AIR LOAD FROM LIGHTING AND ROOF

NOTE: add 40% of the installed watts for lights recessed in a return air ceiling.

Incandescent Lights .... X3.4 = (+). NA

Fluorescent Lights. .... X4.1 = (+). NA

(Roof Load)

NOTE: Use U value & ETD for roof with no ceiling

Sq. Ft.   U-Factor   ETD*

.... X .... X .... = (+). NA

*(ETD correction based on plenum temperature.)

NOTE: Subtract the ceiling load, refer to No. 3.

Ceiling Load Credit .... (−). NA

11. RETURN DUCT HEAT GAIN

Gain factor .... X Line 6 Sensible Gain. .... = NA

12. TOTAL LOADS ON EQUIPMENT (Btuh) — **100,726** | **26,357**

---

13. DESIGN CONDITIONS (HEATING)

Inside db . (-) minus Outside db . = Difference .

14. TRANSMISSION LOSSES

	Exposure	Sq. Ft.	Factor	db Temp Diff	HEATING LOAD Load	HEATING LOAD Notes
Windows			X	X	=	
			X	X	=	
			X	X	=	
Walls			X	X	=	
			X	X	=	
			X	X	=	
			X	X	=	
Doors			X	X	=	
			X	X	=	
Partitions			X	X	=	
Floors			X	X	=	
Roof			X	X	=	
Roof/Ceiling			X	X	=	

15. INFILTRATION

ft³/min .... X db Temp Diff .... x 1.1 =

16. SUBTOTAL HEATING LOAD FOR SPACE

17. SUPPLY DUCT HEAT LOSS

Loss factor .... X Line 16 Sensible Gain .... =

18. VENTILATION

ft³/min .... X db Temp Diff .... x 1.1 =

19. HUMIDIFICATION LOAD   Inside RH (Desired ......)(Max ......)

ft³/min .... ÷ 100 X Btu/hr .... =

(water) gal/day .... X (air) ft³/min .... ÷ 100 = ...

20. RETURN DUCT HEAT LOSS

Loss factor .... X Line 16 Loss .... =

21. TOTAL HEATING LOAD ON EQUIPMENT (Btuh)

407A08.EPS

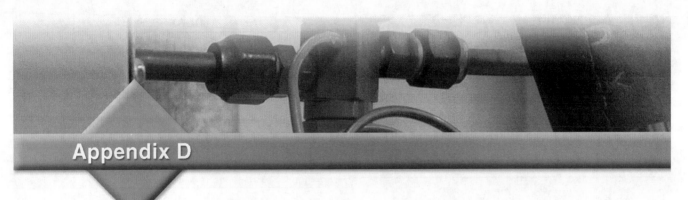

Typical Air Distribution System Duct and Supply Outlet Data

SUPPLY OUTLETS

FLOOR OUTLETS - PERIMETER

CFM	SIZE (IN)	APPROX. SPREAD (FT)	FACE VELOCITY (FPM)	FREE AREA (SQ IN)
70	2¼ × 10	9	535	18.6
80	2¼ × 12	10	565	21.1
100	2¼ × 14	11	610	23.6
110	4 × 10	10	500	32.4
135	4 × 12	13	500	39.0
175	4 × 14	14	555	45.5

LOW SIDEWALL - PERIMETER

CFM	SIZE (IN)	APPROX. SPREAD (FT)	FACE VELOCITY (FPM)	FREE AREA (SQ IN)
80	10 × 6	13	430	26.7
100	12 × 6	10	440	32.6
120	14 × 6	8	450	38.4

BASEBOARD

CFM	SIZE (FT)	APPROX. SPREAD (FT)	OUTLET VELOCITY (FPM)	FREE AREA (SQ IN)
80	2	7.5	430	26.6

HIGH SIDEWALL

CFM	SIZE (IN)	HORIZ. THROW (FT)	FACE VELOCITY (FPM)	FREE AREA (SQ IN)
80	10 × 4	8	390	29.0
125	10 × 6	10	415	43.3
150	12 × 6	10	410	52.7
165	14 × 6	9.5	375	62.1

ROUND CEILING OUTLETS

CFM	SIZE (IN)	HORIZ. THROW (FT)	OUTLET VELOCITY (FPM)	FREE AREA (SQ. IN)
45	6	3	500	12.2
105	8	5	580	26.1
185	10	7	580	43.8
285	12	8.5	575	65.7
425	14	10.5	600	91.9

SQUARE CEILING OUTLETS

CFM	SIZE (IN)	HORIZ. THROW (FT)	OUTLET VELOCITY (FPM)	FREE AREA (SQ IN)
50	6 × 6	3.5	450	15.4
135	8 × 8	5	550	35.1
250	10 × 10	6	620	58.1
325	12 × 12	7	550	85.1

Tables based on Lima Register Co. Catalog Data, @ .028 in w.c. drop across outlet.

DUCT SIZING

CFM	SUPPLY (IN)		RETURN (IN)		CFM
	ROUND	REC-TANGULAR	ROUND	REC-TANGULAR	
50	5	8 × 6	6	8 × 6	50
100	6	8 × 6	7	8 × 6	100
150	7	8 × 6	8	8 × 6	150
200	8	8 × 8	10	10 × 8	200
300	9	10 × 8	10	12 × 8	300
400	10	12 × 8	12	14 × 8	400
500	12	14 × 8	12	16 × 8	500
600	12	16 × 8	14	18 × 8	600
700	12	18 × 8	14	20 × 8	700
800	14	20 × 8	14	24 × 8	800
900	14	22 × 8	16	26 × 8	900
1000	16	24 × 8	16	30 × 8	1000
1100	16	26 × 8	18	34 × 8	1100
1200	16	28 × 8	18	36 × 8	1200
1300	16	30 × 8	18	28 × 10	1300
1400	18	32 × 8	18	30 × 10	1400
1500	18	28 × 10	20	32 × 10	1500
1600	18	30 × 10	20	34 × 10	1600

Table based on 0.08 inch w.c./100 ft for supply ducts and 0.05 inch w.c. for return ducts.

RETURN AIR GRILLES

CFM	SIZE (IN)	FREE AREA (SQ IN)
100	10 × 6	36.4
125	12 × 6	44.4
170	12 × 8	61.0
145	14 × 6	52.4
200	14 × 8	72.0
245	24 × 6	89.6
335	24 × 8	122.0
310	30 × 6	110.8
425	30 × 8	152.0

VERTICAL STACKS

SUPPLY (CFM)	STACK SIZE (IN)	RETURN (CFM)
100	3¼ × 10	75
125	3¼ × 12	90
150	3¼ × 14	110

2¼" stacks = 55% of 3¼" stack capacity

PANNED JOIST (16 IN O.C.)

RETURN CFM	NOMINAL JOIST DEPTH (IN)	ACTUAL JOIST DEPTH (IN)
260	6	5½
375	8	7½
525	10	9½

407A09.EPS

Appendix E

Duct Friction Chart

FRICTION LOSS FOR ROUND DUCT

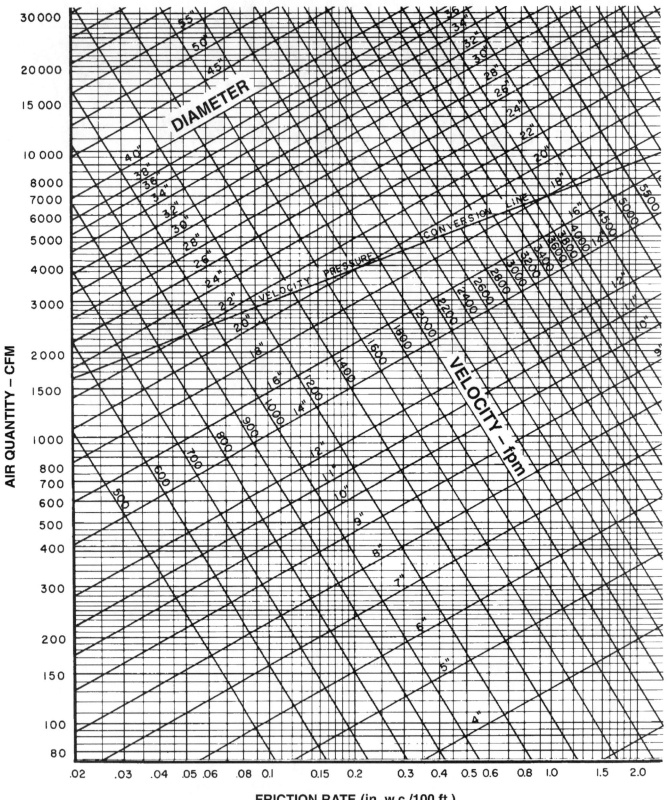

FRICTION RATE (in. w.c./100 ft.)

407A10.EPS

Additional Resources

This module is intended to be a thorough resource for task training. The following reference works are suggested for further study. These are optional materials for continued education rather than for task training.

Air Distribution System Selection. York, PA: International Corporation.

HVAC Duct Construction Standards – Metal and Flexible. Chantilly, VA: Sheet Metal and Air Conditioning Contractors National Association (SMACNA).

Manual D, Duct Design for Residential Winter and Summer Air Conditioning. Washington, DC: Air Conditioning Contractors of America (ACCA).

Manual G, Selection of Distribution Systems. Washington, DC: Air Conditioning Contractors of America (ACCA).

Manual J, Load Calculation for Residential Winter and Summer Air Conditioning. Washington, DC: Air Conditioning Contractors of America (ACCA).

Manual N, Load Calculation for Commercial Winter and Summer Air Conditioning. Washington, DC: Air Conditioning Contractors of America (ACCA).

Residential Air System Design. Syracuse, NY: Carrier Corporation.

Figure Credits

Manual J, Load Calculation for Residential Winter and Summer Air Conditioning. Air Conditioning Contractors of America, 407F02, 407F03, 407F06, 407F12, 407F15–407F18, Appendix A, Appendix B

Manual N, Load Calculation for Commercial Winter and Summer Air Conditioning. Air Conditioning Contractors of America, Appendix C

Carrier Corporation, 407F04, 407F05, 407F11, 407F14, 407F19, 407F20, 407F24, 407F28, 407F50, Appendices D and E

ASHRAE Handbook of Fundamentals, 1993, Chapter 25, Table 3. © American Society of Heating, Refrigerating and Air Conditioning Engineers, Inc., www.ashrae.org, 407F13

National Fenestration Rating Council (NFRC), 407SA01

Hart & Cooley, Inc, 407F27

Trane, 407F49

NCCER makes every effort to keep these textbooks up-to-date and free of technical errors. We appreciate your help in this process. If you have an idea for improving this textbook, or if you find an error, a typographical mistake, or an inaccuracy in NCCER's Contren® textbooks, please write us, using this form or a photocopy. Be sure to include the exact module number, page number, a detailed description, and the correction, if applicable. Your input will be brought to the attention of the Technical Review Committee. Thank you for your assistance.

Instructors – If you found that additional materials were necessary in order to teach this module effectively, please let us know so that we may include them in the Equipment/Materials list in the Annotated Instructor's Guide.

Write:	Product Development and Revision
	National Center for Construction Education and Research
	3600 NW 43rd St, Bldg G, Gainesville, FL 32606
Fax:	352-334-0932
E-mail:	curriculum@nccer.org

Craft

Module Name

Copyright Date

Module Number

Page Number(s)

Description

(Optional) Correction

(Optional) Your Name and Address

HVAC Level Four

03408-09

Commercial and Industrial Refrigeration Systems

03408-09
Commercial and Industrial Refrigeration Systems

Topics to be presented in this module include:

Overview

In an earlier level, you were introduced to the types of self-contained product refrigerators and freezers used in retail applications such as markets and convenience stores. Businesses such as supermarkets, warehouses, ice cream and ice-making plants, food and beverage distributors, food packaging plants, and food transportation firms require larger and more complex systems. In scale, these systems can be compared to the air conditioning systems found in large office buildings, while the retail refrigeration equipment is more comparable to residential and light commercial air conditioning systems. Unlike the types of equipment found in retail applications, systems used in commercial and industrial refrigeration applications are likely to be custom engineered for the structure and may involve complex control systems. Systems that use ammonia as a refrigerant are fairly common in commercial and industrial work, whereas they are not likely to be found in retail applications. This module examines the equipment, control systems, and refrigerants used in commercial and industrial refrigeration applications.

Objectives

When you have completed this module, you will be able to do the following:

1. Identify different types of refrigerated coolers and display cases and describe each one's common application.
2. Compare the basic components used in commercial/industrial refrigeration systems with those used in retail refrigeration systems.
3. Identify single, multiple, and satellite compressor systems. Describe the applications, installation considerations, and advantages and disadvantages of each type.
4. Identify packaged condensing units and unit coolers. Describe their applications, operation, and installation considerations.
5. Identify two-stage compressors and explain their operation and applications.
6. Identify the various accessories used in commercial refrigeration systems. Explain why each is used and where it should be installed in the system.
7. Identify the various refrigeration control devices. Explain the purpose of each type and how it works.
8. Compare the components used in ammonia systems with those used in halocarbon-based refrigerant systems.

Trade Terms

Anhydrous	Latent heat defrost
Commodities	Lyophilization
Condensing unit	Satellite compressor
Cooler	Secondary coolant
Cryogenics	Sublimation
Cryogenic fluid	Total heat rejection
Deck	(THR) value
Immiscible	Unit cooler

Required Trainee Materials

1. Pencil and paper
2. Appropriate personal protective equipment

Prerequisites

Before you begin this module, it is recommended that you successfully complete *Core Curriculum*; *HVAC Level One*; *HVAC Level Two*; *HVAC Level Three*; and *HVAC Level Four*, Modules 03401-09 through 03407-09.

This course map shows all of the modules in the fourth level of the *HVAC* curriculum. The suggested training order begins at the bottom and proceeds up. Skill levels increase as you advance on the course map. The local Training Program Sponsor may adjust the training order.

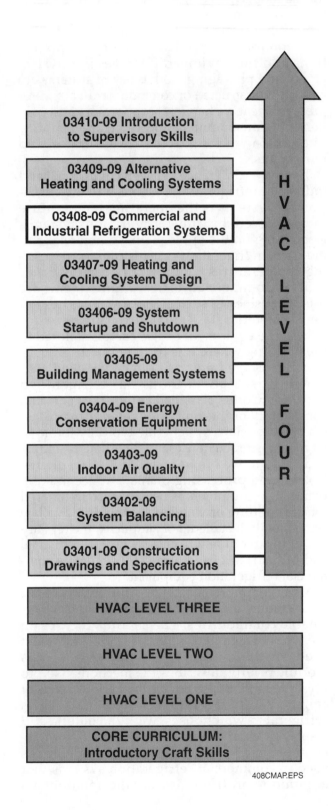

03410-09 Introduction to Supervisory Skills

03409-09 Alternative Heating and Cooling Systems

03408-09 Commercial and Industrial Refrigeration Systems

03407-09 Heating and Cooling System Design

03406-09 System Startup and Shutdown

03405-09 Building Management Systems

03404-09 Energy Conservation Equipment

03403-09 Indoor Air Quality

03402-09 System Balancing

03401-09 Construction Drawings and Specifications

HVAC LEVEL THREE

HVAC LEVEL TWO

HVAC LEVEL ONE

CORE CURRICULUM: Introductory Craft Skills

HVAC LEVEL FOUR

408CMAP.EPS

1.0.0 ◆ INTRODUCTION

Commercial and industrial refrigeration systems and comfort cooling systems all have a common purpose: removing heat. Refrigeration systems are mainly used to remove heat from substances for the food, chemical, and manufacturing industries. Air-conditioning systems are used to remove heat from buildings and vehicles to provide comfort control. Excluding their applications and the obvious differences in physical layout and packaging, both types of systems are basically constructed of common hardware: compressors, heat exchangers (evaporators and condensers), fans, pumps, pipe, duct, and controls. Their main working fluids are air, water, and refrigerants.

The major difference between commercial and industrial refrigeration systems and air-conditioning systems is the operating temperature range served by the refrigeration systems. Depending on the application, commercial and industrial refrigeration systems can have operating temperatures ranging between 60°F and –250°F. Commercial refrigeration systems are used in businesses such as grocery stores, restaurants, cafeterias, fast-food stores, and beverage stores. The equipment can be custom tailored to the building where it is installed or a self-contained packaged unit that can be installed anywhere it is needed. Industrial refrigeration systems are often classified as those systems that consistently or intermittently require an on-the-job operator. Such systems are typically found in large food storage houses, packing houses, industrial plants, ice cream manufacturing plants, frozen food processing plants, and ice making plants. The equipment is usually custom tailored for the building or warehouse in which it will be permanently installed. For simplicity, the remainder of this module will classify both commercial and industrial refrigeration equipment as commercial refrigeration equipment.

Most commercial refrigeration involves the refrigeration and freezing of food and other perishable **commodities**. This module focuses on the types of commercial refrigeration equipment used for this purpose. The first part of this module briefly describes the methods used to preserve food. It also covers the various kinds of commercial refrigeration equipment, such as refrigerated warehouses, walk-in **coolers**, and display cases. The second part of the module covers the refrigeration methods and components used in commercial refrigeration systems that are unique or different from the components used in comfort cooling systems. Also covered are some special refrigeration applications and systems.

2.0.0 ◆ REFRIGERATION AND THE PRESERVATION OF FOOD PRODUCTS

The field of food processing starts at the point of harvest and extends to consumption (*Figure 1*). Throughout this process, the food must be properly prepared and stored in order to maintain its freshness and quality. Refrigerated storage is defined as any space where refrigeration is used to provide controlled storage conditions. It can be an entire building such as a warehouse, a section of a building dedicated to storage, or a single stand-alone unit. Refrigerated storage areas are often divided into several sections, with each section being cooled at a different temperature. The specific temperature used is determined by the safe preservation temperature of the commodities involved.

In addition to storage, the movement of products from one point in the food processing chain to the next requires refrigerated transport. Products intended for export may spend from one to three weeks in some type of refrigerated transport unit.

2.1.0 Cold Storage

Cold storage is used to preserve perishable food in its fresh, wholesome state for extended periods. This is done by controlling the food's temperature and humidity during storage. Cold storage conditions are determined by the type of food being stored. They are also determined by the length of time the food is to be held in storage.

Foods placed in cold storage can be divided into two broad groups based on their required temperature range: perishable or chilled temperature range and frozen temperature range.

- *Perishable or chilled temperature range* – This group includes commodities that must be refrigerated below a specified temperature but not allowed to freeze. For some products in this category, such as fruits, vegetables, and flowers, living processes (such as the growth of shoots or ripening) continue during storage. Exposure to cold temperatures only retards these processes, with the degree of retardation depending on the temperature level being

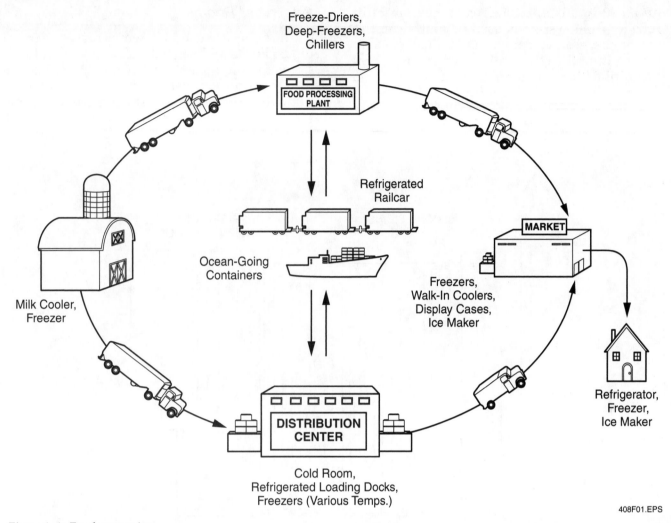

Figure 1 ◆ Food processing

maintained. Airflow is crucial to removing the heat generated during the respiratory cycle of many products in the perishable temperature range. With high air flow rates also comes the potential for higher evaporation rates from many types of produce. This can cause premature dehydration. Avoiding dehydration requires that humidity be maintained at the proper level for each type of product. For example, celery is best stored at temperatures between 32°F and 36°F, with a relative humidity of 98 to 100 percent. Pumpkins, on the other hand, are best stored at temperatures between 50°F and 55°F, and a relative humidity of 50 to 70 percent. The storage temperatures and humidity requirements for the different kinds of produce are so varied that no generalizations can be made. Some require a curing period before storage. Other products, such as certain kinds of potatoes, require different storage conditions depending on their intended use.

• *Frozen temperature range* – This group includes those products whose temperature must not rise above freezing, such as ice cream and frozen prepared foods, meats, meat products, and fish. Airflow is not as critical for frozen products as it is for nonfrozen products.

Cold storage plants (*Figure 2*) are used to process and/or store large quantities of foods intended for distribution to supermarkets and restaurants. The plant or warehouse food handling methods and storage requirements determine the building design. Refrigerated storage plants and warehouses that handle a variety of foods are divided as required into several cooler rooms (chill rooms) and freezer rooms to preserve the foods involved. Each refrigerated room is separated from the other rooms based on the storage temperature needed. Typically, the temperatures in a warehouse can range from 35°F to 60°F with humidity control, and to –20°F without humidity control. Generally, a custom-built refrigeration system is used to cool the entire warehouse and maintain the individual temperatures within the refrigerated rooms. The larger the application, the more likely that ammonia will be employed as the refrigerant. Ammonia system construction and installation becomes more cost-effective as size increases.

2.1.1 Chill Rooms

Chill rooms are used for temporary storage of foods. Fruits and vegetables are the main foods stored in chill rooms. They are normally placed in the chill rooms in their freshly harvested condition, after the necessary sorting and cleaning. Medium-range temperatures from 32°F to 60°F are used in chill rooms. The specific temperature

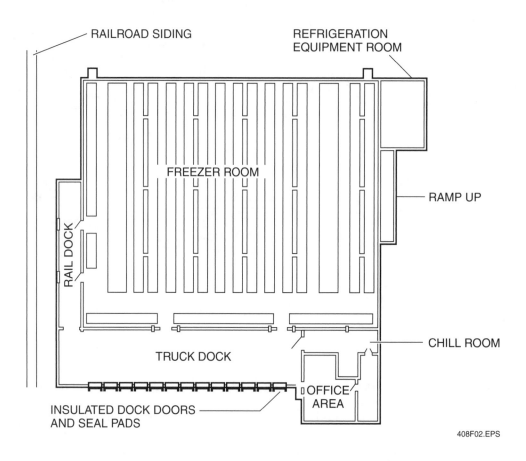

408F02.EPS

Figure 2 ◆ Floor plan for a typical refrigerated warehouse.

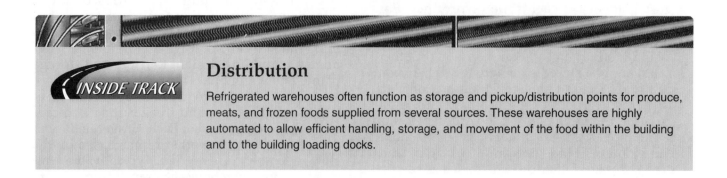

Distribution

Refrigerated warehouses often function as storage and pickup/distribution points for produce, meats, and frozen foods supplied from several sources. These warehouses are highly automated to allow efficient handling, storage, and movement of the food within the building and to the building loading docks.

maintained in any one room is determined by the storage temperature and humidity level needed for the type of food being stored. The list of specific temperatures and relative humidities required for the various fruits and vegetables is extensive. To try to reproduce that information here is beyond the scope of this module. However, this information is readily obtained in the books referenced in the back of this module.

Foods stored in chill rooms are typically kept at temperatures just above their freezing point to maintain top product quality. The required temperature must be precisely maintained. In some cases, variations as small as 2°F to 3°F can result in loss of product quality. Obviously a loss of refrigeration would seriously jeopardize products throughout the facility, resulting in huge financial losses. As a result, adding redundant, or back-up, equipment is common.

Most chill rooms also have a means of controlling the relative humidity to within 3 to 5 percent of the desired level. This is often done by controlling the temperature of the evaporators using spray nozzles or by using precise humidifier controls. Air is circulated within the room to maintain an even temperature.

2.1.2 Freezer Rooms

Prolonged storage of food requires that the food be frozen, allowing the storage time to be lengthened from days or weeks to months. Freezer rooms held at temperatures below freezing are used to store frozen foods. Temperatures are typically maintained at levels ranging between 10°F and −10°F. Rooms used to store ice cream are often maintained at a temperature of −20°F. Frozen foods can deteriorate during the period between production and consumption. The most important factors contributing to the deterioration of frozen foods are the storage temperature

and storage time. The amount of protection provided by the package containing the frozen food is also important. Some of the bacteria in frozen foods may be killed during freezing and frozen storage, but all the bacteria are never completely destroyed. When defrosting, foods are still subject to bacterial decomposition.

2.2.0 Commercial Freezing Methods

Some cold storage plants and warehouses are used to both process and freeze foods. The process of freezing reduces the temperature of a food product from the ambient level to the storage level and changes most of the water in the product to ice. As shown in *Figure 3*, the freezing process has three phases. The first phase removes the product's sensible heat. During the second phase, the product's latent heat of fusion is removed and the water in the product is changed to ice crystals. In phase three, continued cooling of the product removes the sensible heat below the freezing point and reduces the temperature to the required frozen storage temperature.

NOTE

The values for the specific heat, freezing point, and latent heat of fusion vary for the different food products. This information is readily obtained in the books referenced in the back of this module.

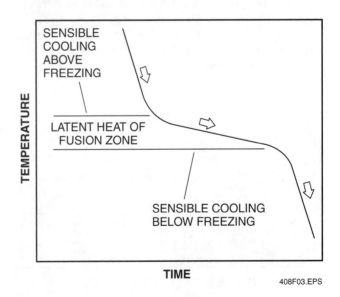

Figure 3 ◆ Typical freezing curve.

Freezing is a time-temperature related process. The longest part of the process is normally the latent heat of fusion as the water turns to ice.

Several methods are used for freezing food products. These methods are briefly described in the following sections.

2.2.1 Air Blast Freezing

Air blast freezing employs cold air circulated around the product at high velocities, maximizing the heat exchange process. The air removes heat from the product via convection and releases it to an air-to-refrigerant heat exchanger coil. One advantage of the blast freezing approach is the ability to freeze objects of inconsistent or irregular size and shape. Evaporator coil air discharge temperatures are typically –20°F to –40°F, but varies depending on the product.

The two main types of blast freezing applications are batch freezing and continuous freezing. In batch freezing applications, product is loaded into the freezer and allowed to remain in place until the process is complete. The product is often palletized or placed in moveable racks, then placed in a room equipped with the necessary evaporator coils and high-velocity fans. To ensure that the air makes adequate contact with the product, the individual product packages are often separated on the top and sides to allow free air flow. As shown in *Figure 4*, the evaporator coils are often separated from the product by a horizontal partition, directing discharge air from the coils through the product load and back to the coil inlet. The arrangement can be series flow, with air flowing along the length of the load as shown in *Figure 4*, or crossflow with the air moving perpendicular to the length of the load. In continuous freezing, product moves by conveyor or other means through the refrigerated enclosure at a constant speed appropriate for the product type.

2.2.2 Contact Freezing

Contact freezing produces packaged or unpackaged frozen products by pressing them between cold metal plates, or by contact with a moving belt exposed to the refrigerated surface. This process is typically used for products of consistent size and shape, to ensure that proper contact is made with individual pieces. Heat is extracted by direct conduction through the plate surfaces and into the refrigerant circulating through channels or tubes integrated into the plate. Contact freezing is considered to be one of the most efficient approaches, because contact transfers heat more rapidly. One other advantage is that there is little or no loss of moisture from the product, as opposed to freezing methods using high volumes of air. Moisture loss means product weight loss and possible quality degradation, significantly impacting the value of the end product.

Figure 5 shows on example of a horizontal plate freezer. Product such as fish is loaded between the plates. Once loaded, the plates are pressed down on the product and the freezing process begins. Plate freezers can also use vertical plate arrangements.

Belt freezers, as shown in *Figure 6*, are generally used for unpackaged products. The film is

408F05.EPS

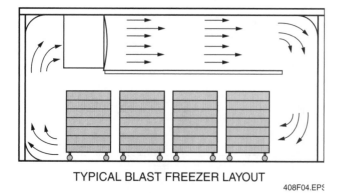

TYPICAL BLAST FREEZER LAYOUT

408F04.EPS

Figure 4 ◆ Blast freezer configuration.

Figure 5 ◆ Horizontal plate freezer.

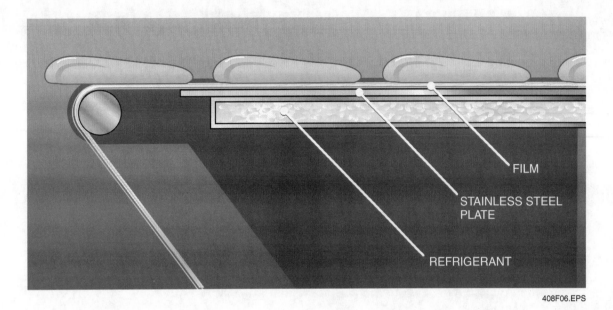

FILM

STAINLESS STEEL
PLATE

REFRIGERANT

408F06.EPS

Figure 6 ◆ Belt freezer design.

pulled with rollers, moving the product along the surface. Used film is disposed of, ensuring good hygiene. Soft, sticky and wet products (scallops, for example) can be a problem with contact freezers, as product can adhere to the refrigerated surface. The special and extremely thin film used with belt freezers like the one shown does not allow product adhesion, and does not significantly impede heat transfer.

2.2.3 Cryogenic Freezing

In the refrigeration industry, **cryogenics** is generally accepted to define the freezing of product using fluid (refrigerant) temperatures below –250°F. The boiling points of common **cryogenic fluids**, such as carbon dioxide, nitrogen, and helium, which are normally in their gaseous state, are below this temperature. The boiling points of common refrigerants such as HFCs and HCFCs are above this temperature. Liquid nitrogen is most commonly used, while liquid helium provides the lowest temperature.

Cryogenic freezing is generally done by injecting the fluid into the freezing chamber. In some cases, only the surface is to be frozen cryogenically. The product is then transferred to a blast freezer to complete the process. The frozen surface reduces moisture loss from the product. This process is known as cryomechanical freezing.

Figure 7 shows one example of a cryogenic freezing unit. Cryogenics can be used in spiral-style or tunnel-belt freezers, as well as batch freezers. In other cases, the product is immersed in the cryogenic fluid, floating in a tank maintained at a specific temperature and pressure. In all cases, the cryogenic fluid absorbs heat from the product, causing the fluid to vaporize. This change of state, a latent heat process, creates the highest heat exchange rate. This is the case in all common refrigeration applications regardless of the refrigerant used.

2.2.4 Freeze-Drying

Freeze-drying, also known as **lyophilization**, is generally a cryogenic process, relying first on the extremely rapid freezing action of this method. The product is then subjected to high levels of vacuum and sufficient heat to dehydrate the product, removing all moisture through **sublimation**. The result is a product that no longer requires refrigeration for long-term preservation, is much lighter in weight, and can be returned to its original state by simply adding water. *Figure 8* is an example of a typical commercial unit.

Some unique products you may have seen, such as astronaut ice cream, are created through this process. Some products, such as watermelon, are poor candidates for freeze-drying due to their extremely high water content. In addition to foods, freeze-drying is used to preserve pharmaceuticals and biological samples, as well as in the recovery of valuable water-damaged documents.

Figure 7 ◆ Cryogenic spiral freezer.

408F07.EPS

408F08.EPS

Figure 8 ◆ Freeze-drying unit.

3.0.0 ◆ REFRIGERATED TRANSPORT UNITS

Refrigerated transport units maintain product temperature during shipment, thereby keeping products fresh until they reach distant markets. Without the ability to control product temperature in transit, consumers would not be able to enjoy many of the foods now found in local stores. Refrigerated transport equipment can be grouped into three methods of shipping: refrigerated shipboard containers, refrigerated truck and trailer units, and refrigerated railcars.

3.1.0 Refrigerated Shipboard Containers

Exported commodities are usually shipped in container units. These heavily insulated steel or aluminum boxes are typically 8' wide by 8' high by 20' or 40' long, with a refrigeration unit installed at one end. The refrigeration equipment is a packaged unit that contains all the components necessary to remove heat from the cargo area. This includes the compressor, condenser, metering device, evaporator, and all electrical controls. *Figure 9* shows a typical refrigerated container.

The uniform size of containers allows them to be stacked aboard ship and on the docks. The loaded containers can be transported to and from the docks via flatbed trailers or railcars. When they arrive, they are either immediately placed on board the ship or held for future shipment in the dock's staging area.

Figure 9 ◆ Refrigerated container.

Most container units use hermetic or semi-hermetic compressors. These units plug into the electrical power panel of the ship or dock facility. When a container is being transported to and from the dock, a portable generator set provides power so the unit can maintain the proper temperature. *Figure 10* shows nose-mount and under-mount generator sets.

A large variety of products can be transported in containers. The unit must be capable of controlling a very wide temperature range, from 90°F down to –25°F. The refrigeration cycle is similar to that of a walk-in cooler, but the unit itself must be extra heavy duty in order to withstand the abuse of traveling on a flatbed, being moved by a forklift, and being lifted in and out of a ship's hold by a crane. Corrosion resistance is also important to the unit's design. Saltwater spray will be present for most of its operational life.

Microprocessor controls allow the unit to monitor system conditions. This provides the best settings for temperature control and energy efficiency. Heating and evaporator defrosting is done with electric heating elements. The condenser is typically air cooled, but most manufacturers offer an optional water-cooled condenser because it allows heat from the inside of the container to be removed, even when the unit is deep inside the ship's hold. Air-cooled condensers can have problems with high ambient temperatures and lack of airflow.

Proper placement of the product in the container is vital if the refrigeration unit is to operate correctly. The correct product temperature must be maintained throughout the container. Product placement must be done in a way that provides

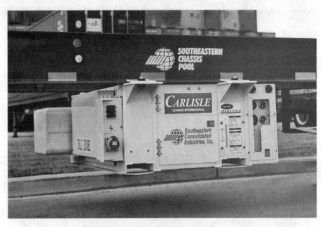

UNDERMOUNT GENERATOR SET

NOSE-MOUNT GENERATOR SET

Figure 10 ◆ Undermount and nose-mount generator sets.

enough space between and above the pallets or boxes. Air must be able to circulate around the product in order to absorb the heat and return to the refrigeration unit. The unit supplies cool air to the product through grooves or T-slots in the floor. The air warmed by the product rises to the top of the container and is drawn into the evaporator by a fan.

Advanced options for containers include humidity control and controlled atmosphere (CA). Humidity controls add moisture to the supply air to protect produce from drying out. CA is a process by which the oxygen, nitrogen, and carbon dioxide levels inside the container are adjusted to minimize ripening and thereby extend the cargo's shelf life.

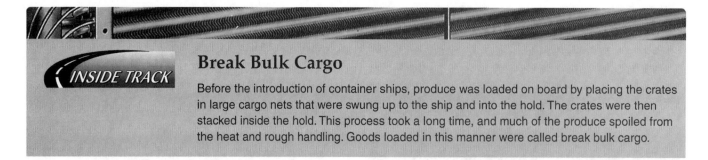

Break Bulk Cargo

Before the introduction of container ships, produce was loaded on board by placing the crates in large cargo nets that were swung up to the ship and into the hold. The crates were then stacked inside the hold. This process took a long time, and much of the produce spoiled from the heat and rough handling. Goods loaded in this manner were called break bulk cargo.

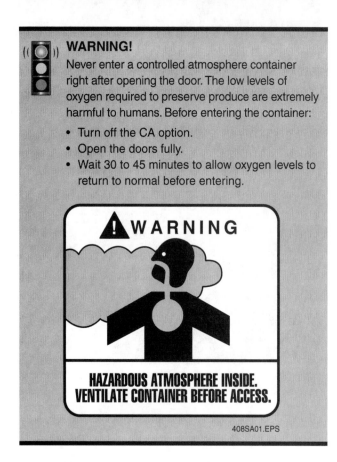

WARNING!

Never enter a controlled atmosphere container right after opening the door. The low levels of oxygen required to preserve produce are extremely harmful to humans. Before entering the container:

- Turn off the CA option.
- Open the doors fully.
- Wait 30 to 45 minutes to allow oxygen levels to return to normal before entering.

⚠ **WARNING**

HAZARDOUS ATMOSPHERE INSIDE. VENTILATE CONTAINER BEFORE ACCESS.

408SA01.EPS

408F11.EPS

Figure 11 ◆ Refrigerated trailer truck.

3.2.0 Trailer and Truck Units

At some point in the supply chain, refrigerated products will be transported by refrigerated trucks. Trucks used to haul perishable and frozen loads come in a variety of sizes and shapes. The most common are over-the-road refrigerated trailers, trucks, and direct-drive delivery vans. *Figure 11* shows a common over-the-road refrigerated trailer truck.

3.2.1 Trailer Units

Refrigerated trailers are used to haul just about every type of temperature-sensitive cargo. These units are used to transport products over long distances, typically across the state or even across the country. Unlike a standard trailer, these units

are insulated, have swing doors with seals, and have a refrigeration unit mounted onto the front.

The refrigeration equipment is a packaged unit that contains all the components necessary to remove heat from the cargo area. These components include the compressor, condenser, metering device, evaporator, and all applicable controls. The trailer's refrigeration unit (*Figure 12*) is heavy duty, allowing it to endure the constant bumps and vibration present during travel. Trailer units may use electric heaters for heat and defrost, but in most cases, heat and defrost come from the use of hot gas. The trailer unit has an integral power supply. Power is provided to an open-drive compressor driven by a small diesel engine connected via a driveshaft. The engine also turns an alternator to supply DC voltage to the unit's electrical components. Fans can be belt driven or direct driven. The unit's controls monitor and protect the cargo, and also protect and control the engine.

Cargo placement within a trailer plays an important role in removing heat from the inside of the trailer. A trailer unit supplies cold air out of the top of the refrigerated space. Sometimes a chute or duct is used to evenly distribute the air along the length of the trailer. The fan forces the air into the spaces around the cargo, where it picks up heat and is drawn back along the floor to the evaporator. Some trailers are equipped

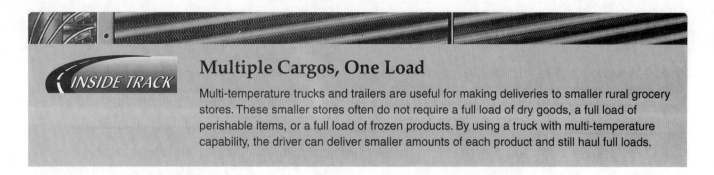

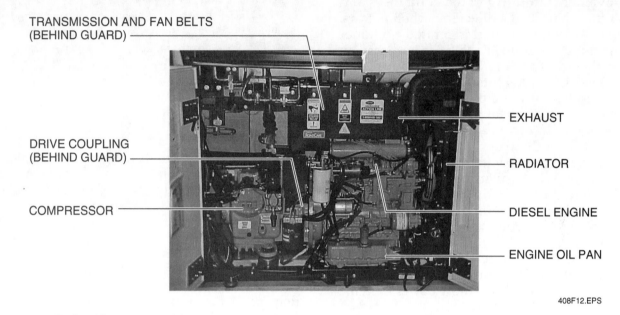

Figure 12 ◆ Trailer refrigeration unit.

with a T-slotted floor, but most use pallets and depend on a clear path through the open pallets. If the cargo is loaded too tightly, the air will not be able to circulate through the load and absorb the heat. For produce with a high rate of respiration, this condition could cause the cargo to spoil.

3.2.2 Truck Units

The primary difference between a refrigerated truck (*Figure 13*) and a refrigerated trailer is that a trailer requires a separate drive vehicle (cab) to tow it from place to place, while a truck is self-contained. Truck units also differ in size and cooling capacity. Trucks are not as large as trailers and are typically used for local delivery of cargo in cities or between adjacent cities. Truck units have a packaged refrigeration unit mounted on the front of the cargo area, just above the driver's cab.

Both refrigerated trucks and trailers may be configured with multi-temperature capabilities. A multi-temperature truck or trailer consists of a refrigeration unit supplying two or more

evaporator sections connected in parallel. Refrigeration systems using multiple evaporators are covered later in this module.

Direct-drive refrigeration units are used in trucks of lower capacity used for local delivery. They differ from the standard truck unit in that their compressor is a much smaller automotive-type compressor located in the engine compartment instead of being packaged in the refrigeration unit. These units carry

Figure 13 ◆ Refrigerated truck.

smaller loads or operate with higher cargo temperature ranges. Power to drive the compressor is transferred from the engine by a belt. Electrical power is supplied by the vehicle's alternator. Direct-drive systems may combine the evaporator and condenser section into one frame or they may use remote evaporators.

3.3.0 Refrigerated Railcars

Improvements to refrigerated containers and trailer units have had a direct impact on railcars. In many cases, the same refrigerated container used to transport cargo in a ship can be placed on a flatbed railcar to continue its trip inland. These containers either plug into a power source built into the railcar or use a portable generator set mounted onto the nose of the container.

Refrigerated railcars also use technological advances from the truck/trailer industry. By mounting a high-capacity trailer refrigeration unit onto a heavily insulated boxcar, larger loads can be delivered at reduced delivery times. Refrigerated railcars of this kind are capable of protecting 130,000 pounds of cargo with a temperature range of –20°F to 87°F. Another way of using railroads to transport perishable products is to load refrigerated trailers onto flatbed rail cars (*Figure 14*). The trailers draw power from the train while in transit.

4.0.0 ◆ REFRIGERATION SYSTEMS AND COMPONENTS

Many types of mechanical refrigeration systems are used to cool or freeze foods and other substances. The operation and components in a refrigeration system used to cool food and other products are basically the same as those used in comfort cooling systems. *Figure 15* shows a basic refrigeration system.

408F14.EPS

Figure 14 ◆ Refrigerated trucks on railcars.

The basic refrigeration system components are the compressor, evaporator, condenser, and expansion device. What changes from system to system is the type of refrigerant, the system operating temperatures, the size and style of the components, and the installed locations of the four components and the lines. The main working fluids are air, water, and refrigerants. The events that take place within the mechanical refrigeration system happen again and again, in the same order. This series of events is called the refrigeration cycle. The operation of the basic refrigeration system and its components are described in detail in the *HVAC Level One* module, *Introduction to Cooling*. You should review the material in that module.

The function performed by each component is as follows:

- *Evaporator* – This is a heat exchanger in which the heat from the area or item being cooled is transferred to the refrigerant.

- *Compressor* – This creates the pressure differences in the system needed to make the refrigerant flow and the refrigeration cycle work.

- *Condenser* – This is a heat exchanger in which the heat absorbed by the refrigerant is transferred from the refrigerant to the cooler outdoor air or another cooler substance, such as water.

- *Expansion device* – This provides a pressure drop that lowers the boiling point of the refrigerant just before it enters the evaporator. The expansion device is also commonly called the metering device.

Also shown in *Figure 15* is the piping used to connect the basic components in order to provide

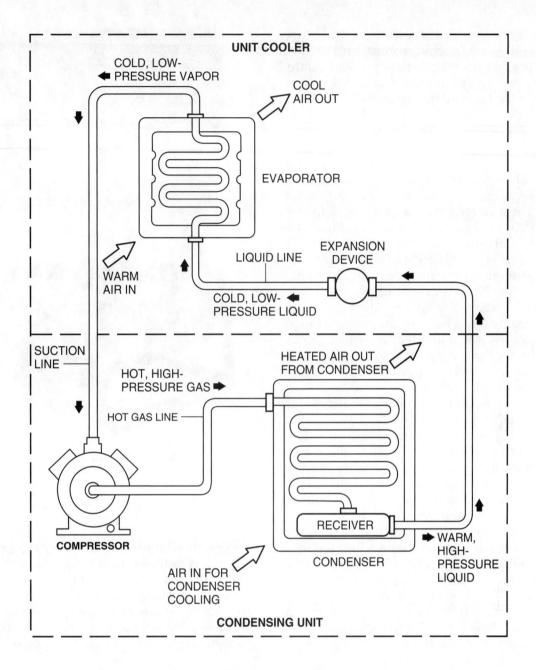

UNIT COOLER

COLD, LOW-
PRESSURE VAPOR

COOL
AIR OUT

EVAPORATOR

WARM
AIR IN

LIQUID LINE

EXPANSION
DEVICE

COLD, LOW-
PRESSURE LIQUID

SUCTION
LINE

HOT, HIGH-
PRESSURE GAS

HEATED AIR OUT
FROM CONDENSER

HOT GAS LINE

RECEIVER

WARM,
HIGH-
PRESSURE
LIQUID

COMPRESSOR

AIR IN FOR
CONDENSER
COOLING

CONDENSER

CONDENSING UNIT

TEMPERATURE RANGE

| −40 | −17 | −1 | 7 | 16 | 21 | 27 | °C |
| −40° | 0 | 30 | 45 | 60 | 70 | 80 | °F |

← REFRIGERATION →|← AIR CONDITIONING →

408F15.EPS

Figure 15 ◆ Basic refrigeration system.

the path for refrigerant flow. Together, the components, accessories, and lines form a closed refrigeration system. The tubing lines are:

- *Suction line* – This carries heat-laden refrigerant gas from the evaporator to the compressor.

- *Hot gas line* – This carries hot refrigerant gas from the compressor to the condenser. It is also referred to as the discharge line.

- *Liquid line* – This carries liquid refrigerant, formed in the condenser, to the expansion device.

In refrigeration systems used to cool warehouses, cold storage (chill) rooms, and walk-in coolers, the compressor, condenser, and related components are commonly packaged into an assembly called a **condensing unit** (*Figure 16*). In practice, condensing units can be either air cooled or water cooled. They may contain one or more compressors and/or fans (air cooled). Condensing units may be installed in equipment or mechanical rooms, mounted on the roof or other outdoor location, or both. The evaporator, expansion device, and fans are also packaged into an assembly that is installed in the individual chill room or walk-in cooler. This assembly is commonly called a **unit cooler**. Some large refrigeration systems may consist of individual components piped together at the job site into a customized system. Self-contained reach-in coolers and display cases have all the system components packaged into a unit that is enclosed in the cabinet. In-depth coverage of systems used in the retail sector can be found in *HVAC Level Three, Retail Refrigeration Systems*.

Refrigeration equipment manufacturers classify equipment according to its cooling class, which is based on the evaporator operating temperature. Within each classification, there are a variety of models to cover the capacity requirements of various applications. *Table 1* shows common refrigeration applications and cooling classes assigned by one manufacturer.

Refrigeration systems used in supermarkets usually have the most demanding cooling requirements (*Figure 17*). A typical supermarket might use one or more medium-temperature refrigeration systems with parallel compressors for the meat, deli, dairy, and produce refrigerators and walk-in coolers. The system may have a separate compressor for the meat or deli refriger-

LARGE UNIT COOLER

DUAL-COMPRESSOR, AIR-COOLED CONDENSING UNIT

408F16.EPS

Figure 16 ◆ Typical refrigeration system components.

ators, or all units may have a single compressor. The low-temperature refrigerators and coolers may be grouped on one or more systems with parallel compressors. The ice cream refrigerators might be on a **satellite compressor** or single compressor. Meat cutting and food preparation rooms are usually placed on a single unit. Parallel

Table 1 Cooling Applications and Temperature Class

Application	Storage Temp. Range	Cooling (Evaporator Temp. Class)		
		Low −40°F to −10°F	Medium −10°F to +30F	High +30°F to +50°F
Meat cutting and packaging, fresh fruit fruit and vegetable storage, floral boxes	+35°F to +60°F			X
Meat storage rooms and extended storage of many fruits and vegetables	+28°F to +34°F			X
Low ceiling coolers for all applications	+35°F and above			X
Low ceiling freezers for all applications	−20°F to +34°F	X	X	
Walk-in coolers and warehouses for meat, dairy, and produce	+35°F and above			X
Walk-in freezers	−30°F to +34°F	X	X	

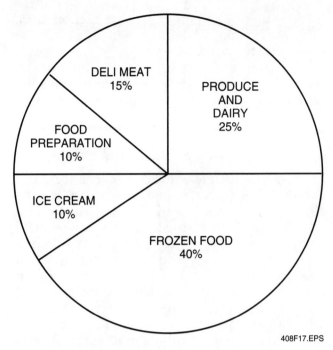

Figure 17 ◆ Distribution of cooling requirements in a typical supermarket.

and satellite compressors are discussed in detail later in this section.

An important reason to group the various refrigeration circuits with similar temperature requirements is to improve compressor efficiency. The lower the required system temperature, the lower the compressor energy efficiency ratio (EER), thus the more expensive it is to operate.

EERs for systems operating at extremely low temperatures can be as low as 3, as opposed to ratings at or above 14 with systems operating at or above 40°F evaporator temperatures.

4.1.0 Compressors

Reciprocating, rotary, screw, and centrifugal compressors are all used in refrigeration systems. The particular type used is determined by the refrigeration load and application. Hermetic and semi-hermetic reciprocating compressors are the most widely used, with the semi-hermetic being the most common. Direct-drive and/or belt-drive open reciprocating compressors are used in mobile refrigeration units and on older stationary refrigeration systems. The operating principles for these compressors are described in detail in the *HVAC Level One* module, *Introduction to Cooling*, and the *HVAC Level Three* module, *Compressors*. Review the material contained in these modules.

4.1.1 Single-Compressor Applications

Single compressors are found in about half of the refrigeration systems in use. The compressor sizes commonly range between 0.5 and 30 horsepower (hp). A single compressor is often used to cool multiple evaporators in a line of display cases or coolers (*Figure 18*) or in multi-temperature mobile refrigeration units. One advantage of using a single compressor to supply multiple

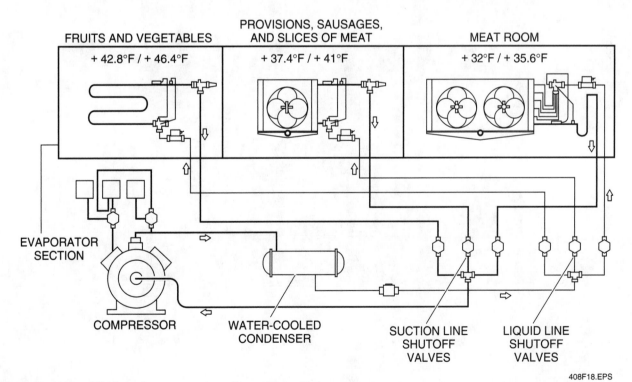

Figure 18 ◆ Simplified single-compressor system with multiple evaporators.

evaporators is the relatively low cost. Another is that the heat from the equipment can be more easily captured in a heat recovery unit, then used to heat the store or the store's domestic hot water supply.

With multiple evaporators connected to a single compressor, and each evaporator experiencing different loads throughout the day, the use of compressor unloaders is generally recommended. This approach allows the compressor to adjust its capacity as necessary to accommodate the load at any given time, greatly reducing the number of start-stop cycles. A single compressor can also be used to serve multiple evaporators operating at different temperatures. In this application, as shown in *Figure 18*, the compressor must be set to operate at the suction pressure and corresponding evaporator temperature needed by the coldest evaporator in the system. To prevent the other evaporators operating at higher temperatures from falling to this pressure and temperature, evaporator pressure regulators

(EPRs) are installed on only those suction lines. EPRs are examined later in this module. In more sophisticated applications, the compressor's suction pressure setpoint can be automatically controlled and adjusted as evaporator loads are satisfied, preventing it from operating at lower setpoints than are necessary.

4.1.2 Multiple-Compressor Applications

The use of multiple compressors connected in parallel allows greater system capacities and the ability to meet varying load conditions more effectively. Connecting two or three smaller compressors in parallel normally results in a higher Btu-per-horsepower capacity than when one larger compressor is used. Compressors in the 5hp, 7.5hp, and 10hp size are typically used for this purpose. Compressors connected in parallel can also provide a system backup in the event that one of the other compressors breaks down.

INSIDE TRACK

Supermarket Parallel Rack Systems

Packaged racks of compressors, complete with controls and safety device, are readily available for installation. Features and available options include integral oil reservoirs with all required oil management devices, a wide selection of compressor types, and alarm systems that visually and audibly indicate a loss of refrigerant or oil as well as mechanical component failure.

408SA02.EPS

Parallel compressors are normally operated with a large receiver and multi-tapped liquid, suction, and discharge (hot) gas manifolds. These manifolds or headers provide the point where the evaporators located in each of the various display cases and/or coolers are connected to the system. The condenser used with parallel compressors may be part of the compressor assembly or rack, or it may be remotely located. *Figure 19* shows a basic system using two compressors connected in parallel.

The compressors may all be the same size, or they can be different sizes. In parallel compressor systems, the compressors can be cycled on and off as needed for capacity control. Compressors may be controlled or staged based on a drop in the system suction pressure. If the compressors are of equal size, one or more mechanical or electronic methods of capacity control can be used to cycle the compressors on and off, while the unit maintains one economical pressure range. Parallel compressors of different sizes can be staged to

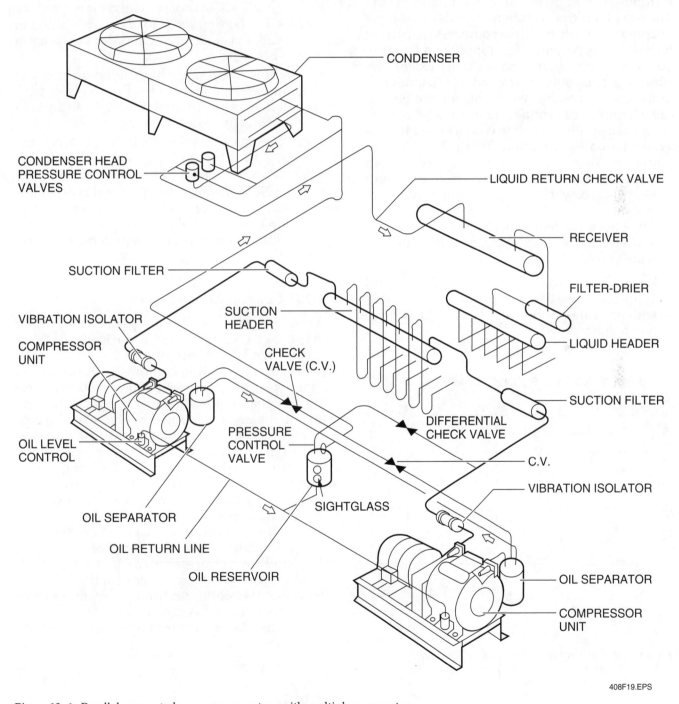

Figure 19 ◆ Parallel-connected compressor system with multiple evaporators.

408F19.EPS

obtain more steps of capacity than the same number of equally sized compressors. *Figure 20* shows an example of capacity control that can be obtained by using three parallel compressors of 5hp, 7.5hp, and 10hp.

In systems using multiple compressors, attention must be given to the return of the oil to the compressors, compressor protection, and the effect on the remainder of the system when some of the compressors are in the off cycle. Oil must return to the compressors under all operating conditions. The use of multiple compressors introduces more places to trap and hold the oil. The use of multiple condensers and/or multiple evaporators with multiple compressors further increases this potential for trapping and storing both liquid refrigerants and oil. To the extent possible, piping layouts must avoid oil accumulation in the portions of the system that are not operating. Compressors connected in parallel require crankcase and hot gas line equalization to provide adequate protection. This is because unequal pressures in multiple compressors can cause oil to be blown or drawn from one compressor crankcase to another, creating mechanical problems.

Some other guidelines that should be observed when selecting and installing parallel compressors are as follows:

• If possible, standardize on 5hp, 7.5hp, and 10hp compressors. Equip each compressor with pressure controls that satisfy all refrigeration requirements.

• Group similar types of equipment that operate at approximately the same suction pressure on the same circuit. For example, group frozen food cases together and ice cream cases together. Never combine medium-temperature or high-temperature equipment designed to cool nonfrozen foods with low-temperature equipment designed to cool frozen foods.

• If the compressor size needed for a group of ice cream cases is less than 5hp, the ice cream cases should be combined with frozen food cases that can be defrosted at the same time.

• Combine walk-in coolers. If necessary, produce cases and cutting room loads can be added to the walk-in cooler circuit in order to make use of 5hp to 10hp compressors.

• Combine multi-shelf high-temperature and medium-temperature cases. Produce and dairy coolers may be combined with these units to optimize compressor sizing.

• Put open meat cases on their own compressor. If they must be combined with other equipment, use individual EPR valves for each case.

• Use EPR valves instead of solenoids and thermostats whenever possible. Use solenoids and multiple circuit time clocks only when combining systems that have different defrost requirements.

4.1.3 Satellite Compressors

A satellite compressor is a separate compressor that uses the same source of refrigerant as that used by a related group of parallel compressors. However, the satellite compressor functions as an independent compressor connected to a cooling area that requires a lower temperature level than that being maintained by the related parallel compressors. Systems operating at a lower temperature run lower suction pressures and are less efficient. The use of a satellite compressor allows the parallel compressors in the same system to operate at a higher, more efficient suction pressure. The satellite compressor can be a remote unit, or it can be installed on the same rack as a group of parallel compressors (*Figure 21*). Satellite compressors can be used in both medium- and low-temperature applications. For example, it is common for a low-temperature frozen food cooling system to use a satellite compressor to cool the lower-temperature ice cream freezers.

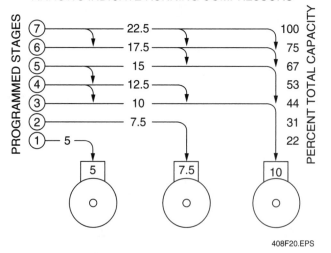

ARROWS INDICATE RUNNING COMPRESSORS

408F20.EPS

Figure 20 ◆ Capacity control by staging unequally sized compressors.

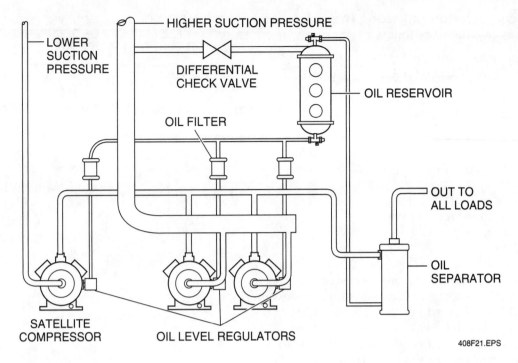

Figure 21 ◆ Satellite compressor.

4.1.4 Use of Two-Stage Compressors

A conventional compressor uses a single-stage process to compress the refrigerant gas in one or more of its cylinders. Single-stage compression means the refrigerant gas is drawn into the suction valve of the compressor cylinder(s) on the piston downstroke, and then is compressed to the required pressure on the compression up-stroke of the piston. The compressed gas is then discharged from the compressor cylinder(s) for routing to the condenser. Two-stage compressors are used in ultra low-temperature systems to pump very low-pressure suction line vapor up to the condensing pressure and temperature conditions. They are also used in some refrigeration applications where the compression ratio of a single-stage compressor would exceed 10 to 1. You will recall that compression ratio is calculated as absolute discharge pressure divided by absolute suction pressure. The absolute pressure (psia) is obtained by adding gauge pressure (psig) to atmospheric pressure (14.7 psi, sometimes rounded to 15 for simplicity).

Two-stage compressors (*Figure 22*) compress the refrigerant gas in a two-step process involving two sequential cylinders or compressors. The system suction gas is applied to the first stage, usually a large cylinder, where it is compressed to about the midpoint of the compression curve. The gas is then discharged into the suction valve

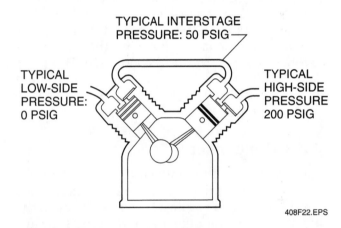

Figure 22 ◆ Two-stage compressor.

of a second cylinder. In the second cylinder, the refrigerant gas is compressed again and raised to the final condensing pressure and temperature conditions. It is then discharged for routing to the condenser. As shown, a typical interstage pressure between the first and second cylinders (suction pressure for the second cylinder) is 50 psig. The use of a two-stage compressor results in a lower compression ratio per stage than that with a single-stage compressor. In the example shown, the compression ratio of the first stage is 4.4 to 1, and for the second stage, it is 3.3 to 1. If a single-stage compressor were used with the same suction and discharge pressures, the compression ratio would be 14.6 to 1.

The compression ratio for a two-stage compressor is calculated as follows:

Compression ratio =

$$\frac{\text{gauge discharge pressure} + 14.7}{\text{gauge suction pressure} + 14.7}$$

$$\text{Stage 1} = \frac{50 + 14.7}{0 + 14.7} = 4.4$$

$$\text{Stage 2} = \frac{200 + 14.7}{50 + 14.7} = 3.3$$

For a single-stage compressor:

$$\frac{200 + 14.7}{0 + 14.7} = 14.6$$

One disadvantage of using two-stage compression is the high temperatures developed between the first and second stages as a result of compressing the refrigerant in the first stage. One method used to eliminate this problem is to desuperheat the gas. This is done by allowing a certain amount of liquid refrigerant from the system receiver to be metered through an expansion valve into the compressor interstage piping (*Figure 23*). As a result, the refrigerant flash gas acts to cool the discharge gas from the first stage before it is applied to the second stage. This is the same approach used in some hot gas bypass arrangements to desuperheat the suction gas to prevent compressor overheating.

4.2.0 Condensers

Some commercial and industrial refrigeration systems use evaporative condensers or water-cooled condensers with cooling towers. However, the most common condensing apparatus is the air-cooled condenser (*Figure 24*). A refrigeration system air-cooled condenser can be an integral part of an indoor or outdoor condensing unit assembly, or it can be a separate unit installed at a remote location from the compressor(s). The remote condenser can be placed outdoors, or it may be placed indoors to heat portions of the building in winter. Both single-circuit and multiple-circuit condensers are used.

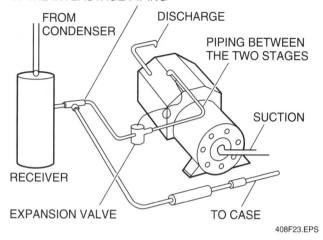

SMALL AMOUNT OF LIQUID REFRIGERANT DIVERTED FOR THE PURPOSE OF DESUPERHEATING THE GAS IN THE INTERSTAGE PIPING

408F23.EPS

Figure 23 ◆ Desuperheating method used in a two-stage compressor.

408F24.EPS

Figure 24 ◆ Air-cooled condenser.

4.2.1 Condenser Ratings

Condensers used in refrigeration systems are rated by their **total heat rejection (THR) value**. THR is the total heat removed in the desuperheating, condensing, and subcooling of the refrigerant as it flows through the condenser. The rating of an air-cooled condenser is based on the temperature difference (TD) between the dry-bulb temperature of the air entering the condenser coil and the saturated condensing temperature corresponding to the pressure at its input. Normally, the condenser

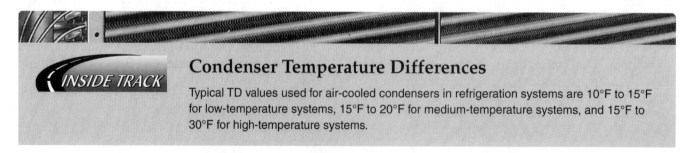

Condenser Temperature Differences

INSIDE TRACK

Typical TD values used for air-cooled condensers in refrigeration systems are 10°F to 15°F for low-temperature systems, 15°F to 20°F for medium-temperature systems, and 15°F to 30°F for high-temperature systems.

picked to work with a particular compressor is selected by matching their THR ratings. The THR capacity of a condenser is considered proportional to its TD. For example, the capacity of a condenser at 30°F is about 50 percent greater than that of the same condenser at a TD of 15°F. Condensers can also be rated in terms of net refrigeration effect (NRE), also known as net heat rejection (NHR). NHR is the total heat rejection minus the heat of compression added to the refrigerant in the compressor.

4.2.2 Control of Air-Cooled Condensers

For a refrigeration system to operate properly, the condensing pressure and temperature must be maintained within certain limits. An increase in condensing temperature causes a loss in capacity. If extreme enough, it may overload the compressor. Low condensing pressures can reduce the flow of refrigerant through conventional expansion valves, resulting in a starved evaporator. Some medium-temperature and low-temperature refrigeration systems use electronically controlled, low-pressure drop expansion valves to ensure that a sufficient supply of refrigerant is fed to the evaporator even when condensing pressures are very low. Conventional thermostatic expansion valves (TXVs) generally require a specific pressure drop to operate at their rated capacity, based on the refrigerant used. For example, HCFC-22 expansion valves are manufacturer-rated at a 100 psig pressure drop. With conventional TXVs, the discharge pressure must be maintained high enough to produce a pressure drop in this range in order to maintain full refrigeration capacity.

To prevent excessively low head pressures that can occur during winter operation, refrigerant-side or air-side control methods are commonly used with condensers. Refrigerant-side control can be done by adjusting the amount of active condensing surface used in the condenser coil. This is done by the controlled flooding of the condenser coil with liquid refrigerant. This method requires the use of a receiver and a larger charge of refrigerant. It also involves the use of temperature-actuated and/or pressure-actuated valves to meter the proper amount of refrigerant needed to flood the condenser in response to the variable loads.

Another common method of refrigerant-side control is to use a condenser with two equal parallel sections, each handling 50 percent of the load during normal summer operation. During the winter, only half of the condenser is used. Solenoid or three-way valves are used to shut off one of the condenser sections as well as any pump-down circuits and fans.

Due to its low initial cost, air-side control is one of the most common methods of head pressure control. It is accomplished by using one of three methods, or a combination of the three: fan cycling, modulating condensing unit dampers, or fan speed control.

Condenser fan cycling can be used even on condensing units with a single fan. As head pressure falls, a pressure switch opens and cycles the fan off. Once the head pressure rises to a predetermined setpoint, the switch closes and the fan restarts. Although simple and inexpensive, this method often results in constant and rapid head pressure changes. This, in turn, causes the expansion valve to constantly change its position in an attempt to regulate the flow of refrigerant and cope with the changing pressure differential. Fan cycling provides better performance when used with larger condensers having multiple fans. Although the head pressure will continue to change up and down between the pressure settings, it is generally a much slower process and the expansion valve responds to the changes more smoothly. With condensers using multiple fans, one fan is often left to run constantly while the others can be cycled as needed.

Some condensers are fitted with dampers on the discharge side of the fans. The damper position is usually controlled directly by refrigerant pressure, with the damper modulating toward the closed position as the head pressure falls. Although this generally results in a more consistent head pressure, the refrigerant-operated damper actuators expose the system to a greater potential for refrigerant leakage. Systems using electronic monitoring of head pressure and positioning of an electric damper actuator are available, but are more expensive. With either approach, air flow through the condenser coil can be modulated from 0 percent to 100 percent. This method is sometimes combined with fan cycling; one fan is fitted with a damper while the other fans are cycled on and off.

The use of variable-speed fans is another method of head pressure control that provides consistent pressures. This method has gained popularity as the cost and reliability of solid-state motor speed controls has improved. As shown in *Figure 25*, one type of control monitors liquid line temperature and uses this input to determine fan speed. As a general rule, fan motors using ball bearings generally perform better than those using impregnated sleeve bearings in speed control applications. This is due to lubrication issues, because sleeve bearings often do not produce proper lubrication at low speeds. This type of

408F25.EPS

Figure 25 ◆ Temperature-controlled condenser fan speed control.

controls allows for use of sleeve-bearing motors by incorporating a minimum speed setpoint. It also has the advantage of simple installation, since the refrigerant circuit is not accessed. *Figure 26* presents an example of a condenser fan speed control using a pressure input. This provides a more direct method of speed control than monitoring liquid line temperature.

4.2.3 Multiple Condensers

Hot gas line connections to multiple air-cooled condensers must maintain roughly equal pressure drop to each condenser. Otherwise, there is a tendency for liquid refrigerant to back up in the condenser with the lower pressure. One piping

408F26.EPS

Figure 26 ◆ Pressure-controlled condenser fan speed control.

method used to help achieve equal pressures uses a slightly oversized hot gas line or header to supply the condensers. This oversized header is essentially frictionless. This allows the takeoffs feeding each of the condensers to be individually sized to provide the required pressure drop or friction loss across the related condenser.

4.2.4 Increasing Liquid Line Refrigerant Subcooling

Subcooling is used in low-temperature and other refrigeration systems to reduce the refrigerant temperature in the liquid line below the saturation temperature. Remember that there is a loss of capacity in most systems at the outlet of the metering device. This is due to the volume of refrigerant that is vaporized to reduce the temperature of the remaining refrigerant to the saturation temperature corresponding to the evaporator pressure. Refrigerant that is vaporized at the outlet of the metering device cannot contribute its latent heat capacity to the cooling process in the evaporator. Subcooling the liquid refrigerant reduces the temperature difference across the metering device, and less refrigerant is vaporized in the process. This can result in capacity increases of roughly 0.5 percent per 1°F of additional subcooling provided at the same suction and discharge pressures.

One method used to increase subcooling is to add a smaller coil section in the air entry side of the system condenser. With this arrangement, the liquid refrigerant that leaves the main condenser coil is routed to the receiver as in the basic system. However, from the receiver, the liquid refrigerant is routed back through the small subcooling coil in the condenser, where it receives additional subcooling before being applied to the system expansion valve(s).

Subcooling may also be increased by integrating a subcooling section within the condenser's main coil assembly. There, the coil's circuitry at the liquid area of the coil is constructed to provide more passes through fewer tube circuits.

Additional subcooling can also be provided by using a water-cooled condenser in addition to the air-cooled condenser. The best approaches for this in the commercial/industrial sector use sufficiently cool waste water from other processes. This method uses water that is otherwise wasted to reduce energy consumption and increase refrigeration capacity.

Suction-to-liquid heat exchangers (*Figure 27*) can also be added to systems to increase subcooling, although the value can be questionable. They are simple double-walled heat exchangers, with suction gas flowing through one side and liquid refrigerant flowing through the other. There is no question that the liquid refrigerant does experience additional subcooling from the cool suction vapor, but it is argued that the heat energy never leaves the system since the heat is absorbed by the suction gas. The increased pressure drop in both refrigerant streams can also add to system energy consumption. An added advantage is that the process increases the superheat of the suction vapor, helping to ensure that liquid refrigerant droplets do not return to the compressor. This is an important consideration because commercial and industrial refrigeration systems generally operate at lower superheat values than comfort cooling systems. A variety of studies indicate that the overall effect is positive with some refrigerants and negative with others, so it is suggested that some research be done before applying suction-to-liquid heat exchangers.

Another common method used to increase subcooling in the liquid line does not involve the condenser. This method uses a higher-temperature system, such as an air-conditioning system, to cool the liquid line in a lower-temperature system. This subcools the liquid refrigerant flowing in the low-temperature system liquid line.

408F27.EPS

Figure 27 ◆ Suction-to-liquid heat exchangers.

This method is popular because a high-temperature system can remove heat with much more efficiency than could be achieved using one of the condenser-related subcooling methods described earlier.

4.2.5 Guidelines for Installing Remote Air-Cooled Condensers

The most important factor to take into account when installing air-cooled condensers or condensing units is the need for a supply of ambient air to the condenser and the removal of heated air from the condensing unit or remote condenser area. Always follow the manufacturer's recommendations concerning clearances from walls and obstructions that should be maintained around the top, bottom, and sides of the unit. If the manufacturer's requirements are not followed, higher system head pressures can result. This can cause poor system operation and possible failure of the equipment. It is also important that adequate space be provided around the unit so that the components can be accessed for removal and/or servicing. Condenser units must not be located in the vicinity of steam, hot air, or exhaust fumes.

Ideally, condenser units should be mounted over corridors, utility areas, restrooms, and other areas where high levels of sound are not important. Roof-mounted units should be installed level on steel channels or I-beam frames capable of supporting the weight of the unit. Mount the condenser over building columns or load-bearing walls. The condenser should be mounted away from noise-sensitive areas and have adequate support to avoid vibration and the transmission of noise into the building. Vibration-absorbing pads or springs should be installed between the condensing unit legs or frame and the roof mounting assembly. Some additional guidelines

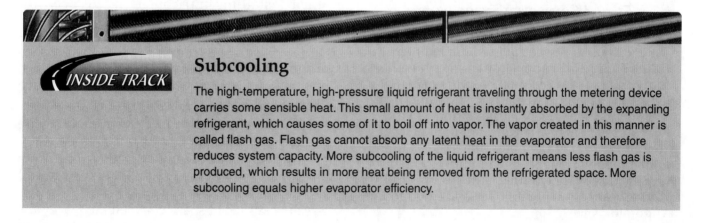

INSIDE TRACK

Subcooling

The high-temperature, high-pressure liquid refrigerant traveling through the metering device carries some sensible heat. This small amount of heat is instantly absorbed by the expanding refrigerant, which causes some of it to boil off into vapor. The vapor created in this manner is called flash gas. Flash gas cannot absorb any latent heat in the evaporator and therefore reduces system capacity. More subcooling of the liquid refrigerant means less flash gas is produced, which results in more heat being removed from the refrigerated space. More subcooling equals higher evaporator efficiency.

that should be observed when installing remote air-cooled condensers are as follows:

- Never use the condenser coil manifolds, control panels, or return piping bends to lift or move the condenser unit.
- Use the building plans or condenser manufacturer's specifications to identify which set of condenser circuits should be connected to a specific compressor.
- Route and support all piping so that vibration and the stress caused by thermal expansion and building settling are minimized.
- Construct a weatherproof enclosure over all the piping access openings in the roof.
- Route all discharge lines away from the control panel.
- Route all liquid return lines so that they do not protrude more than a few inches in front of the condenser before dropping to the roof.
- Insulate all discharge piping near the control box to protect workers from burns or electrical shock.
- Build a wooden platform or other nonconductive structure for workers to stand on when servicing electrical devices.

4.3.0 Evaporators

Many kinds of evaporators are used in commercial refrigeration. They can be divided into two main groups. One group includes the evaporators used for cooling air, which in turn cools the contents of coolers or display cases. The other group includes liquid-cooling evaporators that are used to cool liquids such as water, soft drinks, other beverages, and brines. The remainder of this section will focus on air-cooling evaporators. Evaporators used for cooling air operate by forced convection or natural convection.

4.3.1 Forced Convection Evaporators (Unit Coolers)

The unit coolers (*Figure 28*) used in food preparation rooms, chill rooms, and walk-in coolers incorporate a forced convection evaporator. It is a packaged assembly usually consisting of the evaporator coil, blower fan, expansion valve, and defrost mechanism. The size and shape of the storage area generally determines the type and number of unit coolers used and their location(s). Forced convection cooling of the room results from the fan(s) forcing air over and through the evaporator coil.

408F28.EPS

Figure 28 ◆ Typical unit cooler.

Unit coolers can cool a refrigerated room or cabinet quickly; however, they have a tendency to cause rapid dehydration or drying of the foods unless care is taken to match the unit with its intended use. When storing many varieties of fresh produce, the humidity in the storage area should be kept high to maintain product quality. Conversely, the humidity should be kept low in a room used to handle and chill warm, red meat carcasses. This is necessary to avoid the formation of fog and to prevent condensing moisture from dripping onto the meat.

Unit coolers used in areas requiring high humidity levels normally have large coils operating with small air-to-refrigerant temperature differences (4°F to 8°F). The rate of airflow across the evaporator coil is high to provide the required refrigeration with the small drop in temperature. When unit coolers are used in areas requiring low humidity, the evaporator coil is small and operates with higher air-to-refrigerant temperature differences (20°F to 30°F). The rate of airflow across the evaporator is low. The evaporator coil(s) in unit coolers that operate at saturated suction temperatures below 30°F must be defrosted periodically. The method used can be air defrost, electric defrost, or hot gas defrost. These defrost methods are described later in this module.

The most important factor to take into account when installing unit coolers is the need for adequate airflow across the evaporator. Always follow the manufacturer's recommendations concerning clearances from walls and obstructions around the top, bottom, and sides of the unit. Some guidelines that should be observed when installing unit coolers are as follows:

- Make sure that the air pattern covers the entire room or area.
- Avoid installing the unit directly above doors and door openings in low-temperature and medium-temperature applications.
- Allow sufficient space between the rear of the unit cooler and the wall to permit free return of the air.

- Always trap drain lines individually to prevent vapor migration.
- Traps on low-temperature units must be outside of the refrigerated enclosure. Traps subject to freezing temperatures must be wrapped with heat tape and insulated.
- In coolers or freezers with glass display doors, be sure that the cooler air discharge blows above, not directly at, the doors. If the doors extend above the blower level, use a baffle to properly direct the air.

4.3.2 Natural Convection Evaporators

Natural convection evaporators are used in many types of open display cases. Air circulation in these evaporators depends on thermal circulation, where warm air rises and cool air descends. These evaporators can be grouped into three classes: frosting, defrosting, and nonfrosting.

Frosting evaporators are commonly used in frozen food storage cases. They continuously build up frost from the moisture in the air. The refrigeration system must be shut down periodically to allow the evaporator to defrost. The defrost cycle can be either automatic or manual. If the refrigerant temperature is below 28°F, heat energy of some kind must be used to defrost the coil; otherwise, the evaporator must be turned off longer than the normal cycle.

Defrosting evaporators run on a defrosting cycle. While the system is operating and the condensing unit is running, the temperature of the evaporator is low. This results in a buildup of frost on the evaporator. When the compressor shuts off, the coil warms above 32°F and the frost melts. This defrosting process is called air defrosting. One disadvantage with some defrosting evaporators is that when the top of the evaporator defrosts, moisture may flow down the evaporator surface. If this happens before it can properly drain, the moisture can freeze around the lower parts of the evaporator, eventually blocking air circulation and interfering with proper refrigeration.

Nonfrosting evaporators operate at temperatures just at or above freezing. When properly operating, frost does not form on these evaporators.

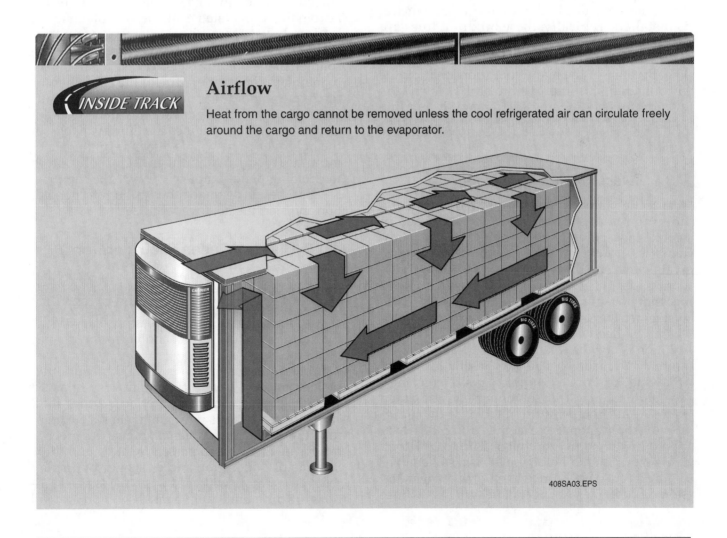

INSIDE TRACK

Airflow

Heat from the cargo cannot be removed unless the cool refrigerated air can circulate freely around the cargo and return to the evaporator.

408SA03.EPS

4.3.3 Multiple Evaporators

Many different configurations can exist in a multiple evaporator system. Each coil normally has its own metering device. Frequently, each coil is also controlled by a solenoid shutoff valve. While each coil acts more or less as an independent coil, be sure that each expansion valve is sensitive to the condition of the coil it controls and is not affected by the conditions in the other coils. Make sure that the control bulb of the lower expansion valve is mounted at a location that is not affected by refrigerant slopover from the coil above it. Because of the numerous conditions that can be encountered with expansion valves when installed with multiple evaporators, it is important to install the expansion valves as recommended by the manufacturer.

The physical location of the suction outlets of different evaporator coils also varies widely. Many are at the bottom of the coil, thus creating a natural self-draining condition for any oil in the coil. Also, many coils are headered, thus creating a potential trap at the bottom of the header unless a drain is provided to drain the oil. If drains are not provided by the coil manufacturer, special oil drain lines must be built into the system when it is installed. Typically, these oil drain lines are positioned so that the oil flows by gravity into the refrigerant suction line or the TXV equalizing line. Because of the many conditions that can be encountered, it is important to always install the evaporators and related suction and liquid line piping as recommended by the evaporator manufacturer.

4.3.4 Eutectic Plate Evaporators

A eutectic plate evaporator, commonly called a holding plate, is a type of evaporator used in mobile land-based and marine refrigeration systems where continuous power is not readily available to run the refrigeration system on a 24-hour basis. The holding plate is a box-like assembly that houses coiled tubing through which the system refrigerant gas is passed, surrounded by a chamber filled with a eutectic solution. A eutectic solution is one that maintains a constant temperature while changing state from a solid to a liquid (thawing) or from a liquid to a solid (freezing). The particular eutectic solution used is designed to freeze at a temperature suitable for the refrigeration application. For example, 0°F to 10°F may be required for a freezer unit or 24°F to 26°F for a refrigerator unit. A holding plate is installed in or on the interior walls or ceiling of the refrigerated storage container. Refrigeration systems that use holding plates instead of standard evaporators are called holdover refrigeration systems.

The use of a holdover refrigeration system in a truck, railcar, or boat enables the refrigeration system equipment to run for only part of the day and still maintain stable refrigeration cold box temperatures for up to 24 hours or more. For example, in a truck holdover refrigeration system, the system compressor is typically energized and allowed to run at night. This allows the refrigerated box to be precooled and the eutectic solution contained in the holding plate to completely freeze. This is done by circulating the system refrigerant gas through the evaporator tubing inside the holding plate. When the compressor is energized, the cycle time is controlled by a temperature probe that monitors the temperature of the holding plate and cycles the system as required. The compressor in these systems can be driven in one of several ways, such as directly from an engine or by an AC or DC electric motor.

The holdover refrigeration system compressor remains de-energized while the truck is being used to make delivery runs during the day. During this interval, called the holdover period, the frozen eutectic solution in the holding plate absorbs the heat in the refrigerated container, maintaining the loaded chilled or frozen cargo at the proper temperature. This is accomplished because the frozen solution in the holding plate acts like a block of ice to maintain the temperature of the container for the required holdover period. Note that the surface area of the holding plate and number of holding plates used determines the time available for the holdover period. This depends on the maximum heat load. When the eutectic solution in the holding plate completely thaws, the compressor must be run again to refreeze the eutectic solution and start the holdover cycle again.

4.4.0 Display Cases (Refrigerators)

Open, self-service display cases (*Figure 29*) that operate at medium and low temperatures are used extensively in food markets. Display cases are used to maintain the temperature of a product that has previously been cooled to the proper temperature.

Display case operation is significantly affected by temperature, humidity, and movement of the surrounding air. The refrigeration load for food store display cases is normally rated by the manufacturer based on ambient summer design conditions of 75°F and 55 percent relative humidity. Satisfactory operation and efficiency of display case refrigerators will vary when the store conditions are different than the rated conditions.

MULTISHELF PRODUCE MERCHANDISER

FROZEN FOOD ISLAND

408F29.EPS

Figure 29 ◆ Open, self-service display cases.

Sources of radiant heat, such as display lighting, ceiling temperatures, product packaging, and density of display case loading, all affect display case operation. If not properly controlled, all of these heat sources can raise the surface temperatures of the products being displayed, thereby shortening their shelf life. For example, an 80°F ceiling temperature can raise the surface temperature of meat in a display case 3°F to 5°F.

At 100°F, the surface temperature of meat can be raised 4°F to 8°F. The surface temperature of a loosely wrapped package of meat with an air space between the film and surface may be 2°F to 4°F above the ambient temperature because of the greenhouse effect within the packaging.

Display cases used in market installations are normally an endless construction type, which allows a group of individual displays to be joined together to form a continuous display bank. Separate end sections are provided for the first and last units in the display bank. All the display cases are constructed with surface zones of transition between the refrigerated area and the room atmosphere. Thermal breaks separate the zones to minimize the amount of surface on the refrigerator below the dew point. The evaporators, air distribution system, and method of defrost built into each type of display case are highly specialized and designed to produce the particular display results desired. The defrost methods used with display cases are discussed in detail later in this module.

A **deck** is a shelf, pan, or rack that supports the refrigerated item. Display cases are often classified by the number of decks they contain:

- Multiple-deck dairy display
- Multiple-deck produce display
- Top discharge produce display
- Multiple-deck fruit and produce display
- Multiple-deck frozen food display
- Back-to-back, single-deck frozen food/ice cream display

The display cases listed above can all be classified as open, self-service display cases. The open display case refrigerates the food using a blanket of cold air. The foods are stored below the top level of the cold blanket of air and are usually wrapped in clear plastic to protect the food from dust and germs in the air and from contamination by customer handling. The evaporator in an open display case is normally located in the bottom of the cabinet with a fan used to circulate the air over the coil and through the case.

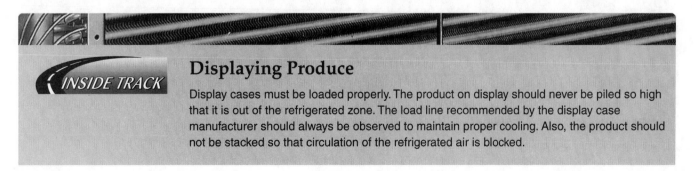

INSIDE TRACK

Displaying Produce

Display cases must be loaded properly. The product on display should never be piled so high that it is out of the refrigerated zone. The load line recommended by the display case manufacturer should always be observed to maintain proper cooling. Also, the product should not be stacked so that circulation of the refrigerated air is blocked.

When compared to closed coolers, open, self-service display cases need additional refrigeration capacity because of their higher refrigeration loss. The refrigeration losses in an open display case are great enough that about twice the amount of refrigeration capacity is needed to cool a product than would be needed to cool the same product in a closed case. Warm, moist air from outside the cabinet mixes with the cold, dry air in the case. Customers reaching into the case to remove stored products cause air movement and additional mixing of warm and cold air. This adds a substantial load to the refrigeration cycle because it must cool the additional air and condense the additional moisture. To minimize the mixing of cold and warm air, open display cases should be located away from all externally induced air circulation devices, such as fans, entrance doors, heating duct openings, and air blasts from unit heaters.

Open frozen food display cases are used to display packaged ice cream and other frozen foods. Because they operate at temperatures of 0°F and below, a substantial difference exists between the room air and the cold air within the case. Because of this greater temperature difference, more moisture-laden air infiltrates the case, increasing the refrigeration losses. These losses, coupled with the lower operating temperatures, require that the condenser used with a frozen food case be larger than that used with a medium-temperature unit.

It is important to note that the dramatic refrigeration losses described here do have a positive effect elsewhere. Although the losses create higher refrigeration loads, the comfort cooling load of the facility itself can be significantly reduced. As a result, many supermarkets have significantly less comfort cooling equipment installed, helping to offset the investment in refrigeration apparatus. In some cooler geographic areas, the cooling effect of the open refrigeration cases, offset by the heat gain of the facility, can nearly or completely eliminate the need for comfort cooling systems. Conversely, refrigeration losses to the space can increase the winter heating load significantly, resulting in substantial energy costs during the winter months, as the two systems are working against each other. The process of calculating heat gains and losses for food markets must therefore be done with great care and attention to detail to ensure accuracy and proper performance of all mechanical systems as a combined unit.

Some open frozen food cases use vertical plate evaporators. These evaporators are placed about 12" apart along the full length of the storage compartment. The packaged frozen foods are stored between the plates. In some cases, frozen foods cannot be stored in the upper 2" or 3" of the plate compartments due to the warm air that penetrates through that top layer of cold air. The evaporator plates in these display cases are marked to show the level beyond which the products

Supermarket Energy and Global Warming

GOING GREEN

Supermarkets are among the highest energy-consuming types of commercial facilities in the world, with the refrigeration systems consuming the lion's share at 50 percent or more. Huge amounts of heat must be rejected, and proportionally large amounts of refrigerant are present in the facility due to the required refrigeration capacity and lengthy piping runs. As a result, the Total Equivalent Warming Impact (TEWI) of grocery operations are very high.

Numerous studies have been done on TEWI reduction in supermarket operations and many strategies are employed to reach this goal. For example, rejected heat is captured from refrigerants and used to heat water for food preparation and cleaning activities, as well for use in comfort heating. A reduction of refrigerant charge results in a reduction of financial risk associated with refrigerant loss, and a reduction in the potential for environmental damage when leakage does occur. One method used to reduce the total refrigerant charge is to use secondary coolants. These are other fluids that have no significant environmental impact and function as a heat transfer agent. These coolants can be chilled by the system inside the mechanical room environment, then pumped to the various refrigeration loads throughout the store. This strategy can reduce the required primary refrigerant charge by thousands of pounds in a single facility.

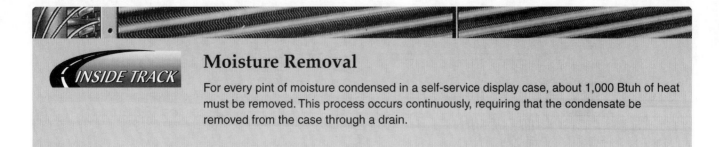
should not be stored. Some display cases use a method of forcing a cold air blanket over the top of the evaporator plates to permit storing food up to the top edge of the plates. This is done by using a forced-air fan, an auxiliary coil built into the structure above the case, or an extra plate evaporator placed lengthwise above the top edge of the plates in the freezing compartment. A blanket of cold air at least 2" or 3" thick must separate the frozen food and the warm, moist air above the storage area in the display case.

4.5.0 Accessories

Accessories (*Figure 30*) are added to the basic refrigeration system in order to improve safety, endurance, efficiency, or servicing. This is especially true in commercial refrigeration systems. Some of these components are factory installed, while others may be installed in the field. This section briefly describes the most common accessories. They include the following:

- Filter-driers
- Sightglass/moisture-liquid indicators
- Suction line accumulators
- Crankcase heaters
- Oil separators and oil control systems
- Receivers

- Manual shutoff and service valves
- Fusible plugs, relief valves, and relief manifolds
- Check valves
- Compressor mufflers
- Vibration isolators

4.5.1 Filter-Driers

No matter how many precautions are taken during the assembly, installation, and service of a refrigeration system, some moisture and other contaminants will enter the system. However, leaks in the system are the main cause of such contamination. Water vapor and foreign matter in sufficient quantities will cause problems in any refrigeration system. Synthetic polyolester (POE) oils increase the chances of a system moisture problem. POE oils attract and absorb moisture from the ambient surroundings faster and in greater quantities than conventional mineral oils. Because they absorb moisture, using POE oils increases the amount of moisture in the system. Moisture may freeze within the orifice of the expansion valve, cause corrosion of metal parts, and/or wet the motor windings of hermetic and semi-hermetic compressors. In the presence of system heat, the moisture will react with the system oil and refrigerant to produce corrosive acid.

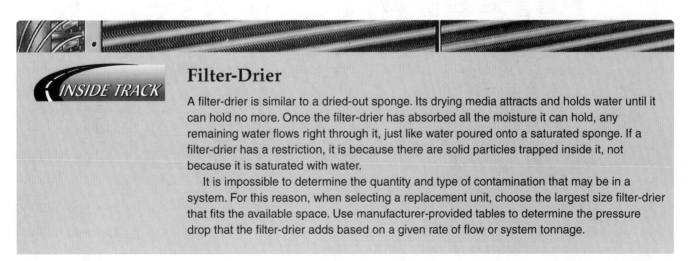

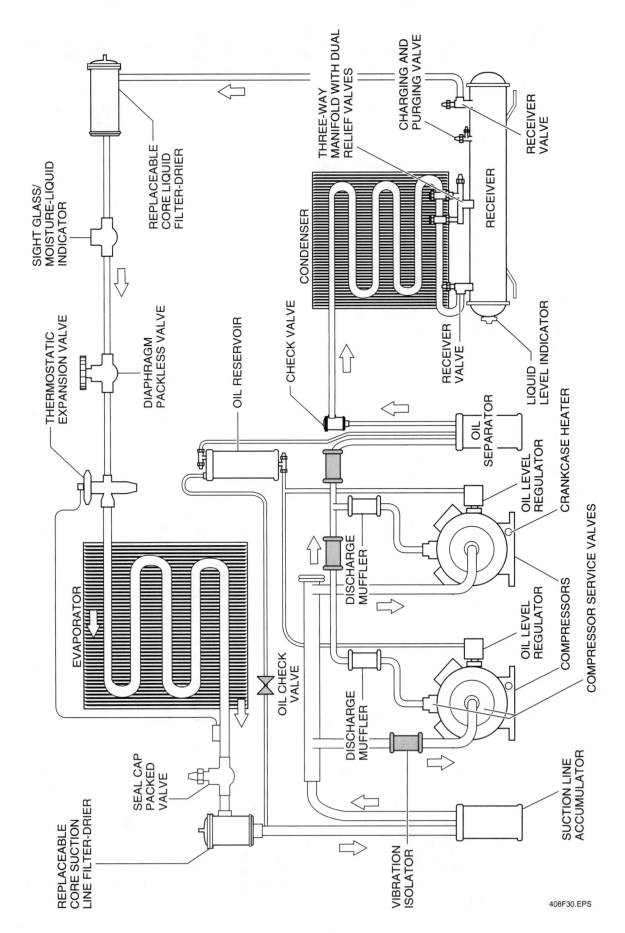

Figure 30 ◆ Parallel compressor refrigeration system with accessories.

408F30.EPS

Foreign matter can contaminate the compressor oil and become lodged in the various mechanical parts of the system.

Filter-driers (*Figure 31*) are used to remove water vapor and foreign matter from the refrigerant stream. Suction line filter-driers are installed upstream of the compressor service valve, accumulator, or other accessories that may be installed. A filter-drier combines the functions of a refrigerant filter and a refrigerant drier in one device. Its use protects the metering device,

evaporator, and compressor from foreign matter such as dirt, scale, or rust. The drier portion removes moisture from the system and traps it. Filter-driers are installed downstream of the condenser or receiver outlet service valve and upstream of the liquid line solenoid valve (if so equipped). On some occasions, especially in the suction line, a simple filter is used in lieu of the filter-drier. This is appropriate when moisture is not considered to be an issue. Filters often cause lower pressure drops than filter-driers.

CROSS SECTION SHOWING DESICCANT
BALLS AND SOLID-PARTICLE SCREEN

CROSS SECTION SHOWING
SOLID CORE

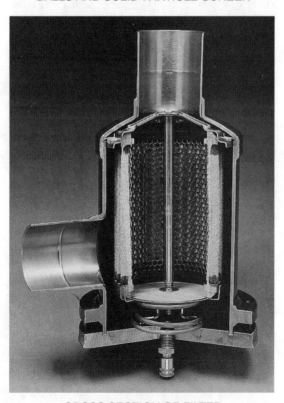

CROSS SECTION OF FILTER
WITH REPLACEABLE CORE

REPLACEABLE CORE

408F31.EPS

Figure 31 ◆ Filter-driers.

Filter-driers are available both as sealed (throwaway) units and with replaceable cores. Some are equipped with Schrader-type access valves that allow measurement of the pressure drop across the filter. This allows restricted filters or cores to be identified quickly so they can be replaced when the pressure drop is excessive. Normally, filter-driers are replaced immediately after a major system repair.

4.5.2 Sightglass/Moisture-Liquid Indicators

The sightglass (*Figure 32*) is like a window that allows you to view the condition of the system refrigerant at the sightglass location. It is typically used when checking the refrigerant charge. The normal location for the sightglass is in the liquid line, downstream of the condenser or receiver outlet and the filter-drier, and immediately ahead of the expansion valve. The appearance of bubbles in the sightglass generally indicates a refrigerant shortage, but it can also signify a restriction in the liquid line. Frequently, the sightglass also serves as a moisture indicator. This type of sightglass indicates the presence of moisture in the system by the color of the moisture-sensing element. Typically, blue indicates a dry and safe system, light violet cautions that some moisture is present, and pink indicates that the amount of moisture in the system is at a dangerous level. Sightglass/moisture-liquid indicators normally require no service. However, in cases of extreme acid formation in a system after a compressor burnout, the acid may damage the sensing element or etch the glass. This would require that the device be replaced.

4.5.3 Suction Line Accumulators

The suction line accumulator is a trap that prevents the compressor from taking in slugs of liquid refrigerant or compressor oil. If liquid refrigerant is allowed to enter the compressor, noisy operation, high consumption, and compressor damage may result. Accumulators are installed in the suction line as close to the compressor suction inlet as possible. At this location, any quantities of liquid refrigerant or oil will be trapped temporarily in the accumulator. The trapped refrigerant remains in the accumulator until it is evaporated. Some accumulators have heaters that help to vaporize the refrigerant liquid.

4.5.4 Crankcase Heaters

Crankcase heaters are installed on most compressors to prevent refrigerant from migrating into the system and mixing with oil when the compressor is off. All heaters evaporate refrigerant from the oil. Heaters are typically fastened to the bottom of the crankcase or inserted directly into the compressor crankcase (immersion type). Band-type heaters that encircle the outside shell of welded hermetic compressors are frequently used. Some of these use PTC (positive temperature coefficient) technology to regulate the output of the heater; as the compressor warms, the heater

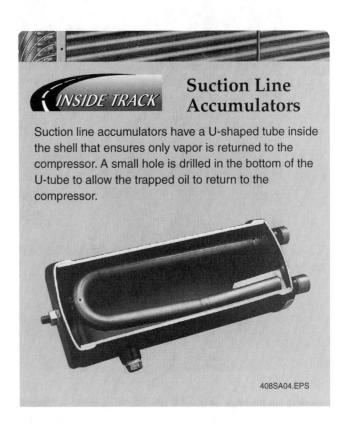

INSIDE TRACK

Suction Line Accumulators

Suction line accumulators have a U-shaped tube inside the shell that ensures only vapor is returned to the compressor. A small hole is drilled in the bottom of the U-tube to allow the trapped oil to return to the compressor.

408SA04.EPS

408F32.EPS

Figure 32 ◆ Sightglass/moisture-liquid indicator.

produces less heat. Band-type heaters are inappropriate for use with semi-hermetic and open drive compressors. *Figure 33* shows a variety of clamp-on and insertion type crankcase heaters.

4.5.5 Oil Separators and Oil Control Systems

Oil coats the inside of every component through which it passes. It reduces the heat transferability and efficiency of the evaporator and condenser. Oil separators minimize the amount of oil that circulates through a refrigeration system. They also slow down the accumulation of oil in places from which it is difficult to return. The oil separator (*Figure 34*) is one of the primary components of any oil control system. They are typically installed in the hot gas discharge line as close to the compressor as practical. Separators usually have a small sump to collect the trapped oil. A float valve in the sump maintains a seal between the high-pressure and low-pressure sides of the system. As the oil level in the sump rises, the float raises and allows high pressure refrigerant vapor to push the oil back to the low-pressure crankcase of the compressor. When used with a single compressor, the oil separator generally functions without additional system components except for an oil filter (*Figure 35*), which captures any debris.

408F34.EPS

Figure 34 ◆ Oil separator.

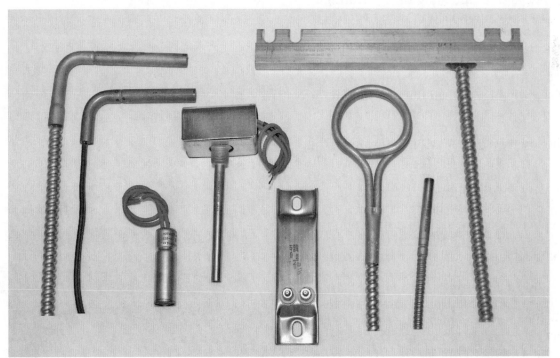

408F33.EPS

Figure 33 ◆ Crankcase heaters.

Figure 36 ◆ Oil reservoir.

Figure 35 ◆ Oil filter.

In parallel compressor systems, the correct oil level must be maintained in each of the compressors, regardless of operating conditions. The amount of oil leaving the individual compressors through the discharge line can vary dramatically due to age/wear and operating time. Additional components are included, along with the oil separator, to maintain the oil level. The oil separator sends the accumulated oil to an oil reservoir (*Figure 36*), since it does not have a large holding capacity. To reduce the pressure in the reservoir to a level slightly higher than the pressure in the compressor crankcase, high pressure is relieved through the oil differential check valve (*Figure 37*) back to the low pressure side of the system. The valve shown is preset to an appropriate value to ensure that adequate pressure is maintained in the reservoir to push oil into the crankcase and relieve any pressure in excess of this value.

Oil is delivered back to the individual compressor crankcases through the oil level control (*Figure 38*). One of the flanged ports shown on the control is used to mount it to the compressor crankcase, while the other is covered with a sightglass for viewing. As the oil level drops in the crankcase, the internal float of the control allows oil to enter. The oil level is adjustable in this particular device from ¼ sightglass to ½ sightglass.

Figure 39 is an example of an oil control system.

Figure 37 ◆ Oil differential check valve.

Figure 38 ◆ Oil level control.

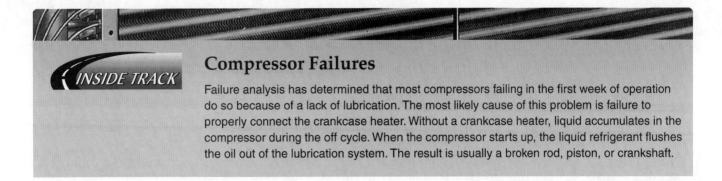

Compressor Failures

Failure analysis has determined that most compressors failing in the first week of operation do so because of a lack of lubrication. The most likely cause of this problem is failure to properly connect the crankcase heater. Without a crankcase heater, liquid accumulates in the compressor during the off cycle. When the compressor starts up, the liquid refrigerant flushes the oil out of the lubrication system. The result is usually a broken rod, piston, or crankshaft.

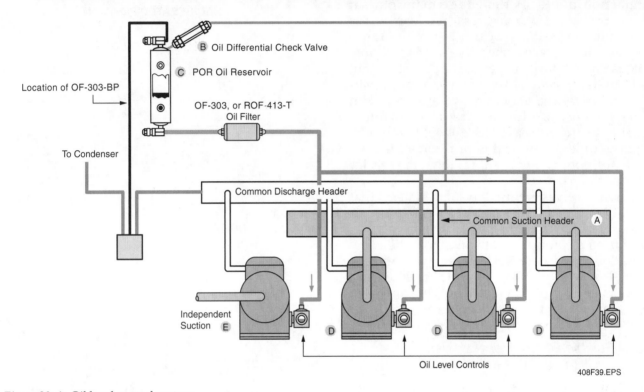

Figure 39 ◆ Oil level control system.

4.5.6 Receivers

The liquid receiver is a tank or container used to store liquid refrigerant in the refrigeration system. This storage is needed in air-cooled refrigeration systems that have large load variations where the system is not required to remove heat at a constant rate. These variations in load allow the refrigerant in the condenser to freely drain into and accumulate in the receiver for temporary storage. The receiver can also be used to store the system charge during system service procedures or prolonged shutdown periods. Note that most water-cooled refrigeration systems do not use a receiver because it is an integral part of the water-cooled condenser. The receiver is installed in the liquid line between the condenser and the metering device. Receivers have inlet and outlet connections and a relief port. They normally contain a sightglass or test cocks that can be used to determine the amount of liquid refrigerant in the receiver.

4.5.7 Manual Shutoff and Service Valves

Various types of manual shutoff and service valves are installed in refrigeration systems. They enable you to seal off parts of the system while installing gauges. They also provide access to the closed system for servicing. Service valves can be installed in any line that may have to be valved off, usually at a place that is easily accessible to the technician. Common valves include charging valves installed in the liquid line or the downstream side of a receiver, receiver outlet and inlet shutoff valves, and compressor shutoff and service valves.

Valves can have packing around their valve stems, or they can be packless. Many are provided with back-seating construction; that is, when the valve is fully opened, the valve disc seats against a second seat. This arrangement seals the packing from the system pressure, preventing the leakage of refrigerant between the packing and the valve stem. Back-seating valves frequently have a port located between the packing and the back seat of the valve. This service port can be used for refrigerant charging or as a connection point for the gauge manifold set when measuring system pressures. The port is opened to the system by closing (front-seating) the valve or by placing the valve in any intermediate position between open and closed. It is shut off from the system by fully opening (back-seating) the valve. Sealed valve caps are used with packed valves as a further precaution against leakage. To prevent leaks, these should always be in place, unless removed in order to adjust the valve position. *Figure 40* shows the three service valve positions.

Schrader valves are used in some systems that are not equipped with service shutoff valves. A Schrader valve uses the same type of valve as those used to pump air into an automobile tire. Schrader valves are self-sealing and have a spring-loaded core that closes when released and opens when depressed. They provide access to the system when the core is depressed.

4.5.8 Fusible Plugs, Relief Valves, and Relief Valve Manifolds

The fusible plug is a device related to temperature and pressure that melts when a specific temperature that corresponds to a given refrigerant pressure is reached. Once the plug melts, the entire refrigerant charge is released. For this reason, the use of fusible plugs is being eliminated in newer systems.

Relief valves are safety devices designed to relieve pressure, preventing the buildup of excessive refrigerant pressure within a system. They

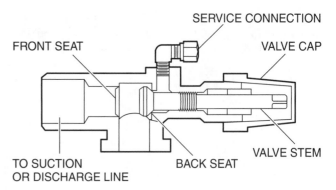

BACK-SEATED POSITION

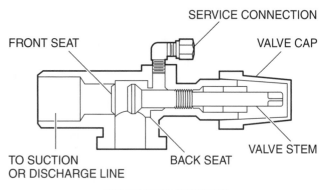

MID-SEATED POSITION

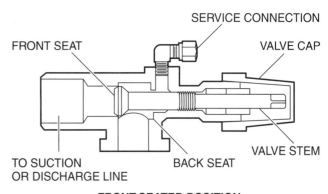

FRONT-SEATED POSITION

408F40.EPS

Figure 40 ◆ Service valve positions.

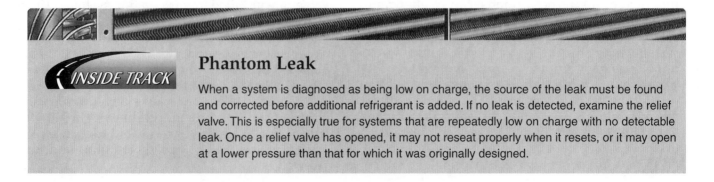

Phantom Leak

When a system is diagnosed as being low on charge, the source of the leak must be found and corrected before additional refrigerant is added. If no leak is detected, examine the relief valve. This is especially true for systems that are repeatedly low on charge with no detectable leak. Once a relief valve has opened, it may not reseat properly when it resets, or it may open at a lower pressure than that for which it was originally designed.

are normally installed in the condenser or liquid receiver. In addition, some codes require that a relief valve be installed on the low-pressure side of the system. Relief valves reset after the excessive pressure has been released, trapping the remaining refrigerant charge.

As shown in *Figure 30*, relief valve manifolds enable the installation of two relief valves at the same piping position. This allows for one of the two relief valves to be safely isolated for service or replacement, while the second valve remains active.

4.5.9 Check Valves

Check valves allow refrigerant to flow in one direction only. They are used to prevent the refrigerant in the system from flowing into system components or piping where it is not wanted. A typical application is when a compressor is located in a much cooler place than the condenser. With this condition, it is possible for refrigerant to migrate from the condenser and condense at the compressor discharge connection during periods of shutdown. If the compressor valves are not tight, the refrigerant can enter the cylinders where it will cause severe slugging when the compressor is restarted. The installation of a check valve in the compressor discharge line prevents this problem.

4.5.10 Compressor Mufflers

Mufflers (*Figure 41*) are used mainly in systems with open or semi-hermetic reciprocating compressors. Reciprocating compressors generate audible pulsations that transmit into and travel along the system piping. A muffler installed in the discharge line, as near the compressor as practical, is used to remove or dampen these pulsations. Mufflers lower the system noise and prevent pos-

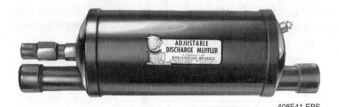

Figure 41 ◆ Compressor muffler.

sible damage to piping and components caused by vibration. Mufflers should add only a minimal pressure drop when in the discharge line.

4.5.11 Vibration Isolators (Absorbers/Dampers)

Vibration isolators (*Figure 42*), also commonly called vibration absorbers or dampers, are flexible tube devices used to prevent the carryover of vibrations generated by one system component to other parts of the system. They are commonly installed in the compressor suction and discharge

Figure 42 ◆ Vibration isolator.

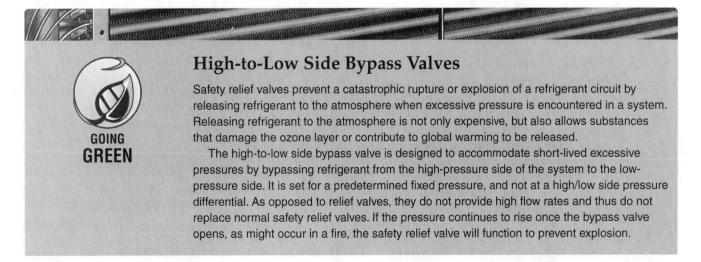

High-to-Low Side Bypass Valves

GOING GREEN

Safety relief valves prevent a catastrophic rupture or explosion of a refrigerant circuit by releasing refrigerant to the atmosphere when excessive pressure is encountered in a system. Releasing refrigerant to the atmosphere is not only expensive, but also allows substances that damage the ozone layer or contribute to global warming to be released.

The high-to-low side bypass valve is designed to accommodate short-lived excessive pressures by bypassing refrigerant from the high-pressure side of the system to the low-pressure side. It is set for a predetermined fixed pressure, and not at a high/low side pressure differential. As opposed to relief valves, they do not provide high flow rates and thus do not replace normal safety relief valves. If the pressure continues to rise once the bypass valve opens, as might occur in a fire, the safety relief valve will function to prevent explosion.

lines near the compressor to reduce the transmission of any condensing unit vibrations. For best sound absorption, two vibration isolators can be used in each line, one mounted vertically and the other horizontally.

5.0.0 ◆ REFRIGERATION SYSTEM CONTROL DEVICES

Control devices used to regulate the operation of a refrigeration system may be either manual or automatic. They may respond to temperature and/or pressure.

5.1.0 Crankcase Pressure Regulating Valves

Crankcase pressure regulating (CPR) valves are installed in the suction line ahead of the compressor (*Figure 43*). The valve controls the maximum

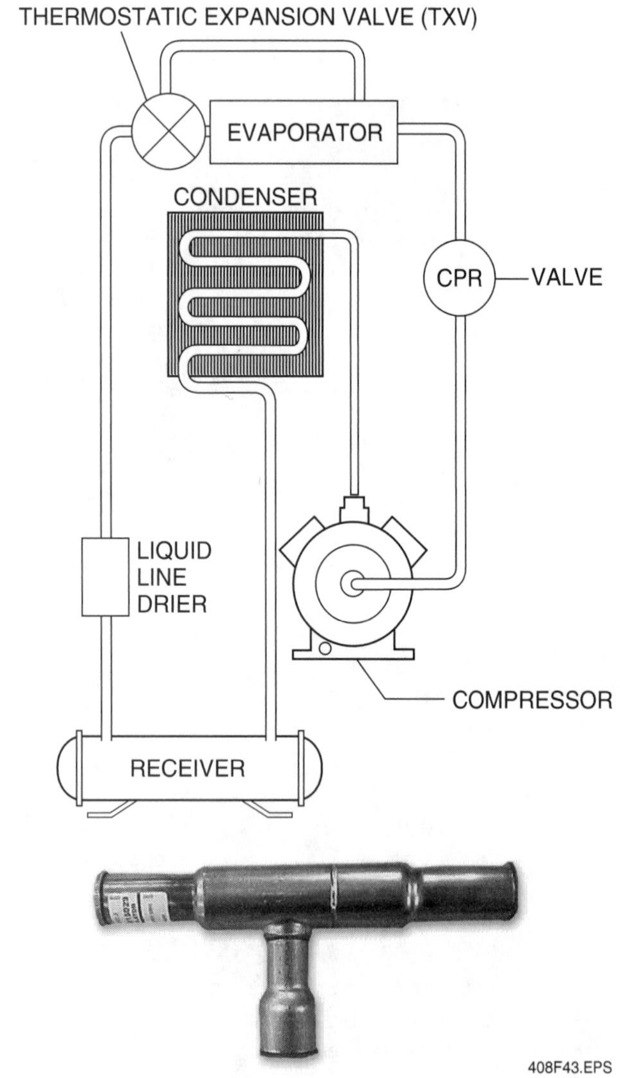

Figure 43 ◆ Crankcase pressure regulating (CPR) valves.

pressure at the compressor suction line and provides overload protection for the compressor motor. The valve is adjusted for the maximum pressure specified by the manufacturer of the condensing unit when available.

Compressors in commercial and industrial refrigeration systems commonly experience an overload during two conditions. One condition occurs during initial startup or a restart of the system after an extended shutdown, especially in low temperature systems. At this time, the compressor is likely operating well outside its selected operating range, and the load must be managed to prevent compressor overheating and/or motor damage. The other condition follows a defrost period, when the warm evaporator places an additional load on the compressor for a short period of time.

When an overload occurs, the valve modulates to prevent suction gas pressures from exceeding the pressure for which the valve is adjusted. When the overload passes and the pressure drops below the valve setting, the valve opens wide.

The valve setting is determined by a pressure spring and the valve closes on a rise in suction line pressure. One common field adjustment method is to operate the system under what is perceived to be an excessive load while monitoring the compressor amperage with an ammeter. The valve is then adjusted toward the closed position until the compressor amperage falls just below the compressor rated load amps (RLA) as stated on its data plate.

5.2.0 Evaporator Pressure Regulating Valves

Evaporator pressure regulating (EPR) valves are installed in the suction line between an evaporator and the compressor (*Figure 44*). They are designed to maintain the desired minimum pressures, and therefore temperatures, within close limits. These valves are used to control the evaporator temperature on systems that use multiple evaporators operating at different temperatures or on systems where the evaporating temperature cannot be allowed to fall below a predetermined level.

The EPR valve responds to the inlet pressure of the evaporator and opens when the inlet pressure equals or exceeds the opening setpoint. As the evaporator temperature continues to rise, the valve opening increases, permitting an increased flow of refrigerant. The valve closes off when the inlet pressure drops below the setpoint. Consequently, the evaporator temperature does not drop below this established point.

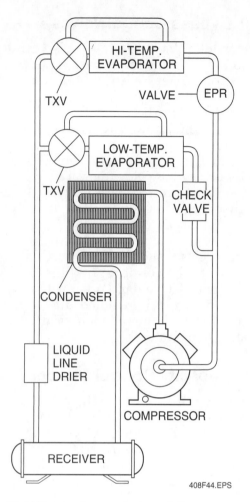

Figure 44 ◆ Evaporator pressure regulator.

408F44.EPS

On multiple evaporator systems where different evaporator temperatures are required, EPR valves are used to hold the saturation pressure at the required setpoint above the common system suction pressure. The EPRs prevent the lowering of temperatures in the warmer evaporators below that desired, while the compressor continues operating to satisfy the temperature requirements needed for the colder evaporators.

5.3.0 Air-Cooled Condenser Pressure Regulator

The condenser pressure regulator (head pressure control) maintains proper condenser pressures when the ambient temperature of the outdoor air is low. The head pressure control is a three-way modulating valve (*Figure 45*). It is controlled by the discharge (head) pressure. When the outdoor air temperatures are high enough to cause the compressor discharge pressure to be above the head pressure control valve setpoint, refrigerant flows through the system in the normal manner.

It flows from the compressor through the condenser and valve ports C and R to the receiver, then through the metering device and into the evaporator.

As the ambient outdoor air temperatures fall, there is a corresponding decrease in the discharge pressure. When the discharge pressure falls below the head pressure control valve setpoint, the compressor discharge gas is routed through

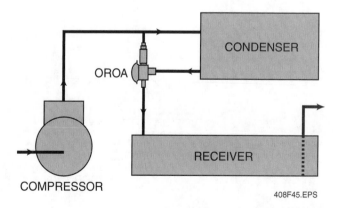

Figure 45 ◆ Condenser pressure regulator (three-way valve).

408F45.EPS

port B of the valve into the receiver, creating a higher pressure at the condenser outlet. The higher pressure at the condenser outlet reduces the flow from port C and causes the level of condensed liquid to rise in the condenser. This flooding of the condenser with liquid refrigerant reduces the available condensing surface. The result is to increase the pressure in the condenser and maintain an adequate high-side pressure. Although the three-way valve is popular for small systems, its capacity is limited. Larger systems accomplish the same task by using by using two separate valves to control refrigerant flow and maintain head pressure. *Figure 46* provides an example of this arrangement.

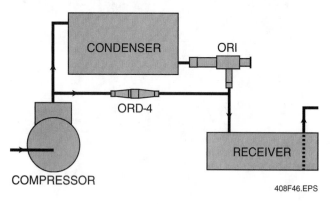

Figure 46 ◆ Condenser pressure regulation (two valves).

5.4.0 Bypass Control Valves

Bypass control valves for air-cooled condensers function as condenser bypass devices on air-cooled condensers with winter start control. These types of valves are installed in the condenser bypass line between the compressor discharge and the liquid line receiver (*Figure 47*). These devices operate in response to receiver pressure. The bypass valve opens when the receiver pressure is below the valve setpoint.

As each compressor starts on cycle, high-pressure discharge gas is admitted directly to the receiver. The pressure quickly builds up in the receiver; when the valve operating pressure is reached, it closes and the compressor discharges directly into the condenser. Without a bypass control, some delay would occur before adequate receiver pressure is reached due to the time required for the larger condenser mass to reach operating temperatures.

5.5.0 Capacity Control Devices

Capacity control refers to the method used to control a system so that its heat removal ability (capacity) matches changing system load conditions. In commercial refrigeration systems, the load on the evaporator can vary widely, ranging from full load to a small percentage of full load.

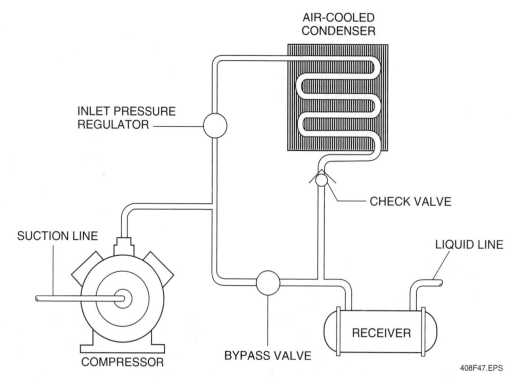

Figure 47 ◆ Bypass control valve.

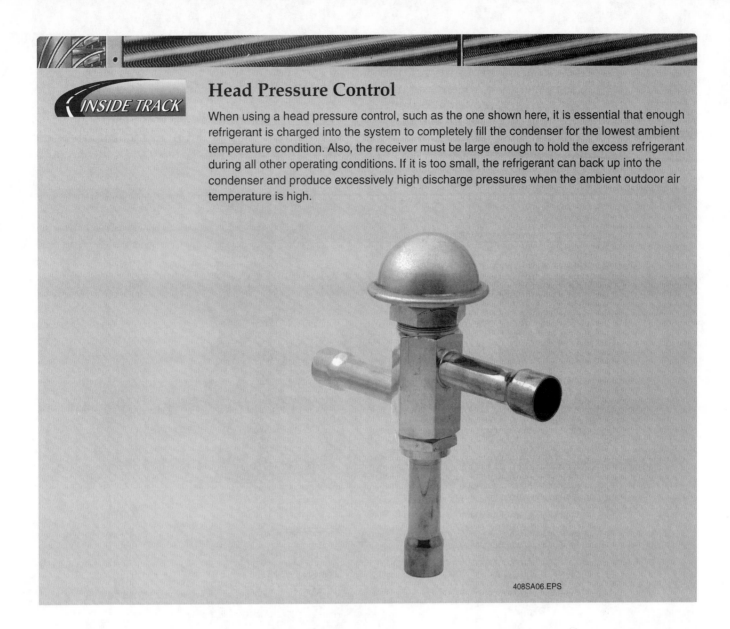
Differences in the load can be caused by the type of product being cooled, the quantity of product being cooled, and the ambient temperature. At system loads less than full load, less refrigerant is boiled off within the evaporator, causing the compressor suction temperature and pressure to reduce. When a system operates at or near minimum load conditions for an extended interval, problems can occur that affect the operating life of the compressor. Low suction pressures and temperatures result in reduced refrigerant gas flow, which can cause the motor in hermetic compressors to overheat. It can also result in poor oil return, which can cause compressor lubrication problems.

Various forms of capacity control have been developed to regulate the capacity of the system compressor to match changing load conditions.

Some very common methods of capacity control include compressor on-off, compressor cylinder unloading, and multi-speed compressors. As these methods have been discussed in previous modules and are directly related to the compressor itself, several other methods related primarily to the refrigerant side of the system will be examined in this module. These methods include hot gas bypass and suction modulation.

5.5.1 Hot Gas Bypass

In cases where short cycling the compressor or using cylinder unloaders is not satisfactory or where energy management is not a factor, you can control compressor capacity using a hot gas bypass. The hot gas bypass system is simply a method of bypassing the condenser with the compressor discharge gas to prevent the compressor suction pressure from falling below a predetermined setpoint, thereby limiting the saturated suction temperature of the evaporator. This prevents the system from reducing the temperature of the refrigerated space any further, since the evaporator coil temperature cannot be reduced further.

All hot gas bypass valves (*Figure 48*) open in response to a decrease in pressure on the low side of the system. They modulate as necessary to maintain the predetermined low-side pressure, admitting as much high-pressure vapor as needed. A solenoid valve is typically installed upstream of the regulator, or incorporated into the valve itself, to allow this part of the circuit to

408F48.EPS

Figure 48 ◆ Hot gas bypass valves.

be disabled, such as when the load is satisfied and the compressor enters a pump down mode. The solenoid also prevents any refrigerant migration during the off cycle.

On close-connected systems using a single evaporator, the hot gas is bypassed into the evaporator immediately after the expansion valve. *Figure 49* provides one example of this application. Special T fittings (*Figure 50*) are available from the manufacturers of hot gas bypass regulators and accessories to introduce the hot gas between the expansion valve and the refrigerant distributor assembly. Some distributors and T fittings are also available as a single assembly. This special fitting allows the hot gas to enter the refrigerant stream without disturbing the operation of the expansion valve itself. Its design also ensures good mixing of the hot gas and the liquid/vapor refrigerant mixture for smooth flow into the evaporator coil. A common piping T should not be used for this application. The distinct advantage of this method is that the expansion valve continues to do its job. As superheat begins to rise as a result of the introduction of hot gas, the expansion valve responds by opening further to maintain the proper superheat value. This prevents the compressor from overheating due to high superheat. The suction pressure and refrigeration capacity remain at the desired level.

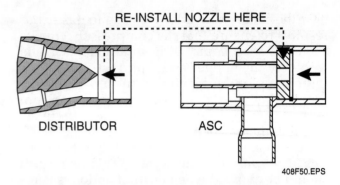

Figure 50 ◆ Typical hot gas connector installation.

On systems where the condensing unit and evaporator are distant from each other, or when multiple evaporators are involved, the hot gas can be introduced into the suction line upstream of the compressor (*Figure 51*). Although functional, this is not the preferred method, as additional steps must be taken to ensure the compressor does not overheat. First, the elevated superheat from the hot gas must be reduced. This is done by using a separate expansion valve known as the desuperheating expansion valve. This valve must be carefully selected by the manufacturer for each application, rather than from a typical catalog. Mixing liquid refrigerant together with hot gas vapor in the suction line, especially

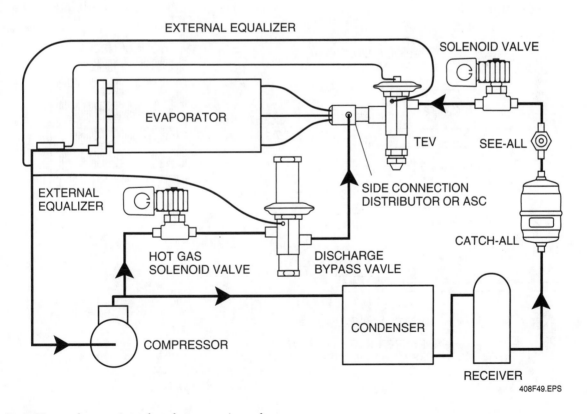

Figure 49 ◆ Hot gas bypass, introduced at expansion valve.

near the compressor inlet, can lead to disastrous results. To ensure that the compressor remains protected from liquid refrigerant floodback, and to ensure the hot gas and liquid refrigerant are thoroughly mixed together, a suction accumulator is usually installed downstream of the connection from both devices. This design effectively provides the function of capacity control through hot gas bypass, and must be used for other practical reasons on some systems. However, it should be avoided when the first option is feasible.

5.5.2 Suction Modulation

Suction modulation capacity control is used on equipment with advanced microprocessor controls. It uses a series of sensors to input system pressure and temperature information to the microprocessor on a continuous basis. To control capacity, the microprocessor sends a signal to a drive module that engages a stepper motor. The stepper motor valve is located in the suction line just before the compressor and opens or closes to regulate the flow of refrigerant through the suction line. The suction line is large enough to allow the valve to handle significant demands for cooling such as in deep freeze applications or quick temperature pull-downs. *Figure 52* shows a suction modulation valve typical of those used in a container refrigeration unit.

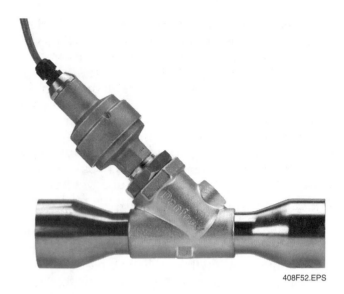

408F52.EPS

Figure 52 ◆ Suction modulation valve.

5.6.0 Pump-Down Control

Refrigeration systems usually equalize during the off cycle, which means the high-side pressure and the low-side pressure reach equilibrium at some point between the normal operating pressures. This equalization results in a mixture of gas and liquid refrigerant throughout the system in most direct expansion systems. However, on some occasions refrigerant gas migrates to the compressor and condenses into a liquid in the compressor

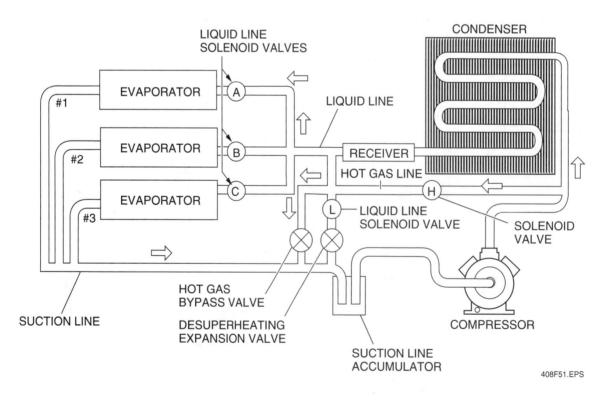

408F51.EPS

Figure 51 ◆ Hot gas bypass valve with multiple evaporators.

crankcase. Another phenomenon somewhat unique to refrigeration, especially in low-temperature applications, is that a significant amount of refrigerant vapor left motionless in the evaporator during an off cycle quickly condenses to liquid, flooding back to the compressor at startup. Both conditions can result in lubricating oil loss, scored compressor bearings, broken piston rods, and valve damage. To alleviate this problem, a pump-down cycle can be used. The advantages of using a pump-down cycle include the following:

- Reduced flooded starts
- Refrigerant pumped into the condenser/receiver during the off cycle
- Quicker cooling during the on cycle

During the pump-down cycle, the compressor pumps the refrigerant charge into the condenser and liquid line or into a receiver at the end of each operating cycle. This keeps the liquid refrigerant out of the low side of the system, which includes the evaporator, suction line, and compressor crankcase.

Crankcase or sump heaters usually prevent refrigerant from condensing in the crankcase during the off cycle. However, systems with larger-than-normal refrigerant operating charges can have liquid refrigerant in the compressor crankcase after a long off cycle. A suction line accumulator (*Figure 53*) and a pump-down cycle will normally protect the compressor from liquid slugging during startup after a prolonged off cycle. The suction line accumulator offers system protection during the running cycle, whereas the pump-down cycle offers protection during the off cycle.

When a pump-down cycle is used, a thermostat controls a liquid line solenoid valve that opens or closes in response to the thermostat's setpoint. By closing the liquid line, the refrigerant is pumped into the condenser and receiver. The liquid line solenoid valve should be located just upstream from the expansion valve and as close to the evaporator coil inlet as possible. This allows for maximum storage volume of the liquid refrigerant during pump-down and prevents the liquid line from sweating. Compressor operation is controlled by a low-pressure cutout control that is set for a suction pressure below any value expected during normal operation.

5.7.0 Defrost Systems

As long as there is a supply of moisture inside the refrigerated space, it will condense on the surface of the evaporator, just as is the case in comfort

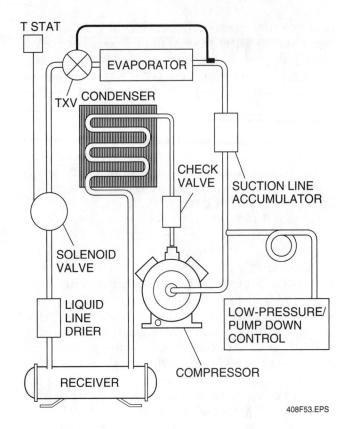

Figure 53 ◆ Pump-down system.

408F53.EPS

cooling applications. However, when the coil surface is below freezing, frost and ice will build, impeding and eventually blocking airflow through the coil. The coil must be defrosted for it to resume proper operation.

In commercial and industrial refrigeration applications, the primary defrost approaches are as follows:

- Off-cycle defrost
- Timed defrost
- Electric defrost
- Hot gas defrost

Regardless of the defrost approach used, there are several common factors regarding frost accumulation on the evaporator coils. Fixtures that were selected and designed to operate at high air-to-refrigerant temperature differences (such as those that are not required to maintain higher humidity levels) will generally build frost more rapidly as their coil temperature will likely be well below the dew point for the box. This encourages greater moisture condensation on the coil. Fixtures that have low air-to-refrigerant temperature differences will not accumulate condensate and frost as quickly.

Another factor in frost accumulation is related to the use of the fixture and the infiltration of outside air. Many refrigerated enclosures are expected to experience high consumer or operator traffic. In hot, humid weather, a fixture that is opened many times each day for product access will constantly be exposed to hot, moist air. This significantly increases the volume of airborne moisture in the enclosure. In addition, the infiltration of hot air replacing refrigerated air significantly increases the required refrigeration operating time as well. Conditions such as this can overwhelm and interfere with fixture performance in spite of defrost cycles. Consider the potential difference in performance of an identical refrigeration system installed in two different locations, one in a remotely located store with little consumer traffic in Arizona, and the other in a consistently busy, 24-hour refrigerated distribution center in coastal Florida.

It is important to note that all heat added to the refrigerated enclosure during the defrost cycle, regardless of the source, must again be removed once the process is complete. This fact has a significant impact on the refrigeration load, and the time a system is projected to spend in its defrost mode must be subtracted from the available time for refrigeration. For example, if the total refrigeration load for a freezer were to be calculated as 96,000 Btus over a 24-hour period (4,000 Btuh), but the unit is expected to operate in defrost for 3 hours per day, then the refrigeration equipment must be able generate this capacity in 21 hours instead of 24 hours. This would require 4,571 Btuh of actual refrigeration capacity.

5.7.1 Off-Cycle Defrost

Off-cycle defrost is the simplest and most passive of the defrost approaches. In fixtures that maintain temperatures at or above 36°F, the coil is simply allowed to defrost during the normal off cycle. Fixtures in this category will usually operate at a coil/saturated suction temperature of 16°F to 31°F, depending on the desired humidity levels for the stored product and the box design. At these temperatures, the coil will certainly build frost, but should defrost naturally once the refrigeration apparatus has cycled off, because the fixture temperature is above freezing.

One assist to this natural defrost approach is to ensure that the fixture temperature is not set lower than needed. Maintaining the box even a few degrees colder than necessary can seriously impede or defeat the natural defrost process, resulting in frozen coils and rising box temperatures. Once this occurs, the fixture will require shutdown for an extended period to rid the coil of accumulated ice.

5.7.2 Timed Defrost

Some systems that operate at slightly lower temperatures than those using simple off-cycle defrost will require a longer period to clear accumulated ice than a normal off cycle provides. Fixtures that operate at 32°F to 36°F often benefit from this approach. Typically, a 24-hour timer (*Figure 54*) with normally closed contacts, placed in series with the fixture's operating controls, is set for a reasonable period of time yet significantly longer than the normal off cycle. The fixture temperature may rise a few degrees above normal during the extended period, but will quickly recover. As is true with all defrost scenarios, setting the timer for a defrost period that coincides with the fixture's lightest period of use (late at night, for example) helps prevent the fixture temperature from rising beyond an acceptable level. Several cycles per day may be required.

5.7.3 Electric Defrost

Electric defrost systems are prevalent in high- and medium-temperature systems that operate at temperatures too cold for off-cycle or timed

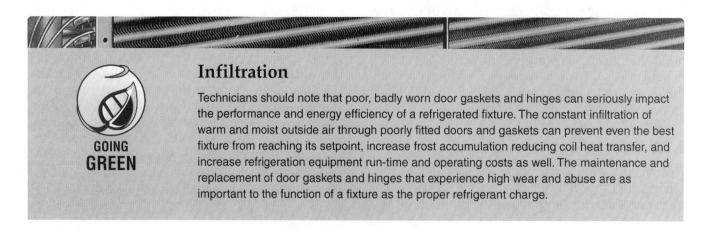

Infiltration

Technicians should note that poor, badly worn door gaskets and hinges can seriously impact the performance and energy efficiency of a refrigerated fixture. The constant infiltration of warm and moist outside air through poorly fitted doors and gaskets can prevent even the best fixture from reaching its setpoint, increase frost accumulation reducing coil heat transfer, and increase refrigeration equipment run-time and operating costs as well. The maintenance and replacement of door gaskets and hinges that experience high wear and abuse are as important to the function of a fixture as the proper refrigerant charge.

GOING GREEN

Figure 54 ◆ Typical 24-hour time clock.

proper operation, the heater must be in good thermal contact with the coil fins to effectively distribute heat and minimize the required defrost period. During the defrost period, a time clock stops the refrigeration equipment and energizes the electric heater. Depending on the unit design, the evaporator fans may continue to operate through the defrost cycle. In many cases, the size and shape of the defrost heater will be unique to a given evaporator coil or manufacturer. This requires access to original parts when replacement becomes necessary. Some coil designs require that the heater be placed deep inside the coil assembly, making repair and/or replacement more challenging, but resulting in a highly effective and rapid defrost cycle.

You may also find some units equipped with radiant electric heaters (*Figure 56*). The resistance heater wire is spirally wound and enclosed inside a quartz tube, with the wires passed through the ends of the sealed tube cap. Using infrared radiant heat, defrost can be accomplished without the heater being in contact with the coil itself. They are capable of reaching a high temperature very quickly and are corrosion-resistant. The quartz enclosure can be a bit fragile, however.

When electric defrost is required, the condensate removed from the coil must also be maintained above freezing until it has completely exited the refrigerated environment. Most condensate collection pans, either as part of a unitary

defrost to be effective, as well as in commercial freezers. Electric defrost components are relatively inexpensive and reliable, and are simple to troubleshoot and repair.

Figure 55 shows a typical arrangement for resistive defrost heaters, in this case installed on both the face and back of an evaporator coil. For

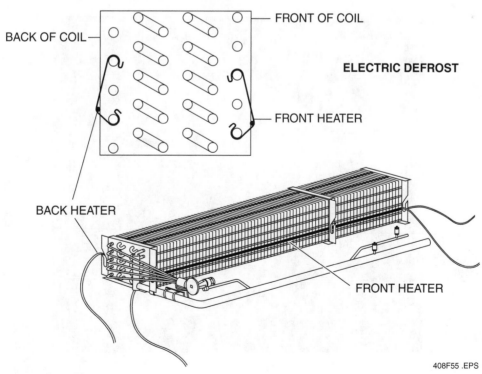

Figure 55 ◆ Evaporator electric heat arrangement.

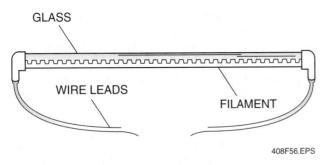

Figure 56 ◆ Infrared quartz heater assembly.

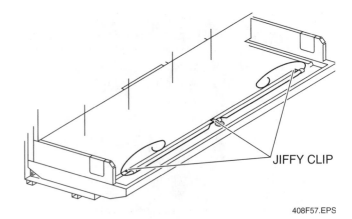

408F57.EPS

Figure 57 ◆ Condensate pan heater.

evaporator coil assembly or as a separate collection point inside the fixture, must also be equipped with electric heaters such as the one shown in *Figure 57*. Some installations, where the drain line is routed through a freezing area, may require that the condensate drain line be kept above freezing as well. This is usually done by attaching electric pipe heating cable to the outside of the line, or by using an appropriate waterproof heat cable routed on the inside of the condensate drain line.

Electric defrost systems are generally controlled by a more sophisticated time clock than a simple 24-hour model. Electromechanical clocks have a dial assembly such as the one shown in *Figure 58*, allowing the timer to be configured for the time, duration, and number of defrost periods during a 24-hour period. The inner defrost dial sets the duration of the defrost period. Defrost is terminated after the period has elapsed. The

timer contacts then return to their normal position and the refrigeration cycle resumes.

Electric defrost can also be terminated by a temperature sensor or line-mounted thermostat. By monitoring the temperature of the coil, or the refrigerant suction line adjacent to the coil, the thermostat provides an indication that the coil has reached the appropriate temperature. Defrost can be terminated before the time clock has reached the allowed maximum time. The proper setpoint for temperature termination may differ among products, as it is based on the manufacturer's chosen location and laboratory testing. Line-mounted thermostats are generally nonadjustable, so it is important that the proper device be chosen for replacement.

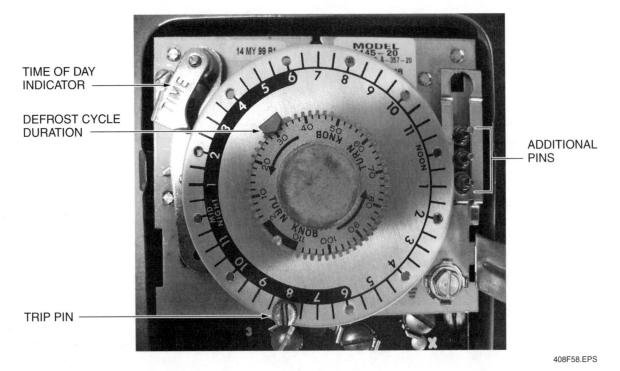

408F58.EPS

Figure 58 ◆ Typical defrost timer dial.

The final choice of defrost control and termination is often an option provided to the end user, either as a standard feature or at an added cost. Defrost termination can therefore be time-only, temperature-only, or a combination of the two.

5.7.4 Hot Gas Defrost

Hot gas defrost is widely considered to be the fastest and most efficient approach. Instead of having only the surface area contact on the coil provided by electric heating elements, the entire surface of the coil tubing is incorporated into the defrost process. However, there are more complex issues in the refrigeration circuit to address with this method. Due to the additional refrigerant circuit components required, and the increased potential for compressor damage when defrost components do not operate properly, great care must be taken in its application.

Hot gas defrost, in its simplest form, sends discharge gas from the compressor directly to the evaporator, bypassing the metering device through the use of several solenoid valves. The coil temperature is allowed to rise to a predetermined pressure and the refrigerant vapor rapidly condenses in the cold evaporator as the process begins. The latent heat of condensation, not the temperature of the discharge gas, is the primary factor in removing enough heat to bring the coil surface above freezing.

One problem for this approach is that little or no heat is being added to the refrigeration circuit beyond compressor motor heat and the heat of compression, since the evaporator is now condensing refrigerant. As a result, the system may not be able to develop sufficient heat to complete the defrost. This can be avoided to some degree by more frequent, but short defrost periods. Another, more serious problem is the potential for condensed liquid refrigerant to return to the compressor in significant volume. For hot gas defrost to work well then, refrigerant leaving the evaporator as a liquid must be re-vaporized fully and a source of heat must be incorporated. This can be done in several ways.

One approach works well with multiple evaporators, often three or more. This method is sometimes referred to as **latent heat defrost**. One or more evaporators remain in service, absorbing heat and transferring it to the refrigerant, while one coil is being defrosted. Hot gas is ported to the defrosting coil through a series of solenoid valves. This approach works especially well on large supermarket systems where many evaporators are connected to a compressor rack system.

Another approach for small systems is very much like a conventional heat pump; the evaporator becomes the condenser and the condenser becomes the evaporator.

One very effective approach uses a reservoir of water, also known as a thermal bank, to absorb and store heat during the normal operating cycle. When the defrost cycle is initiated, this stored heat is used to re-evaporate the refrigerant. This results in a very efficient and rapid defrost with limited potential for liquid floodback to occur.

During the normal operating cycle, hot gas refrigerant is sent through a coil submerged inside the thermal bank. The water is heated and the refrigerant is pre-cooled before moving on to the condenser. Once the defrost cycle begins, hot gas is first ported to the evaporator drain pan and on to the evaporator coil, where the ice melts and the refrigerant gas is condensed to a liquid. As this liquid or liquid/vapor refrigerant mixture exits the evaporator, it passes through the thermal bank, absorbing heat from the water and re-evaporating the refrigerant. The water in the bank is often frozen during the process, adding more latent heat to the process.

6.0.0 ◆ AMMONIA SYSTEMS

The choice of the refrigerant used in a low-temperature industrial system is mainly between HCF404A and ammonia. Obviously, there is a choice only when the federal, state, and local codes permit the use of ammonia. Codes in many locations restrict the use of ammonia systems. In general, it is only used in applications that are physically separated from public access. Ammonia systems are typically custom designed for a particular application. These systems can span a wide range of evaporating and condensing conditions. They may be used to cool a food freezing plant operating at temperatures from +50°F to –50°F or to cool a distribution warehouse requiring multiple temperatures for the storage of meat, produce, ice cream, and other frozen foods.

6.1.0 Ammonia Refrigerant (R-717)

Anhydrous ammonia, called refrigerant R-717, has excellent heat transfer qualities and is commonly used in industrial refrigeration systems. Ammonia is a chemical compound of nitrogen and hydrogen (NH_3). Under ordinary conditions, it is a colorless gas. Its boiling point at atmospheric pressure is –28°F and its melting point from a solid is –108°F. Because of its low boiling point, refrigeration can occur at temperatures considerably below 0°F without the need for below atmospheric pressures in the evaporator. It has a latent heat value of 565 Btu/lb at 5°F. This enables large refrigeration effects while using relatively small equipment. In the presence of moisture, ammonia chemically reacts with and attacks copper and bronze; therefore, steel piping and fittings are used in ammonia systems.

> **WARNING!**
> The term *anhydrous* refers to the lack of water in ammonia for commercial refrigeration use. Ammonia is highly hydroscopic, seeking water from any available source. A volume equal to 1,300 gallons of ammonia vapor can be absorbed into a single gallon of water. Its attraction to water places moist surfaces of the human body, such as eyes, nasal passages, mouth and lungs, at risk when exposed to excessive amounts of ammonia vapor.

Ammonia is an **immiscible** refrigerant. This means that it does not dissolve in oil. In mechanical (compressor-driven) ammonia refrigeration systems, this causes the system mineral-based oil to separate from the ammonia and collect in the evaporator or low side of the system. This oil must be removed, or it will coat the heat transfer surfaces of the evaporator, reducing system performance. In some systems, draining is done automatically. In others, it may be done manually.

One straightforward advantage to using ammonia as a refrigerant is that leaks are self-alarming; the odor alone gets immediate attention and alerts system operators to the presence of a leak. Concentrations in air as small as 3 to 5 parts per million (ppm) can be detected by humans. Small leaks can be pinpointed by using small handheld wicks coated with sulfur. The wick is lit and produces a relatively colorless smoke. In the presence of ammonia vapors, the smoke becomes white and quite dense. Since ammonia is one of the oldest refrigerants in use today, the simple sulfur stick has been used to locate ammonia leaks for over 100 years.

6.2.0 Ammonia Refrigeration Systems and Components

Ammonia systems fall into two classes: mechanical and nonmechanical. In the mechanical system, a compressor functions to lower the pressure and temperature of the ammonia (refrigerant) in the evaporator, allowing it to boil and absorb heat from the item being cooled. The pressure and temperature of the ammonia in the condenser are raised, allowing it to condense and give up its heat to either air or water. The compressor also maintains a pressure difference between the high-pressure and low-pressure sides of the system. This pressure difference causes the ammonia to flow throughout the system. Nonmechanical ammonia systems are called absorption systems. They do not have a compressor. Instead, they use a chemical cycle to provide cooling. Absorption systems have many applications and are used widely in recreational vehicle refrigeration systems. They are also used in residential refrigeration applications where the use of electrical devices is limited. The principles of operation of absorption systems are described in detail in many of the references listed in the back of this module. The remainder of this section will provide a brief description of some basic mechanical ammonia systems.

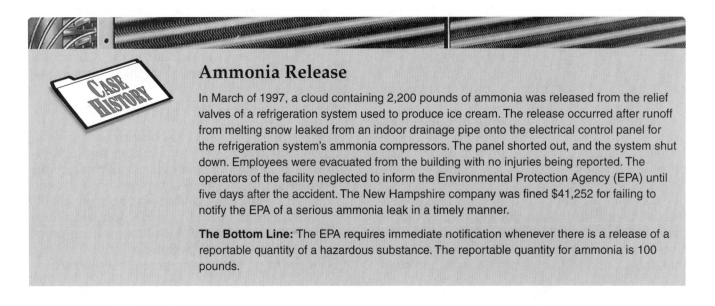

Ammonia Release

In March of 1997, a cloud containing 2,200 pounds of ammonia was released from the relief valves of a refrigeration system used to produce ice cream. The release occurred after runoff from melting snow leaked from an indoor drainage pipe onto the electrical control panel for the refrigeration system's ammonia compressors. The panel shorted out, and the system shut down. Employees were evacuated from the building with no injuries being reported. The operators of the facility neglected to inform the Environmental Protection Agency (EPA) until five days after the accident. The New Hampshire company was fined $41,252 for failing to notify the EPA of a serious ammonia leak in a timely manner.

The Bottom Line: The EPA requires immediate notification whenever there is a release of a reportable quantity of a hazardous substance. The reportable quantity for ammonia is 100 pounds.

6.2.1 Single-Stage and Compound Ammonia Refrigeration Systems

Mechanical ammonia refrigeration systems can be single-stage or multi-stage (compound) systems. Compound systems compress the gas from the evaporator to the condenser in several stages. Typically, compound systems are used when producing temperatures of –15°F and below. Single-stage ammonia systems consist of the same basic components as used in a single-stage refrigeration system that operates with a halocarbon-based refrigerant.

Figure 59 shows a basic compound (two-stage) ammonia system. As shown, it has two compressors: a low-stage compressor and a high-stage compressor. It also has two cooling loads: a low-temperature evaporator and a medium-temperature evaporator.

During operation, low-pressure vapor returned from the evaporators is applied to the suction input of the low-stage compressor. There, the vapor is compressed, then discharged at an intermediate pressure into an intercooler unit. This unit is located between the two compressors. The intercooler acts to cool or desuperheat the discharge gas from the low-stage compressor to prevent the overheating of the high-stage compressor. Desuperheating is done in the intercooler by bubbling the discharged gas from the low-stage compressor through liquid refrigerant contained in the unit. Another method mixes the liquid refrigerant normally entering the intercooler with the discharge gas. Desuperheated intermediate-pressure vapor flows from the intercooler into the suction input of the high-stage compressor. There, the vapor is compressed, then discharged to the system condenser. The hot high-pressure gas cools in the condenser and the resulting liquid refrigerant flows from the condenser into the receiver. From the receiver, the liquid refrigerant flows to and through the evaporator metering devices for application to the evaporators. It also flows from the receiver into the intercooler.

6.2.2 Ammonia Liquid Recirculation Systems

Ammonia liquid recirculation systems (*Figure 60*) are liquid overfeed systems. In this type of system, ammonia is supplied from a low-pressure receiver to the system evaporators. The low-pressure receiver is a vessel that stores liquid ammonia at low pressure and is used to supply the evaporators with liquid ammonia, either by gravity or circulated by a pump. The compressor suction line is connected to the low-pressure receiver, maintaining the pressure and, in turn, the temperature of the liquid refrigerant being recirculated. Large suction line accumulators are often incorporated as well to prevent compressor damage in the event the low pressure receiver overfills with liquid refrigerant. At startup, there is also the potential for liquid carryover into the suction line if the pressure is reduced too quickly and the ammonia liquid boils violently. As a result, the

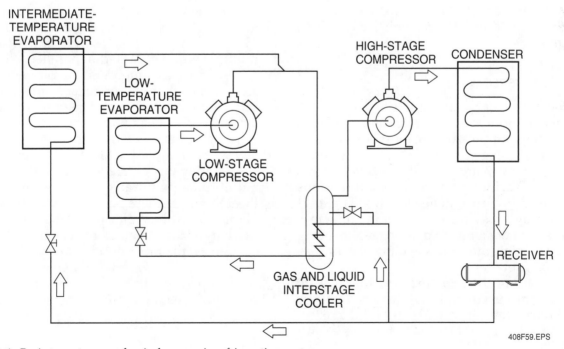

Figure 59 ◆ Basic two-stage mechanical ammonia refrigeration system.

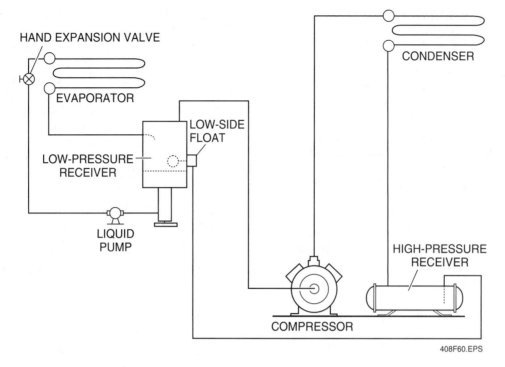

Figure 60 ◆ Simplified ammonia liquid recirculation system.

pressure should be reduced slowly in the low-pressure receiver when starting a warm system.

The low-pressure receiver also takes the suction gas from the evaporators and separates the gas from the liquid. Because the amount of ammonia fed into the evaporators is usually several times the amount that is actually evaporated there, liquid ammonia is always present in the suction return to the low-pressure receiver. Makeup ammonia is supplied to the low-pressure receiver from the high-pressure receiver through the metering device, which is usually a float valve.

6.2.3 Compressors

Reciprocating, rotary vane, and screw compressors are commonly used in ammonia systems. Reciprocating compressors are most often used in smaller single-stage or multiple-stage systems. Screw compressors are typically used in systems over 100hp. Various compressor combinations may be used in multi-stage systems. The low-pressure stage may be a rotary vane or screw compressor where large volumes of gas are being moved. The high-pressure stage may be a reciprocating or screw compressor.

When reciprocating compressors that require oil cooling are used, external heat exchangers using a refrigerant or secondary cooling are frequently used to cool the oil. Oil for screw compressors is usually cooled either by the direct injection of liquid refrigerant into the compressor to cool the discharge gas, external cooling by a heat exchanger, or external cooling using refrigerant from the condenser in a shell-and-tube heat exchanger.

6.2.4 Condensers and Receivers

Water-cooled shell-and-tube and evaporative condensers are commonly used with ammonia systems. Air-cooled condensers are seldom used. The condenser can be a single unit, or it can be multiple units connected in parallel. When using shell-and-tube condensers, the piping should be arranged so that the condenser tubes are always filled with water. The condenser should also have air vents on its head with hand valves for manual purging. Receivers fed by the condenser should always be mounted below the condenser so that the condensing surface is not flooded with ammonia. Evaporative condensers must be located where there is sufficient airflow and no obstructions that can cause recirculation of the discharged air. When using a single evaporative condenser, the receiver must always be operated at a lower pressure than the condensing pressure. It also operates at a cooler temperature than the condensing temperature.

6.2.5 Evaporators and Evaporator Control

Several types of evaporators can be used in ammonia systems. Fan coil, direct-expansion evaporators can be used, but their use is limited to systems with suction temperatures of 0°F or higher. Flooded shell-and-tube evaporators are often used in ammonia systems where indirect or secondary cooling fluids such as water, brine, or glycol must be cooled. The various metering devices used to control ammonia system evaporators are basically the same kind and operate the same way as the metering devices used in halocarbon-based refrigerant systems. The same thing is true for the various methods used to defrost the evaporators in ammonia systems that operate below 35°F. Ammonia evaporator coils are made of either aluminum or steel tubing.

6.3.0 Ammonia Safety Considerations

Ammonia has a harsh effect on the respiratory system. Only very small quantities can be breathed safely. Five minutes at 50 ppm is the maximum exposure allowed by OSHA. Ammonia is hazardous to life at 5,000 ppm and is flammable at 150,000 to 270,000 ppm. Ammonia has an odor that can be smelled at 3 ppm to 5 ppm. This odor gets very irritating at 15 ppm. While it is dangerous to human health, ammonia is considered environmentally safe.

Rusting pipes and vessels in older systems containing ammonia can create a safety hazard. Also, cold liquid refrigerant should not be confined between closed valves in a pipe where the liquid may warm because the warm liquid can expand, causing the pipe to burst. Slight smells of ammonia sometimes can occur because of leakage at valve packings or pump seals, or venting from recently drained oil or from an air-purging device.

Any escape of ammonia must be prevented from causing personal injury. When an ammonia leak occurs where people are present, the odor provides a warning. In unoccupied areas, such as refrigerated storage rooms or unmanned machine rooms, automatic sensors are used to provide a warning of a potential problem. These sensors include detector tubes and solid-state electronic, electrochemical, or infrared devices. The sensitivity of each type of sensor varies with its design and intended use. The solid-state sensor works well for initiating alarms in the range of ammonia concentration of several hundred ppm, but is not very sensitive in the range of 20 ppm to 50 ppm. Electrochemical units provide better protection for occupied areas because many are sensitive to

ammonia concentrations in the range of 0.7 ppm to 1,000 ppm.

OSHA requires that maintenance and service personnel who are expected to respond in any way to an ammonia release or other emergency regarding ammonia receive appropriate training in accordance with its regulations. Having experience working with ammonia is not sufficient to satisfy the OSHA requirement, and the training must be appropriate for the duties the individual is expected to perform in an emergency situation. End users of ammonia, as well as contractors, have a responsibility to ensure that the proper training has been received and documented. Contractors should take responsibility for general safety and emergency response training, while the end user should ensure that all contractor personnel who work at its site, as well as its own maintenance employees, receive site-specific training.

Before working with ammonia, become familiar with the information contained on its MSDS. You should also know the first aid procedures used for exposure to ammonia. In addition, the safety guidelines listed below should also be observed:

- Wear the proper protective equipment, such as a respirator mask, face shield, and gloves when working with ammonia to avoid inhalation and contact with your skin and eyes.
- If bodily contact with ammonia is made, flush any affected areas with water.
- Remove clothing unless it is so saturated with liquid ammonia that it is frozen to the skin.
- Do not wear contact lenses while working with ammonia.
- If a spill has occurred and no one is in the room or in danger, start ventilation equipment in the room before entering it, even if you are wearing a respirator mask.

7.0.0 ◆ SECONDARY COOLANTS

A **secondary coolant** is any cooling liquid that is used as a heat transfer fluid. It changes temperature without changing state as it gains or loses heat. For the lower temperatures of refrigeration, this requires a coolant with a freezing point below that of water. Systems that use brine or water as a medium to transfer heat from a space or substance to be cooled to the evaporating coil of a refrigeration system are called secondary coolant systems. *Figure 61* shows a simple secondary coolant system, where the substance

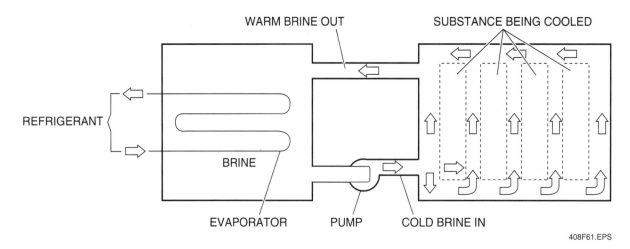

Figure 61 ♦ Simple secondary coolant system.

being cooled is placed in containers and the brine (secondary coolant) is circulated around it. The heat is transferred from the substance to the brine, then is transferred from the brine to a cooler refrigerant flowing through the evaporator coil of the primary refrigeration system.

Water is commonly used as the heat transfer medium at temperatures above 32°F. Water is an ideal medium because of its fluidity and heat-carrying capacity. However, it cannot be used at temperatures below 32°F. Brines are used where temperatures below 32°F are needed. Brines may consist of water mixed either with inorganic salts such as sodium chloride (table salt) or calcium chloride, or with organic compounds such as alcohols or glycols. Fluorinated or chlorinated hydrocarbon and halocarbon solutions may also be used. When a water solution with alcohol or glycol is used, it usually consists of methanol and water, ethanol and water, ethylene glycol and water, or propylene glycol and water.

Sodium chloride and calcium chloride are the most commonly used brines. Both of these brines have the advantage of low cost. Sodium chloride is cheaper than calcium chloride and is preferred where contact with unsealed foodstuffs may occur. However, it cannot be used satisfactorily below 6°F. Commercial grade calcium chloride can be used down to a temperature of –40°F. However, both of these brines are corrosive and require that an inhibitor be used. Sodium dichromate is an economical and satisfactory inhibitor.

Systems using salts are less susceptible to leakage than systems using water solutions of alcohol or glycol. Alcohol is used mainly in industrial processes where its flammability may not be a problem and where temperatures down to –40°F may be needed. The toxicity of wood alcohol (methanol) is a disadvantage, whereas ethanol and denatured grain alcohol are nontoxic, which may be an advantage in some systems.

Water solutions of glycol are often used in commercial applications rather than in industrial processing. Both ethylene and propylene glycol are corrosive, and an inhibitor must be used with either of these solutions. Galvanized surfaces are very prone to corrosive attack by glycols and are therefore not used in these systems. Potable water and an appropriate inhibitor should be used for making up the glycol brines.

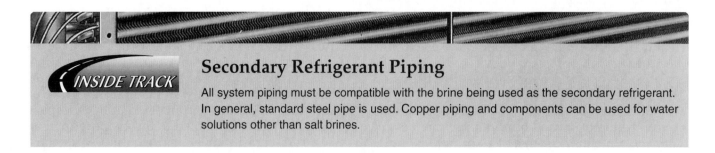

Secondary Refrigerant Piping

All system piping must be compatible with the brine being used as the secondary refrigerant. In general, standard steel pipe is used. Copper piping and components can be used for water solutions other than salt brines.

The freezing point of brine is the temperature at which crystallization begins. The lowest temperature to which sodium chloride brine can be brought without crystallizing is just above –6°F, with a 23 percent sodium chloride concentration by weight. The temperature limit for calcium chloride is –59.8°F, with a 28.5 percent concentration by weight. A 70 percent concentration by volume of alcohol in water will result in a freezing point of –164°F. A 65 percent concentration by volume of ethylene glycol will depress the freezing point to –56°F, but a higher concentration will raise the freezing point of the solution. A 60 percent solution by volume of propylene glycol will depress the freezing point of the solution to –57°F. Naturally, the freezing point of the brine solution will be lower than the effective system operating temperature. For example, while the crystalliza-tion of sodium chloride occurs at –6°F, the lowest effective operating temperature is +6°F.

The important factors to consider for secondary coolants other than water for conditioning purposes are specific gravity, viscosity, conductivity, specific heat, and freezing and boiling points. High rates of heat transfer can be achieved with solutions having high values of specific gravity, conductivity, and specific heat, but with low values of viscosity.

Glycols can be used effectively in systems at relatively high temperatures when they are combined with proper stabilizers. Ethylene glycol is more toxic than propylene glycol; therefore, propylene glycol is preferred for food freezing processes. Hydrocarbon and halocarbon heat transfer mediums are expensive and are used only in very low-temperature processes.

Review Questions

1. The main difference between commercial refrigeration systems and air-conditioning cooling systems is the _____.
   a. type of compressors used
   b. operating temperature range
   c. type of evaporators used
   d. type of condenser used

2. The temperature and humidity maintained in cold storage are determined by the type of food being stored and the _____.
   a. local climate
   b. size of the storage room
   c. number of pallets
   d. length of time it is to be held in storage

3. If the stored contents of a walk-in cooler become prematurely dehydrated, the cause may be that _____.
   a. the temperature in the cooler is too high
   b. the cooler door seal is not airtight
   c. the circulating air velocity is too high
   d. the urethane insulation is too thin

4. To maintain top quality, fruits and vegetables stored in chill rooms are normally stored _____.
   a. at temperatures just below their ripening temperature
   b. in airtight bins
   c. at a temperature just above their freezing point
   d. between vertical plate evaporators

5. Freezers used to store ice cream often operate at _____.
   a. –20°F
   b. 0°F
   c. 20°F
   d. 32°F

6. The commercial freezing method by which the product to be frozen is pressed between plates is called _____.
   a. freeze drying
   b. contact freezing
   c. air blast freezing
   d. cryogenic freezing

7. Lyophilization produces food products that _____.

    a. contain all the moisture originally found in the product

    b. require freezing until the point of consumption

    c. are lightweight and require no refrigeration

    d. must always be cooked before consumption

8. Which of the following are typically used for local delivery trucks?

    a. Direct-drive refrigeration units

    b. Trailer refrigeration units

    c. Cryogenic freezers

    d. Large insulated ice blocks

9. The energy efficiency ratio (EER) of a compressor is the lowest (least efficient) when used to cool _____.

    a. food preparation rooms

    b. meat cutting rooms

    c. meat refrigerators

    d. ice cream freezers

10. Parallel-connected multiple compressors are used mainly to _____.

    a. cool multiple evaporators in a line of display cases

    b. cool multi-temperature mobile refrigeration units

    c. increase system capacity and meet varying load conditions

    d. operate a system at extremely low temperatures

11. The compression ratio for a compressor operating at a 260 psig discharge pressure and a 32 psig suction pressure would be _____.

    a. 0.17

    b. 5.8

    c. 8.1

    d. 292

12. A refrigerant-side method to control air-cooled condensers in response to varying loads is _____.

    a. cycling the fans on and off

    b. modulating the unit dampers

    c. changing the fan speed

    d. adjusting the amount of active condensing surface in the condenser

13. Unit coolers used in areas that require high humidity levels normally have _____.

    a. larger evaporator coils that operate with a small air-to-refrigerant temperature differential

    b. smaller evaporator coils that operate with a small air-to-refrigerant temperature differential

    c. smaller evaporator coils that operate with a large air-to-refrigerant temperature differential

    d. larger evaporator coils that operate with a large air-to-refrigerant temperature differential

14. An assembly typically installed in a walk-in cooler that consists of an evaporator, metering device, and fans is called a(n) _____.

    a. condensing unit

    b. ventilation unit

    c. unit cooler

    d. air circulation unit

15. The refrigeration load for food store display cases is normally rated by the manufacturer based on the _____.

    a. summer design conditions

    b. number of customers per day

    c. thermal breaks between zones

    d. thickness of the cold air blanket

16. A device used to capture oil leaving one or more compressors is a(n) _____.

    a. oil separator

    b. oil reservoir

    c. oil differential check valve

    d. crankcase pressure regulator

17. It is best to install two _____ in each line, one mounted vertically and the other horizontally.
    a. vibration isolators
    b. sightglass/moisture-liquid indicators
    c. evaporator pressure regulator valves
    d. compressor mufflers

18. One method of preventing compressor overload from excessive refrigeration load is accomplished using a _____.
    a. crankcase pressure regulator
    b. hot gas bypass regulator
    c. compressor discharge modulator
    d. condenser fan cycling

19. The evaporator pressure regulator (EPR) valve is used to _____.
    a. control the maximum pressure at the compressor suction line in order to protect the compressor motor from overload
    b. control the evaporator temperature in systems with multiple evaporators that operate at different temperatures
    c. control system capacity in addition to limiting evaporator temperature
    d. maintain proper evaporator pressures when the adjacent ambient temperatures are low

20. The three-way condenser pressure regulator elevates head pressure by _____.
    a. adding hot gas directly into the evaporator
    b. varying the speed of one or more condenser fans
    c. causing liquid refrigerant to flood the condenser, reducing its active surface area
    d. throttling the flow of discharge gas from the compressor

21. Hot gas bypass is a method of controlling compressor capacity often used when _____.
    a. rapid compressor cycling is not a problem
    b. energy costs are a significant concern
    c. compressor short-cycling is unsatisfactory
    d. variable speed compressors are available

22. Which of the following is an *incorrect* statement regarding the features of pump-down controls?
    a. The potential for flooded starts is reduced.
    b. The refrigerant is pumped from the low side to the high side.
    c. A pressure switch controls the compressor instead of a thermostat.
    d. Refrigerant migrates from the low side to the high side continuously during the off cycle.

23. A defrost method used with multiple evaporators where one evaporator is being defrosted while the others continue refrigerating is called _____.
    a. latent heat defrost
    b. time defrost
    c. hot gas bypass
    d. off-cycle defrost

24. Ammonia is _____.
    a. a chemical compound of nitrogen and hydrogen
    b. harmful to the environment
    c. a miscible refrigerant
    d. safe for humans

25. A secondary refrigerant commonly used at temperatures above 32°F is _____.
    a. a solution of sodium chloride and water
    b. a solution of methanol and water
    c. water
    d. a solution of calcium chloride and water

Summary

Commercial and industrial refrigeration systems are used mainly to cool substances for the food, chemical, and manufacturing industries. Excluding their applications and the obvious differences in the physical layout and packaging, refrigeration systems operate in the same way as comfort air-conditioning systems. Both types of systems have the same basic hardware: compressors, heat exchangers (evaporators and condensers), fans, pumps, pipe, duct, and controls. Their main working fluids are air, water, and refrigerants. The principal use of commercial refrigeration systems involves the refrigeration and freezing of food.

The major difference between commercial/industrial refrigeration systems and comfort air-conditioning systems is the operating temperature range that the refrigeration systems serve.

Depending on the specific application, commercial/industrial refrigeration systems can have operating temperatures that range between 60°F and –250°F. Some other differences include the following:

- The number of evaporators connected to a single condensing unit
- The number of compressors used
- The number of condensers used
- The variety of refrigerants used
- The variety of defrosting methods used
- More than one temperature often provided by a single system

Notes

Anhydrous: Containing no water.

Commodities: Commercial items such as merchandise, wares, goods, and produce.

Condensing unit: A packaged refrigeration system assembly containing the compressor, condenser, and any related components.

Cooler: A refrigerated storage device that protects commodities at temperatures above 32°F.

Cryogenics: Refrigeration that deals with producing temperatures of –250°F and below.

Cryogenic fluid: A substance that exists as a liquid or gas at ultra low temperatures of –250°F and below.

Deck: A refrigeration industry trade term that refers to a shelf, pan, or rack that supports the refrigerated items stored in coolers and display cases.

Immiscible: A condition in which a refrigerant does not dissolve in oil and vice versa.

Latent heat defrost: A method used in hot gas defrost applications that uses the heat added in one or more active evaporators as a source of heat to defrost another evaporator coil.

Lyophilization: A dehydration process which first incorporates freezing of a material, then a reduction of the surrounding pressure with simultaneous addition of heat to remove moisture. Moisture leaves the material through sublimation. Also known as freeze-drying.

Satellite compressor: A separate compressor that uses the same source of refrigerant as those used by a related group of parallel compressors. However, the satellite compressor functions as an independent compressor connected to a cooling area that requires a lower temperature level than that being maintained by the related compressors.

Secondary coolant: Any cooling liquid that is used as a heat transfer fluid. It changes temperature without changing state as it gains or loses heat.

Sublimation: The change in state directly from a solid to a gas, such as the changing of ice to water vapor, without passing through the liquid state at any point.

Total heat rejection (THR) value: A value used to rate condensers. It represents the total heat removed in desuperheating, condensing, and subcooling a refrigerant as it flows through the condenser.

Unit cooler: A packaged refrigeration system assembly containing the evaporator, expansion device, and fans. It is commonly used in chill rooms and walk-in coolers.

Additional Resources and References

Additional Resources

This module is intended to be a thorough resource for task training. The following reference works are suggested for further study. These are optional materials for continued education rather than for task training.

ASHRAE Handbook, Refrigeration Systems and Applications, Latest Edition. Atlanta, GA: American Society of Heating, Refrigerating, and Air Conditioning Engineers.

ASHRAE Handbook, HVAC Systems and Equipment, Latest Edition. Atlanta, GA: American Society of Heating, Refrigerating, and Air Conditioning Engineers.

Figure Credits

A/S Dybvad Stal Industri, 408F05

Linde, Inc., 408F07

GEA Process Engineering Inc., 408F08 www.niroinc.com

Carrier Transicold, 408F09, 408F10, 408F14, 408SA03

Topaz Publications, Inc., 408F11, 408F12, 408F42, 408F43 (photo), 408F58

Grumman Olson Industries, 408F13

Heatcraft Refrigeration Products, 408F16, 408F24

Tyler Refrigeration Division, 408SA02

ICM Controls, 408F25

Johnson Controls. 408F26

Standard Refrigeration Company, 408F27

Krack Corporation, 408F28

Hill Phoenix, 408F29

Emerson Climate Technologies, 408F31 (upper left, bottom left and right), 408F32, 408SA04, 408SA06

Courtesy Sporlan Division – Parker Hannifin, 408F31 (upper right), 408F35–408F39, 408SA05, 408F45, 408F46, 408F48–408F50

Hotwatt Incorporated, 408F33

Temprite Company, 408F34

Refrigeration Research, Inc., 408F41

Danfoss, Inc., 408F52

Intermatic, Inc., 408F54

Hussman Corporation, 408F55, 408F57

NCCER makes every effort to keep these textbooks up-to-date and free of technical errors. We appreciate your help in this process. If you have an idea for improving this textbook, or if you find an error, a typographical mistake, or an inaccuracy in NCCER's Contren® textbooks, please write us, using this form or a photocopy. Be sure to include the exact module number, page number, a detailed description, and the correction, if applicable. Your input will be brought to the attention of the Technical Review Committee. Thank you for your assistance.

Instructors – If you found that additional materials were necessary in order to teach this module effectively, please let us know so that we may include them in the Equipment/Materials list in the Annotated Instructor's Guide.

Write: Product Development and Revision
National Center for Construction Education and Research
3600 NW 43rd St, Bldg G, Gainesville, FL 32606

Fax: 352-334-0932

E-mail: curriculum@nccer.org

Craft _____ Module Name _____

Copyright Date _____ Module Number _____ Page Number(s) _____

Description _____

_____

_____

_____

(Optional) Correction _____

_____

_____

(Optional) Your Name and Address _____

_____

_____

03409-09

Alternative Heating and Cooling Systems

03409-09
Alternative Heating and Cooling Systems

Topics to be presented in this module include:

Overview

The 21st century has brought in higher energy costs and a new awareness of the damage that the unrestricted burning of fossil fuels can have on the environment. Experts agree that the operation of HVAC equipment consumes significant quantities of electrical power as well as fossil fuels, such as natural gas and oil. Improvements have been made to increase the energy efficiency of conventional HVAC equipment. At the same time, there is increased interest in alternative equipment and technologies, combined with a review of the use of some older technologies, that can be used to save energy without negatively impacting the environment. This module will introduce some of the new and old technologies that are being used today to address these problems.

Objectives

When you have completed this module, you will be able to do the following:

1. Describe alternative technologies for heating, including:
   - In-floor
   - Direct-fired makeup unit (DFMU)
   - Solar
   - Air turnover
   - Corn or wood pellet burners
   - Waste oil/multi-fuel
   - Fireplace inserts
2. Describe alternative technologies for cooling, including:
   - Ductless system (DX/hydronic)
   - Computer room
   - Chilled beams
   - Multi-zone

Trade Terms

Active solar heating system
Air stratification
Air turnover unit
Catalytic element
Chilled-beam cooling system
Creosote
Cooled equipment enclosure
Direct-fired makeup air unit
Ductless split-system air conditioner
Evaporative cooler

Evaporative pre-cooler
Geothermal heat pump
Hot aisle/cold aisle configuration
Indirect solar hydronic heating system
Infiltration
Passive solar heating system
Radiant heating system
Thermosyphon system
Type HT vent
Type PL vent
Valance cooling system

Required Trainee Materials

1. Pencil and paper
2. Appropriate personal protective equipment

Prerequisites

Before you begin this module, it is recommended that you successfully complete *Core Curriculum*; *HVAC Level One*; *HVAC Level Two*; *HVAC Level Three*; and *HVAC Level Four*, Modules 03401-09 through 03408-09.

This course map shows all of the modules in the fourth level of the *HVAC* curriculum. The suggested training order begins at the bottom and proceeds up. Skill levels increase as you advance on the course map. The local Training Program Sponsor may adjust the training order.

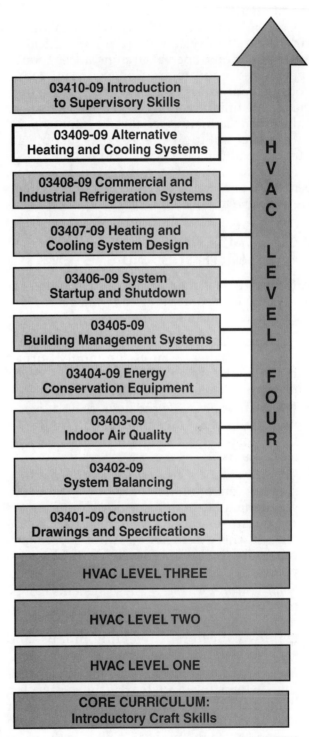

03410-09 Introduction to Supervisory Skills

03409-09 Alternative Heating and Cooling Systems

03408-09 Commercial and Industrial Refrigeration Systems

03407-09 Heating and Cooling System Design

03406-09 System Startup and Shutdown

03405-09 Building Management Systems

03404-09 Energy Conservation Equipment

03403-09 Indoor Air Quality

03402-09 System Balancing

03401-09 Construction Drawings and Specifications

HVAC LEVEL THREE

HVAC LEVEL TWO

HVAC LEVEL ONE

CORE CURRICULUM: Introductory Craft Skills

HVAC LEVEL FOUR

409CMAP.EPS

1.0.0 ◆ INTRODUCTION

In addition to conventional heating systems, there are many alternative systems available that can be used to heat a structure. These alternative systems are used because the application is either not suited to a conventional system, or the availability of a low-cost fuel or energy source makes the alternative system attractive.

Cooling a structure with a conventional ducted air distribution system is often impractical. In other situations, special circumstances require specialized equipment. For example, rooms housing computers and other electronic equipment have unique cooling requirements that can only be satisfied with specialized equipment.

In this era of high energy costs, people are turning to alternative systems to reduce energy consumption without sacrificing comfort or other benefits. Many innovative products and systems are available to satisfy this need.

2.0.0 ◆ ALTERNATIVE HEATING METHODS AND SYSTEMS

Alternative heating methods and systems can take many forms. Methods and systems covered in this module include:

- *Appliances that burn solid fuel, such as stoves, furnaces, and boilers* – Fuels commonly burned include, but are not limited to, wood, wood pellets, shelled corn, and coal.
- *Waste-oil heaters* – These devices burn waste motor oil, used transmission and hydraulic fluid, waste vegetable oil, and other petroleum-based fluids.
- *Geothermal and water-source heat pumps* – Instead of removing heat from the outdoor air, **geothermal heat pumps** remove heat from the temperature-stable earth, resulting in increased energy efficiency. Water-source heat pumps extract heat from well water, lakes, and ponds.
- *Solar heating systems* – These systems can be passive or active. Passive systems have few or no moving parts or controls. Active systems employ collectors, circulating pumps, and sophisticated control systems.
- *In-floor radiant heating systems* – Hydronic heating systems use a system of tubes carrying heated fluid. The tubes are imbedded in or installed under a floor. The floor is then heated, which in turn heats objects in the room by radiation. Electric heaters installed in or under the floor provide the same result.

- *Direct-fired makeup units (DFMU)* – These units are designed to heat air being brought into a building. The heated air is used to replace air being exhausted from the building.

3.0.0 ◆ SOLID-FUEL APPLIANCES

Solid fuels such as wood and coal have been used to provide heat since prehistoric times. In the United States, coal and wood were widely used for home heating until the middle of the 20th century. At that time, they were replaced by cleaner and more convenient to use fuels such as natural gas and fuel oil. By the 1970s, rising energy costs forced people to take another look at solid fuels as an energy source. As a result, solid-fuel burning stoves, furnaces, and boilers started to be used more, especially in rural areas where this type of fuel was plentiful and inexpensive.

3.1.0 Wood-Burning Stoves

Wood-burning stoves (*Figure 1*) and fireplace inserts are used to provide area heating or to supplement a conventional heating system. Wood-burning stoves can burn chunks of wood, wood pellets, or shelled corn. If so designed, coal may be burned in some wood-burning stoves. Stoves

409F01.EPS

Figure 1 ◆ Installed wood stove.

are typically constructed of welded steel plates or cast iron. The combustion chamber may be lined with firebrick for enhanced combustion and heat retention. Modern stoves are designed to burn the fuel through controlled combustion. This can be accomplished using a manual air damper, a thermally activated air damper, or by a thermostatically controlled, electrically operated combustion air blower or air damper.

Wood stoves and fireplace inserts sold in the United States since 1988 must meet EPA smoke emission standards. If the stove is equipped with a **catalytic element**, smoke emissions cannot exceed 4.1 grams per hour. Non-catalytic stoves use a number of construction techniques such as internal baffles to enhance combustion and reduce smoke emissions. Non-catalytic stoves cannot emit more than 7.5 grams of smoke per hour. Stoves provide heat through radiation. Some stoves can be fitted with an optional circulating fan that moves air over the surface of the stove to pick up additional heat.

Only properly seasoned wood should be burned in a wood-burning stove or furnace. Wood must be seasoned for several months to remove moisture. Hardwoods are preferred because they burn slower. Improperly seasoned wood (called green wood) has a high moisture content. When green wood is burned, the water vapor given off combines with other combustion products to form **creosote** in the chimney.

WARNING!
Creosote can catch fire in the chimney, causing a dangerous condition.

3.2.0 Wood-Burning Furnaces

Wood-burning forced-air furnaces are used as the primary source of heat in some structures. They are typically installed in two ways: as a stand-alone heating appliance or as a supplement to an existing forced-air furnace. Some wood-burning furnaces have dual-fuel capability in one cabinet. The furnace has a common circulating fan that moves air over the heat exchanger regardless of whether heat is being supplied by wood or by an integral gas or oil burner.

A wood-burning furnace used as a supplemental heater (*Figure 2*) is basically a wood stove with a sheet metal enclosure that allows air to circulate over the hot fire box. This type of furnace usually does not have a circulating air blower. Instead, the blower in the companion gas or oil furnace is used to supply the air. To accomplish this, the

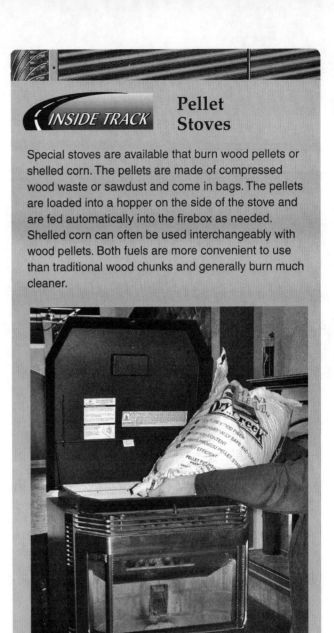

INSIDE TRACK

Pellet Stoves

Special stoves are available that burn wood pellets or shelled corn. The pellets are made of compressed wood waste or sawdust and come in bags. The pellets are loaded into a hopper on the side of the stove and are fed automatically into the firebox as needed. Shelled corn can often be used interchangeably with wood pellets. Both fuels are more convenient to use than traditional wood chunks and generally burn much cleaner.

409SA01.EPS

supply air plenum has to be modified so that the air from the gas or oil furnace blower is diverted through the cabinet of the wood-burning furnace and then into the main supply air duct (*Figure 3*). In effect, the wood-burning furnace is in series with the gas or oil furnace. There are several ways to control blower operation in the wood heat mode. When the wood-burning furnace is installed in series with the gas or oil furnace, the fan selector switch on the room thermostat sub-base can be set for continuous operation.

409F02.EPS

Figure 2 ◆ Add-on wood-burning furnace.

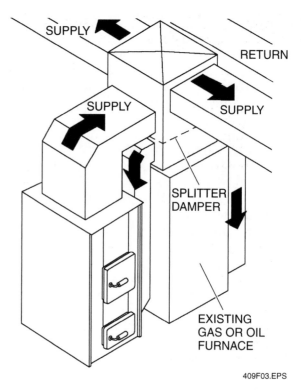

409F03.EPS

Figure 3 ◆ Add-on furnace ducting.

A disadvantage of this arrangement can be cold drafts coming from the supply air register if the wood fire dies out.

Another arrangement is to install a separate fan switch in the plenum directly above the wood-fired furnace (*Figure 4*). This fan switch is connected in parallel with the fan switch in the gas or oil furnace. That way, the fan switch in the gas or

oil furnace can control the fan when the gas or oil furnace is supplying heat and the fan switch on the wood-burning furnace can control the fan when wood is being burned.

Supplemental wood-burning furnaces can also be equipped with a circulating air blower and installed in parallel with an existing gas or oil furnace. By installing a blower, it can also be

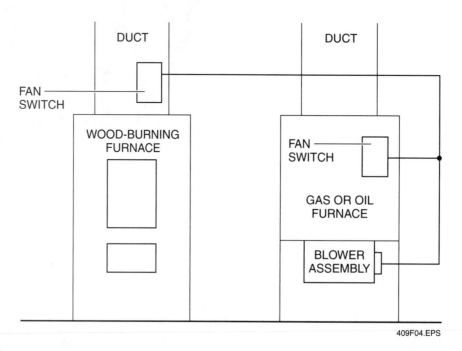

409F04.EPS

Figure 4 ◆ Add-on furnace fan switch wiring.

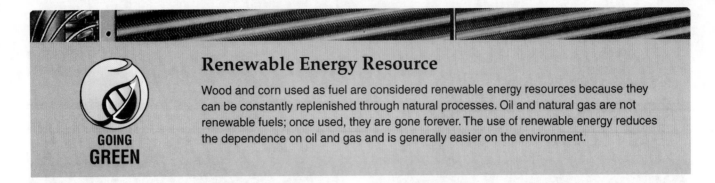

Renewable Energy Resource

Wood and corn used as fuel are considered renewable energy resources because they can be constantly replenished through natural processes. Oil and natural gas are not renewable fuels; once used, they are gone forever. The use of renewable energy reduces the dependence on oil and gas and is generally easier on the environment.

installed as the primary heat source (no gas or oil furnace). Supplemental wood-burning furnaces have the flexibility to adapt to a wide variety of existing forced-air heating systems. During operation, the fan switch in the supplemental furnace will keep the circulating fan operating almost constantly as long as there is fire in the firebox. As the fire dies out, the fan may begin to cycle on and off. Typically, the room thermostat of the gas or oil furnace is set low enough so that the gas or oil furnace does not operate while the wood-burning furnace is supplying heat. When the fire goes completely out, room temperature will drop low enough for the room thermostat to call for heat from the gas or oil furnace.

A dual-fuel, wood-burning furnace can best be described as a gas or oil furnace and a wood-burning furnace combined in one package. Electric heat versions are also available. Some models are also able to burn coal. A dual-fuel furnace is used as the primary heat source in a structure and the duct system is installed the same as any other forced-air furnace. Their controls tend to be more sophisticated than those found on a simpler supplemental wood-burning furnace. Some even allow for the installation of air conditioning.

Two room thermostats are used in the furnace configuration shown in *Figure 5*. One controls the oil burner and the other controls combustion air for the wood. The wood burner room thermostat

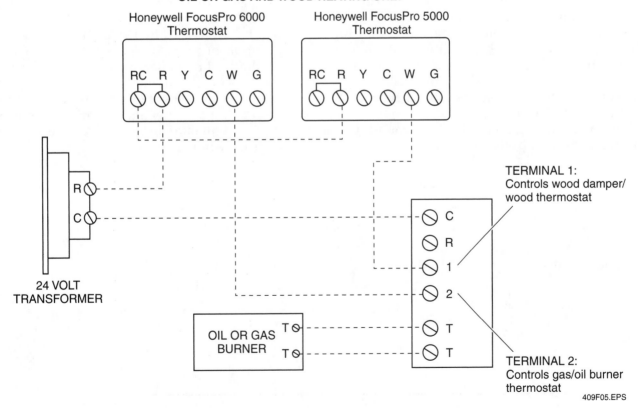

Figure 5 ◆ Typical dual-fuel furnace thermostat wiring.

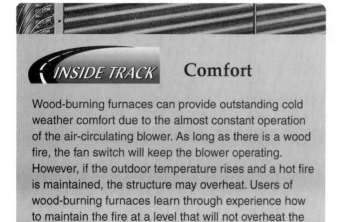

is set higher than the oil burner thermostat. As the fire dies down or goes out, room temperature will fall and the room thermostat controlling the oil burner will activate the burner to maintain room temperature. On a typical dual-fuel furnace, the gas or oil burner can be used to light the wood or the wood can be lit in the conventional manner.

3.3.0 Wood-Burning Boilers

Wood-burning boilers, like forced air furnaces, are used as the primary source of heat in a structure. They are typically installed in two ways: as a stand-alone heating appliance or as a supplement to an existing boiler. Some wood-burning boilers have dual-fuel capability in one cabinet. The boiler has a common circulating pump that moves water through the boiler sections regardless of whether heat is being supplied by wood or by an integral gas or oil burner.

A wood-burning boiler (*Figure 6*) can provide all the heat for a building, thereby leaving the

409F06.EPS

Figure 6 ◆ Add-on wood-burning boiler.

existing gas- or oil-fired boiler as a backup. The pump on the wood boiler operates continuously, circulating the hot water from the wood boiler through the existing boiler or high-efficiency exchanger. When there is a demand for heat, the existing boiler can remain unfired, unless the wood boiler is not refueled and the fire dies out. The wood-burning boiler is installed in series with the existing boiler. If the water temperature is maintained high enough, the burner on the existing boiler will not operate. If the water temperature drops further and a heat demand is present, the aquastat in the existing boiler will energize the burner (*Figure 7*).

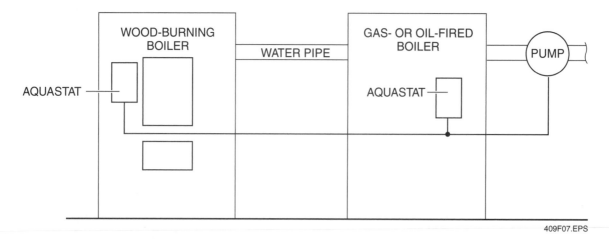

409F07.EPS

Figure 7 ◆ Add-on boiler aquastat wiring.

Supplemental wood-burning boilers can be installed indoors next to the existing boiler. They can also be installed outdoors in a self-contained weatherproof structure, complete with its own chimney (*Figure 8*). The heated water is pumped through insulated pipes buried underground and connected to the existing hydronic heating system inside the structure. Advantages of this system are that it is cleaner (no ashes or wood debris inside the structure) and no new chimney or vent has to be installed in the structure.

Like its forced-air counterpart, a dual-fuel, wood-burning boiler is a gas or oil boiler and a wood-burning boiler combined in one package (*Figure 9*). It offers the same advantages as a dual-fuel furnace. A dual-fuel boiler is used as the primary heat source in a structure and the piping and radiation is installed the same as any other hydronic heating system.

3.4.0 Wood-Burning Appliance Installation and Maintenance

Heating appliances that burn wood or other solid fuel have unique installation and maintenance requirements including:

- Clearance to combustibles
- Combustion air
- Venting
- Field wiring, piping, and ductwork
- Fuel storage
- Cleaning and maintenance

Before installing any wood-burning appliance, always thoroughly read the manufacturer's installation and service literature before proceeding.

3.4.1 Clearance to Combustibles

Any fuel-burning appliance, such as furnaces, will always carry specifications for installation clearances from combustible materials such as floors, walls, and building structural members.

409F08.EPS

Figure 8 ◆ Outdoor wood-burning boiler.

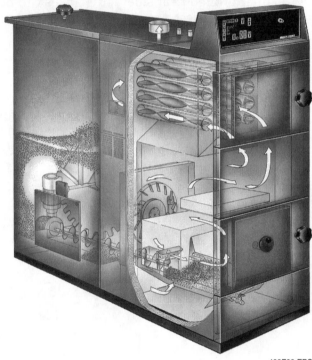

409F09.EPS

Figure 9 ◆ Dual-fuel burner.

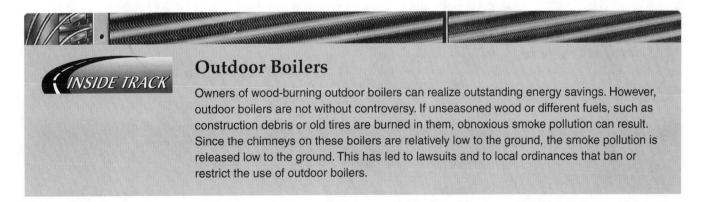

Outdoor Boilers

Owners of wood-burning outdoor boilers can realize outstanding energy savings. However, outdoor boilers are not without controversy. If unseasoned wood or different fuels, such as construction debris or old tires are burned in them, obnoxious smoke pollution can result. Since the chimneys on these boilers are relatively low to the ground, the smoke pollution is released low to the ground. This has led to lawsuits and to local ordinances that ban or restrict the use of outdoor boilers.

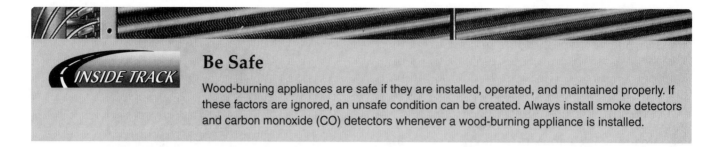
Wood-burning appliances are no different. However, since wood-burning stoves and furnaces operate at higher temperatures than conventional gas or oil furnaces, those distances tend to be much greater. For example, wood stoves must be placed on an approved, non-combustible base. When installed in a room, the stove must maintain a 36" clearance from the top, sides, back, and front to any combustibles. If the stove is equipped with an approved heat shield, distances can be reduced. The closest a stovepipe can be to combustibles is three times the diameter of the pipe. In other words, a 6" stovepipe must be at least 18" away from combustibles (*Figure 10*).

Check the manufacturer's literature for clearance requirements when installing wood-burning stoves, furnaces, and boilers. For the most up-to-date information on all aspects of installing wood-burning appliances, consult National Fire Protection Association (NFPA) codes.

3.4.2 Combustion Air

Stoves and furnaces that burn solid fuel must be supplied with adequate air for combustion. In a loosely-built structure, air entering through normal **infiltration** may be adequate to supply all combustion air needs. If the structure is tightly built, outside air will have to be brought into the structure to support combustion. Outside air can be brought in through a duct into the area near the stove or furnace (*Figure 11*). As a rule of thumb, the duct size should be equal to or greater than the size of the chimney or vent of the appliance.

If the stove or furnace is installed in a confined space such as a utility room, combustion air and ventilation air grilles must be installed in a wall or door that is open to the rest of the structure. The combustion air grille is installed near the floor and the ventilation air grille is installed near the ceiling. Each grille must have a free air opening of one square inch for each 1000 Btuh of input capacity. For example, assume a combination wood/oil furnace has a maximum input capacity of 144,000 Btuh. Two grilles having at least 144 square inches of free area would be required for the utility room. Be aware that free air area is not the same as the area of the face of the grille. Consult the grille manufacturers literature for the free air area for any given grille.

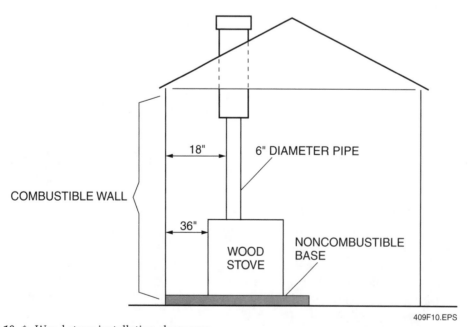

Figure 10 ◆ Wood stove installation clearances.

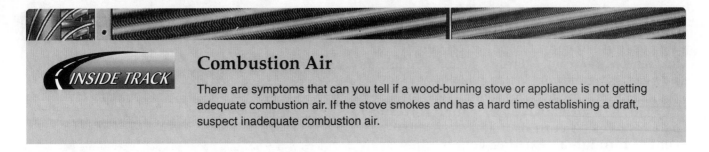
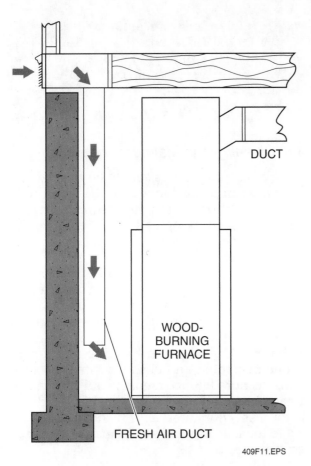

DUCT

WOOD-BURNING FURNACE

FRESH AIR DUCT

409F11.EPS

Figure 11 ◆ Wood-burning appliance combustion air supply.

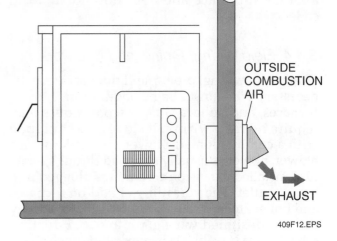

OUTSIDE COMBUSTION AIR

EXHAUST

409F12.EPS

Figure 12 ◆ Direct venting of corn or pellet stove.

409F13.EPS

Figure 13 ◆ Type HT vent.

Corn and pellet stoves can be equipped with a direct vent that also serves as the combustion air source. The vent draws in combustion air and exhausts the products of combustion through a special through-the-wall fitting (*Figure 12*).

3.4.3 Venting

Wood-burning stoves, furnaces, and boilers must be vented, either through a correctly sized, tile-lined masonry chimney or through a correctly sized all-fuel Type HT metal vent (*Figure 13*). **Type HT vents** are rated to handle temperatures as high as 1000°F. When installing a Type HT vent, do not mix components from a different manufacturer's Type HT vent. Mixing components may create an unsafe condition. The agency rating of the assembled vent is only valid if rated components from the same manufacturer are used. Never vent a wood-burning stove, furnace, or boiler through a chimney or vent serving a fireplace or other fuel-burning appliance such as a gas or oil furnace. When installing a Type HT vent, follow the manufacturer's installation instructions and all appropriate local and national codes. Corn and wood pellet stoves generally do not have flue temperatures that are as high as those of conventional wood stoves. For that reason, they can be directly vented through a sidewall using the vent kit supplied by the stove manufacturer. Special pellet venting pipe or **Type PL vent**, is also available for venting corn and pellet stoves. A corn or pellet stove may be vented through an existing vent or fireplace chimney as long as it is not being used to vent another appliance and does not violate local codes.

3.4.4 Field Wiring, Piping, and Ductwork

Changes to wiring, piping, and ductwork may be necessary when installing wood-burning stoves, furnaces, and boilers, Simple wood stoves may require no field wiring. If stoves are equipped with a combustion air blower or circulating air blower, power is usually supplied through a factory-installed extension cord that is plugged into a wall outlet. The versatility of add-on furnaces and boilers means that an almost infinite variety of field-designed wiring, piping, and ducting schemes are possible. When installing such a system, make sure that any modifications you make are workmanlike and comply with the equipment manufacturer's recommendations and all appropriate national and local codes.

> **NOTE**
> Dual-fuel appliances should be installed strictly in accordance with the manufacturer's instructions and all local and national codes.

3.4.5 Fuel Storage

Solid fuels, such as wood chunks and coal, are dirty and can take up a lot of space. Piles of firewood can serve as a home for insects and rodents. For those reasons, solid fuel should not be stored indoors in any quantity. Seasoned firewood should be stored outdoors, off the ground, and covered to protect it from the weather. Bring in only enough firewood for one or two days use.

Coal used in stoves and furnaces is typically sold in bags. Dirt can be kept to a minimum by removing the coal only as needed. Shelled corn and wood pellets are inherently clean and sold in bags. The corn or pellets are poured directly into the hopper on the stove and fed into the stove as fuel is consumed.

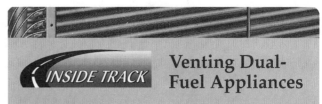

Venting Dual-Fuel Appliances

Dual-fuel appliances such as wood/oil or wood/gas furnaces and boilers often use a common vent. If installing such a furnace or boiler as a replacement for a conventional gas or oil furnace or boiler, the existing chimney or vent may not be adequate to handle the higher temperature of the replacement appliance. When installing a dual-fuel appliance, use a Type HT vent or a correctly sized and lined masonry chimney to ensure a safe installation.

3.4.6 Cleaning and Maintenance

Wood-burning appliances should be cleaned and maintained in accordance with the manufacturer's instructions. Maintenance items include oiling motors, checking belts, and replacing filters. When wood burns, it can create creosote that can build up on the internal surfaces of the stove, furnace, and chimney (*Figure 14*). Creosote is combustible, and can ignite if left to accumulate. Once ignited, creosote can burn with the intensity of a blowtorch, permanently damaging the stove and/or vent and possibly causing a structure fire. For that reason, all wood-burning appliances must be inspected and cleaned when creosote buildup is noted. Since creosote buildup varies between installations, there is no hard and fast rule for inspection intervals. It is not uncommon to have to clean a wood stove and/or chimney every few weeks.

409F14.EPS

Figure 14 ◆ Creosote buildup in a chimney.

The cleaning of a wood stove is very messy and requires special brushes designed to fit specific areas of the stove and chimney (*Figure 15*). A scoop and a bucket may be required to remove and dispose of solids. A vacuum cleaner is required to pick up finer debris. Because creosote is very dirty, wear gloves, eye protection, old clothes and a dust mask to prevent contamination.

Ash is the byproduct of burning solid materials. Ash must be periodically removed from the appliance and stored in a sealed metal container. Once the ash has completely cooled and is free of hot embers, it can be transferred to another metal container for storage or disposal. Never store ashes in non-metallic containers and never put ashes in a trashcan that contains combustibles. Wood ashes can be buried. The minerals in wood ash are beneficial to garden plants and vegetables.

409F15.EPS

Figure 15 ◆ Cleaning brushes.

4.0.0 ◆ WASTE OIL HEATERS

People today realize the importance and necessity of recycling and reusing resources to protect the environment. Heaters that burn waste oil are examples of products that use a resource that, in the past, was often discarded. Waste oil can come from a variety of sources. Automotive repairs shops generate used motor oil, transmission fluid, and other petroleum-based lubricants and fluids. Restaurants and food processing facilities generate vegetable oils that were used in cooking. A waste-oil burner is similar to a conventional gun-type oil burner that has been heavily modified to burn waste oil (*Figure 16*). The burners can be used in forced-air systems or to fire a boiler. Because of the nature of the fuel, waste-oil heaters are likely to be used in commercial and industrial applications and rarely in residential applications.

The quality and physical properties of waste oil can vary significantly. For example, used motor oil is significantly different from used french-fry oil. The fuel handling system and the burner itself must be capable of handling these variables. Filters are required to remove various contaminants in the waste oil. Waste oil burners often have a built-in heater that heats the waste oil to maintain a uniform viscosity. Compressed air, supplied by a built-in air compressor or by an outside source is required to help atomize the fuel. The atomized fuel is typically ignited by a high-voltage spark from the ignition transformer. An oil burner primary control using a cad cell flame detector provides a safety shutoff if a flame is not established. Installation and venting of waste oil heaters is similar to conventional oil-fired heaters. Always read the manufacturer's installation literature before proceeding.

> **NOTE**
> Conventional gun-type oil burners all operate in a similar manner and use similar components. That is not the case with burners used in waste oil heaters. Different manufacturers use different engineering and different components to accomplish the same result.

4.1.0 Waste Oil Heating Issues

Burning waste oil can provide significant cost savings but there are negative aspects as well. The Environmental Protection Agency (EPA) has rules that cover the burning of waste oil. For example, regulations govern the conditions under which waste oil containing certain types of contaminants can be burned. Local codes may regulate or restrict the burning of waste oil. Some waste oils will contain excessive contaminants and/or burn dirtier than other oils, resulting in decreased intervals between scheduled maintenance and more frequent equipment breakdowns. The negative aspects as well as the potential cost savings should be carefully weighed before installing a waste oil heater.

409F16.EPS

Figure 16 ◆ Waste oil burner assembly.

5.0.0 ◆ GEOTHERMAL AND WATER-SOURCE HEAT PUMPS

A few feet below the surface of the Earth, the temperature is relatively warm and stable year-round. The water in wells, lakes, and ponds is also relatively warm, even in the winter. All of these sources of clean heat can be used to efficiently heat homes and businesses while minimally impacting the environment.

5.1.0 Ground-Source Heat Pumps

Ground-source or geothermal heat pumps extract heat from the ground. They use coils of tubing containing a non-toxic fluid buried in the ground to absorb the heat (*Figure 17*). The heated fluid is pumped indoors to a refrigerant-to-water heat exchanger where the heat is absorbed by the refrigerant. The heat is then moved to the coil in an air handler where it is rejected into the structure. In summer, when cooling is required, the coil in the air handler acts as an evaporator and absorbs heat in the refrigerant. The refrigerant-to-water heat exchanger acts as a condenser, giving up heat to the fluid. The heated fluid is then pumped through the ground loop where it gives up the heat extracted from the structure to the ground. Geothermal heat pumps that heat water are also available. They can heat water for domestic use, for heating in a hydronic heating system, to heat swimming pools, and to melt snow from driveways. The most common geothermal installation method is to dig a series of four- to six-foot-deep trenches in which coils of tubing are buried (*Figure 18*). If the property is too small for this type of installation, a series of closely spaced deep holes can be drilled into which the tubing can be inserted.

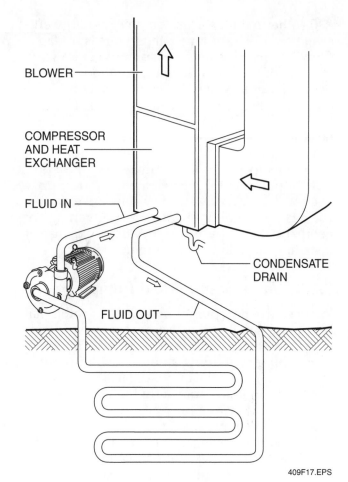

409F17.EPS

Figure 17 ◆ Geothermal heat pump.

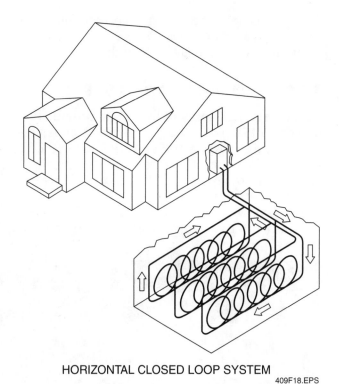

HORIZONTAL CLOSED LOOP SYSTEM

409F18.EPS

Figure 18 ◆ Geothermal heat pump ground loop.

Geothermal heat pumps are very energy efficient because the ground temperature is so stable. Conventional air-source heat pumps have a varying heat output because winter outdoor air temperature varies widely. When it is really cold, the air source heat pump's heat output is low and must be supplemented by electric resistance heaters.

With geothermal heat pumps, the stable ground temperature results in a steady heat output. Supplemental electric heaters are rarely needed in a properly sized system and a defrost cycle is not needed. The major disadvantage of geothermal heat pumps is the added expense of installing the belowground tubing loop.

5.2.0 Water-Source Heat Pumps

Water source heat pumps operate like geothermal heat pumps in that they extract heat from a temperature-stable well, lake, or pond. One form uses tubing containing a non-toxic fluid that is immersed in the lake or pond. Heat is extracted from the water and transferred to the indoor air in the same way as a previously described for a geothermal heat pump (*Figure 19*).

The other water source method directly pumps water from a well, lake, or pond and circulates it through a refrigerant-to-water heat exchanger where the heat from the water is extracted. A disadvantage of this scheme is that if the quality of the water is poor, it can clog or corrode the refrigerant-to-water heat exchanger and any other components in the water side of the system.

6.0.0 ◆ SOLAR HEATING SYSTEMS

Harnessing the energy of the sun to provide heat for comfort is not only energy-efficient, it is very kind to the environment. It is estimated that each square mile of the Earth receives solar energy equivalent to 5 billion kWh each year. This free source of energy is there for the taking. Solar heating systems capture that free energy. The systems fall into two categories: passive and active.

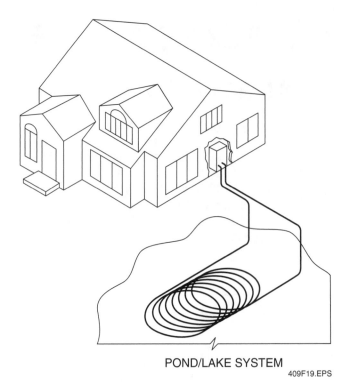

POND/LAKE SYSTEM

409F19.EPS

Figure 19 ◆ Water source heat pump pond or lake water source.

Solar heating systems are popular in southwestern parts of the United States where winter sunshine is plentiful.

6.1.0 Passive Solar Heating Systems

The family cat is aware of passive solar heating. On winter days, cats can be found stretched out on the carpet in front of a south-facing window. The rays of the sun warm both the carpet and the cat. **Passive solar heating systems** (*Figure 20*) both capture and store solar energy. Simple passive systems require two major components: large, south-facing windows, and thermal mass to collect and store the heat. In the United States, the sun is low in the southern sky during winter. Large south-facing windows allow the sun to enter the room where it is captured in a greenhouse effect and heats the thermal mass.

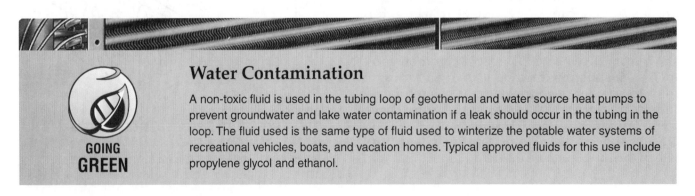

Water Contamination

A non-toxic fluid is used in the tubing loop of geothermal and water source heat pumps to prevent groundwater and lake water contamination if a leak should occur in the tubing in the loop. The fluid used is the same type of fluid used to winterize the potable water systems of recreational vehicles, boats, and vacation homes. Typical approved fluids for this use include propylene glycol and ethanol.

GOING
GREEN

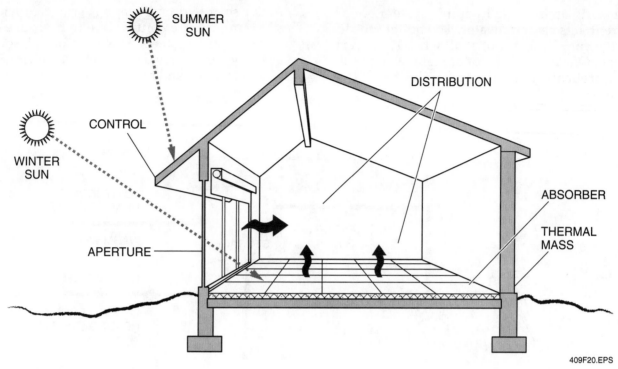

Figure 20 ◆ Passive solar heating system.

The thermal mass often takes the form of dark-colored tile or masonry floors or masonry walls. During the day, the sun's rays heat the thermal mass. After the sun goes down, the heat retained in the thermal mass is gradually released into the space.

In some schemes, the thermal mass is a wall filled with water. There are too many variations of this basic design to fully discuss them all in this module. A common feature includes thermal drapes that automatically close when the sun goes down to prevent heat loss back out the windows. Some passive designs are laid out so that the thermal mass induces gentle convection airflow throughout the room or structure.

Passive solar hydronic heating systems are also available. In this scheme, convection currents move water from a roof-mounted solar collector to a storage tank inside the structure. A **thermosyphon system** (*Figure 21*) is typical of passive systems. It operates by the different densities between the water in the storage tank and the collector. Cool water in the storage tank and hot water in the collector means rapid flow. Hot water in the storage tank and cool water in the collector means little or no flow. This condition could possibly lead to reverse thermosyphoning. A check valve must be installed to prevent this backward flow of water through the system. Thermosyphoning systems do not have temperature controls, so water temperature in the system

will vary with changes in the weather. The warm water in the storage tank is used to preheat water in a conventional hydronic heating system with a boiler. Passive hydronic heating systems are not as well suited for space heating as other types of hydronic solar heating systems.

The major advantage of a passive solar heating system is simplicity. With few or no moving parts, the system heats the structure as long as the sun shines. A major disadvantage of passive systems is that they are prone to overheat the structure on warm sunny days and provide little heat on cloudy days. A conventional heating system is often required to supplement a passive solar heating system.

6.2.0 Active Solar Heating Systems

Active solar heating systems are more complex than passive systems. They use pumps, valves, and other devices to control the flow of fluid through the solar collector and the rest of the system. Indirect systems, often called closed loop systems, use antifreeze such as ethylene glycol as the circulating fluid in the system. The heated antifreeze solution is then passed through a heat exchanger where it gives up heat to water that is circulated through the hydronic heating system pipes and radiation. *Figure 22* shows a typical **indirect solar hydronic heating system**. The storage tank shown would commonly be piped in series with an existing hydronic system boiler where it acts as a preheater. The boiler would operate very little or not at all when the sun is shining. On cloudy days, or at night, when little or no preheating occurs, the boiler heats the structure.

Active solar hydronic heating systems provide better temperature control due to their use of complex control systems. However, like all solar heating systems, they are ineffective on cloudy days and at night, and require some other form of backup heat.

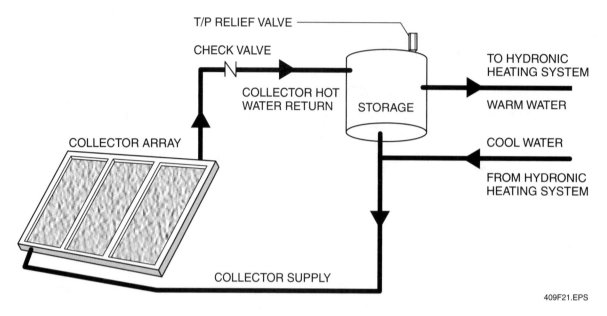

409F21.EPS

Figure 21 ◆ Passive solar thermosyphon hydronic heating system.

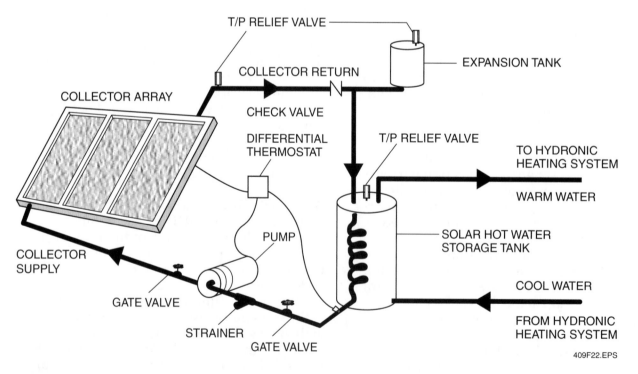

409F22.EPS

Figure 22 ◆ Active solar indirect hydronic heating system.

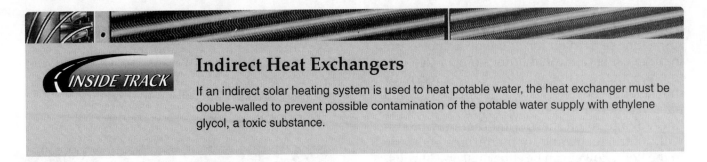
Indirect Heat Exchangers

If an indirect solar heating system is used to heat potable water, the heat exchanger must be double-walled to prevent possible contamination of the potable water supply with ethylene glycol, a toxic substance.

7.0.0 ◆ IN-FLOOR RADIANT HEATING SYSTEMS

In a **radiant heating system**, the object is heated, not the air surrounding the object. Think how warm you can feel if you sit in the sun on a cool day. In-floor radiant heating systems operate on that same principle. The heating elements are imbedded in, or installed below the floor. They can be either electric resistance heating elements or tubes containing hot water (hydronic heat).

7.1.0 Electric Radiant Heating Systems

Electric systems can take the form of mats made of resistive wire or spools of resistive wire. Heating mats are a pre-engineered system that is typically installed as a unit on the underside of a wood floor between the floor joists (*Figure 23*). The mats are low-density heaters, typically about 10 watts (about 34 Btuh) per square foot of surface area. Enough mat has to be installed to satisfy the heat loss of the room at the design temperature. For example, a room with a heat loss of 3,400 Btuh would require about 100 square feet of mat. Since the mats cannot be cut to fit, mats must be carefully selected to fit the area into which they are to be installed. In the example, ten 10-square-foot mats, or four 25-square-foot mats would satisfy the requirement. The mats must be installed strictly in accordance with the manufacturer's instructions and all applicable electrical codes. Many jurisdictions require that an electrician perform the installation.

Heaters are also available in the form of resistive wire that has a watts-per-foot value. Once the heat loss of the room is calculated, the length of

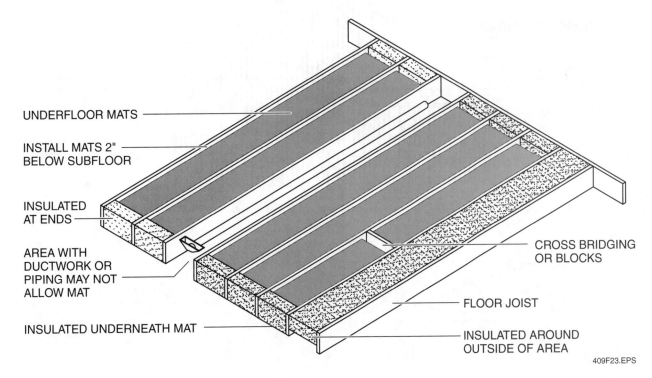

UNDERFLOOR MATS

INSTALL MATS 2" BELOW SUBFLOOR

INSULATED AT ENDS

AREA WITH DUCTWORK OR PIPING MAY NOT ALLOW MAT

INSULATED UNDERNEATH MAT

CROSS BRIDGING OR BLOCKS

FLOOR JOIST

INSULATED AROUND OUTSIDE OF AREA

409F23.EPS

Figure 23 ◆ Radiant electric heating system mat installation.

resistive wire can be determined to satisfy the heat loss. The resistive wire is then laid out in a serpentine pattern on top of the floor and secured in place using dedicated fittings (*Figure 24*). Once in place, the resistive wire can be covered with a thin layer of concrete, ceramic tiles, or wood laminate. Temperature control is usually done with a line-voltage room thermostat that may be used with control relays and/or thermistors embedded in the floor. Power can be either 120 or 230 volts.

Advantages of in-floor electric radiant systems are quiet operation and excellent comfort. However, it can be costly to operate because it is a form of electric heat. Installation can be expensive due the care that must be taken to ensure a safe installation. For example, a carelessly driven nail or staple could penetrate a live circuit, creating a safety hazard.

7.2.0 Radiant Hydronic Heating Systems

Traditional hydronic heating systems use radiators to supply heat to individual rooms. In-floor hydronic radiant loops can be installed in the floor in a manner similar to electric radiant heating systems (*Figure 25*). The radiant loop takes the place of a traditional radiator. Cross-linked polyethylene (PEX) plastic tubing is used in this type of system. Metal reflectors are often installed beneath the tubing to direct the radiant heat upward.

409F25.EPS

Figure 25 ◆ Radiant hydronic heating system tubing installation.

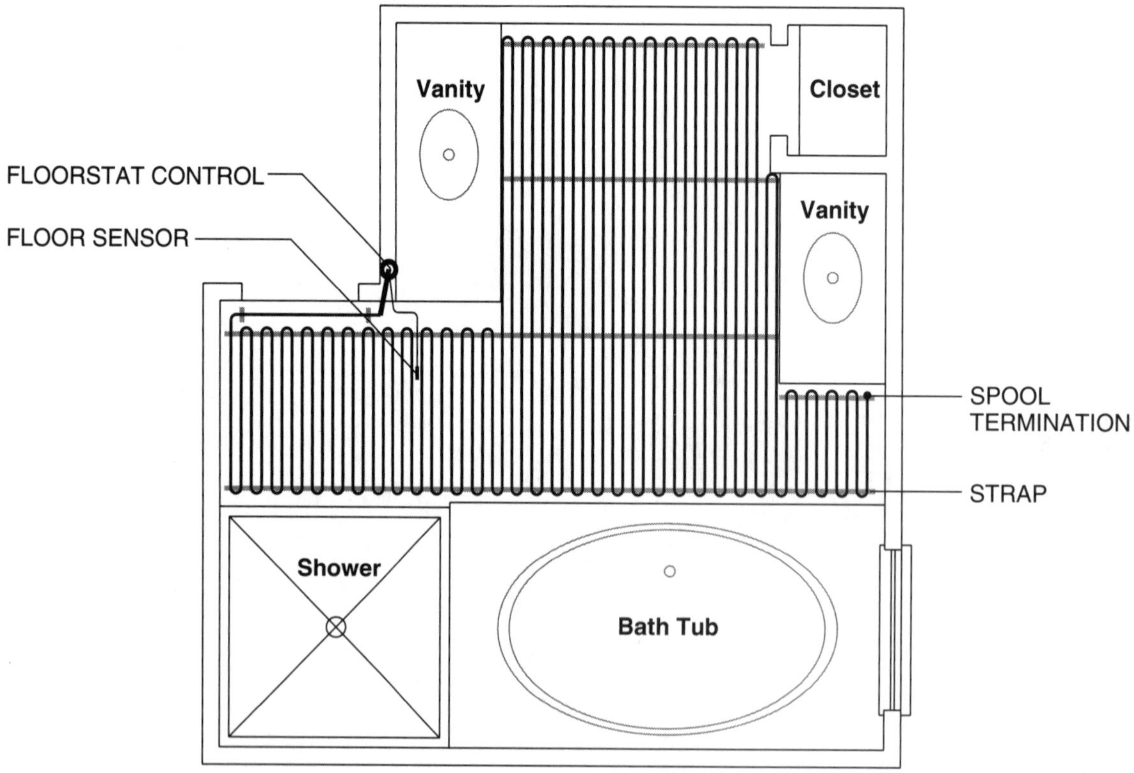

FLOORSTAT CONTROL

FLOOR SENSOR

Vanity

Closet

Vanity

SPOOL TERMINATION

STRAP

Shower

Bath Tub

409F24.EPS

Figure 24 ◆ Radiant electric heating system resistive wire installation.

Low-voltage room thermostats control zone valves that allow heated water to flow through the radiant loop. Other than the type of radiation, an in-floor hydronic radiant heating system is, in all other respects, identical to a conventional hydronic heating system.

Radiant hydronic heating systems offer all the advantages of traditional hydronic heating systems including superior comfort, energy efficiency, and quiet operation. Like all hydronic heating systems, radiant hydronic systems are more expensive to install than other types of heating systems. Radiant heating systems, both electric and hydronic, can also be installed in the ceiling.

8.0.0 ◆ DIRECT-FIRED MAKEUP AIR UNITS

In many commercial and industrial applications, indoor air must be mechanically exhausted to purge harmful contaminants, smoke, and odors. For example, air near a loading dock can be contaminated by vehicle exhaust. A second example is an industrial paint booth, which often requires massive amounts of air to be exhausted.

There are many possible situations that require high volumes of exhaust air, and this exhausted air must be replaced. Further, this makeup air must often be conditioned before entering the space, in order to maintain comfort or to ensure that the industrial process environment remains consistent. Without mechanical equipment to replace exhaust air, negative pressures in the building forces air to enter through any available opening. Air allowed to infiltrate uncontrolled enters without filtration, and also enters at ambient temperature and humidity conditions.

When ensuring that the makeup air is filtered and remains above a certain temperature is the primary consideration (cooling may be unnecessary), a **direct-fired makeup air unit** (*Figure 26*) can be used. This unit replaces the exhausted air, generally using 100 percent air from outdoors, and heats the air as it passes through. These units are typically gas-fired, but contain no heat exchanger or flue vent. The unique feature of the direct-fired unit is that the combustion process takes place directly in the primary airstream. Combustion air is drawn from the same air being supplied to the building, and all byproducts of combustion remain in the airstream entering the building as well. Obviously, it is important that combustion byproducts be sufficiently diluted to

409F26.EPS

Figure 26 ◆ Direct-fired makeup air unit.

ensure the safety of building occupants. Direct-fired units are specifically designed with this key requirement in mind, and the level of hazardous byproducts is sufficiently diluted by the high volume of air flowing through them. In addition, air is being constantly exhausted from the space.

Direct-fired units are generally controlled by a discharge-air temperature sensor arrangement, and use a modulating gas burner design. Unlike a typical residential furnace, which usually produces supply air temperatures of 110°F or more, the discharge temperature of a direct-fired unit is often set at, or slightly above, the desired room temperature. However, they are typically available with rated temperature rises ranging from 60°F to 120°F. Another traditional system feature is a control interlock between the makeup air unit and the related exhaust system – if the exhaust system is not in operation or fails, the makeup air unit is automatically shut down as well. Direct-fired units can be ducted, or designed to simply free-blow into the space they serve.

Figure 27 illustrates a typical direct-fired burner. Most utilize a cast-iron pipe as a manifold, drilled to incorporate the gas orifices, and stainless steel baffles, also known as mixing plates. The burner must be designed to prevent the airstream from disturbing the combustion process, while simultaneously collecting sufficient combustion air to support the flame. One very important feature of having the burner mounted directly into the airstream relates to efficiency. Because all heat produced by the burner remains in the airstream, rather than incorporating a flue vent to atmosphere, direct-fired units generally operate at or near 100 percent combustion efficiency.

409F27.EPS

Figure 27 ◆ Direct-fired burner assembly.

In situations that require the makeup air to be cooled and/or dehumidified as well, direct-expansion (DX) or chilled-water cooling systems can also be incorporated into the direct-fired makeup air unit. This arrangement can provide a packaged system capable of maintaining precise discharge air conditions to the indoor environment utilizing either a combination of return and outdoor air, or 100 percent outdoor air on a year-round basis. Since high entering air temperatures, as well as high wet bulb temperatures, are typically imposed on these systems, evaporator coil and overall refrigerant circuit design may be somewhat different than the typical DX system. Standard cooling systems may easily be overloaded with entering air temperatures in excess of 90°F for extended periods of time, especially when the air is also moisture-laden. Simple applications that do not require heat or precise-discharge air conditions may use evaporative cooling systems, which are discussed later in this module.

9.0.0 ◆ ALTERNATIVE COOLING METHODS AND SYSTEMS

Alternative cooling methods and systems can take many forms. Methods and systems covered in this module include the following:

- Ductless split-system air conditioning systems including both direct expansion (DX) and chilled-water systems.
- Air conditioning systems designed to cool computer rooms.

- Valance cooling systems.
- Chilled-beam cooling systems
- **Evaporative coolers**, commonly called swamp coolers.

10.0.0 ◆ DUCTLESS SPLIT-SYSTEM AIR CONDITIONING SYSTEMS

In some heating and cooling applications, it is difficult or impractical to install a ducted air distribution system. Examples include a home heated with a hydronic system, or a historic building that does not have the space to accommodate ducts without major structural modifications. Other appropriate applications include cooling areas that have unique loads such as highway tollbooths, stand-alone ATMs outside of banks, fast-food restaurant drive-through windows, and add-on sunrooms. Options for cooling in these applications include the use of a room air conditioner or a **ductless split-system air conditioner**. *Figure 28* shows a ductless split system configuration. Ductless split systems are also available as heat pumps.

As the name implies, these are split systems that employ a ductless air handler. The outdoor unit contains the compressor, condenser coil, condenser fan, and controls like any other split-system condenser. The narrow profile and horizontal discharge of air make them ideal for installations where space is limited. The air handlers used in ductless split systems can be mounted on the floor, high on a wall, and either on or in the ceiling. The two components of the system are connected together by refrigerant tubing like any other split system.

Temperature control and control of other system functions is typically done with a handheld remote that is similar to a TV remote (*Figure 29*). Some systems allow the use of a wall-mounted room thermostat that is hardwired to the two major system components.

Most ductless split systems are direct expansion (DX) systems. There are, however, ductless split systems that supply chilled water to the fan coils instead of refrigerant. An air-cooled chiller supplies chilled water that is then circulated to one or more air handlers. The air handlers used in chilled-water systems closely resemble their direct expansion counterparts and are controlled in a similar manner.

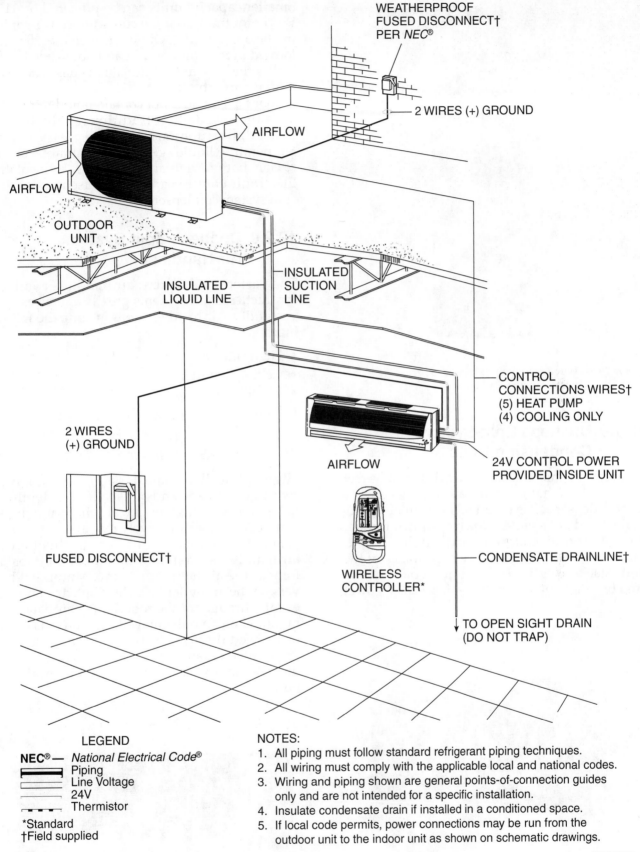

WEATHERPROOF
FUSED DISCONNECT†
PER *NEC*®

2 WIRES (+) GROUND

AIRFLOW

AIRFLOW

OUTDOOR
UNIT

INSULATED
LIQUID LINE

INSULATED
SUCTION
LINE

2 WIRES
(+) GROUND

FUSED DISCONNECT†

CONTROL
CONNECTIONS WIRES†
(5) HEAT PUMP
(4) COOLING ONLY

AIRFLOW

24V CONTROL POWER
PROVIDED INSIDE UNIT

CONDENSATE DRAINLINE†

WIRELESS
CONTROLLER*

TO OPEN SIGHT DRAIN
(DO NOT TRAP)

LEGEND

NEC® — *National Electrical Code*®
Piping
Line Voltage
24V
Thermistor

*Standard
†Field supplied

NOTES:
1. All piping must follow standard refrigerant piping techniques.
2. All wiring must comply with the applicable local and national codes.
3. Wiring and piping shown are general points-of-connection guides only and are not intended for a specific installation.
4. Insulate condensate drain if installed in a conditioned space.
5. If local code permits, power connections may be run from the outdoor unit to the indoor unit as shown on schematic drawings.

409F28.EPS

Figure 28 ◆ Components of a ductless split-system air conditioning system.

Figure 29 ◆ Wireless remote control used with a ductless split-system.

10.1.0 Ductless Split-System Condensing Units

Condensing units (*Figure 30*) are designed as rectangular slabs that resemble a standing suitcase. The condenser coil is mounted horizontally in the cabinet and air is either blown or drawn through the coil and discharged horizontally. Higher capacity single units have the appearance of two units stacked vertically. Units are typically powered by single-phase or three-phase 230V power.

Figure 30 ◆ Ductless split-system condensing unit.

Smaller capacity units can be powered by 115V. Power for the indoor fan coil is typically supplied by the outdoor unit. Often, the metering device is located in the outdoor unit and not inside the fan coil as they are in traditional fan coils. The result is that the tube that carries refrigerant to the indoor fan coil does not contain liquid refrigerant as is the case with a traditional split system. Instead, the tube carries a low-pressure, low-temperature liquid/gas refrigerant mix. For that reason, both refrigerant tubes must be insulated. Electronic controls are widely used in ductless split system condensers.

10.2.0 Ductless Split System Air Handlers

The variety of different air handlers available for ductless split systems greatly enhances their versatility. Air handler types include the following:

- High sidewall
- Ceiling mount
- In-ceiling cassette
- Floor mount

10.2.1 High-Sidewall Air Handlers

High-sidewall air handlers (*Figure 31*) have a rectangular shape and are shallow in depth. By being mounted high on a wall, the fan in the air handler can more effectively move the air into the room. The centrifugal blower in this type of air handler is small in diameter and almost as long as the air handler is wide. Motorized louvers on the air outlet help distribute the air more evenly throughout the room. If the air handler is mounted on an outside wall, the condensate can be drained through the outside wall by gravity. If mounted on an inside wall, an accessory condensate pump can be used for condensate disposal.

Figure 31 ◆ High-sidewall air handler.

High-sidewall air handlers are typically attached to the wall on a mounting plate. The mounting plate also acts as a template for drilling holes for the refrigerant piping, wiring, and condensate drain. Power for the air handler comes from the outdoor unit. Low-voltage DC is used to communicate between the air handler and the outdoor unit. Control of the air handler and outdoor unit is done by way of a wireless handheld remote. Some applications may allow a traditional hardwired room thermostat.

10.2.2 Ceiling-Mounted Air Handlers

Ceiling-mounted air handlers, often called under-ceiling air handlers (*Figure 32*), have many of the same characteristics as high-sidewall air handlers. The main difference is that they are suspended beneath a ceiling. They are often located close to a wall so that refrigerant piping, wiring, and condensate piping can be run through the wall. If located away from an outside wall, an accessory pump must be used to dispose of condensate. Control of the unit is the same as that of a high sidewall unit.

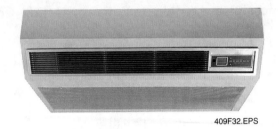

Figure 32 ◆ Universal air handler in the ceiling-mount position.

10.2.3 In-Ceiling Cassettes

In-ceiling cassettes (*Figure 33*) are mounted in drop ceilings. All that protrudes below the ceiling line is the return air intake (located in the center) and the supply air distribution louvers (located around the perimeter). When installed, the unit has a below-the-ceiling appearance of a concentric ceiling diffuser. The blower draws air in through the center of the diffuser and distributes it out through the supply louvers. Refrigerant lines, wiring, and condensate lines are all located above the drop ceiling. A condensate pump is

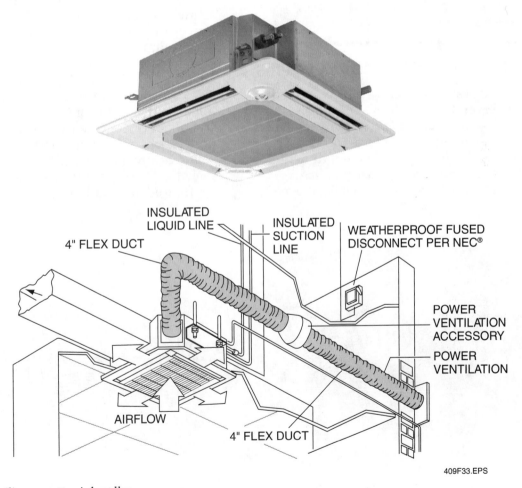

Figure 33 ◆ In-ceiling cassette air handler.

almost always required in this application. Control is similar to that of other ductless split-system air handlers.

In-ceiling cassettes are different from the other types of ductless split-system air handlers in that a limited length and size of ductwork can be attached to the air handler to supply an adjoining space. A knockout on the side of the cassette is provided for attaching the duct. If the ducted air is supplied to an adjoining room, air return must be accomplished using an undercut door or other method. Some cassettes have provisions for the introduction of outside air for ventilation.

10.2.4 Floor-Mounted Air Handlers

On floor-mounted or console air handlers (*Figure 34*), air louvers direct airflow into the room. They are typically mounted against a wall for a through-the-wall connection to the outdoor unit. Condensate disposal and control options are similar to those of the other types of air handlers.

> **NOTE**
> Some manufacturers offer a universal air handler that can be used for floor, low sidewall, or ceiling-mount applications.

10.3.0 Multiple Ductless Split Systems

Ductless split systems are also available in models that allow several air handlers to be attached to a single outdoor unit (*Figure 35*). The different areas where the air handlers are located become, in effect, zones. Each zone can have independent temperature control. The outdoor unit has service valves or other connection points where the refrigerant lines supplying each fan coil are connected. The larger the capacity of the condensing unit, the more fan coils it can supply with refrigerant.

Due to the zoned nature of multiple configurations, there will be times when the full capacity of the outdoor unit will not be required. For example, if only one zone of a three-zone system requires cooling, the compressor will have excess capacity. Different manufacturers deal with compressor capacity control differently. Methods used include dual compressors, two-stage compressors, and variable-speed compressors.

Increased versatility can be realized by using different types of air handlers in different zones. For example, in a four-zone system, the first zone may use a ceiling-mounted air handler, the second and third zones may use high-sidewall air handlers, and the fourth zone may use an in-ceiling cassette air handler.

10.4.0 Ductless Split System Installation and Service

Ductless splits systems are installed similar to conventional split system air conditioners and heat pumps. Outdoor units should be installed to allow proper airflow and access for service. Power for the indoor unit comes from the outdoor unit. Local codes may require a disconnect switch for the indoor unit.

Some ductless split systems do not use a 24V control circuit like those found in conventional split systems. Instead, DC communication signals between the outdoor unit and indoor unit(s) control all functions of the system. Electronic controls are widely used in ductless split systems. These controls may have to be set up or configured at installation to ensure that the out-

409F34.EPS

Figure 34 ◆ Floor-mounted (universal) air handler.

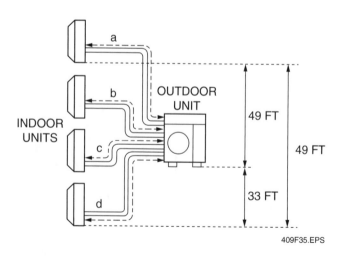

409F35.EPS

Figure 35 ◆ Direct expansion multiple air handler hookup.

door unit is compatible with its various air handlers. Since each manufacturer has its own electronic controls, it is important to read the manufacturer's installation instructions during the initial setup.

Service and troubleshooting of ductless split systems is similar to other split system products. Procedures such as refrigerant charging should be done according to the manufacturer's instructions. Electronic controls often have built-in diagnostics in the form of fault code light-emitting diodes (LEDs). Labels on the equipment and the service literature can help decipher any fault codes.

10.5.0 Chilled-Water Ductless Split Systems

Ductless split system air handlers can be supplied with chilled water instead of refrigerant. In all other respects, chilled-water systems are very similar to their direct-expansion counterparts. The outdoor unit (*Figure 36*) is an air-cooled chiller. It can supply chilled water to one or more air handlers (*Figure 37*). Heat pump versions are also available that heat the water in winter. Capacity control is not as much of an issue with multiple chilled-water fan coil systems as it is with multiple direct expansion systems because the system manages chilled water instead of refrigerant. In the chilled-water system, a chilled-water storage tank can easily absorb excess compressor capacity.

When a fan coil calls for chilled water, chilled water in the storage tank can satisfy some or all of the demand.

409F36.EPS

Figure 36 ◆ Air-cooled chiller used with a ductless split system.

In chilled-water systems, the refrigerant system is sealed. It uses a refrigerant-to-water heat exchanger in place of a conventional evaporator coil. Issues related to refrigerant charging due to refrigerant line length or the addition of refrigerant system components such as a filter-drier do not exist. Piping to the fan coils contains water. It is usually easier to deal with water piping than it is to deal with refrigerant piping.

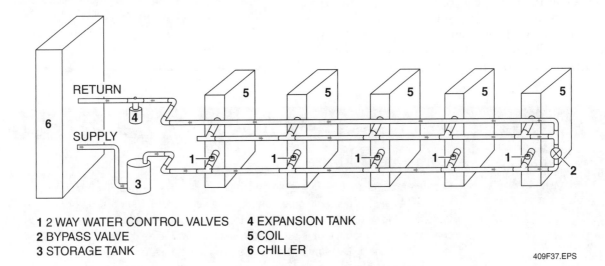

1 2 WAY WATER CONTROL VALVES   **4** EXPANSION TANK
2 BYPASS VALVE   **5** COIL
3 STORAGE TANK   **6** CHILLER

409F37.EPS

Figure 37 ◆ Chilled-water multiple air handler hookup.

11.0.0 ◆ COMPUTER ROOM COOLING SYSTEMS

Computers and other electronic equipment generate a great deal of heat that must be removed to prevent overheating. If allowed to overheat, computers will malfunction or fail. For security reasons, computers and banks of servers are installed in dedicated rooms and closets. The various components of the system are often stacked tightly together, allowing a great deal of heat to build up in a very small area. For those reasons, special equipment is used to cool computers and similar electronic devices. The cooling equipment must be able to reliably handle a constant cooling load as well as a high cooling load.

In most computer and electronic cooling applications, the sensible cooling load represents an even larger amount of the total load than in common comfort applications. Since computers and other electronic devices produce only sensible heat, the only latent load is that which comes from infiltration or from moisture added by humidification systems. Maintaining proper and consistent humidity levels is also important, because electronic systems are sensitive to damage from static electricity. For that reason, humidification accessories and dehumidification cycles with electric reheat are often incorporated into the cooling systems.

The cooling load from the electronics and computer system is typically consistent year-round, so computer room systems must be properly prepared or designed for operation at low ambient conditions. This includes some form of head pressure control during low outdoor tempera-tures, as well as crankcase heaters for compressors located outdoors. Care must also be taken to prevent refrigerant migration outdoors during the off-cycle, generally by using a liquid line solenoid valve. Another feature often found on these systems includes compressor staging and/or hot gas bypass for capacity control to maintain precise control of the environment.

The number and type of systems used to cool computer rooms is great and varied. Some common types of equipment and systems used to cool computers include:

- Raised-floor cooling systems
- Freestanding air handlers
- Liquid chillers
- Cooled equipment enclosures
- Spot coolers

11.1.0 Raised-Floor Cooling Systems

Many computer rooms deliver cool, conditioned air from the space beneath a raised floor (*Figure 38*). The cool air is delivered through perforated grates in the floor. It passes up through equipment cabinets where it picks up heat. The cabinets are arranged in a **hot aisle/cold aisle configuration** where cool air is available in the cold aisle and heated air from the cabinets is dumped into the hot aisles. Return grilles in the ceiling of the room return the air from the hot aisles to the air handler.

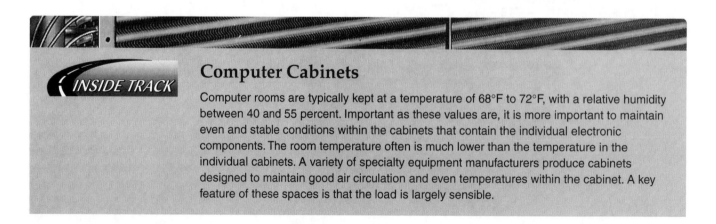

INSIDE TRACK

Computer Cabinets

Computer rooms are typically kept at a temperature of 68°F to 72°F, with a relative humidity between 40 and 55 percent. Important as these values are, it is more important to maintain even and stable conditions within the cabinets that contain the individual electronic components. The room temperature often is much lower than the temperature in the individual cabinets. A variety of specialty equipment manufacturers produce cabinets designed to maintain good air circulation and even temperatures within the cabinet. A key feature of these spaces is that the load is largely sensible.

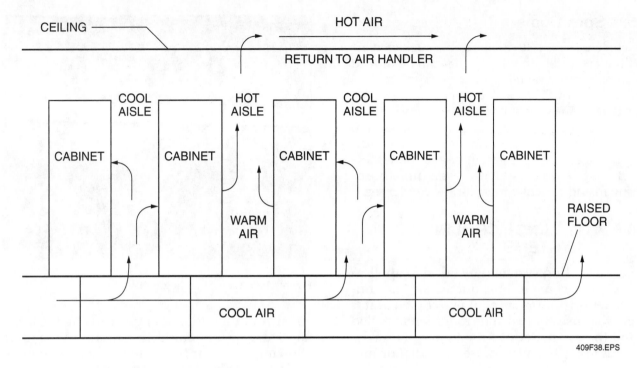

CEILING —

HOT AIR →

RETURN TO AIR HANDLER

| COOL AISLE | | HOT AISLE | | COOL AISLE | | HOT AISLE | |
| CABINET | CABINET | CABINET | CABINET | CABINET |

WARM AIR

WARM AIR

RAISED FLOOR

COOL AIR

COOL AIR

409F38.EPS

Figure 38 ◆ Hot aisle/cold aisle cabinet configuration.

This arrangement does not always allow the heat to be removed evenly from the cabinets, resulting in hot spots in the cabinet. Often, supplemental cooling fans and/or cooling coils are needed to ensure that individual cabinets remain cool. Some cabinets are designed so that the hot exhaust air is ducted directly into the return air plenum and back to the air handler. Compressors must have special controls such as hot-gas bypasses.

11.2.0 Freestanding Air Handlers

Freestanding air handlers are used to cool, dehumidify, and filter the air in equipment rooms. Air is directly discharged from the air handler without the use of any ductwork. The air handler is often supplied with chilled water from a remotely located chiller.

11.3.0 Liquid Chillers

Liquid chillers supply chilled water to air handlers used to cool individual rooms, cabinets, or racks. Chilled water is typically supplied at a temperature above the dew point to prevent condensation that could harm electronic components.

11.4.0 Cooled Equipment Enclosures

Cooled equipment enclosures (*Figure 39*) can best be described as self-contained cooled cabinets designed to house electronic equipment.

Servers or racks of electronic components can be stacked in the cabinet. The self-contained cooling unit, either air-cooled or water-cooled, is designed to distribute cool air evenly across all components in the cabinet to eliminate any hot spots.

409F39.EPS

Figure 39 ◆ Cooled equipment enclosure.

11.5.0 Spot Coolers

Spot coolers (*Figure 40*) are portable packaged air conditioners, often on wheels, that can be moved to an area to provide temporary or supplemental localized cooling. They are equipped with flexible ducts that allow air from outside the conditioned area to be brought in to cool the condenser and to exhaust the condenser air outside of the conditioned space. Spot coolers may also be water-cooled. Water-cooled spot coolers require hoses to bring in and exhaust water for the condenser.

12.0.0 ◆ VALANCE COOLING SYSTEMS

A **valance cooling system** is installed in applications where a conventional ducted air distribution systems is impractical. It is a chilled-water system that uses finned-tube radiation installed around the perimeter of a room, just below the ceiling. It gets its name because the radiation and piping are concealed beneath a decorative valance.

Each valance cooling unit consists of a section or sections of finned-tube radiation, similar to that used in baseboard hydronic heating systems (*Figure 41*). Each unit is equipped with an insulated chilled-water supply and return line. A zone valve may be installed in the supply line. Beneath the finned-tube section is a pan to catch the condensate and drain it away for disposal. The condensate pan is often a part of the decorative valance. In operation, chilled water flows through the coil on a call-for-cooling from the low-voltage room thermostat. The signal energizes the circulator pump and the zone valve, if so equipped. As the radiation cools, it cools the

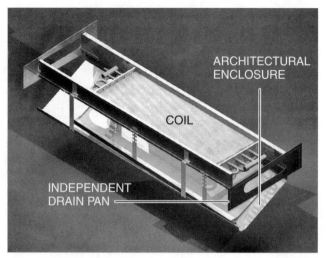

409F41.EPS

Figure 41 ◆ Valance cooling unit.

air surrounding the radiation, causing the air to fall. This sets up natural air currents in the room allowing warm air to rise to the ceiling where it cools, then falls, repeating the cycle. No circulating fans are used in the system. A separate system is normally used to provide ventilation and air filtration.

Because valance cooling systems do not use a circulating fan, the systems are very quiet and dust is kept to a minimum. They are more energy-efficient because the circulating pump consumes much less power than a blower motor. As with other types of hydronic systems, the installed costs of valance cooling systems are much higher than those of conventional forced-air cooling systems. The piping and radiation used in a valance cooling system can serve a dual purpose. The same radiation can be used to heat the structure.

409F40.EPS

Figure 40 ◆ Spot coolers.

13.0.0 ◆ CHILLED-BEAM COOLING SYSTEMS

Chilled-beam cooling systems have been used for several years in Europe and Australia and are now being introduced in the United States. Chilled-beam systems are similar in some respects to valance cooling systems. The name seems to imply that structural members of the building are cooled. In fact, the units used to cool the room are long, finned tube radiators in an enclosure. They resemble beams suspended from the ceiling, thus the name chilled beam. There are two main types of chilled-beam cooling systems: passive and active.

13.1.0 Passive Chilled-Beam Systems

Passive chilled-beam systems (*Figure 42*) resemble valance cooling units in that they both employ finned-tube elements located near the ceiling of a room. Chilled-beam units can be located near outside walls or can be located in other areas of the ceiling. The units may be exposed or concealed behind metal grilles in a suspended ceiling. Passive systems rely on natural convection currents to cool the room. Lighter warm air in the room rises to the chilled-beam unit, where it is cooled. The heavier cooled air then gently falls into the room.

In passive systems, the chilled water supplied to the finned elements has to be slightly warmer than the dew point of the room air to prevent condensation from forming and dripping from the cooling units on the ceiling.

13.2.0 Active Chilled-Beam Systems

Active chilled-beam systems (*Figure 43*) use ducted, conditioned air to help induce additional airflow past the chilled-beam units. Condensation is less likely to occur on the units in an active system.

Chilled-beam systems are noted for their energy efficiency. They also offer quiet operation, excellent indoor air quality, and low maintenance due to their small number of moving parts. Disadvantages include high installation costs compared to more conventional systems. Because this technology is new to the United States, many HVAC contractors may be unfamiliar with it. Those that are familiar with the technology charge a premium for their expertise.

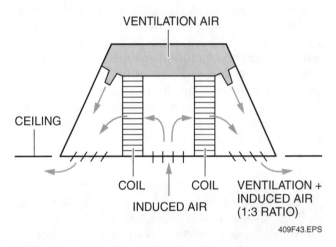

Figure 43 ◆ Active chilled-beam cooling system.

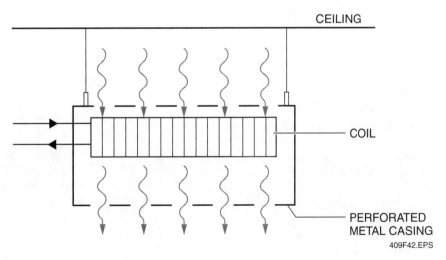

Figure 42 ◆ Passive chilled-beam cooling system.

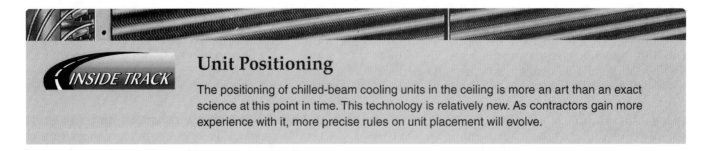

INSIDE TRACK

Unit Positioning

The positioning of chilled-beam cooling units in the ceiling is more an art than an exact science at this point in time. This technology is relatively new. As contractors gain more experience with it, more precise rules on unit placement will evolve.

14.0.0 ◆ EVAPORATIVE COOLERS

Evaporative coolers, often called swamp coolers (*Figure 44*), are used in many parts of the United States for comfort cooling. They offer lower operating costs than conventional air conditioners. Evaporative coolers do not operate using the refrigeration cycle. Instead, they operate on the principle that when water is changed from a liquid to water vapor, heat is absorbed.

An evaporative cooler consists of a louvered cabinet containing a blower assembly, water-absorbing pads, a water pump and water distribution system, and a float valve to control water level in the sump of the unit. In operation, the pump distributes water to the pad or pads to wet them. The blower assembly draws hot, outdoor air across the wet pads. The water on the pads absorbs heat from the air as the water evaporates. This cools and adds moisture to the air.

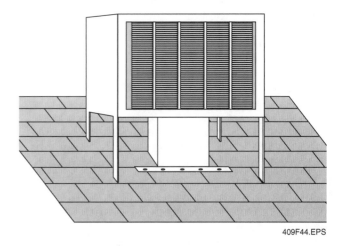

409F44.EPS

Figure 44 ◆ Roof-mounted evaporative cooler.

NOTE

The pads used in evaporative coolers have traditionally been made of shredded aspen wood. Today, pads are available that are made of cellulose or man-made fibers.

This cooled air is then distributed throughout the structure. Evaporative coolers do not require a return duct because they take in outdoor air. To prevent the structure from being pressurized, windows or doors must be opened, or a relief duct provided, to relieve the pressure.

Evaporative coolers are only effective if the outdoor air has a very low relative humidity. That is why they are popular in the hot, dry climate of the southwestern United States. If the relative humidity increases, evaporative coolers become ineffective because the air cannot absorb additional moisture. In those situations, a conventional air conditioner is used for comfort cooling. Sometimes, evaporative coolers are installed so that they share the supply duct system with a conventional air conditioner. Manual isolation

plates are used to prevent the conditioned air from one appliance from being bypassed through another appliance (*Figure 45*).

At one time, a combination heat pump/evaporative cooler was manufactured. It did not prove popular and has since been discontinued.

The major advantage of evaporative coolers is their lower installed cost and lower operating cost compared to a traditional air conditioner. The blower motor and water pump consume much less power than a compressor, condenser fan motor, and evaporator blower motor. However, they do increase water consumption. Disadvantages include poor comfort when humidity is high, as well as increased maintenance.

NOTE

Evaporative coolers are commonly installed on the roof with the supply air duct connected to the bottom of the cooler. Units are available with a side air discharge; other units are available that can be installed in a window like a room air conditioner. Coolers on wheels are available for spot cooling.

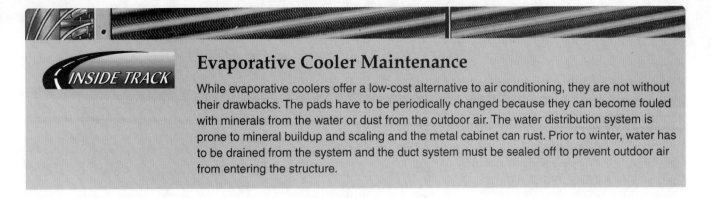

Evaporative Cooler Maintenance

While evaporative coolers offer a low-cost alternative to air conditioning, they are not without their drawbacks. The pads have to be periodically changed because they can become fouled with minerals from the water or dust from the outdoor air. The water distribution system is prone to mineral buildup and scaling and the metal cabinet can rust. Prior to winter, water has to be drained from the system and the duct system must be sealed off to prevent outdoor air from entering the structure.

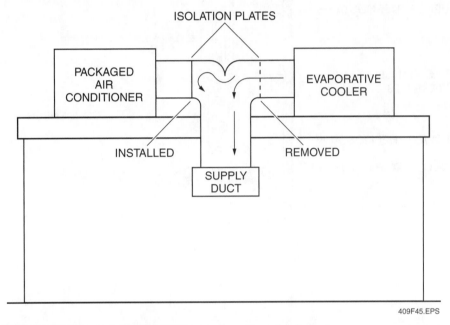

Figure 45 ◆ Air conditioner and evaporative cooler using a common supply air duct.

15.0.0 ◆ ALTERNATIVE ENERGY-SAVING SYSTEMS AND DEVICES

In this time of high energy costs, people are constantly looking for ways to stretch their energy dollar. Many devices and systems save energy by capturing, using, or redistributing waste heat that would otherwise be lost.

Alternative energy-saving systems and devices can take many forms. Systems and devices covered in this module include the following:

- *Heat pump water heaters* – Can be used to heat domestic water or swimming pools.
- *Waste heat water heaters* – Capture the rejected heat from air conditioning or refrigeration systems to heat domestic water.
- *Evaporative pre-coolers* – Used to lower the head pressure and reduce energy consumption of air conditioners.

- *Air turnover systems* – Redistribute stratified air within a space, balancing the room temperature and saving energy.

15.1.0 Heat Pump Water Heaters

Electricity is one of the most expensive forms of energy. If it is the only source of energy available, the cost of comfort heating and domestic water heating can be quite high. The heat pump, when used for comfort heating, can reduce heating costs. The same technology can be used to heat domestic water.

NOTE

Heat pump water heaters used to heat potable water must use a double-walled heat exchanger to prevent contamination of the potable water system if a refrigerant leak should occur in the refrigerant-to-water heat exchanger.

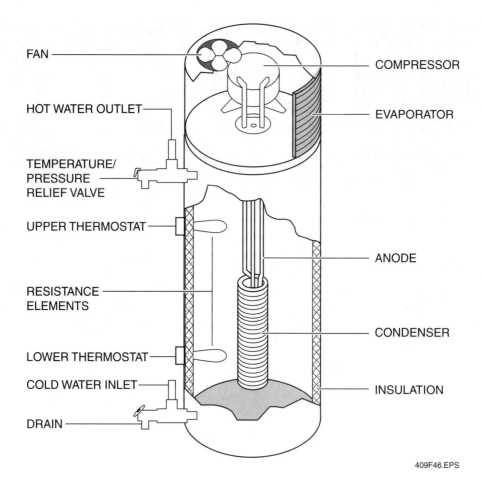

FAN

COMPRESSOR

EVAPORATOR

HOT WATER OUTLET

TEMPERATURE/
PRESSURE
RELIEF VALVE

UPPER THERMOSTAT

ANODE

RESISTANCE
ELEMENTS

CONDENSER

LOWER THERMOSTAT

COLD WATER INLET

INSULATION

DRAIN

409F46.EPS

Figure 46 ◆ Heat pump water heater.

A heat pump electric water heater (*Figure 46*) uses a refrigerant-to-water heat exchanger to heat the water. The heat pump unit is positioned atop a conventional electric water heater tank. It contains a small compressor, evaporator coil, a small evaporator fan, and other components. In operation, the compressor runs and supplies metered refrigerant to the evaporator coil. The evaporator fan draws ambient air across the coil, where it absorbs heat. Refrigerant is returned to the compressor where it is compressed and then sent to the condenser coil. This coil is a refrigerant-to-water heat exchanger immersed in the water storage tank. The heat that was absorbed in the evaporator is rejected to the water, causing it to be heated.

Heat pump water heaters offer excellent energy savings. They are not difficult to install and can be installed indoors or outdoors. Their major disadvantage is high initial cost compared to a conventional electric water heater. That initial installation cost can be offset by the savings realized during operation.

NOTE

In a humid environment, a heat pump water heater will produce condensate. Steps must be taken during installation to ensure that any condensate produced is disposed of properly.

Heat pump water heaters are not limited to heating domestic water. Models are available that heat swimming pool water. The same type used to heat swimming pool water can also be used to preheat water in a conventional hydronic heating system.

15.2.0 Waste Heat Water Heaters

When an air conditioning system runs, the heat removed from the structure is usually lost to the atmosphere. That heat can be captured and recycled for use in industrial processes and to heat water. For example, restaurants and food-processing plants produce a lot of heat as part of their normal operation. They also require a large amount of heat to cook food and heat water.

Heat that is removed by air conditioning and refrigeration equipment used in these types of facilities can be recovered and used to heat water. Recovery is done by using a refrigerant-to-water heat exchanger (*Figure 47*) that removes heat from the discharge gas. Other types of waste heat recovery systems reclaim heat lost out of power exhausts.

15.3.0 Evaporative Pre-Coolers

When a compressor has to work harder, it consumes more energy. High head pressure caused by a high ambient temperature increases the load on the compressor and can make it work harder. If the compressor can be made to work less under the same ambient conditions, power consumption will drop. In hot, dry desert areas, **evapora-tive pre-coolers** can be used to cool the air before it enters the condenser coil. Cooler air entering the condenser causes head pressure to drop. This in turn causes the compressor to consume less power.

An evaporative pre-cooler (*Figure 48*) is nothing more than an evaporative cooler like that discussed earlier in the module. It does not contain a fan, but relies on the air being drawn in through the condenser coil for its airflow. The pre-cooler is positioned in front of the condenser coil. Water is applied from the top to wet the pad. The air being drawn across the condenser coil first passes through the pre-cooler where the air is cooled.

Like all evaporative coolers, pre-coolers are less effective as the relative humidity increases and they can consume a great deal of water. They also have the same maintenance issues that all evaporative coolers have.

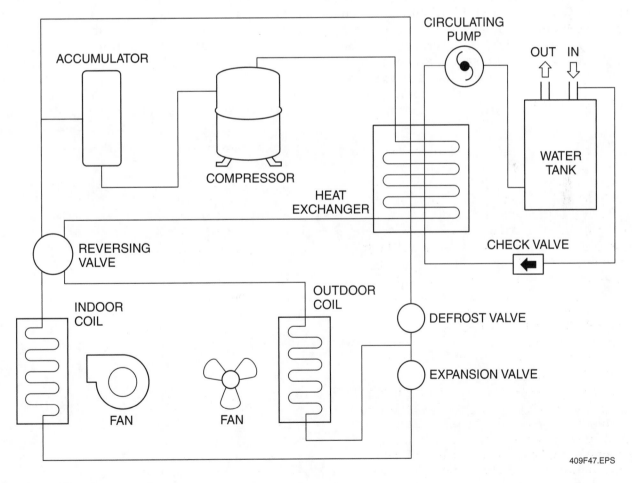

409F47.EPS

Figure 47 ◆ Waste heat water heater.

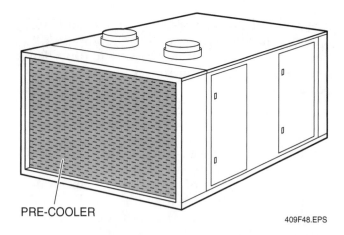

PRE-COOLER

409F48.EPS

Figure 48 ◆ Evaporative pre-cooler installed on a packaged air conditioner.

16.0.0 ◆ AIR TURNOVER SYSTEMS

Air stratification is a characteristic of all rooms with high ceilings. Warm, lighter air rises and is retained near the ceiling. Heavier cold air settles near the floor. This creates different temperature zones between the floor and the ceiling, which can create discomfort. By preventing the air from stratifying, a more even temperature can be maintained, resulting in lower energy costs.

Although this section will focus on systems used to prevent stratification, it should also be noted that air stratification is sometimes em-
ployed as a load-reduction strategy. For example, a light manufacturing operation may be located in a facility with a roof height of 28'. If there is no use of the space above the 12' height, the air distribution system—both supply and return ducts—can be installed from the roof down to the 12' height. This effectively leaves the hot air, which rises toward the roof unconditioned and undisturbed. This reduces the load on the cooling system, often reduces the capacity of the equipment to be installed, and reduces operating costs.

One of the simplest ways to prevent air stratification is to use ceiling-mounted paddle fans. This method has been used for over 100 years and is still used today in residences and light commercial applications such as restaurants.

Paddle fans are ineffective in large industrial and commercial settings, such as in a warehouse with high ceilings. For those applications, an **air turnover unit** is used. Air turnover units are tall vertical air handlers that often contain heating and cooling sections. The fans in air turnover units gently circulate the air, mixing the cool air near the floor with the warmer air near the ceiling. At the same time, the air can be conditioned further by heating or cooling it as it passes through the air handler. Since the air turnover unit is conditioning air that is already heated or cooled to some extent, less energy has to be expended to further condition the air.

1. Wood stoves equipped with a catalytic element cannot have smoke emissions that exceed _____ grams per hour.
   a. 0.4
   b. 4.1
   c. 7.5
   d. 10

2. When installing a wood-burning furnace, the clearance to combustibles requirement is generally the same as that of a gas-fired furnace.
   a. True
   b. False

3. Which of the following vent types would be used to vent a combination gas/wood furnace?
   a. Type B
   b. Type PL
   c. PVC
   d. Type HT

4. Ashes from a wood-burning stove must always be stored in a _____ container.
   a. metal
   b. cardboard
   c. glass
   d. plastic

5. Waste oil heaters must burn oil of a uniform _____ for proper operation.
   a. color
   b. temperature
   c. viscosity
   d. molecular weight

6. Geothermal heat pumps extract heat from _____.
   a. air
   b. the ground
   c. a lake
   d. wells

7. A thermal mass is likely to be found in a passive solar heating system.
   a. True
   b. False

8. The solar heating system that provides the best temperature control is _____.
   a. passive radiant
   b. passive solar hydronic
   c. active solar hydronic
   d. active radiant

9. Which of the following is an advantage of an in-floor radiant electric heating system?
   a. Inexpensive to operate
   b. Simple to install
   c. Inexpensive to install
   d. Excellent comfort

10. The air handler in a ductless split system can be mounted on a wall, on a ceiling, in a ceiling, and on the floor.
    a. True
    b. False

11. A raised-floor cooling system is likely to be found in a _____.
    a. meat packing plant
    b. computer room
    c. hotel lobby
    d. sports stadium

12. Valance cooling systems use a small fan to aid in air circulation.
    a. True
    b. False

13. An evaporative cooler would be most likely found in which of the following United States cities?
    a. Phoenix, AZ
    b. Miami, FL
    c. Boston, MA
    d. New Orleans, LA

14. A cooling system that uses an evaporative cooler does *not* require a _____.
    a. return air duct
    b. water supply system
    c. pressure relief method
    d. blower assembly

15. A simple way to prevent air stratification is to use a _____.
    a. window fan
    b. valance cooler
    c. paddle fan
    d. chilled-water coil

Summary

In addition to conventional heating and cooling systems, there are many alternative systems available that can heat or cool a structure. These alternative systems are used because the application is not suited to a conventional system, or the availability of a low-cost fuel or energy source makes the alternative system attractive. Solid fuels, such as wood, wood pellets, shelled corn, and coal, if they are readily available, are often cheaper than heating fuels such as natural gas, propane, and fuel oil. To be burned as fuel, these solid fuels require special appliances. Solar energy and heat energy from the earth (geothermal energy) are available as sources, but different and often unconventional methods are required to capture and use that type of heat.

Cooling a structure with a conventional ducted air distribution system is often impractical. In those situations, ductless direct-expansion systems or ductless systems using chilled water are used. In hot, dry desert climates, evaporative cooling can be used to cool a structure while at the same time dramatically lowering electrical power consumption. Rooms housing computers and other electronic equipment have unique cooling requirements that can only be satisfied with specialized equipment.

In this era of high energy costs, people are turning to alternative systems to reduce energy consumption without sacrificing comfort or other benefits. Alternative systems include heat pump water heaters used to heat domestic water and swimming pools; water heaters that recover the heat normally rejected by an air conditioning condenser; and evaporative pre-coolers used to cool the air entering a condenser coil. Air turn-over systems move air within a room to even the temperature. This increases comfort while reducing energy usage.

Notes

Active solar heating system: A type of solar heating system that uses fluids pumped through collectors to gather solar heat. It is the most complex solar heating system and provides the best temperature control.

Air stratification: The layering of air in a room based on temperature. The warmest air is concentrated at the ceiling and the coldest air is concentrated at the floor, with varying temperature layers in between.

Air turnover unit: A type of tall air handler that moves stratified air so that the temperature in the room is more even. The air turnover unit may further condition the air in the process.

Catalytic element: A device used in wood-burning stoves that helps reduce smoke emissions.

Chilled-beam cooling system: A cooling system that employs radiators (chilled beams) mounted near the ceiling through which chilled water flows. Passive systems rely on convection currents for cooling. Active systems use ducted conditioned air to help induce additional airflow over the beams.

Cooled equipment enclosure: A type of enclosure or cabinet that contains its own air conditioning system. It is designed to house and cool electronic components.

Creosote: A black, sticky combustible byproduct created when wood is burned in a stove. It must be periodically cleaned from within the stove and chimney before it can build up to the point where it can cause a fire.

Direct-fired makeup air unit: An air handler that heats and replaces indoor air that is lost from a building through exhaust vents.

Ductless split-system air conditioner: A split-system air conditioner in which the indoor air handler is wall- or ceiling-mounted. The air handler blows the conditioned air directly into the room without the use of ductwork.

Evaporative cooler: A comfort cooling device that cools air by evaporating water. It is commonly used in hot, dry climates. Cooling effectiveness drops as the relative humidity of the outdoor air increases.

Evaporative pre-cooler: A device used to pre-cool air entering a condenser coil. Pre-cooling the air reduces the load on the compressor, thus reducing energy costs. It operates on the same principle as an evaporative cooler.

Geothermal heat pump: A heat pump that uses fluid-filled tubing buried in the ground to capture the heat that is always present in the earth.

Hot aisle/cold aisle configuration: A method used to install equipment cabinets in computer rooms to manage the flow of warm and cool air out of and into the cabinets.

Indirect solar hydronic heating system: A type of active solar heating system in which a double-walled heat exchanger is used to prevent toxic antifreeze in the solar collectors from contaminating the potable water system.

Infiltration: The process by which outdoor air enters a structure through cracks, crevices, and other openings in the building.

Passive solar heating system: A type of solar heating system characterized by a lack of moving parts and controls. The sun shining in a window on a winter day is an example of passive solar heat.

Radiant heating system: A type of heating system that heats objects, not the air. In-floor systems using electric heating elements or tubes filled with hot fluid are examples of radiant heating systems.

Thermosyphon system: A type of passive solar heating system in which the difference in temperature of fluids in different parts of the system causes the fluids to flow through the system.

Type HT vent: A metal vent capable of withstanding temperatures up to 1,000°F. It is commonly used to vent wood-burning stoves and furnaces.

Type PL vent: A type of metal vent specifically designed for stoves that burn wood pellets or corn.

Valance cooling system: A type of cooling system in which chilled water is circulated through finned-tube radiators located near the ceiling around the perimeter of a room. Convection currents move the cooled air instead of a blower assembly. A decorative valance conceals the system.

Additional Resources and References

Additional Resources

This module is intended to be a thorough resource for task training. The following reference works are suggested for further study. These are optional materials for continued education rather than for task training.

http://warmair.net

http://www.servicemagic.com/article.show.
Think-Green-when-it-Comes-to-Residential-Heating.15397.html

Figure Credits

Topaz Publications, Inc., 409F01, 409SA01, 409F13

Fire Chief Wood & Coal Furnace - Victorian Sales, 409F02

Alpha American Company, 409F03, 409F05, 409F11

Courtesy of BioHeat USA, 409F06, 409F09

Central Boiler, 409F08

Courtesy of Bob Shipman - Owner - Pinnacle Corn Stoves, 409F12

Chimney Safety Institute of America, 409F14

Schaefer Brush Manufacturing Company, 409F15

Columbia Boiler Company, 409F16

U.S. Department of Energy, Office of Energy Efficiency and Renewable Energy, 409F20, 409F46

Watts Radiant, Inc., 409F23, 409F24

Uponor, Inc., 409F25

Cambridge Engineering, Inc., 409F26

AAON, 409F27

Carrier Corporation, 409F28, 409F33 (line art)

Sanyo HVAC – sanyohvac.com, 409F29, 409F30

EMI – an ECR International Brand, 409F31, 409F32, 409F34

Mitsubishi Electric and Electronics USA, Inc., 409F33 (photo), 409F35

Multiaqua, Inc., 409F36, 409F37

Kooltronic, Inc., 409F39

Portable AC, 409F40

Sigma Corporation, 409F41

ASHRAE Journal, January 2007. © American Society of Heating, Refrigerating and Air Conditioning Engineers, Inc., www.ashrae.org, 409F42, 409F43

NCCER makes every effort to keep these textbooks up-to-date and free of technical errors. We appreciate your help in this process. If you have an idea for improving this textbook, or if you find an error, a typographical mistake, or an inaccuracy in NCCER's Contren® textbooks, please write us, using this form or a photocopy. Be sure to include the exact module number, page number, a detailed description, and the correction, if applicable. Your input will be brought to the attention of the Technical Review Committee. Thank you for your assistance.

Instructors – If you found that additional materials were necessary in order to teach this module effectively, please let us know so that we may include them in the Equipment/Materials list in the Annotated Instructor's Guide.

Write: Product Development and Revision
National Center for Construction Education and Research
3600 NW 43rd St, Bldg G, Gainesville, FL 32606

Fax: 352-334-0932

E-mail: curriculum@nccer.org

Craft _____   Module Name _____

Copyright Date _____   Module Number _____   Page Number(s) _____

Description _____

_____

_____

_____

(Optional) Correction _____

_____

_____

(Optional) Your Name and Address _____

_____

_____

03410-09

Introduction to Supervisory Skills

03410-09
Introduction to Supervisory Skills

Topics to be presented in this module include:

Overview

This module will introduce you to some of the many skills you will have to acquire if you wish to become a supervisor. You will be required to learn many people skills, such as leadership and communication, in addition to the skills of your trade. You will select a management style and learn to apply it. You will also need to learn how to coordinate and handle relations with subordinates, superiors, and customers.

Objectives

When you have completed this module, you will be able to do the following:

1. Describe the skills necessary to be a supervisor.
2. List the characteristics and behavior of effective leaders, as well as the different leadership styles.
3. Explain the difference between problem solving and decision making.
4. Describe ways to deal with common leadership problems, such as absenteeism and turnover.
5. Identify a supervisor's safety responsibilities.
6. Describe the signals of substance abuse.
7. List the essential parts of an accident investigation.

Required Trainee Materials

Pencil and paper

Prerequisites

Before you begin this module, it is recommended that you successfully complete *Core Curriculum*; *HVAC Level One*; *HVAC Level Two*; *HVAC Level Three*; and *HVAC Level Four*, Modules 03401-09 through 03409-09.

This course map shows all of the modules in the fourth level of the *HVAC* curriculum. The suggested training order begins at the bottom and proceeds up. Skill levels increase as you advance on the course map. The local Training Program Sponsor may adjust the training order.

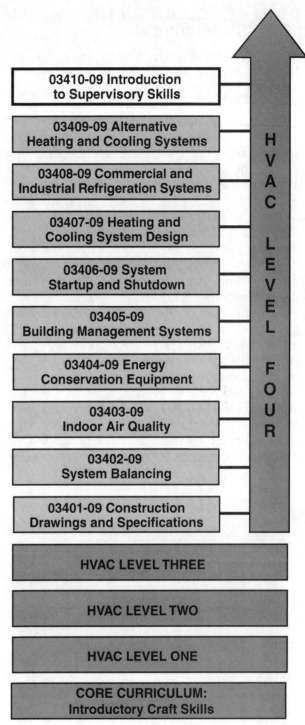

410CMAP.EPS

1.0.0 ◆ INTRODUCTION TO SUPERVISION

Some workers have no desire to progress beyond being an effective and successful craftworker. That is not a defect or a fault; if you feel content with your position, you may be happier as a craftworker. Supervision is not for everyone. For those who wish to become a crew leader or foreman, however, there are many new skills to be learned. The knowledge that you have attained of the craft itself is certainly part of those skills. The foreman or crew leader must know the limitation of the skills shared by their workers as well, in order to properly supervise their work. However, supervision requires many people skills as well. Some of these skills can be acquired simply by working with others in teams. Other skills may require additional specific training.

For the purpose of this module, it is important to define some of the positions to which we will be referring. Craftworkers and foremen depend on the availability of three elements: materials, labor, and technology. The term *craftworker* refers to a person who performs the work of his or her trade(s). The crew leader is a person who supervises one or more craftworkers on a crew. A foreman is essentially a second-line supervisor who supervises one or more crew leaders or front-line supervisors. Finally, a project manager is responsible for managing the construction of one or more construction projects. The organizational chart in *Figure 1* depicts how each of the four positions is related in the chain of command. This module will focus on the role of the crew leader.

Craftworkers and crew leaders differ, in that the crew leader manages the activities that the skilled craftworkers on the crews actually perform. In order to manage a crew of craftworkers, you must have firsthand knowledge and experience in the activities being performed. In addition, you must be able to lead, organize, plan, control, and direct the activities of the various crew members.

Some basic skills of the supervisor include, but are not limited to, the following:

- Determining the training needs of workers and training them
- Maintaining good relationships with and among workers
- Dealing with workers who are not from the same culture, or who are of different genders
- Solving personal problems between workers
- Budgeting time and materials effectively
- Motivating workers to perform efficiently

This module discusses the importance of developing effective leadership skills as a new

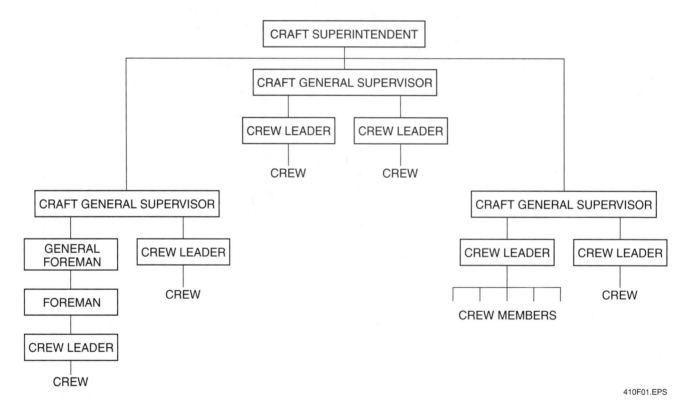

410F01.EPS

Figure 1 ◆ Project organization chart.

supervisor and provides some tips for doing so. Effective ways to communicate with all levels of employees, build teams, motivate crew members, make decisions, and resolve problems will be covered in depth. Finally, an overview of safety supervision is provided.

2.0.0 ◆ BECOMING A LEADER

The crew leader is generally selected and promoted from a work crew or technician status. If you are selected to be a crew leader, that selection will frequently be based on your ability to accomplish tasks, get along with others, meet schedules, and stay within the budget. You must lead the team to work safely and provide a quality product.

Making the transition from a craftworker to a crew leader can be difficult, especially when you find yourself in charge of supervising a group of peers, some of whom you may have known for long periods of time. Crew leaders are no longer responsible for their work alone; rather, they are accountable for the work of an entire crew of people with varying skill levels and abilities, a multitude of personalities and work styles, and different cultural and educational backgrounds. Supervisors must learn to put their personal relationships aside and work for the common goals of the team.

As you move from a craftworker position to the role of a crew leader, you will find that more hours will be spent supervising the work of others than are spent actually performing the technical skill for which you have been trained. *Figure 2* represents the percentage of time craftworkers, crew leaders, foremen, and project managers

spend on technical and supervisory work as their management responsibilities increase.

Your success as a new crew leader is directly related to your ability to make the transition from crew member into a leadership role. Leadership includes possessing the traits and skills to motivate others to follow and perform.

2.1.0 Characteristics of Effective Leaders

A supervisor is, by the nature of the position, a leader. Leadership traits are some of the skills that a supervisor needs in order to be effective. Although the characteristics of leadership are varied, there are some definite commonalities among effective leaders.

First and foremost, effective leaders lead by example. In other words, they work and live by the standards that they establish for their crew members or followers, making sure that they practice what they preach.

Effective leaders tend to have a high level of drive, determination, persistence, and a stick-to-it attitude. In the face of a challenging situation, leaders seek involvement, and work through adversity to achieve their goal. When faced with obstacles, effective leaders don't get discouraged. Instead, they identify the potential problems, make plans to overcome them, and continue to work toward the intended goal. In the event of failure, effective leaders learn from their mistakes and apply that knowledge to their future leadership attempts.

Effective leaders are effective communicators. Accomplishing this may require that the leader overcome issues such as language barriers, gen-

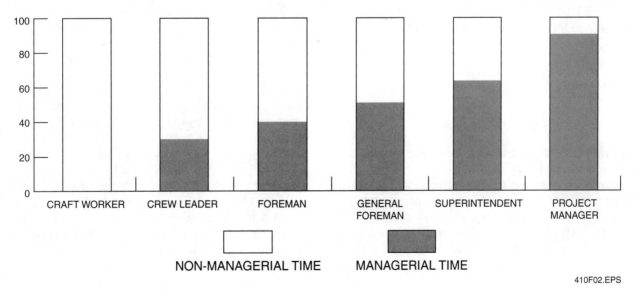

Figure 2 ◆ Percentage of time spent on technical and supervisory work.

der bias, or differences in personalities to ensure that each follower or member of the crew understands the established goals of the project.

Effective leaders have the ability to motivate their followers or crew members to work to their full potential and become effective members of the team. They try to develop their crew members' skills and encourage them to improve and learn as a means to contribute more to the team effort. Effective leaders tend to demand 100 percent from themselves and from their teams, so they work hard to provide the skills and leadership necessary to do so.

Lastly, effective leaders are organized planners. They know what needs to be accomplished, and they use their resources to make it happen. Because they cannot do it alone, leaders enlist the help of their team members to share in the workload. Effective leaders delegate work to their crew members, and they implement company policies and procedures to ensure that the work is completed effectively, efficiently, and safely. When delegating work, they make sure that the crew member understands what to do and the level of responsibility to which they will be held.

Finally, effective leaders have the training and confidence that allows them to make decisions and solve problems. If they are going to accomplish their goals, leaders must learn to take risks when appropriate. Leaders must be able to absorb information, assess courses of action, understand the risks, make decisions, and assume the responsibility for these decisions. Some other major characteristics of effective leaders include the following:

- Planning and organizing skills
- Loyalty
- Fairness
- Enthusiasm
- Ability to teach and learn from others
- Initiative
- Salesmanship
- Good communication skills

2.1.1 Planning and Organizing

Planning and organizing includes an awareness of time and costs. While materials are usually chosen and obtained by the purchaser, as crew leader you should check that all of the components for a particular stage of work are present when that stage is begun. For example, if a turbine is to be installed, you need to know that the fuel pipes and the bolts for anchoring are ready, are the correct size and class, and are on site before they are needed.

You will also need to make certain that your crew is adequate and has appropriately skilled personnel for a particular job before the job begins. If a certain level of welding skills is required, it is your responsibility to determine that those skill levels are available to you. If high level layout and fabrication skills are necessary, be sure that the personnel with those skills will be on hand.

Planning is looking ahead to needs. Organizing is immediate control of resources. Together, planning and organization get the job done efficiently.

2.1.2 Loyalty

As a crew leader, your first loyalty is to the job. That means that you need to think about the requirements of the job all the time. Your long-time friendships with other craftworkers must be kept off the worksite if they interfere with the job. In return, the company owes you the loyalty of supplying you with the tools and personnel to get the job done. The company also owes you support for your appropriate decisions.

2.1.3 Fairness

Fairness is the key to respect. If your decisions are made out of personal likes and dislikes, those decisions will not be respected, either by your subordinates or superiors.

Fairness is not limited to decisions about who gets the dirty jobs, or even to the actual physical job itself. It must exclude all factors in decision making other than the requirements of the job. Gender, color, religious faith, national origin, or political inclination must all be ignored. If you find anyone abusing another crew member, for any reason, your own emotional response must not be a factor in the result. Bring the abuse to an end. At the same time, give rewards to those who earn them, whether you like them or not. The result will be a crew that will work for you.

2.1.4 Enthusiasm

If you are not enthusiastic, none of your workers will be. Work done without enthusiasm is not well done, but rather mediocre. As a leader, you must rise above fatigue and disinterest, and make yourself glad to work each day.

2.1.5 Ability to Teach and Learn from Others

An old proverb says, "The wise man will learn more from the fool than the fool will learn from the wise man." Remember that everyone can teach you something, and everyone can learn from you. Tell your workers about your mistakes, so that they won't repeat them. Learn from the others around you whenever you can.

2.1.6 Initiative

Taking the initiative means taking responsibility for making decisions. It also means keeping the job moving in a positive direction. It doesn't mean waiting for the decisions to be made at the office; it means making the decisions and keeping work moving onward.

2.1.7 Salesmanship

The good leader sells the job to the workers, and to the customer. To sell anything requires belief in the item being sold. This is the point where enthusiasm is necessary, for enthusiastic belief in the job is what sells it. First make sure the job is worth selling, and then sell it.

2.1.8 Good Communication Skills

You must learn to communicate well with everyone you interact with. Learn to listen well, so that you can tell what the speaker thinks you are saying. If you find that the person you are trying to communicate with does not hear what you want to say, try saying it in some other way. Then listen again, until each of you knows what the other is saying.

Remember that not everyone communicates in the same way. Listen for cues to the way people learn and communicate. For example, some people learn and communicate visually. Such a person might respond to an explanation with a visual phrase, such as, "I can't see it" or "I get the picture." Others learn by putting their hands on something, so they might say, "I can work with it" or "It doesn't feel right." Learn to use different communication styles for different people.

2.2.0 Leadership Behavior

Followers have expectations of their leaders. Employees need to feel that their skills and abilities are valued and make a difference. In other words, they crave a feeling of job importance. They also look to their leaders to do the following:

- *Set the example* – Arrive on time, start working when you get there, and keep working until it is time to go home.
- *Suggest and direct work activities* – Workers need to be clear on what you expect them to do on a day-to-day basis.
- *Influence their actions* – Tell and show them clearly what behavior is or is not acceptable.
- *Communicate effectively* – Check to be sure that everybody understands what is expected of them.
- *Make decisions and assume responsibility* – Do not waste time looking for the person responsible for errors and accidents. Assume it is your fault, and fix the problem.
- *Demonstrate loyalty to the group* – You will always walk the thin line between what the workers need, and what the job requires. However, if you take care of your crew, they will take care of you.
- *Abide by company policies and procedures* – Hold the people under you to the rules, and obey them yourself.

2.3.0 Functions of a Leader

The functions of a leader will vary with the environment, the group being led, and the tasks to be performed. However, there are certain functions common to all situations that the leader will be called upon to perform. Some of the major functions are as follows:

- Organize, plan, staff, direct, and control
- Empower group members to make decisions and take responsibility for their work
- Maintain a cohesive group by resolving tensions and differences among its members and between those inside and outside the group
- Ensure that all group members understand and abide by company policies and procedures
- Accept responsibility for the success and failures of the group's performance
- Represent the group to others
- Be sensitive to the differences of a diverse workforce

2.4.0 Leadership Styles

There are three broad categories of leadership style that supervisors can adopt. At one extreme is the autocratic or dictator style of leadership, where the supervisor makes all of the decisions independently, without seeking the opinions of crew members. At the other extreme is the hands-off leadership style, where the supervisor lets the employees make all of the decisions. In the middle is the democratic style, where the supervisor seeks crew members' opinions and makes the appropriate decisions based on their input.

The following are some characteristics of each of the three leadership styles:

Autocratic leaders:

- Expect their crew members to work hard without questioning
- Seldom seek advice from their crew members
- Insist on solving problems alone
- Seldom permit crew members to assist each other
- Praise and criticize on a personal basis
- Have no sincere interest in creatively improving methods of operation or production

Hands-off leaders:

- Believe no supervision is best
- Rarely give orders
- Worry about whether they are liked by their crew members

Democratic leaders:

- Discuss problems with their crew members
- Listen
- Explain and instruct
- Give crew members a feeling of accomplishment by commending them when they do a job well
- Are friendly and available to discuss personal and job-related problems, without using up too much time on idle chatter.

The correct style of leadership for a particular situation or operation depends on the nature of the crew as well as the work it has to accomplish. For example, if the crew does not have enough experience for the job ahead, then a direct, autocratic style may be appropriate. The autocratic style of leadership is also effective when jobs involve repetitive operations that require little decision making.

However, if a worker's attitude is an issue, then democratic action is required. In this case, providing the missing motivational factors may improve performance and result in a better attitude. The democratic style of leadership is also useful when the work is of a creative nature because brainstorming and exchanging ideas with the crew members can be beneficial.

Hands-off leadership is seldom a very effective style. In most cases, either the leadership becomes the responsibility of a crew member, who takes the initiative to assume the role, or chaos results due to lack of formal leadership.

2.4.1 Selecting a Leadership Style

In selecting the most appropriate style of leadership, you should consider the power base from which to work. There are three elements that make up a power base:

1. Authority
2. Expertise
3. Respect

First, the company must give you, the supervisor, enough authority to do the job. This authority must be equal to responsibility, and it must be made known to crew members when they are hired, as well as to those already employed under you, so that they understand who is in charge.

Next, you must have an expert knowledge of the activities to be supervised to be effective. This is important because the crew members need to know they have someone to turn to when they have a question or a problem, when they need some guidance, or when modifications or changes are needed.

Finally, respect is probably the most critical element of power. This derives from being fair to employees, by listening to their complaints and suggestions, and by utilizing incentives and rewards appropriately to motivate crew members. In addition, supervisors who have a positive attitude and a favorable personality tend to gain the respect of their crew members as well as of their peers. Along with respect comes a positive attitude from the crew members.

2.5.0 Ethics in Leadership

The construction and maintenance industry demands the highest standards of ethical conduct. Every day the crew leader has to make decisions that may have ethical implications. If you make an unethical decision as a supervisor, it hurts you, other workers, peers, and the company for which you work. There are three basic types of ethics:

1. Business (legal)
2. Professional (balanced)
3. Situational

Business, or legal, ethics means obeying all laws and regulations related to the issue. Professional, or balanced, ethics is carrying out all activities in such a manner as to be fair to everyone concerned. Situational ethics pertains to specific activities or events that may initially appear to be a gray area. For example, you may ask yourself, "I think this action is not unethical, but how will I feel about myself if my actions are published in the newspaper or if I have to justify my actions to my family, friends, and colleagues?"

Consider the example of professional football. The National Football League has rules by which the game is played. If someone breaks one of the rules, he is acting in an unethical manner from a business or legal sense. It is an accepted standard among professional football players that when they tackle another player it is to take him out of the play and not to purposely injure him. If a player were to tackle another player with the motivation of causing an injury, that is considered unethical.

If you were running down the sideline after catching a pass to score a touchdown, and in your attempt slightly stepped out of bounds (without anyone calling the ball dead at that point) and did not voluntarily acknowledge your action, this is unethical from a situational standpoint. How would you feel the next day if a newspaper printed a picture showing your foot out of bounds and insinuating that your character is questionable because you did not acknowledge breaking the rules?

There are going to be many times when, as a supervisor, you will be put into a situation where you will need to assess the ethical consequences of an impending decision. For instance, should a supervisor continue to keep one of his or her crew working who has broken into a cold sweat due to overheated working conditions, just because the superintendent says the activity is behind schedule? Or should a supervisor who is the only one aware that the reinforcing steel placed by his or her crew was done incorrectly correct the situation before the concrete is placed in the form? If a crew leader is ever asked to carry through on an unethical decision, it is up to him or her to inform the supervisor of the unethical nature of the issue, and if still asked to follow through, refuse to act.

3.0.0 ◆ GENDER AND CULTURAL ISSUES

During the past several years, the construction industry in the United States has experienced a trend in worker expectations and diversity. These two issues are converging at a rapid pace. At no time has there been such a generational merge in the workforce ranging from The Silent Generation (1925–1945), Baby Boomers (1946–1964), Gen X (1965–1979) and the Millennials (1980–2000).

This trend, combined with diversity initiatives by the construction industry, has created a climate that is both exciting and challenging. Today, many construction companies are aggressively seeking a broad base of candidates for the construction industry. To do this effectively, they are using their own resources as well as relying on associations with the government and other construction trade organizations. All current research indicates that this industry will be more dependent on the critical skills of a diverse workforce—a workforce that is both culturally and ethnically fused. Across the United States, the construction industry is aggressively seeking to bring new workers into its ranks, specifically women and racial and ethnic minorities. Diversity is no longer solely driven by social and political issues, but by consumers who need hospitals, malls, bridges, power plants, refineries, and many other commercial and residential buildings.

Some issues relating to gender and race must be addressed on the job site. These include different communication styles of men and women, language barriers associated with cultural differences, the possibility of sexual harassment, and the potential for cases of gender or racial discrimination.

3.1.0 Communication Styles of Men and Women

As more and more women move into construction, it becomes increasingly important that communication barriers between men and women are broken down and differences in behaviors are understood so that they can work together more effectively.

The Jamestown, New York, Area Labor Management Committee (JALMC) offers the following explanations and tips.

1. Women tend to ask more questions than men do. Men are more likely to proceed with a job and figure it out as they go along, while women are more likely to ask questions first.

2. Men tend to offer solutions before empathy; women tend to do the opposite. Both men and women should say what they want upfront—solutions or a sympathetic ear. That way, both genders will feel understood and supported.

3. Women are more likely to ask for help than men. Women are generally more pragmatic when it comes to completing a task. If they need help, they will ask for it. Men are more likely to attempt to complete a task by themselves, even when it might be a good idea to seek assistance.

4. Men tend to communicate more competitively, and women tend to communicate more cooperatively. Both parties need to hear one another out without interruption.

3.2.0 Language Barriers

Language barriers are a real workplace challenge. The U.S. Census Bureau reported that the number of people who speak languages other than English increased to 47 million in 2000. Spanish is now the most common non-English language spoken in the United States. As the makeup of the immigration population continues to change, the number of non-English speakers will rise dramatically, and the languages being spoken will also change. Bilingual job sites are increasingly common.

Companies have the following options to overcome this challenge:

- Offer English classes either at the worksite or through school districts and community colleges
- Offer incentives for workers to learn English

As our workforce becomes more diverse, communicating with people for whom English is a second language will be even more critical. The Knowledge Center's *Manager's Tool Kit* offers the following tips for communicating across language barriers:

- Be patient. If you expect too much, too quickly, you will likely confuse and frustrate the people involved.
- Don't get angry if people don't understand at first. Give them time to process the information in a way that they can comprehend.
- If people speak English poorly but understand reasonably well, ask them to demonstrate their understanding through actions rather than words.
- Avoid humor. Your jokes will likely be misunderstood and may even be misinterpreted as a joke at the worker's expense.
- Don't assume that people are unintelligent simply because they don't understand your language.
- Speak slowly and clearly, and avoid the tendency to yell.
- Use face-to-face communication whenever possible. Over-the-phone communication is often more difficult when a language barrier is involved.
- Use pictures or drawings to get your point across, if necessary.

3.3.0 Cultural Differences

As workers from a multitude of backgrounds and cultures are brought together, there are bound to be differences and conflicts in the workplace. To overcome cultural conflicts, the SHRM suggests the following:

1. Define the problem from both points of view. How does each person involved view the conflict? What does each person think is wrong? This involves moving beyond traditional thought processes to consider alternate ways of thinking.
2. Uncover cultural interpretations. What assumptions are being made based on cultural programming? By doing this, the supervisor may realize what motivated an employee to act in a particular manner. For instance, an employee may call in sick for an entire day to take his or her spouse to the doctor. In some cultures, such an action is based on responsibility and respect for family.
3. Create cultural synergy. Devise a solution that works for both parties involved. The purpose is to recognize and respect other's cultural values, and work out mutually acceptable alternatives.

3.4.0 Sexual Harassment

In today's business world, men and women are working side-by-side in careers of all kinds. There are no longer male occupations or female jobs; therefore, men and women are expected to relate to each other in new ways that are uncharacteristic of the past.

As women make the transition into traditionally male industries, such as construction, the likelihood of sexual harassment increases. Sexual harassment is defined as unwelcome behavior of a sexual nature that makes someone feel uncomfortable in the workplace by focusing attention on their gender instead of on their professional qualifications. Sexual harassment can range from telling an offensive joke or hanging a poster of a swimsuit-clad man or woman in one's cubicle to making sexual comments or physical advances within a work environment. Historically, sexual harassment was thought to be an act performed by men of power within an organization against women in subordinate positions with lesser power. However, as the number of sexual harassment cases have shown over the years, this is no longer the case.

Sexual harassment can occur in a variety of circumstances, including but not limited to the following:

- The victim as well as the harasser may be of the same sex. The victim does not have to be of the opposite sex.
- The harasser can be the victim's supervisor, an agent of the employer, a supervisor in another area, a co-worker, or a non-employee.
- The victim does not have to be the person harassed but could be anyone affected by the offensive conduct.
- Unlawful sexual harassment may occur without economic injury to or discharge of the victim.
- The harasser's conduct must be unwelcome.

The Equal Employment Opportunity Commission (EEOC) enforces sexual harassment laws within industries. When investigating allegations of sexual harassment, the EEOC looks at the whole record including the circumstances and the context in which the alleged incidents occurred. A decision on the allegations is made from the facts on a case-by-case basis.

Prevention is the best tool to eliminate sexual harassment in the workplace. The EEOC encourages employers to take steps necessary to prevent sexual harassment from occurring. Employers should clearly communicate to employees that sexual harassment will not be tolerated. They can do so by developing a policy on sexual harassment, establishing an effective complaint or grievance process, and taking immediate and appropriate action when an employee complains.

Both swearing and off-color jokes are not only offensive to co-workers, but also tarnish a worker's image. Crew leaders need to emphasize that abrasive or crude behavior may affect opportunities for advancement.

3.5.0 Gender and Minority Discrimination

With the increase of women and minorities in the workforce, there is more room for gender and minority discrimination. Consequently, many business practices, including the way employees are treated, the organization's hiring and promotional practices, and the way people are compensated, are being analyzed for equity. More attention is being placed on fair recruitment, equal pay for equal work, and promotions for women and minorities in the workplace.

The EEOC requires that companies be equal opportunity employers. This means that organizations hire the best person for the job, without regard for race, sex, religion, age, etc. Once traditionally a male-dominated industry, construction companies are moving away from this notion and are actively recruiting and training women, younger workers, people from other cultures, and workers with disabilities to compensate for the shortage of skilled workers.

Despite the positive change in recruitment procedures, there are still inadequacies when it comes to pay and promotional activities associated with women and minorities. Individuals and federal agencies continue to win cases in which women and minorities have been discriminated against in the workplace, thus demonstrating that discrimination persists.

To prevent discrimination cases, employers should develop valid job-related criteria for hiring, compensation, and promotion. These measures should be used consistently for every applicant interview, employee performance appraisal, etc. Therefore, all employees responsible for recruitment and selection, supervision, and evaluating job performance should be trained on how to use the job-related criteria legally and effectively.

4.0.0 ◆ PROBLEM SOLVING AND DECISION MAKING

Like it or not, problem solving and decision making are a large part of every supervisor's daily work. The supervisor's first job is to get the job done on time, within budget, and safely. There are always problems to be resolved and decisions to be made, especially in fast-paced, deadline-oriented industries. Sometimes, the difference between problem solving and decision making is not clear. Decision making refers to the process of choosing a course of action in a manner appropriate to the situation. Problem solving involves determining the difference between the way things are and the way things should be, and finding out how to bring the two together. The two activities are interrelated because in order to make a decision, you may also have to use problem-solving techniques.

4.1.0 Types of Decisions

Some decisions are routine or simple. These types of decisions are often based on past experiences. An example is deciding how to get to and from work. If you've worked at the same place for a long time, you are already aware of the options for traveling to and from work (for example, take the bus, drive a car, carpool with a coworker, take a taxi). Based on past experiences with these options, you can make a decision on how best to commute to work.

On the other hand, some decisions are nonroutine or more difficult. These types of decisions require more careful thought on how to carry out an activity: by using a formal problem-solving technique. An example is planning a trip to a new vacation spot. If you are not sure how to get there, where to stay or what to see, one option is to research the area to determine possible routes, hotel accommodations, and attractions. Then, you can make a decision about your best options without the benefit of direct experience.

4.2.0 Formal Problem-Solving Techniques

To make nonroutine decisions, the following procedure can be used as a part of formal problem solving.

Step 1   Recognize that a problem exists. A problem is the difference between the way things are and the way you want them to be. Solving a problem means eliminating the differences between what exists and what is desired. Consider the following example:

It is Friday afternoon, and you are passing the scrap bin. To your surprise, you notice a number of pieces of tubing sticking out of the bin, all with burrs on one end, but otherwise quite useable. Some appear as much as eight feet long. You climb up the ladder and peer into the bin. Several of the tubes have been cut very badly on a miter.

You approach the trade supervisor, Mary, and mention the odd scraps. She returns to the scrap bin with you, and she agrees that the tubes could have been used. You both wonder why the miters are so sloppily and unevenly cut. Mary mentions that one of the mechanics had been sick, and missed several days as a consequence, leaving his helper, Rob, to do the work.

After a little more research, you find out that Rob is fresh out of high school. He is a first year student in the HVAC curriculum. According to his instructor, he is a good student, willing to learn and a hard worker. However, the class has so far only done training in math, basic print reading, hand tools, and safety.

Rereading the scenario, the first thing to do to solve the problem is to identify the signals (statements) that indicate that a problem exists. In this case, the main signal is that a lot of good looking, useable tubing is in the scrap bin.

There may be other areas related to the one noted above. If so, what are they? As the reader can see, the signal is the thing that will motivate the supervisor to find a solution to the problem.

Step 2   Determine what you know or assume you know about the problem, and separate facts from nonfacts.

This second step involves answering questions about the problems. Specifically, the *what*, *when*, *where*, and *who* should be determined. Once these questions have been considered, the information must be categorized as factual or non-factual.

Referring to the example above, some responses to an analysis of the problem may be as follows:

What?

- There is too much good, useable tubing in the scrap pile.
- There is too much waste of materials on this job site.
- The job has gone over budget because of this waste.

When?

- The problem was discovered on Friday afternoon.

Where?

- The problem is the scrap bin near the mechanics' shop.

Who?

- One of the mechanics was off sick for two days, and he left a helper in charge.
- The mechanic acted incorrectly by leaving the helper in charge.
- The helper, Rob, doesn't know how to do his job.

Based on this analysis, some of the statements are facts while the others are assumptions, judgments, expressions of frustration, and/or blame. To solve the real problem, you must eliminate the non-factual items and deal only with the facts.

Facts:

- There is too much good, useable tubing in the scrap bin.
- The problem was discovered on Friday afternoon.
- The problem is in the scrap bin near the mechanics' shop.
- The mechanic in charge was off sick for two days, and he left a helper in charge.

Nonfacts:

- There is too much waste of materials on this job site.
- The job has gone over budget because of this waste.
- The mechanic acted incorrectly by leaving the helper in charge.
- The helper doesn't know how to do his job.

Step 3 State the problem.

Using only facts, formulate problem statements. A fact is a statement of the problem when it meets the following criteria:

- A fact notes the difference between what is and what should be.
- A fact indicates that the present situation has potential for change.

Using the factual statements from the example case, the problem statements are as follows:

- There is much good tubing in the scrap bin that could be used in the construction of the job.
- The mechanic's helper may not know how to cut tubing to minimize waste.

Step 4 Develop objectives that will eliminate the problem.

Describing the desired end result (an objective) is a turning point in the problem-solving sequence. At this point, decision making comes into play. It involves choosing to do one of two things: to act on only one idea that comes into mind or to act on what appears to be the most effective of several alternatives.

Written or stated objectives are the basis of the decision-making process. The more precise the objective, the easier it will be to choose the most effective and efficient plan of action.

Continuing with the previous example, some of your objectives may be the following:

- By the end of business on Monday, I will find out whether or not Rob knows how to cut tubing in a way that reduces waste.
- By the end of business on Tuesday, I will discuss with Mary that training helpers is a part of her job.
- By the end of business on Wednesday, Mary will prepare an on-the-job training program for Rob.
- By the end of business on Friday, Rob will begin an on-the-job training program on how to cut tubing efficiently.

Step 5 Develop alternate solutions, and select one that solves the problem.

At this point, you should list all possible actions to accomplish the objectives, thereby solving the problem. When developing the list, do not make judgments about each alternative. Instead, list each possible action, and then weigh the alternatives to make a decision.

The previous example might have the following list of alternatives:

- Talk to Rob to determine whether or not he is aware of how to cut tubing efficiently; if not, provide training to enable him to do so.
- Provide an on-the-job training program.
- Discuss the problem with Mary and determine the best solution together.
- Inform Mary that it's her responsibility as a lead person to provide training where and whenever needed. Therefore, instruct her to develop and implement a training program.
- Follow up after the program to see if, in fact, Rob can now perform satisfactorily.

Step 6 Develop a plan of action.

Once the solution to the problem has been determined, the next step is to decide who is responsible for carrying out the plan and the deadline for accomplishing it, if not already stated in the objective(s).

For this example, a definite plan of action is as follows:

1. Meet with Mary to discuss the problem. Get a commitment from her that by Monday she has discussed the problem with her helper and reported back with the outcome.
2. Upon meeting with Mary the second time, obtain a commitment from her to develop a week-long, on-the-job training program for Rob. Inform Mary that it is her duty to implement the training program and follow up on it. The plan is due by Tuesday.
3. Meet with Mary to review and make any needed revisions to her plan.
4. Follow up next week to ensure that Rob has successfully completed the training program.

Once all six steps of the formal problem-solving technique are completed, you must follow up to ensure the action plan was completed as intended. If not, take corrective measures to see that the plan is carried out. In addition, take steps to prevent the same problem from occurring in the future.

4.2.1 The Pareto Principle

Sometimes, the number of problems seems overwhelming. The Pareto principle, also called the 80–20 rule, can help you to pick the problems to focus on. The principle, really a rule of thumb, says that the vast majority (80 percent) of problems are the result of a limited number (20 percent) of causes. If you can identify the causes of the most common problems, you can focus on fixing those causes first, thus efficiently solving the majority of your problems. Having done so, take a look at the remaining problems, and identify the most common causes for the majority of the remaining problems. In this way, improvement is most efficiently carried on continuously.

4.2.2 Involving the Employees in Decision Making

In many cases, as the leader, you will make the decisions. However, it is important to involve the employees in the decision-making process, both to empower them to make decisions on their own, and to take responsibility for their work.

Brainstorming is one way to involve the employees. Bring the group involved in an aspect of the work together, and ask them how to solve a problem related to that part of the job. Write down all answers without judgment, even if it seems ridiculous, like inviting gremlins to carry off all of the waste material. When everybody has suggested one or more solutions, stop and hold a vote on which solutions to keep. Now have a discussion of the pros and cons of each idea, followed by a second vote on which ideas to keep. Even if the idea that has received the most votes isn't one you can apply, one or more of the other ideas may be. Now you can make a decision that has some support among the workers who have to carry out the idea, and the workers will feel that they have at least expressed their concerns.

4.2.3 Budgeting and Cost Control

Budgeting properly does not mean telling your superiors what you think they want to hear. Do a full takeoff of materials and equipment needed, and estimate labor as accurately as you can. You may not agree with the company estimator, but you can give management a second opinion on work costing.

Part of the crewleader's attention must be devoted to cost control. A project is budgeted at a certain level, and excess cost due to waste material, wasted work, and reworks directly decreases the profit margin on the job. One way of affecting costs is by ensuring that all material is onsite when it is needed. Another is to be found by proper scheduling, to be certain that the correct personnel are on hand at the right time.

4.3.0 Dealing with Leadership Problems

When you are responsible for leading others, it is inevitable that you will encounter problems and be forced to make decisions about how to respond. Some problems will be relatively simple to handle, such as covering for a sick crew member. Other problems will be complex and much more difficult. Common problems include poor attitudes toward the workplace, an employee's inability to work with others, and absenteeism and turnover.

Each employee is accountable for his or her job. Any problem that interferes with the job must be resolved. The supervisor's responsibility is to help the employee resolve such problems. Disciplinary procedures are the tools by which the supervisor can help the employee. Nobody wants to discipline employees, so systems exist for helping the employee to resolve the problems.

Discipline is nearly always progressive. The first incident of a problem is usually dealt with by giving the employee a verbal warning. The fact that this is called a verbal warning does not mean that the incident and the warning are not recorded. Keep a record of every disciplinary interaction with employees.

The second occurrence of a violation is usually given in the form of a written warning. Most companies use a standard form for the written warning. Again, be sure to keep a record of this warning also. Further violations will be met by sanctions, ultimately leading to discharge of the employee.

4.3.1 Poor Attitude toward the Workplace

Sometimes, employees have poor attitudes toward the workplace because of bad relationships with their fellow employees, negative feelings toward their supervisor, or a dislike of the job in general. Whatever the case, it is important that in your role as supervisor you determine the cause of the poor attitude.

The best way to determine the cause is to talk with that employee one-on-one, listening and asking questions to uncover information. Once the facts have been assessed, you can determine how to correct the situation and turn the negative attitude into a positive one.

If you discover that the problem stems from factors in the workplace or the surrounding environment, you have several choices. First, you can move the worker from the situation to a more acceptable work environment. Next, you can change that part of the work environment found to be causing the poor attitude. Finally, you can take steps to change the employee's attitude so that the work environment is no longer a negative factor.

4.3.2 Inability to Work with Others

Sometimes you will encounter situations where an employee has a difficult time working with others on the crew. This could be a result of personality differences, an inability to communicate, or some other explanation. Whatever the reason, you must address the issue for the effectiveness of the team.

The best way to determine why individuals do not get along or work well together is to speak openly with the parties involved. You should speak openly with the employee as well as the other individual(s) to find out why. Once you

discover the source of the conflict, you can determine how to respond and get the workers communicating and working as a team again. On the other hand, it is possible nothing can be done to resolve the situation. In this case, you would either have to transfer the employee to another crew or have the problem crew member terminated. This latter option should be used as a last measure and should be discussed with your superiors or human resources department.

4.3.3 Absenteeism and Turnover

Timesheets or cards, as documentation of an employee's record of timeliness, are an absolute necessity. This is especially significant because the record is needed in order to establish a paper trail in case of disciplinary actions. All documentation should be prepared in a timely fashion.

Documentation of different types will vary from company to company. Documentation of safety violations, job hazards, discipline, and other issues depend on context, type of work, and the destination of reportage. Forms may be simple or very complex, but must be completed in a timely and correct fashion.

Absenteeism and turnover are big problems. Without workers available to do the work, jobs are delayed and money is lost.

Absenteeism refers to workers missing their scheduled work time on a job. Absenteeism has many causes, some of which are inevitable. For instance, people get sick, they have to take time off for family emergencies, and they have to attend family events such as funerals. However, some causes of absenteeism can be prevented by the supervisor.

The most effective way to control absenteeism is to make the company's policy clear to all employees. Companies that do this find that chronic absenteeism diminishes as a problem. Explain the policy to new employees, including the number of absences allowed and what type of sick or personal days can be taken. In addition, inform workers on how to notify their supervisors when they miss work and the consequences of exceeding the number of sick or personal days allowed.

Once the policy on absenteeism is explained to employees, you must be sure to implement it consistently and fairly. If the policy is administered equally, employees are likely follow it. However, if it is not administered equally and some employees are given exceptions, the rate of absenteeism will increase.

Despite having an absenteeism policy, some employees will be chronically late or absent. In cases where employees abuse the policy, confirm that they understand the company's policy and insist that they comply with it. If an employee's absenteeism continues, disciplinary action may be in order.

Turnover refers to the loss of an employee initiated by that employee. In other words, the employee quits and leaves the company to work elsewhere or is fired for cause. Termination for cause is considered to have been initiated by the employee, since some action of the employee has caused the termination to be required.

Some of the major causes of turnover include the following:

- *Uncompetitive wages and benefits* – Workers may leave one company to go to another that pays higher wages and/or offers better benefits. They may also leave to go to another industry that pays more.
- *Lack of job security* – Workers leave to find more permanent employment.
- *Unsafe working conditions* – Workers leave for safer conditions.
- *Unfair/inconsistent treatment by their immediate supervisor* – Workers leave in pursuit of more pleasant relationships.
- *Poor working conditions* – Workers may want to work inside instead of outside, or in a quieter environment.

Essentially, the actions for controlling absenteeism are also effective for turnover. Past studies have shown that maintaining positive relationships on the job site go a long way in reducing both turnover and absenteeism. This takes effective leadership on the part of the supervisor.

5.0.0 ◆ SUPERVISORS AND SAFETY

Accidents at work sites have high costs, both in human misery and monetary loss. Supervisors must try to prevent accidents by carrying out the company's safety program and making sure workers perform their tasks safely. Ultimately, the responsibility for personal safety lies with the individual worker. Are you wearing the appropriate PPE for the job? Are you following correct procedures? For the leader, these same questions must be asked with respect to each employee. Workers who are using saws or grinders and are not wearing full eye protection risk getting shavings or particles in their eyes. It is the responsibility of the leader to notice whatever tool is being used and to insist on compliance with company safety policy.

There are safe ways of performing any task, and it is the leader's responsibility to ensure that those ways are followed so that no workers are injured. Any time you notice unsafe procedures or behavior, it is your responsibility to stop them. If there is an accident and it can be proven you were present, observed an unsafe act and took no action to stop it, you may be held legally responsible for any injuries. To be successful, the crew leader must:

- Be aware of the costs of accidents
- Understand all federal, state, and local governmental safety regulations
- Be involved in training workers in safe work methods
- Conduct safety meetings
- Be involved in safety inspections, accident investigations, and fire protection and prevention

Providing your team with a safe working environment by preventing accidents and enforcing safety standards will help you finish jobs on time and within budget.

5.1.0 Safety Responsibilities

Each employer must set up and manage a workplace safety and health program to reduce injury, illness, and fatalities. The program must be appropriate to the workplace conditions and consider the number of workers employed and potential work hazards.

To be successful, the safety and health program must have management leadership and employee participation. In addition, training and informational meetings play an important part in effective programs. Being consistent with safety policies is the key.

5.2.0 Safety Program

The supervisor plays a key role in the successful implementation of the safety program. The supervisor's attitudes toward the program set the standard for how crew members view safety. Therefore, the supervisor must follow all program guidelines and require crew members to do the same.

Safety programs consist of the following:

- Safety policies and procedures
- Hazard identification and assessment
- Safety information and training
- Safety record system
- Accident investigation procedures
- Appropriate discipline for not following safety procedures

5.3.0 Safety Policies and Procedures

Employers are responsible for following OSHA and state safety standards. Usually, they incorporate OSHA and state regulations into a safety policies and procedures manual. Basic safety requirements are presented to new employees during their orientation to the company. If the company has a safety manual, the new employee is required to read it and sign a statement indicating that was understood. If the employee cannot read, the employer can have someone read it to the employee, answer any questions that arise, and have the employee sign a form stating that the information is understood.

It is not enough to tell employees about safety policies and procedures on the day they are hired and never mention them again. Rather, supervisors must constantly emphasize and reinforce the importance of following all safety policies and procedures. In addition, encourage employees to play an active role in determining job safety hazards and prevention.

5.4.0 Hazard Identification and Assessment

Safety policies and procedures have to be specific to the company and clearly present the hazards of the job. Supervisors also need to identify and assess potential worksite hazards and comply with OSHA and state standards. OSHA recommends that employers conduct inspections of the workplace, monitor safety and health information logs, and evaluate new equipment, materials, and processes for potential hazards before they are utilized.

Both supervisors and employees play important roles in identifying hazards. It is the supervisor's responsibility to establish which working conditions are unsafe and inform employees of the hazards and their locations. Also, crew members should be encouraged to notify supervisors about hazardous conditions. To accomplish this exchange of information, supervisors must be present and available on the job site.

The supervisor needs to make employees aware of the built-in hazards within the construction industry so they can be avoided. Examples of hazards include working on high buildings, in tunnels that are deep underground, on caissons, in huge excavations with earthen walls buttressed by heavy steel I-beams, and other naturally dangerous projects. The supervisor can also take safety measures, such as installing protective railings to prevent workers from falling from buildings, as well as scaffolds, platforms, and shoring.

5.5.0 Safety Information and Training

It is important for employers to provide periodic information and training to new and long-term employees and as often as necessary until all employees are adequately trained. Present special training and informational sessions when safety and health information changes or workplace conditions create new hazards.

Whenever a supervisor assigns an experienced employee a new task, he or she must ensure the employee is capable of doing the work in a safe manner. This can be accomplished by providing safety information one-on-one or in group training. The supervisor can provide the following information:

- Define the task
- Explain how to do the task safely
- Explain what tools and equipment to use and how to safely use them
- Identify the necessary personal protective clothing
- Explain the nature of the hazards in the work and how to recognize them
- Stress the importance of personal safety and the safety of others
- Hold regular safety meetings with the crew's input
- Review material safety data sheets that may be necessary

In many cases, a standard procedure is the production and posting of a job hazard analysis (JHA)—a document describing the hazards potentially associated with a particular job or piece of equipment. The JHA is posted at the point where the operator would see it, to protect anyone from injury due to ignorance. In some organizations, the JHA forms must be signed by the operator before using the equipment.

6.0.0 ◆ SUPERVISOR INVOLVEMENT IN SAFETY

To be effective leaders, supervisors must be actively involved in the safety program. Supervisory involvement includes conducting safety meetings and inspections, promoting first aid and fire protection and prevention, preventing substance abuse on the job, and investigating accidents.

6.1.0 Safety Meetings

A safety meeting may be a brief informal gathering of a few employees or a formal meeting with instructional films and talks by guest speakers. The audience size and topics to be addressed determine the format of a meeting. Small, informal safety meetings are typically conducted weekly.

Plan safety meetings in advance, and communicate the information to all employees affected. In addition, make the topics timely and practical. For example, if upcoming work is confined space work, a review of confined space safety is a good safety meeting topic.

6.2.0 Inspections

Supervisors must make frequent inspections to prevent accidents as well as control the number of accidents that occur. They need to inspect the job sites where their workers perform their tasks. It is recommended that this be done before the start of work each day and during the day at different times.

Supervisors must correct or protect workers from existing or potential hazards in their work areas. Sometimes, supervisors may be required to work in areas controlled by other contractors. Here, the supervisors may have little control over unsafe conditions. In such cases, they should immediately bring any hazards to the attention of the contractor at fault, to their own superior, and to the person responsible for the job site.

Supervisors' inspections are only valuable if action is taken to correct what is wrong or hazardous. Consequently, supervisors must be aware of unsafe acts in their work sites. When an employee performs an unsafe action, the supervisors must explain to the employee why the act was unsafe, ask that it not be done again, and request cooperation in promoting a safe working environment. The supervisor may decide to document what happened (*Figure 3*) and what the employee was asked to do to correct the situation. It is then very important that supervisors

follow up to ensure the employee is complying with the safety procedures. Never allow a safety violation to go uncorrected. Forms used will be dependent on context.

6.3.0 Substance Abuse

Unfortunately, drug and alcohol abuse is a reality within our country and in the workplace. Drug abuse means inappropriately using drugs, whether they are legal or illegal. Some people use illegal street drugs, such as cocaine or marijuana, or abuse prescription drugs, such as Ritalin or Vicodin. Others consume alcohol to the point of intoxication.

It is essential that supervisors enforce company policies and procedures about substance abuse. Supervisors must work with management to deal with suspected drug and alcohol abuse and not handle these situations themselves. These cases are normally handled by human resources or a designated manager. There are legal consequences of drug and alcohol abuse and the associated safety implications. That way, the business and the employee's safety are protected. If you suspect that an employee is suffering from drug or alcohol abuse, you should immediately contact your supervisor and/or human resources for assistance.

Construction supervisors have to deal with immediate job safety because of the dangers associated with construction work. It is the supervisor's duty to make sure that safety is maintained at all times. This may include removing workers from a work site where they may be endangering themselves or others. For example, suppose several crew members go out and smoke marijuana or have a few beers during lunch. Then they return to work to erect scaffolding for a concrete pour. If you smell marijuana on the crew members' clothing or alcohol on their breath, you must step in and take action. Otherwise, they might cause an accident that could delay the project or cause serious injury or death.

It is often difficult to detect drug and alcohol abuse. The best way is to search for identifiable effects, such as those mentioned above or sudden changes in behavior that are not typical of the employee. Some examples include the following:

- Unscheduled absences; failure to report to work on time
- Significant changes in the quality of work
- Unusual activity or lethargy
- Sudden and irrational temper flare-ups
- Significant changes in personal appearance, cleanliness, or health

Incident Report Form

(Please print legibly or type)

Date of Incident:_____   Report Date:_____

Name(s) of personnel involved:

Name:			
Employee Number:			
Address:			
Phone:			

Witness(es):

Name:			
Employee Number:			
Address:			
Phone:			

Location of Incident: _____

Time: _____ A.M./P.M.   Supervisor: _____

Details of Incident: _____

_____

_____

_____

_____

_____

_____

_____

_____

_____

_____

Person Submitting Report: _____   **Title:** _____

Address: _____

_____

Office Use Only:
Date Report Received:_____

410F03.EPS

Figure 3 ◆ Sample incident report.

There are other more specific signs which should arouse suspicions, especially if more than one is exhibited. Among them are as follows:

- Slurring of speech or an inability to communicate effectively
- Shiftiness or sneaky behavior, such as an employee disappearing to wooded areas, storage areas, or other private locations
- Wearing sunglasses indoors or on overcast days to hide dilated or constricted pupils
- Wearing long-sleeved garments, particularly on hot days, to cover marks from needles used to inject drugs
- Attempting to borrow money from co-workers
- The loss of an employee's tools and company equipment

6.4.0 Accident Investigations

There are two times when a supervisor is involved in an accident investigation. The first time is when an accident, injury, or report of work-connected illness takes place. If present on site, the supervisor must proceed immediately to the location where the incident occurred to see that proper first aid is being provided and ensure that other safety and operational measures are taken to prevent another incident.

After the incident, the supervisor needs to make a formal investigation and submit a report. An investigation looks for the causes of the accident by examining the circumstances and talking to the people involved. Investigations are perhaps the most useful tool in accident prevention.

For years, a prominent safety engineer was confused as to why sheet metal workers frequently fractured their toes. The supervisor had not made a thorough accident investigation, and the injured workers were embarrassed to admit how the accidents really occurred. It was later discovered they used the metal reinforced cap on their safety shoes to hold the sheet metal vertically in place when they fastened it. The sheet metal was inclined to slip and fall behind the safety cap onto the toes and cause a fracture. With a proper investigation, further accidents could have been prevented.

The four major parts of an accident investigation are as follows:

- Describing the accident
- Determining the cause of the accident
- Determining the parties involved and the part played by each
- Determining how to prevent recurrences

1. A crew leader differs from a craftworker in that a _____.
   a. crew leader need not have direct experience in those job duties that a craftworker typically performs
   b. crew leader can expect to oversee one or more craftworkers in addition to performing some of the typical duties of the craftworker
   c. crew leader is exclusively in charge of overseeing, since performing technical work is not part of this role
   d. crew leader's responsibilities do not include being present on the job site

2. A crew leader is a person who supervises one or more _____.
   a. craftworkers
   b. foremen
   c. supervisors
   d. construction projects

3. Selection as a crew leader is based on your _____.
   a. age
   b. ability to meet schedules
   c. education level
   d. popularity

4. Strong supervisors lead by _____.
   a. paying more money
   b. watching
   c. explaining
   d. example

5. Among the many traits effective leaders should have is _____.
   a. the ability to communicate the goals of a project
   b. the drive necessary to carry the workload by themselves in order to achieve a goal
   c. a perfectionist nature, which ensures that they will not make useless mistakes
   d. the ability to make decisions without needing to listen to others' opinions

6. To accomplish their goals, leaders must learn to _____.
   a. blame others
   b. assume responsibility
   c. conceal their errors
   d. suppress motivation

7. As a crew leader, you will need to make certain that your crew has _____ for a particular job.
   a. plenty of time available
   b. appropriate skilled personnel
   c. transportation to the scene
   d. another supervisor

8. For the crew leader, first loyalty is to the _____.
   a. supervisor
   b. paycheck
   c. job
   d. customer

9. To learn to communicate well, learn to _____.
   a. speak loudly
   b. talk slowly
   c. use short words
   d. listen well

10. In the autocratic style of leadership, the supervisor _____.
    a. makes all the decisions
    b. leaves the decisions up to the workers
    c. makes the decisions after receiving the workers' input
    d. doesn't need decisions made

11. Democratic leaders discuss problems with the crew, listen, explain, and instruct.
    a. True
    b. False

12. The three elements of a power base are _____.
    a. popularity, efficiency, and pressure
    b. authority, expertise, and respect
    c. money, personal influence, and secrecy
    d. openness, thoughtfulness, and sincerity

13. Decision making refers to the process of choosing _____ in a manner appropriate to the situation.
    a. what you want
    b. inaction
    c. a course of action
    d. a goal

14. A problem is the difference between the way things are and the way you want them to be.
    a. True
    b. False

15. Written or stated _____ are the basis of the decision-making process.
    a. questions
    b. penalties
    c. objectives
    d. problems

Summary

In order to move into supervisory roles, the worker must learn many skills. In addition to craft skills, the crew leader must learn to deal with relationships in a different way. The key to functioning relationships as a crew leader is fairness to everyone and open communication with all. Cultural and gender issues must be irrelevant to decisions about personnel, only specific personal skills and strengths being the deciding factors. Loyalty to the job and to the crew are both necessary in order to accomplish the work. Finally, the crew leader must help to create an environment oriented toward safety and fair practices.

Notes

NCCER makes every effort to keep these textbooks up-to-date and free of technical errors. We appreciate your help in this process. If you have an idea for improving this textbook, or if you find an error, a typographical mistake, or an inaccuracy in NCCER's Contren® textbooks, please write us, using this form or a photocopy. Be sure to include the exact module number, page number, a detailed description, and the correction, if applicable. Your input will be brought to the attention of the Technical Review Committee. Thank you for your assistance.

Instructors – If you found that additional materials were necessary in order to teach this module effectively, please let us know so that we may include them in the Equipment/Materials list in the Annotated Instructor's Guide.

Write: Product Development and Revision
              National Center for Construction Education and Research
              3600 NW 43rd St, Bldg G, Gainesville, FL 32606

Fax: 352-334-0932

E-mail: curriculum@nccer.org

Craft   Module Name
_____

Copyright Date   Module Number   Page Number(s)
_____

Description
_____
_____
_____
_____

(Optional) Correction
_____
_____
_____

(Optional) Your Name and Address
_____
_____
_____

Glossary of Trade Terms

Active solar heating system: A type of solar heating system that uses fluids pumped through collectors to gather solar heat. It is the most complex solar heating system and provides the best temperature control.

Air stratification: The layering of air in a room based on temperature. The warmest air is concentrated at the ceiling and the coldest air is concentrated at the floor, with varying temperature layers in between.

Air turnover unit: A type of tall air handler that moves stratified air so that the temperature in the room is more even. The air turnover unit may further condition the air in the process.

Algorithm: A mathematical equation consisting of a series of logic statements used in a computer or microprocessor to solve a specific kind of problem. In HVAC applications, algorithms are typically used in microprocessor-controlled equipment to control a wide range of control function operations based on the status of various system sensor input signals.

Anhydrous: Containing no water.

Application-specific controller: A digital controller installed by a manufacturer on a specific product at the factory.

Arrestance efficiency: The percentage of dust that is removed by an air filter. It is based on a test where a known amount of synthetic dust is passed through the filter at a controlled rate, then the weight of the concentration of dust in the air leaving the filter is measured.

Atmospheric dust spot efficiency (dust spot efficiency): The percentage of dust that is removed by an air filter. It is the number that is normally referenced in the manufacturer's literature, filter labeling, and specifications. The dust spot efficiency of a filter is based on a test where atmospheric dust is passed through a filter, then the discoloration effect of the cleaned air is compared with that of the incoming air.

Atmospheric pressure: The pressure exerted on all things on the surface of the earth as the result of the weight of the atmosphere.

Baud: The rate at which information is transmitted across communication lines.

Biological contaminants: Airborne agents such as bacteria, fungi, viruses, algae, insect parts, pollen, and dust. Sources include wet or moist walls, duct, duct liner, fiberboard, carpet, and furniture. Other sources include poorly maintained humidifiers, dehumidifiers, cooling towers, condensate drain pans, evaporative coolers, showers, and drinking fountains.

Bit: Short for binary digit. The smallest element of data that a computer can handle. It represents an off or on state (zero or one) in a binary system.

Bleed-off: A method used to help control corrosion and scaling in a water system. It involves the periodic draining and disposal of a small amount of the water circulating in a system. Bleed-off aids in limiting the buildup of impurities caused by the continuous addition of makeup water to a system.

Boyle's law: With a constant temperature, the pressure on a given quantity of confined gas varies inversely with the volume of the gas. Similarly, at a constant temperature, the volume of a given quantity of confined gas varies inversely with the applied pressure.

Building management system (BMS): A centralized, computer-controlled system for managing the various systems in a building. Also known as a building automation systems (BAS).

Building-related illness: A situation in which the symptoms of a specific illness can be traced directly to airborne building contaminants.

Bus: A multi-wire communication cable that links all the components in a hard-wired computer network.

Catalytic element: A device used in wood-burning stoves that helps reduce smoke emissions.

Charles' law: With a constant pressure, the volume for a given quantity of confined gas varies directly with its absolute temperature. Similarly, with a constant volume of gas, the pressure varies directly with its temperature.

Chilled-beam cooling system: A cooling system that employs radiators (chilled beams) mounted near the ceiling through which chilled water flows. Passive systems rely on convection currents for cooling. Active systems use ducted conditioned air to help induce additional airflow over the beams.

Glossary of Trade Terms

Commodities: Commercial items such as merchandise, wares, goods, and produce.

Condensing unit: A packaged refrigeration system assembly containing the compressor, condenser, and any related components.

Conductance: A measurement of the heat flow through nonhomogeneous materials such as glass blocks, hollow tiles, and concrete blocks. Specifically, it is the heat flow rate through one square foot of a nonhomogeneous material of a given thickness when there is a 1°F temperature difference between the two surfaces of the material.

Conductivity: The ability of a material to conduct heat.

Control point: The name for each input and output device wired to a digital controller.

Cooled equipment enclosure: A type of enclosure or cabinet that contains its own air conditioning system. It is designed to house and cool electronic components.

Cooler: A refrigerated storage device that protects commodities at temperatures above 32°F.

Coordination drawings: Elevation, location, and other drawings produced for a project by the individual contractors for each trade to prevent a conflict between the trades regarding the installation of their materials and equipment. Development of these drawings evolves through a series of review and coordination meetings held by the various contractors.

Creosote: A black, sticky combustible byproduct created when wood is burned in a stove. It must be periodically cleaned from within the stove and chimney before it can build up to the point where it can cause a fire.

Cryogenics: Refrigeration that deals with producing temperatures of –250°F and below.

Cryogenic fluid: A substance that exists as a liquid or gas at ultra low temperatures of –250°F and below.

Cubic feet per minute (cfm): A measure of the amount or volume of air in cubic feet flowing past a point in one minute. Cubic feet per minute can be calculated by multiplying the velocity of air, in feet per minute (fpm), by the area it is moving through, in square feet. It can also be measured with various test instruments.

Cut list: An information sheet that is derived from shop drawings. It is the shop guide for fabricating duct runs and fittings.

Dalton's law: The total pressure of a mixture of confined gases is equal to the sum of the partial pressures of the individual component gases. The partial pressure is the pressure that each gas would exert if it alone occupied the volume of the mixture at the same temperature.

Data collection: The collection of trend, runtime, and consumable data from the digital controllers in a building.

Deadband: In a chiller, the tolerance on the chilled-water temperature control point. For example, a 1°F deadband controls the water temperature to within ±0.5°F of the control point temperature (0.5°F + 0.5°F = 1°F deadband).

Deck: A refrigeration industry trade term that refers to a shelf, pan, or rack that supports the refrigerated items stored in coolers and display cases.

Desiccant: A material that has a high capacity for absorbing moisture; for example, calcium chloride.

Detail drawing: A drawing of a feature that provides more elaborate information than is available on a plan.

Dew point: The temperature at which water vapor in the air becomes saturated and starts to condense into water droplets.

Diffuser: An outlet that discharges supply air into a room in various directions and planes.

Digital controller: A digital device that uses an input module, a microprocessor, and an output module to perform control functions.

Direct digital control (DDC): The use of a digital controller is usually referred to as direct digital control or DDC.

Direct-fired makeup air unit: An air handler that heats and replaces indoor air that is lost from a building through exhaust vents.

Ductless split-system air conditioner: A split-system air conditioner in which the indoor air handler is wall- or ceiling-mounted. The air handler blows the conditioned air directly into the room without the use of ductwork.

Glossary of Trade Terms

Elevation view: A view that depicts a vertical side of a building, usually designated by the direction that side is facing; for example, right, left, east, or west elevation.

Environmental tobacco smoke (ETS): A combination of sidestream smoke from the burning end of a cigarette, cigar, or pipe and the exhaled mainstream smoke from the smoker.

Ethernet: A family of frame-based computer networking technologies for local area networks (LANs). The name comes from the physical concept of the ether. It defines a number of wiring and signaling standards.

Evaporative cooler: A comfort cooling device that cools air by evaporating water. It is commonly used in hot, dry climates. Cooling effectiveness drops as the relative humidity of the outdoor air increases.

Evaporative pre-cooler: A device used to pre-cool air entering a condenser coil. Pre-cooling the air reduces the load on the compressor, thus reducing energy costs. It operates on the same principle as an evaporative cooler.

External static pressure: The total pressure loss of the duct system ductwork and components external to the supply fan assembly.

Fan brake power: The actual total power needed to drive a fan to deliver the required volume of air through a duct system. It is greater than the expected power needed to deliver the air because it includes losses due to turbulence, inefficiencies in the fan, and bearing losses.

Firmware: Computer programs that are permanently stored on the computer's memory during a manufacturing process.

Floor plan: A building drawing indicating a plan view of a horizontal section at some distance above the floor, usually midway between the ceiling and the floor.

Formaldehyde: A colorless, pungent byproduct of hydrocarbons that can cause irritation of the eyes and upper air passages.

Friable: The condition in which materials can release particulates into the air.

Gateway: A link between two computer programs allowing them to share information by translating between protocols.

Geothermal heat pump: A heat pump that uses fluid-filled tubing buried in the ground to capture the heat that is always present in the earth.

Green building: A sustainable structure that is designed, built, and operated in a manner that efficiently uses resources while being ecologically friendly.

Grille: A louvered covering for any opening through which air passes.

High-efficiency particulate air (HEPA) filter: An extended media, dry-type filter mounted in a rigid frame. It has a minimum efficiency of 99.97 percent for 0.3-micron particles when a clean filter is tested at its rated airflow capacity.

Hot aisle/cold aisle configuration: A method used to install equipment cabinets in computer rooms to manage the flow of warm and cool air out of and into the cabinets.

Hypertext transfer protocol (http): The base protocol used by the world wide web.

Immiscible: A condition in which a refrigerant does not dissolve in oil and vice versa.

Indirect solar hydronic heating system: A type of active solar heating system in which a double-walled heat exchanger is used to prevent toxic antifreeze in the solar collectors from contaminating the potable water system.

Indoor air quality (IAQ): Measure of the quality of interior air that could affect health and comfort of a building's occupants.

Induction unit system: An air conditioning system that uses heating/cooling terminals with circulation provided by a central primary air system that handles part of the load, instead of a blower in each cabinet. High-pressure air (primary air) from the central system flows through nozzles arranged to induce the flow of room air (secondary air) through the unit's coil. The room air is either cooled or heated at the coil, depending on the season. Mixed primary and room air is then discharged from the unit.

Infiltration: The process by which outdoor air enters a structure through cracks, crevices, and other openings in the building.

Integrated building design: A collaborative decision-making process that uses a project design team that extends from a project's inception through its design and construction phases.

Internet protocol (IP): A data-oriented protocol used for communicating data across a packet-switched internetwork.

Internet protocol address (IP address): A unique address (computer address) that certain electronic devices use in order to identify and communicate with each other on a computer network utilizing the internet protocol (IP) standard.

Internet: A worldwide communication network used by the public to interface computers.

Interoperability: The ability of digital controllers with different protocols to function together accurately.

Latent heat defrost: A method used in hot gas defrost applications that uses the heat added in one or more active evaporators as a source of heat to defrost another evaporator coil.

Layup: An industry term referring to the period of time a boiler is shut down.

Load-shedding: Systematically switching loads out of a system to reduce energy consumption.

Local area network (LAN): A server/client computer network connecting multiple computers within a building or building complex.

Local interface device: A keypad with an alphanumeric data display and keys for data entry. It is connected to the digital controller with a phone-type cable that provides power and communication.

Longitudinal section: A section drawing where the "cut" is made along the long dimension of the building.

Lyophilization: A dehydration process which first incorporates freezing of a material, then a reduction of the surrounding pressure with simultaneous addition of heat to remove moisture. Moisture leaves the material through sublimation. Also known as freeze-drying.

Microbial contaminants: See *biological contaminants*.

Micron: A unit of length that is one millionth of a meter, or about 1/25,400 of an inch.

Modem: Short for modulator-demodulator. A device that links a computer to communication lines. It converts voice to electronic signals and vice versa.

Monel®: An alloy made of nickel, copper, iron, manganese, silicon, and carbon that is very resistant to corrosion.

Multiple chemical sensitivity (MCS): A medical condition found in some individuals who are vulnerable to exposure to certain chemicals and/or combinations of chemicals. Currently, there is some debate as to whether or not MCS really exists.

Network: A means of linking devices in a computer-controlled system and controlling the flow of information among these devices.

New building syndrome: A condition that refers to indoor air quality problems in new buildings. The symptoms are the same as those for sick building syndrome.

Off-gassing: The process by which furniture and other materials release chemicals and other volatile organic compounds (VOCs) into the air.

Ozone: An unstable, poisonous oxidizing agent that has a strong odor and is irritating to the mucus membranes and the lungs. It is formed in nature when oxygen is subjected to electric discharge or exposure to ultraviolet radiation. It is also generated by devices such as photocopiers, electronic air cleaners, and other equipment that uses high voltages.

Passive solar heating system: A type of solar heating system characterized by a lack of moving parts and controls. The sun shining in a window on a winter day is an example of passive solar heat.

Personal digital assistant (PDA): An electronic device which can include some of the functionality of a computer, cell phone, music player and camera.

Plan view: The overhead view of an object or structure.

Pontiac fever: A mild form of Legionnaire's disease.

Product-integrated controller (....

Radiant heating system: A type of heating system that heats objects, not the air. In-floor systems using electric heating elements or tubes filled with hot fluid are examples of radiant heating systems.

Radon: A colorless, odorless, radioactive, and chemically inert gas that is formed by the natural breakdown of uranium in soil and groundwater. Radon exposure over an extended period of time can increase the risk of lung cancer.

Recycle shutdown mode: A chiller mode of operation in which automatic shutdown occurs when the compressor is operating at minimum capacity and the chilled-water temperature has dropped below the chilled-water temperature setpoint. In this mode, the chilled-water pump remains running so that the chilled-water temperature can be monitored.

Register: A grille equipped with a damper or control valve.

Retort: A container in which substances are cooked, distilled, or decomposed by heat.

Riser diagram: A one-line schematic depicting the layout, components, and connections of a piping system or electrical system.

Runaround loop: A closed-loop energy recovery system in which finned-tube water coils are installed in the supply and exhaust airstreams and connected by counterflow piping.

Satellite compressor: A separate compressor that uses the same source of refrigerant as those used by a related group of parallel compressors. However, the satellite compressor functions as an independent compressor connected to a cooling area that requires a lower temperature level than that being maintained by the related compressors.

Schedules: Tables that describe and specify the types and sizes of items required for the construction of a building.

Secondary coolant: Any cooling liquid that is used as a heat transfer fluid. It changes temperature without changing state as it gains or loses heat.

Section drawing: A drawing that depicts a feature of a building as if there were a cut made through the middle of it.

Sensible heat recovery device: An air-to-air recovery device that transfers only sensible heat between the supply and exhaust airstreams. It does not exchange latent heat (heat contained in water vapor) between the supply and exhaust airstreams.

Shop drawing: A drawing that indicates how to fabricate and install individual components of a construction project. A shop drawing may be drafted from the construction drawings of a project or provided by the manufacturer.

Sick building syndrome: A condition that exists when more than 20 percent of a building's occupants complain during a two-week period of a set of symptoms, including headaches, fatigue, nausea, eye irritation, and throat irritation, that are alleviated by leaving the building and are not known to be caused by any specific contaminants.

Site plan: A construction drawing that indicates the location of a building on a land site.

Software: Computer programs transferred to the computer from various media and stored in an erasable memory.

Specific density: The weight of one pound of air. At 70°F at sea level, one pound of dry air weighs .075 pound per cubic foot.

Specific volume: The space one pound of dry air occupies. At 70°F at sea level, one pound of dry air occupies a volume of 13.33 cubic feet.

Static pressure: The pressure exerted uniformly in all directions within a duct.

Sublimation: The change in state directly from a solid to a gas, such as the changing of ice to water vapor, without passing through the liquid state at any point.

Sustainable: Designed to reduce impact on the environment.

Takeoff: The process of surveying, measuring, itemizing, and counting all materials and equipment needed for a construction project, as indicated by the drawings.

Temperature differential: The difference in air temperature on two sides of an object.

Glossary of Trade Terms

Tenant billing: The ability to charge a building tenant for after-hours use of the building's HVAC system.

Thermal conductivity: The ability of a given substance to conduct heat; specifically, it is the heat flow per hour (Btuh) through one square foot of one-inch thick homogeneous material when the temperature difference between the two faces is 1°F.

Thermosiphon: A passive heat exchange process in which liquid is circulated by means of natural convection.

Thermosyphon system: A type of passive solar heating system in which the difference in temperature of fluids in different parts of the system causes the fluids to flow through the system.

Throw: The horizontal or vertical axial distance an airstream travels after leaving a supply outlet before the maximum stream velocity is reduced to a specific terminal velocity.

Total heat recovery device: An air-to-air recovery device that can transfer both sensible and latent heat (heat contained in water vapor) between supply and exhaust airstreams.

Total heat rejection (THR) value: A value used to rate condensers. It represents the total heat removed in desuperheating, condensing, and subcooling a refrigerant as it flows through the condenser.

Total heat: Sensible heat plus latent heat.

Total pressure: The sum of the static pressure and the velocity pressure for any cross section of an air duct. It determines how much energy must be supplied to the system by the fan to maintain airflow.

Transients: Short-duration interference signals that are coupled to and transmitted on power lines, communication lines, and/or computer network bus lines. Transient signals can be caused by natural or man-made electrical or electromagnetic disturbances, signals, or emissions. When present in computer network bus lines, transients can cause computer program glitches and/or interrupt operational sequences.

Transverse section: A section drawing where the "cut" is made along the short dimension of the building.

Traverse readings: A series of velocity readings taken at several points over the cross-sectional area of a duct or grille.

Type HT vent: A metal vent capable of withstanding temperatures up to 1,000°F. It is commonly used to vent wood-burning stoves and furnaces.

Type PL vent: A type of metal vent specifically designed for stoves that burn wood pellets or corn.

U-factor: The heat flow per hour through one square foot of material when the temperature difference between the two surfaces of the material is 1°F.

Unit cooler: A packaged refrigeration system assembly containing the evaporator, expansion device, and fans. It is commonly used in chill rooms and walk-in coolers.

User interface module: A keypad with an alphanumeric data display and keys for data entry. It is connected to a network communication bus.

Valance cooling system: A type of cooling system in which chilled water is circulated through finned-tube radiators located near the ceiling around the perimeter of a room. Convection currents move the cooled air instead of a blower assembly. A decorative valance conceals the system.

Velocity: The pressure in a duct due to the movement of the air. It is the difference between the total pressure and the static pressure.

Volatile organic compounds (VOCs): A wide variety of compounds and chemicals found in such things as solvents, paints, and adhesives, that are released as gases at room temperature.

Volume: The amount of air in cubic feet flowing past a given point in one minute (cfm).

Web browser: A software application which enables a user to display and interact with text, images, videos, music and other information typically located on a web page at a web site on the world wide web or a local area network.

Web page: A resource of information that is suitable for the world wide web and can be accessed through a web browser.

Wide area network (WAN): A server/client computer network spread over a large geographical area.

World wide web: A system of interlinked hypertext documents accessed via the internet. Commonly shortened to "the web."

Index

Index

Velocity pressure, 2.15, 7.28
Velometers, 2.19-2.21
Ventilation
    demand-control, 3.31, 5.44
    exchange rate, 3.4, 4.2
    mechanical, 3.4
    natural, 3.4
    night-time free cooling, 5.43-5.44
    outside air for ventilation, 3.16, 4.20, 9.8, 9.30
    purge modes, 3.20
    system inspection, 3.14
    wood-burning appliances, 9.9-9.10
Ventilation control, 3.16-3.17, 3.20
Ventilators, energy and heat recovery, 4.2-4.6
Vibration isolators, 8.37-8.38
Volatile organic compound (VOC) sensors, 3.30-3.31
Volatile organic compounds (VOCs), 3.5, 3.6
Volume, 7.28
    defined, 2.15
    measuring, 2.19-2.20, 2.23
    specific, 2.5, 2.10
    temperature- pressure relationships, 2.3-2.5, 2.11
Volume balancers, 2.20
Volume dampers, 7.45-7.46

W

Walls, load factors, 7.16-7.19
Waste heat water heaters, 9.32-9.33
Waste oil heaters, 9.2, 9.12
Water heaters, alternative, 9.13, 9.32-9.33
Water recovery system, 4.27-4.29
Water-side economizers, 4.22
Water treatment
    boilers, 6.12-6.13
    cooling tower water systems, 6.26
Web browser, 5.25
Web browser system integration, 5.29-5.30
Web page, 5.25
Web page server, 5.25, 5.38
Wet air filters, 3.21
Wet-bulb temperature, 2.9
Wet-bulb temperature lines, psychrometric chart, 2.10
Wide area network (WAN), 5.24, 5.24-5.25
Window glass
    energy performance ratings, 7.18
    load factor, 7.15
Wood-burning appliances
    boilers, 9.6-9.7
    furnaces, 9.3-9.6
    installation and maintenance, 9.7-9.10
    stoves, 9.2-9.3
World wide web, 5.29

Y

Year-round air conditioning units (YAC), 6.36-6.38

Z

Zone control
    interoperability strategies, 5.45
    residential, 5.12-5.13
    VAV systems, 5.17-5.18, 5.41-5.44
    VVT systems, 5.15-5.17, 5.18, 5.39-5.41
Zone dampers, 5.15-5.16, 5.40-5.45
Zone management
    demand-controlled ventilation, 5.44
    heating, 5.42-5.44
    humidity control, 5.44
    morning warm-up, 5.43
    occupancy period scheduling, 5.41
    occupied cooling, 5.40, 5.41-5.42
    occupied heating, 5.40-5.41
    real-time HVAC usage, 5.45
    simultaneous cooling and heating, 5.41, 5.43
    smoke control, 5.44-5.45
    unoccupied period, 5.41, 5.43